BFH5763

Physiology and Biochemistry of Metal Toxicity and Tolerance in Plants

Physiology and Biochemistry of Metal Toxicity and Tolerance in Plants

Edited by

M.N.V. Prasad
*University of Hyderabad,
Hyderabad, India*

and

Kazimierz Strzałka
*Jagiellonian University,
Krakow, Poland*

KLUWER ACADEMIC PUBLISHERS
DORDRECHT / BOSTON / LONDON

A C.I.P. Catalogue record for this book is available from the Library of Congress.

ISBN 1-4020-0468-0

Published by Kluwer Academic Publishers,
P.O. Box 17, 3300 AA Dordrecht, The Netherlands.

Sold and distributed in North, Central and South America
by Kluwer Academic Publishers,
101 Philip Drive, Norwell, MA 02061, U.S.A.

In all other countries, sold and distributed
by Kluwer Academic Publishers,
P.O. Box 322, 3300 AH Dordrecht, The Netherlands.

Printed on acid-free paper

QK
753
.M47
P49x
2002

All Rights Reserved
© 2002 Kluwer Academic Publishers
No part of the material protected by this copyright notice may be reproduced or
utilized in any form or by any means, electronic or mechanical,
including photocopying, recording or by any information storage and
retrieval system, without written permission from the copyright owner.

Printed in the Netherlands.

Contents

Preface   vii

Contributors   ix

Metal Permeability, Transport and Efflux in Plants   1
    E. CSEH

Glutathione and Thiol Metabolism in Metal Exposed Plants   37
    B. TOMASZEWSKA

Metal Chelating Peptides and Proteins in Plants   59
    L. SANITÀ DI TOPPI, M.N.V. PRASAD
    AND S. OTTONELLO

Metabolism of Organic Acids and Metal Tolerance in
    Plants Exposed to Aluminum   95
    H. MATSUMOTO

Physiological Responses of Non-Vascular Plants to
    Heavy Metals   111
    N. MALLICK AND L.C. RAI

Physiological Responses of Vascular Plants to Heavy Metals   149
    F. FODOR

Proline Accumulation in Heavy Metal Stressed Plants:
    An Adaptive Strategy   179
    P. SHARMILA AND P. PARDHA SARADHI

Influence of Metals on the Biosynthesis of Photosynthetic
    Pigments   201
    B. MYŚLIWA-KURDZIEL AND K. STRZAŁKA

Heavy Metal Influence on the Light Phase of Photosynthesis 229
   B. MYŚLIWA-KURDZIEL, M.N.V. PRASAD AND
   K. STRZAŁKA

Gas Exchange Functions in Heavy Metal Stressed Plants 257
   ELŻBIETA ROMANOWSKA

Heavy Metal Interactions with Plant Nutrients 287
   Z. KRUPA, A. SIEDLECKA, E. SKÓRZYNSKA-POLIT
   AND W.MAKSYMIEC

Functions of Enzymes in Heavy Metal Treated Plants 303
   A. SIEDLECKA AND Z. KRUPA

Heavy Metals and Nitrogen Metabolism 325
   GRAŻYNA KŁOBUS, MAREK BURZYŃSKI AND
   JÓZEF BUCZEK

Plant Genotypic Differences Under Metal Deficient and
   Enriched Conditions 357
   S. LINDBERG AND M. GREGER

Genotoxicity and Mutagenicity of Metals in Plants 395
   BRAHMA B. PANDA AND KAMAL K. PANDA

Metal Detoxification Properties of Phytomass:
   Physiological and Biochemical Aspects 415
   MIGUEL JORDAN

Subject index 427

Biodiversity index 431

Preface

Metals impose great influence on various aspects of life at all levels of its organization. Several metals are the essential constituents of biomolecules and participate in many important biological processes. However metals can be harmful when present in excess amounts, while a few of them are toxic to biota even at extremly low concentration.

Phytotoxicity of metallic compounds dates back to 1896, when the French farmer applied "Bordeaux mixture" (copper sulphate, lime and water) to control fungal pests. A few years later, spraying of iron sulphate solution on a cereal crop dominated with weeds resulted in killing of only the weeds.

Plant-metal interactions are complex and depend on many factors. The most important being the plant species exposed, its developmental stage and the chemistry of metal including its concentration. Plant-metal interactions may have beneficial, harmful or will have no effect on the plant, depending on metal species and concentration.

Various physiological and biochemical processes in plants are affected by metals. The contemporary physiological, biochemical and molecular investigations on toxicity and tolerance in metal stressed plants are prompted by the growing metal pollution in the environment. In order to understand as how plants survive and accumulate metals from the contaminated and polluted environment, the physiology, biochemistry and molecular biology of plants under metal stress need critical investigation to ascertain answers about the underlying processes of acclimation (short term) and adaptation (long term).

Metal toxicity to plants has great impact and relevance not only for plants but also to the ecosystem in which the plants form an integral component. Plants growing in metal polluted locations exhibit altered metabolism, growth reduction, lower biomass production and metal accumulation and these functions are of human health concern. Edible plants with high doses of accumulated toxic metals are harmful not only to humans but also for the animals when used as animal feed.

Metals affect numerous biochemical and physiological processes in plants. On the other hand, plants developed various defence mechanisms to counteract metal toxicity. Only detailed study of these processes and mechanisms would allow scientists to understand the complex plant-metal interactions. Therefore, the aim of this book is to give an overview of the most important aspects of physiological and biochemical basis for metal toxicity and tolerance in plants. Thus, the contents of the book encompass the mechanisms of metal uptake and transport in plants, influence of metals on vital processes such as biosynthesis of the photosynthetic pigments, light and dark phases of photosynthesis, respiration and gas exchange, nitrogen metabolism and interaction of metals with plant nutrients. Special emphasis is given to genotoxicity and mutagenicity of metals, importance of plant genotype and their interaction with metals, capability of selected plant genotypes to hyperaccumulate certain metals, and the function of enzymes in metal exposed plants. Similarities and differences between vascular and non-vascular plants in their responses to metals are also dealtwith. The defence mechanisms against metal toxicity, including the role of organic acids, glutathione, metal chelating peptides and proteins as well as adaptive strategy involving proline accumulation are also included. The last chapter deals with the role of phytomass in metal detoxification.

The book was intended to provide the reader with the state-of-the-art-knowledge about the selected topics *supra vide*. In different chapters, the experimental data and the current trends were reported and some general conclusions were also drawn.

It is believed that this book might serve as reference for plant physiologists, plant biochemists and molecular biologists who have interest in the biology of heavy metal toxicity and tolerance. It will also be of great interest to university and college teachers and students of environmental biotechnology, environmental botany, agriculture, horticulture, silviculture, soil science and plant ecophysiology.

M.N.V.P. is grateful to Padma Bhushan Professor P. Rama Rao, Vice-Chancellor, University of Hyderabad; Professor T. Suryanarayana, Dean, School of Life Sciences for encouragement and supporting academics. K.S. acknowledges the financial support of the Polish Committee for Scientific Research (KBN) under the project No 6 P04A 028 19. The editors are extremely thankful to the DST-KBN Indo-Polish programme of mutual cooperation (DST/INT/POL/P-4/1 dt.19.3.01) which catalyzed the scientific collaboration not only between the editors but also synergized active participation of the leading scientists in the participating nations. Thanks are also due to all contributors for the comprehensive and critical reviews. The editors thank Dr Jacco Flipsen for his keen interest and punctuality in its release.

Editors

Contributors

Editors

M.N.V. Prasad
Department of Plant Sciences
University of Hyderabad
Hyderabad 500046, India

K. Strzałka
Department of Plant Physiology and Biochemistry
The Institute of Molecular Biology
Jagiellonian University
7 Gronostajowa Street, PL 30-387 Kraków, Poland

Contributors

E. Cseh
Department of Plant Physiology
Eötvös University
Budapest, H-1445 P.O.Box 330, Hungary

M. Greger
Department of Botany
Stockholm University
Stockholm, S-106 91, Sweden

M. Jordan
Departamento de Ecología, Facultad deCiencias Biológicas
Pontificia Universidad Católica de Chile, Casilla 114-D
Santiago, Chile

G. Kłobus
Marek Burzyński and Józef Buczek
Department of Plant Physiology, Institute of Botany
Wrocław University, PL 50–328 Wrocław, Poland

Z. Krupa
Department of Plant Physiology
Maria Curie-Skłodowska University
Akademicka 19, PL 20-033 Lublin, Poland

S. Lindberg
Department of Plant Biology
Swedish University of Agricultural Sciences
Uppsala, S-75007, Sweden

W.Maksymiec
Department of Plant Physiology
Maria Curie-Skłodowska University
Akademicka 19, PL 20-033 Lublin, Poland

N. Mallick
Department of Agriculture and Food Science
Indian Institute of Technology
Kharagpur India

H. Matsumoto
Research Institute for Bioresources
Okayama University,
Chuo, Kurashiki, Okayama, 710-0046, Japan

B. Myśliwa-Kurdziel
Department of Plant Physiology and Biochemistry
The Institute of Molecular Biology
Jagiellonian University
7 Gronostajowa Street, PL 30-387 Kraków, Poland

S. Ottonello
Department of Biochemistry and Molecular Biology
University of Parma, Parma, Italy

B.B. Panda
Genecology and Tissue Culture Laboratory
Department of Botany
Berhampur university, Berhampur 760 007, Orissa, India

K.K. Panda
Genecology and Tissue Culture Laboratory
Department of Botany
Berhampur university, Berhampur 760 007, Orissa, India

L.C.Rai
Department of Botany
Banaras Hindu University
Varanasi-221005, India

E. Romanowska
Department of Plant Physiology
University of Warsaw
02–096 Warszawa, Miecznikowa 1, Poland

L. Sanità di Toppi,
Department of Evolutionary and Functional Biology
Section of Plant Biology
University of Parma, Parma, Italy

P. Pardha Saradhi
Department of Environmental Biology
University of Delhi
Delhi 110 007, India

P. Sharmila
Department of Environmental Biology
University of Delhi
Delhi 110 007, India

A.Siedlecka
Department of Plant Physiology
Maria Curie-Skłodowska University
Akademicka 19, PL 20-033 Lublin, Poland

E. Skórzynska-Polit
Department of Plant Physiology
Maria Curie-Skłodowska University
Akademicka 19, PL 20-033 Lublin, Poland

K. Strzałka
Department of Plant Physiology and Biochemistry
The Institute of Molecular Biology
Jagiellonian University
7 Gronostajowa Street, PL 30-387 Kraków, Poland

B. Tomaszewska
Adam Mickiewicz University
Institute of Molecular Biology and Biotechnology
Department of Plant Metabolism
Fredry 10, 61–701 Poznań, Poland

CHAPTER 1

METAL PERMEABILITY, TRANSPORT AND EFFLUX IN PLANTS

E. CSEH
*Department of Plant Physiology, Eötvös University
Budapest, H-1538 P.O.Box 120, Hungary*

1. INTRODUCTION

The alleviation of heavy metal stress affecting higher plants include the fields of uptake, translocation and efflux. In order to understand the arising problems first we should go through the present knowledge on these basic processes.

On the basis of the accumulated data we can agree that in higher plants the greater part of heavy metals is adsorbed in the roots and their transport to the shoot is prevented. This transition in the distribution ratio may be very significant: in cucumber 86.7% (Fodor et al., 1998), in stinging nettle about 98% (Fodor et al., 1996) of Pb remains in the root. In tomato this value falls between 81.7 – 90.1%, in sunflower between 77.6–56.5% (Antosiewicz, 1993). *Silene maritima* excludes Zn from the shoot (Baker, 1978). Cd also accumulates in the root (Cutler and Rains, 1974; Jarvis et al., 1976; Cataldo et al., 1981; Barceló et al., 1988; Baker, 1981) in spite of that it is a much more mobile element than Pb, Fe or Cu.

2. STRUCTURE OF CELL WALLS: STRUCTURAL CARBOHYDRATES

It is highly questionable where the heavy metals are captured in the root tissues because a large portion of ^{59}Fe – 80-90% after a 60 minute uptake period and 50% after one week consequent growth – can be removed from the roots by the method of Bienfait et al., (1985) (Fig. 1). After a 2-hour uptake period half of the amount of ^{64}Cu can be replaced by inactive Cu or 66% can be removed by washing with Na_2EDTA which unambiguously tha t they are located in the apoplast (Fig. 2). The cell wall can be

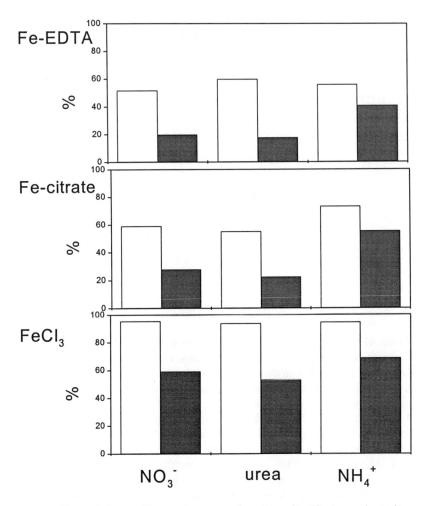

Figure 1. Removable iron from cucumber roots after 60 minutes (empty) and one week (filled) uptake period on different N-forms

considered as a large-surfaced ion exchanger on the basis of its structure, mainly due to the OH⁻ and COO⁻ groups of polysaccharides constituting the matrix, so a high cation exchange capacity (C.E.C.) may be expected in the root. In the primary cell walls the cellulose chains embedded in the matrix account for ~20% of the

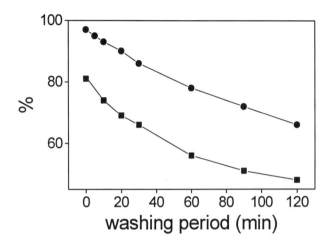

Figure 2. Exchangeable (with 10^{-6} M Cu ■) and removable (with 10^{-4} M Na₂EDTA •) copper (^{64}Cu) from wheat roots after 2 hour uptake period.

wall material and they exhibit crystalline and paracrystalline order (Reid, 1997) enclosing the intermicellar and interfibrillar spaces (AFS), (Briggs and Robertson, 1957) where the passively absorbed ions are located. Since the free space constitutes far from negligible portion of the root cell walls it is possible that the fairly mobile cations can be found in the water free space in ionic form, too. Three principal groups of compounds constituting plant cell wall matrix can be distinguished: the pectins, the hemicelluloses and glycoproteins. All three groups of compounds are extremely complicated and variable in composition. Based on this two distinct types of primary cell walls are distinguished: type 1 is found in dicotyledonous plants and monocotyledonous plants with large leaves while type 2 is associated with the grasses (Carpita and Gibeaut, 1993; Carpita, 1996). The main constituent of type 1 cell walls is mostly xyloglucan that consists of very regular molecular fragments and give 20% of the wall. The xyloglucan chain is split at the glucose residues without substituents by the endo-β(1→4)-D-glucanase. The resulting molecular fragments, oligosaccharides (Fry et al., 1993) are growth regulators because of their fucose content. Consequently, any effect increasing the activity of endoglucanase will secondarily influence growth.

Type 2 cell walls contain xylan-rich hemicellulose (glucurono-arabino-xylan, GAX) with side chains containing acidic groups. This is important concerning the binding of cations because graminaceous plants have cell wall with little pectin. Besides, a major feature of Poales is the enrichment of aromatic substances in the primary cell wall (Carpita, 1996; Ishii, 1997), that are otherwise characteristic to the lignified secondary walls.

Most of them – the esters of hydroxycinnamates, ferulates and p-coumarates – bind to galacturonoarabinoxylan. Ferulates are able to cross-link two GAX chains leading to the formation of a network. Observing the possibility of cation adsorption it is important that type 2 cell walls, the primary wall of graminaceous plants, contain little pectin. In type 1 walls xyloglucans are oriented along with cellulose microfibrils while in type 2 it is not possible because of the side-chains of the glukuronoarabinoxylan molecule.

Concerning the binding of heavy metals the most important component of the cell wall is the pectin, because the backbone of the acidic pectins are composed of D-galacturonic acid residues connected by α (1→ 4) - linkages in case of homogalacturonan and rhamnogalacturonan I. In case of rhamnogalacturonan II the galacturonic acid and glucuronic acid content is about 30% (Reid, 1997). The homogalacturonan fragments, similarly to the oligosaccharides made up of xyloglucan, play a role in cell growth via the induction of a phytoalexins (Aldington and Fry, 1993), influence membrane polarization and through it the fluxes of various substances and ions so that they are able to induce potassium efflux (Mathieu et al., 1998). Rhamnogalacturonan II that is composed of at least 12 different monosaccharide residues, several of which are unusual, is attributed to have a similar bioactive role. It inhibits the membrane transport of certain amino acids (Aldington and Fry, 1994).

These chain-forming acidic fragments provide the opportunity for the high adsorption of cations in dicots and large-leaved monocots. The pectin content of Poaceae members is much less resulting in significantly lower cation exchange capacity (Keller and Deuel, 1957). It is interesting that it took almost 50 years to give an explanation to the special behaviour of Poaceae.

2.1. The glycoproteins of the cell wall

Cell wall proteins may be claimed to play a regulatory role in the development of plants on the basis of their varied composition and their expression in different tissues. Cell wall proteins are encoded by multigene families, the members of which are stress-inducible and consequently their regulation may be cell type specific (Reiter, 1998). They may be present in a considerable amount reaching 10% of the cell wall. They are distinguished on the basis of their amino acid composition: hydroxyproline-rich (HRGPs = extensin), glycine-rich (GRPs), proline-rich (PRPs) and arabinogalactan (AGPs) proteins. Besides these chimeric proteins exist that contain extensin-like domains. These cell wall proteins are accumulated in different cell types, so they may have different functions. Certain amino acid sequences may characterise the proteins listed above: $Ser(Hyp)_4$ in HRPGs, $(Pro)_3$ XYLys in PRPs, $Ala(HyP)_n$ in AGPs and glycine makes up 70% of GRPs (Cassab, 1998; Reiter, 1998; Sommer-Knudsen et al.,1998; Showalter,1993; Keller, 1993).

Besides $Ser(HyP)_4$ extensin contains valine, lysine, histidine and tyrosine in varying amounts. The formation of the extensin network and the rigidity of the cell wall can be attributed to the isodityrosine linkages between two Tyr residues of different molecules (Fry, 1982). Extensin is of basic nature because of its lysine content.

For translocation it is important that, besides the plasmodesmata of primary and secondary cell walls and the bordered pits of the phloem elements and xylem walls, what kind of cell wall proteins are synthesised and how they influence transport rates via the alteration of the permeability of plasmodesmata. Extensin may be peroxidised under stress conditions, e.g. in the presence of H_2O_2 and Fe II ions. It is assumed that the accumulation of extensin in the cell wall results in the decline of growth (Ye and Varner, 1991). Extensin also plays a part in defence against pathogens.

PRPs are mainly present in the xylem which may imply these proteins are involved in xylem differentiation or in lignification because the phenolic molecules are able to bind to prolyl residues (Luck et al., 1994). The majority of the AGPs contain more than 90% carbohydrate and they differ from the rest of cell wall proteins in that AGPs are soluble and can be found in the extracellular space rather than in the cell wall.

2.2. The role of the cell wall in adsorption of heavy metals

On the basis of the composition of the cell wall, adding the fact that the substances forming the cell wall should be produced and transported out of the cell through the plasmalemma under strict regulation, it is hard to accept the view that heavy metals once bound to the cell wall of the root do not influence the metabolism of the plant and so it escapes from their damaging effect and attributing the damage to those accumulating in the cytoplasm or cell organelles. What is the validity of this hypothesis? May the widely sought primary effects originate from the binding of heavy metals in the cell wall? Can the binding of expansins responsible for the extension of the cell wall change if they meet heavy metals?

For the extension of the cells the molecule chains should slip on each other. During the investigation of the molecular background of "acid growth mechanism" of plant tissues new proteins called expansins were discovered that are responsible for the extension of the cell wall (McQueen-Mason et al., 1992; Cosgrove, 1997; 1999). They occur in very small amount in the non-crystalline regions of cellulose. They do not exhibit hydrolytic activity and any enzymatic activity. The protein consists of two domains: one for the binding to cellulose and another with putative catalytic activity likely to function by disrupting the non-covalent bonding between wall polysaccharides (Cosgrove, 2000; McQueen-Mason and Rochange, 1999). The functioning of expansins is essential for the uptake of substances because without the extension of the cells there is no water uptake and only the exchange of substances may occur. Expansins were found in all studied plants. They comprise a large superfamily with at least two major branches, α-expansins and β-expansins.

The amino acid sequences of various α-expansins are similar in 70-90%, the β-expansins are much more variable. Its accumulation was shown in the cortex of corn roots near the endodermis. It was also found in the cell wall and intercellular spaces while in the cell probably the secretory vesicles contain expansins (Zhang and Hasenstein, 2000).

Our present knowledge on the extensibility of the cell wall supports the view that heavy metals induce rigidity in the cell wall resulting in the inhibition of cell expansion and consequently growth. The process of synthesis and extension of the cell wall can be separated. Incorporation of new substances without the possibility of extension causes only the thickening of the wall. Naturally it occurs during the formation of the secondary cell wall. For the expansion of the wall the complicated molecular network should loosen. For this, two structures should necessarily change: glycoproteins (pectins) where the molecules are attached by Ca-bridges and covalent bonds at negative charges and the cellulose – xyloglucan network, where the xyloglucans are linked to the surface of cellulose microfibrils by hydrogen bonds joining several microfibrils together (Cosgrove, 1993; McQueen-Mason et al., 1993). Xyloglucans often have numerous side chains which hinder the attachment to the cellulose microfibrils, so "debranching" stimulates their binding. The decrease in the methylation and acetylation of pectin also favours the formation of a more rigid structure (Cosgrove, 1997).

What is the heavy metal concentration range which supports the view that heavy metal retaining ability of the root leads to tolerance? Based on the results of uptake experiments it can be postulated that a peak can be recorded i.e. exchangeable potassium can be measured in the apparent free space if there is no Ca^{2+} in the solution. In the presence of calcium the uptake is linear i.e. no potassium is adsorbed in the cell wall and all binding sites are occupied by Ca^{2+}. In plant cells at optimal Ca supply at least 50% of Ca is in the cell walls (Marschner, 1995). On a dry matter basis 120-130 µM Ca is in the cell wall. Compared to this the concentration of other heavy metals, except for Mn and Fe, in non-inhibited situation, hardly reach 1 µM. Among microelements, in a tolerant plant, only Mn reaches the level of Ca, but it will not accumulate in the cell wall because of its high mobility. These data give evidence for the possibility of toxic heavy metals may have enough space in the cation binding sites of the matrix. However, it does not mean they are ineffective once bound to the cell wall. In onion root tips it was found that Pb increased the amount of polysaccharides and often irregularly thickening cell walls (Wierzbicka, 1998). Simultaneously the surface of the cell membrane increased, plasmotubules were formed, which may substantially increase the uptake surface and the number of binding sites.

The protoplast of living and growing cells maintain dynamic balance with the surrounding cell wall because it continually supplies substances required for the growth of the cell wall and the cell wall in turn provide the possibility for water and ion uptake from the external space due to its extensibility. When the cell looses its ability to expand net water transport immediately stops but the influx and efflux of substances may go on because of the metabolism of the cell. In growing cells the plasma membrane is appressed against extracellular material. It is assumed that turgor pressure is responsible for this appression (Kohorn, 2000). The plasmalemma should contain all the proteins responsible for the transport and synthesis of cell wall materials. One and most important of these is the cellulose synthase. This enzyme is a complex protein consisting of several subunits electronmicroscopically appearing as a rosette connecting to saccharose synthase at the inner side of the membrane synthesize a microfibril containing ~36 chains. Saccharose synthase generates UDP-glucose units which form cellobiose constituting the $\beta(1\rightarrow 4)$-glucan chain of cellulose (Delmer and Amor, 1995; Albersheim et al., 1997; Delmer, 1999; Brown Jr. and Saxena, 2000). But this enzyme responsible for cellulose synthesis also synthesizes $\beta(1\rightarrow 3)$-glucans, callose as a response to wounding or different biogenic or abiogenic stresses (Kudlicka and Brown,

Jr., 1997). Aluminum was proven to induce callose synthesis in the cell wall next to plasmodesmata which consequently causes the closure of plasmodesmata resulting in the perturbation of cell-to-cell communication (Sivaguru et al., 2000). The other enzyme sensitive to stress factor is the WAK i.e. the receptor kinase connected to the cell wall which has a cytoplasmic Ser/Thr protein kinase domain, spans the plasma membrane and extends a domain into the cell wall (He et al., 1996, 1999). Mutations in WAKs demonstrate that they are essential for plant development and required during the pathogen response (He et al., 1998). Large amount of WAK is also covalently linked to pectin. There is small population of WAK that is soluble with detergents.

3. ROLE OF PLASMALEMMA IN HEAVY METAL TOLERANCE

Plasmalemma, may exclude toxic ions or their toxic concentrations if the highly specific transport proteins, others regulating efflux and the proton symport-antiport systems, i.e. its integrity is preserved. The integrity of the plasmalemma can be checked by measuring the efflux of potassium ions. Membrane integrity may be lost by SH oxidation, the inhibition of H^+ATPase and lipid peroxidation due to free radicals (Meharg, 1993). Plasmalemma is damaged first of all by Cu ions. In copper tolerant species Zn, Cd, Pb tolerance may be observed.

In order to understand the role of the plasma membrane the transporters should be discovered. It is still a question of debate, whether heavy metals penetrate the membrane through a common transporter as it was hypothesized by Clarkson and Lüttge (1989) or they have specific transporters.

3.1. Transport ATPases

The primary motive force for membrane transport is provided by the proton transport ATPases (H^+ATPase) and the vacuolar H^+PPase. They carry protons through the membranes, creating the electrochemical potential and pH gradients, that is the proton motive force ($\Delta p = \Delta\mu_H^+ / F = \Delta\Psi - 59\Delta pH$). The H^+ATPase located in the plasmalemma generate -120-160 mV (inside negative) membrane potential and a pH gradient of 1.5-2 magnitude. This is the driving force for all metabolite movement through the membrane, and it determines the direction of transport, too. The negative charge inside the membrane promotes the influx of cations by facilitated diffusion, through channels or by uniport, and the symport of anions and organic molecules (sugars, amino acids) by one or more protons. The energy for the movement against the concentration gradient is supplied by the H^+ATPase. The hyperpolarization caused by proton extrusion leads to opening of specific cation channels, while depolarization causes their closure, at the same time the efflux channels are opened (K^+_{in} and K^+_{out} channels of stomatal guard cells). Anion channels are regulated similarly (Czempinski et al., 1999).

The plasmalemma H^+ATPase belongs to the vanadate sensitive phosphoenzyme family. It is a P-type ATPase which produces phosphorylated intermediate during the ATP hydrolysis that is induced by K^+ (Serrano, 1989). Its substrate is MgATP. The polypeptide is active either as a 100 kDa monomer or as a homodimer. The monomer has 10 transmembrane domains, both the C and N terminal and a large hydrophilic loop are located at the cytoplasmic side of the membrane (Sze et al., 1999). Its expression is regulated by a large, multigene family which consists of 10 different genes in *Arabidopsis* (Sussman, 1994). There are isoforms of the P-ATPase synthesized in plants, that are specific (different isoforms, e.g. epidermis or phloem companion cell) or several genes are expressed differently among plant organs (See reviews by Baudouin and Boutry, 1995; Sze et al., 1999). The H^+ATPase was the first protein to be cloned (Harper et al., 1989; Pardo and Serrano, 1989; Boutry et al., 1989).

The P-type ATPases can be found in almost all organisms and they can carry numerous ions besides protons (K^+, Na^+, Mg^{2+}, Ca^{2+}, Cu^{2+}, Cd^{2+}) which is one of their most important tasks in plants. P-type ATPases often show very low similarity to each other. But those parts of the polypeptides that are believed to be involved in communication between ATP hydrolysis and conformational change, in ion binding and in ATP binding, respectively, are common between all P-type ATPases. Some of them appeared during the early stages of evolution (in bacteria) others appeared much later (H^+ATPase) (Palmgren and Axelsen, 1998). Most organisms appear to contain a variety of P-type ATPases. Yeast, for example, has 16 P-type ATPases belonging to all five major families. In *Arabidopsis,* more than 40 type ATPases are predicted based on cloned genes. P-type ATPases belong to five families: heavy metals-, phospholipids-, type A and B Ca^{2+}ATPases-, H^+ATPase and no assigned specificity ATPases (Palmgren and Harper, 1999).

3.1.1. Ca-ATPases, Ca-channels
Another relevant viewpoint for heavy metal effect – mostly because there are easily mobilized metals like Cd, Ni, Mn, Zn – is the role of Ca^{2+} and the mode of its transport. Ca concentration may not exceed 200 nM in the cytosol. Besides, it prevalently binds to calmodulin. Since the free Ca^{2+} is a signal, the regulation of its concentration is a key factor to the proper function of metabolism. Ca ions are transported by 3 different systems: the Ca-ATPase, the Ca-channel and the H^+/Ca^{2+} antiport. The antiporter protein was isolated in 1996 (Hirschi et al., 1996) and it occurs in both vacuolar and cell membranes, moreover it was shown in thylakoids in the chloroplast (Ettinger et al., 1999). The showing little similarity in different organisms. It is a relatively small molecule (~ 48 kD) consisting of a few hundreds of amino acids. Its Ca-transporting activity is considerable, greater than that of Ca^{2+}-ATPase (Ueoka-Nakanishi and Maeshima, 2000).

Plant Ca-ATPases, similarly to Ca-ATPases found in animal cells can be divided into two groups (IIA and IIB). The IIA Ca-ATPases can be found in the ER and are not stimulated by calmodulin. The IIB ATPase is stimulated by calmodulin and is located in different endomembranes. The genes for both of them have been identified in plants, too (Palmgren and Harper, 1999). Ca-ATPases are also capable for Mn^{2+} translocation (Liang et al., 1997). Ca channels located in the plasma membrane have two major classes (White, 1998). One of them is relatively non-selective, the other type is more selective voltage dependent channel. Voltage dependent channels are activated by membrane depolarization, caused various effects, for example the increase of Ca^{2+} concentration. In the vacuolar membrane, there are at least 4 different Ca channels, two of which are ligand gated and two are voltage gated. Ca channels exist in the endomembranes, too.

3.1.2. P-type heavy metal pumps

It is of great interest for all ions, but particularly for heavy metals how their uptake from the apoplast, their efflux from the cytoplasm and their transport to a specific compartment is regulated. In addition the borderline of concentration between essentiality and toxicity is narrow. Despite there are evidences for the transport of copper, zinc, cadmium and silver by P-type ATPase, at present none of the plant heavy metal pumps have been biochemically characterised. The C and N terminals of these proteins have regulatory domains with heavy metal binding motifs, but their role in the transport is unknown (Palmgren and Harper, 1999). In human tissues with abnormal Cu metabolism genes causing illnesses have been isolated and sequenced, and they turned out to be coding for P-type Cu-ATPases that are membrane integral proteins. In yeast the analogs of these genes have also been identified that also code for P-type Cu-ATPase and transport Cu out of the cytosol. It is not yet discovered how the ATPase transport copper but it seems possible that the transport form is Cu^+. The heavy metal binding domain and the carrier domain of the molecules may not be identical (reviewed in Harris, 2000).

3.1.3. ABC transporters (ATP-binding casette)

The main characteristics of ABC transporters is that they are energized directly by MgATP, so they do not gain energy through the chemiosmotic way by proton movement like other transport processes. So far more than 100 ABC proteins have been isolated from living organisms. All transporter proteins consist of two basic structural units: the hydrophobic integral domain (four or six transmembrane spans) and the ATP-binding domain facing the cytosol. Functional proteins can also result from the association of two homologous polypeptides. The substrate for plant ABC transporters is MgATP and not ATP. ATP can be replaced by GTP and UTP to a certain extent, but not inorganic pyrophosphate. They are sensitive to vanadate. They are located in the membranes in eucaryotic cells where they catalyse the efflux of a number of compounds but they can also be found in the membranes of peroxisomes, mitochondria, endoplasmic reticulum and the vacuole. They can carry a remarkable variety of substances: ions, carbohydrates, lipids, xenobiotics, anti-cancer drugs, pigments and high molecular mass peptids. In addition, some ABC proteins can act as ion channels or regulators of ion channels. They basically influence metabolism, they may take part in cell detoxification by extrusion of toxic materials, e.g. heavy metals or by compartmentalisation (reviewed in Rea, 1999; Davies and Coleman, 2000). It has been proven that among organic molecules they transport glutathione-S conjugates and oxidised glutathione. It was demonstrated with yeast (*S. pombe*) mutant that the glutathione derivative phytochelatins are transported to the vacuole in both Cd-complexed or non-complexed form by ABC transporters.

4. HIGH AND LOW-AFFINITY TRANSPORT SYSTEMS

The first evidence for the existence of high and low-affinity systems was provided by the example of potassium (Epstein, 1972). In general, high-affinity systems function at low external concentrations, while the low-affinity systems function at high external concentrations. The threshold concentration is at about 0.4-1 mM. The high-affinity system acts with carrier proteins with high energy consumption, transports relatively few molecules and has saturation characteristics. The low-affinity system acts with channel proteins with low energy consumption, transports a large number of molecules, and its concentration dependence is linear. The affinity of the two systems differs with 3-4 orders of magnitude. The transport characteristics originally found with the largely mobile K^+ and Cl^- later turned out to be valid for all transported substances. High and low-affinity systems work in both the plasmalemma and the tonoplast. Only in the last two years, after the genetic background of transporters was revealed, the carriers (high-affinity system) were proven to be able to function at both low and high concentrations in case of potassium.

As it was mentioned before, in case of microelements probably due to their very low amount in the external solution and the cell, a single multisubstrate system was expected and the existence of transport systems with different affinity seemed unlikely, too. In case of the macroelements the external concentration ranges between 5-6 orders of magnitude to which roots should acclimate. Concerning the toxic nature of microelements this can be shifted only towards deficiency. Eide (1998) also begins his review on the heavy metal transport processes in yeast that "this field was filled with dragon up to the last years". In understanding the transport of essential and toxic microelements only small steps are made, because their movement does not occur only in ionic form and the stability constants of their complexes vary according to the composition of their environment and the pH. In the soil, due to the microbial activity and at the root surface due to the heterogeneity of the exuded complex-forming substances, it is a hard task to determine the real amount of substances eventually reaching the cell wall and the uptake surfaces. Measurements made with wheat and cucumber roots showed that the adsorption of ^{59}Fe-EDTA, ^{59}Fe-citrate and ^{59}FeCl$_3$ at the root surfaces was different in orders of magnitude (Cseh et al., 1994). In addition the complex may penetrate the membrane causing further changes in metabolism. Microelements that are present in low amount in the cell require carrier molecules in order to get to the specific molecules where they are needed. Their uptake, transport and assimilation should happen in a strictly regulated manner.

4.1. The genetic background of uptake mechanisms: our knowledge on the uptake

The description of the genetic background of uptake mechanisms made it clear that each metal ion has multicomponent, specific uptake system. At low external concentration when deficiency symptoms appear the expression of transporter genes increases, high-affinity transport systems are induced or activated. In normal supply conditions the low-affinity systems work. The most detailed information is available about the uptake of Fe perhaps because there are many similarities between the cells of yeast and higher plants and it is much easier to work with yeast mutants compared to higher plant cells. So the results achieved with yeast are always a few steps further.

Plants are divided to two great groups on the basis of iron uptake: Strategy I and II. Grass species belong to Strategy II (Marschner et al., 1986). They give off iron complexing phytosiderophors – compounds belonging to the mugineic acid family – to the environment (Takagi et al., 1984) which complex Fe(III) facilitating its uptake. Dicotyledonous plants and monocots with large leaves belong to Strategy I. In these plants iron must be reduced before the uptake (Chaney et al., 1972). Plants respond to iron deficiency by an increasing electron transport and proton extrusion. The considerable acidification of the root surface promotes the solubilization of iron otherwise insoluble below pH 4. All plant cells have a reducing capacity which is due to a low activity, constitutive electron transport, and a turbo electron transport induced mainly by iron or other microelement deficiencies (Bienfait, 1988) that increase reducing capacity generally 5-8-fold, occasionally 20-fold. Redox systems located in the plasma membrane play a part not only in iron uptake but also in blue light utilisation (response), defence against pathogens and the cell wall synthesis.

4.1.1. Redox proteins
It is generally accepted that the electron donor of redox proteins is NAD/P/H (Crane et al., 1991) and their artificial electron acceptor is ferricyanide, while naturally occurring acceptors may be iron-chelates (Fe-citrate, Fe-malate etc.). It has long been argued (Holden et al., 1991) whether ferricyanide reductase and iron-chelate reductase are the same or not (Brüggeman et al. 1990). Some researchers declare that the constitutive reductase can only reduce ferricyanide (Bienfait, 1985) and a new enzyme is synthesized under iron deficiency or the activity or quantity of the constitutive reductase increases and reduce Fe-chelates (Holden et al., 1991). Anyhow, the enzyme of the turbo electron transport has not been identified, yet.

Eide (1998) summarises the results achieved with yeast cells. *S. cerevisiae,* - similarly to other cells – has two Fe uptake systems and it also exhibits an increased reducing capacity under iron deficiency capable of reducing Fe(III) in various chelates. In yeast *Fre1* gene and protein was isolated which are responsible for the reduction. Some parts of this protein resembles cytochrome b_{558} subunit that is able to bind FAD and NADPH. The Fre1 is a glycosylated flavocytochrome. In the reduction and uptake of iron *Fre1-7* gene is involved which has an unrevealed role. It is in turn, clear that Fre1 protein is not enough for the reduction, it is supplemented by the Fre2 and Utr1 proteins. Utr1 is located on the cytoplasmic side, it is a hydrophilic, soluble protein involved in Fe^{3+}/Cu^{2+} reduction. Under iron deficiency a high affinity system is switched on – in yeast cells its activity increases 200-fold – which is regulated by the *Fet3* gene and it is responsible for the uptake of Fe^{2+}. *Fet3* encodes a protein with remarkable similarity to a family of enzymes known as multicopper oxidase. It needs 6 Cu atoms for its activity. They catalyse four single-electron oxidations of substrate, an action that is followed by a four-electron reduction of O_2 to generate two molecules of water. Multicopper oxidase is an integral membrane protein. The other subunit of the high affinity uptake system is synthesized by the *Ftr1* gene, which acts as a permease and transports Fe(III) into the cell. The high affinity system possibly functions as follows:

Fre1 reduces iron, than it is reoxidised by the multicopper oxidase directly passing it over to the Fe^{3+} binding site of the Ftr1 permease and it is transported to the cell.

The low-affinity system is specific for Fe^{2+} over Fe^{3+} but may be capable of transporting other metal ions (Ni^{2+}, Cd^{2+}, Co^{2+}, Cu^{2+}) as well. The *Fet4* gene encodes the transport protein of low-affinity system. It is interesting that the amount of this transport protein also increases under iron deficiency.

To date, none of the genes responsible for Fe uptake has been isolated from dicotyledonous plants, except for the *Irt1* gene from *Arabidopsis thaliana* (Eide et al., 1996), which is not equivalent either to *Fet3* or *Fet4*. Irt1 accumulates under iron deficiency and it is inhibited by other divalent cations (Cd, Co, Zn, Mn) (Briat and Lobréaux, 1997). A modified version of *Fre1* gene originated from yeast increased root reducing capacity in transgenic tobacco plants (Oki et al., 1999). Best results were achieved with transgenic tobacco plants in which both Fe reducing genes of the yeast (*Fre1* and *Fre2*) were expressed. Not only iron but also Zn, Cu and Mn concentration increased in the leaves of the plants (Mok et al., 2000).

Many details have already been revealed about the proteins constituting the transmembrane electron transport chain, but the chain itself still remain obscure. It was noticed that most, if not all, plasmalemma preparations contain NADH-ferricyanide oxidoreductase and NADH- cytochrome c oxidoreductase. Undoubtedly, there are b-type cytochromes, too, and they are capable of the reduction of ascorbate. Cyt b_{561} protein isolated and partially purified from bean hypocotyl is reduced by ascorbate. It is a glycosylated molecule with 50-55 kDa molecular mass. It has been suggested that electrons can be transferred from cytosolic ascorbate to apoplastic monodehydro-ascorbate – that remains in reduced state – via cyt. b_{561} in the plasmalemma. The plasma lemma associated monodehydroascorbate reductase might recycle the monodehydroascorbate to ascorbate in the cytosol. Monodehydroascorbate reductase is so far the only redox enzyme purified to homogeneity and identified in the plasmalemma. It has a molecular mass of 45 kDa. It prefers NADH, contains one FAD and is specific for the β-hydrogen on NADH. This enzyme shows high activity with ferricyanide *in vitro* (Bérczi and Møller, 1998, 2000). However, it seems to be proven that the enzyme is not a Fe^{3+}-chelate reductase.

The Fe-chelate reductase enzyme localised in the plasmalemma is capable of reducing Fe-EDTA and Fe-citrate. In the last decade many efforts were taken to isolate the enzyme. A 31 kDa polypeptide was purified from onion roots by Serrano et al. (1994) simultaneously with the 27 kDa quinone reductase, which showed high redox activity with NADH and Fe^{3+}-citrate as substrates. A ferric-chelate reductase with similar characteristics was also purified from maize and tomato root microsomal fractions (Bagnaresi and Pupillo, 1995; Bagnaresi et al., 1997*)*. Very recently Robinson et al. (1999) identified the *Fro2* gene in *Arabidopsis*. The *Fro2* gene encodes a protein with a predicted molecular mass of 82 kDa and having a FAD binding domain. Ferric-chelate reductase activity was fully restored in deficient mutants when the *Fro2* gene was inserted. It is interesting that the need for iron reduction is wide accepted, the ferricyanide and Fe-EDTA reduction of various leaf and root tissues were measured, the different behaviour of graminaceous species and other plants was clearly identified and in spite of these the redox systems found in the plasmalemma are still not satisfactorily connected to the reduction of iron, especially to the induced activity under iron deficiency. It is well evidenced that NAD(P)H oxidase, b-type cytochromes, quinones (vitamin K_1), flavins, metals (Fe, Cu, Mn) can be found in the plasmalemma. Different research laboratories isolated redox systems (malate dehydrogenase, quinone reductase, nitrate reductase, Fe-chelate reductase) and they localized them in the outer or inner side of the membrane (see: Bérczi and Møller, 2000) but they failed to put them in specific order to form a complete system.

4.1.2. Ascorbic acid
Ascorbic acid which can accumulate to millimolar concentration in plant tissues might deserve much larger concern. More than 10% of soluble carbohydrates is ascorbate and it found in highest concentration in the chloroplast (Noctor and Foyer, 1998). Glutathione and ascorbate can act as antioxidants in either enzymatic or non-enzymatic way. Ascorbate is able to react directly with free radicals: the oxygen, the hydroxyl radicals, superoxide and singlet oxygen. It prevents the activity of enzymes that have prosthetic transition metals. Two transition metals, Cu and Fe, catalyse the formation of OH radicals (Haber-Weiss reaction) which can be prevented by ascorbic acid.

Two protons and two electron are released when monodehydroascorbate (also called as ascorbate free radical) and dehydroascorbate is produced. Both dehydroascorbate reductase and monodehydroascorbate reductase were isolated in plant cells. The former one is a component of the glutathione cycle (Noctor and Foyer, 1998) while the latter was identified by Bérczi and Møller (1998) as the ferricyanide oxido-reductase. Ascorbate is present not only within the cell but also in the apoplast. In the cell wall it has a definite role in the elimination of the peroxide radicals thus facilitating cell expansion. Ascorbate is required for the hydroxylation of proline (it keeps iron in reduced state), which is one of the precursors of extensin. Peroxidases act in the formation of isodityrosine bonds of the chains constituting extensin which increase the rigidity of the wall decreasing the expansibility. Caffeic acid, ferulic acid, coniferyl alcohol molecules involved in lignification may be oxidized and thus connect hemicellulose chains (de Cabo et al., 1996). The high ascorbic acid level in the apoplastic fluid would prevent the formation of diphenyl bridges at the same time as it consumes hydrogen peroxide necessary for peroxidase action. The decrease in the apoplastic ascorbic acid level with age will decrease the hydrogen peroxide consumption by ascorbic acid oxidation and it will be used by phenolic cross-linking leading to cell wall stiffening. Since phenolic radicals react much faster with ascorbate than each other (Takahama, 1998; Zancani and Nagy, 2000) the possibility of formation of a more rigid matrix network decreases, the cell is capable of further growth. The ascorbic acid content in the cell wall decreases with aging and thus growth is decreasing (Sánchez et al., 1997).

Neither reduced ascorbic acid nor oxidized dehydroascorbate freely permeates through biological membranes. Ascorbate is a hydrophilic compound that does not readily cross lipid bilayers by simple diffusion (Horemans et al., 1997). Dehydroascorbic acid is transported into the cells. Ascorbic acid appears to be synthesized in the cytosol and is exchanged with dehydroascorbate by facilitated diffusion (Rautenkranz et al., 1994., Horemans et al., 1998), so the constant presence of the oxidized and reduced form is ensured in the apoplast. As early as 1991 Askerlund and Larsson pointed out that right side vesicles filled with ascorbate are capable of reducing Fe-citrate so the ascorbate acts as electron donor in the transplasmamembrane electron transport. Ascorbate and monodehydroascorbate may be permanently present in the apoplast. Since ascorbate acts as an electron donor in the elimination of active oxygen species monodehydroascorbate is produced, e.g. during the reduction of H_2O_2 by ascorbate peroxidase. The produced ascorbate free radical (MDHA) may be reduced again by the electron carrier cyt b_{561} in the apoplast (Horemans et al., 1994), but it can also be transported to the cytosol where ascorbate is formed (Bérczi and Møller, 2000). In cucumber roots which are extremely sensitive to iron deficiency and exhibit the highest known Fe-chelate reducing capacity ascorbate and dehydroascorbate content increased by 71% and 216%, respectively, under iron deficiency compared to iron sufficient conditions (Cseh and Fodor, 1997; Zaharieva et al., 1999). The activity of the membrane-bound monodehydroascorbate reductase (3-fold) and ascorbate-peroxidase increased, too. Therefore, iron deficiency significantly influence ascorbate metabolism. Consequently, the key to iron reduction may be found in the ascorbate metabolism, taking into account that that ascorbate is capable of the *in vitro* reduction of Fe^{3+} even if it is complexed. No doubt, ascorbate and dehydroascorbate maintain a constant redox state in the apoplast.

Taking all these into consideration the only convincing fact is that Fe^{3+} is reduced before arriving at its transporter in the majority of plants. Except for the Irt1 Fe^{2+} transporter, we can only guess about the transport of iron, its translocated form and its fate in the cell.

4.1.3. Uptake of copper, manganase and zinc
In case of yeast copper, manganase and zinc uptake involves a low and a high affinity uptake system. The transporter proteins and their genetic background has been described. Similarly to Fe, the high affinity uptake system transports Cu^+ instead of Cu^{2+} (see Eide et al., 1996). The *copt1* gene isolated from *Arabidopsis* putatively encodes the Cu transport protein (Kampfenkel et al., 1995). Mn^{2+} may substitute Ca^{2+} and Mg^{2+} e.g. in the P-type ATPases so it may be transportable by them. Mn concentration exceeds that of Fe in cucumber roots, xylem sap and leaves which is an unresolved question at present (Cseh et al., 2000; unpublished). Investigating the sequence identity of zinc transporters in yeast it was revealed that together with the Irt and other transporter proteins these proteins contain a putative metal-binding domain. This was supported by the observation that in iron deficient plants Cd uptake remarkably increased which could not have been explained by either the increased proton efflux or the stimulation by the redox state. It was hypothesized that the induction of the *Irt1* gene plays a role in the uptake of other divalent cations (Cohen et al., 1998). Plant metal transporters identified to date are reviewed by Clemens (2001).

4.1.4. The intracellular transport
All heavy metals that form complexes with high stability constant would have a toxic effect in ionic state and would not be transported to its specific place. Consequently, there must be specific transporter proteins moving between the particles following their uptake. In *Arabidopsis thaliana* and other plants a novel protein family was discovered (Dykema et al., 1999), the isoprenylated, farnesylated or geranylgeranylated proteins containing putative metal-binding motifs that can bind Cu^{2+}, Ni^{2+} and Zn^{2+}. This finding has an interesting implication because the farnesyl- and the geranylgeranyl-PP are the precursors of terpenoid biosynthesis and the structural units of plant growth hormons.

5. THE IRON UPTAKE OF GRAMINACEOUS PLANTS: THE MUGINEIC ACID FAMILY AND THE NICOTIANAMINE

Graminaceous plants opposite to other plants are able to chelate Fe^{3+} by the phytosiderophores released from their roots and reabsorb the complex. The phytosiderophore mugineic acids were discovered by Japanese researchers (Takagi, 1976; Sugiura and Nomoto, 1984; Takemoto et al., 1978; Mori and Nishizawa, 1987) and presently their biosynthesis also has been revealed. S-adenosyl-methionine is synthesized from L-methionine using ATP, from which nicotianamine is produced by trimerisation catalysed by the nicotianamin-synthase. Here the pathway forks between graminaceous and other plants because only the Poaceae has the nicotianaminotransferase enzyme that synthesizes the first mugineic acid, the 2'-deoxymugineic acid (Higuchi et al., 1995a,b; Kanazawa et al., 1995). This compound occurs mostly in wheat but in smaller amount in other plants, too. The highest concentration of mugineic acids was found in barley while the lowest is produced in rice and maize (Kawai et al., 1988). In general a mixture of mugineic acids is synthesised in the plants but there is always a species-specific, dominant type.

Mugeneic acids are secreted from the apical zone of the roots together with potassium ions in a ratio of 1:1. They were shown to be transported along the potassium concentration gradient by antiport from the cortical cells. They may be secreted as monovalent anions through anion channels (Sakaguchi et al., 1999) and are synthesized permanently in the cells but the amount released to the environment is regulated by light intensity (Cakmak et al., 1998). The release starts after 3 hours of illumination and lasts only a few hours (Takagi et al., 1984). The reasons for the light regulation have not yet been revealed. The synthesis is remarkably increased by iron deficiency, it may be increased 200-fold within a few days. The uptake of Fe(III)-mugineic acid complex requires energy, but it exceeds the uptake from artificial chelators 1000-fold (Römheld and Marschner, 1986). It is capable of solubilising iron over pH 8. Neither calcium nor magnesium phosphate hinders the uptake. The stability constant is 18,1. It forms complex with other heavy metals: Cu, Zn, Co (Zhang et al., 1991a) and its secretion is induced by other divalent cation deficiencies: Mn, Zn and Cu (Ma and Nomoto, 1996). In durum wheat the phytosiderophore secretion increased in Zn deficiency (Cakmak et al., 1996) and the mobilization of Zn pools in the cell wall under iron deficiency (Zhang et al., 1991b). Not only the Fe(III)-complex is transported through the membrane but also the Zn-phytosiderophore complex (von Wirén et al., 1996). In natural conditions, mainly on calcareous soils inducing iron deficiency, it was pointed out that the increase of phytosiderophore secretion increases not only Fe uptake but also the uptake of Zn, Ni and Cd up to 200 % (Römheld and Awad, 2000). In heavy metal contaminated soils this should be taken into consideration, mainly in case of graminaceous plants releasing phytosiderophores.

Nicotianamine, the metabolite of phytosiderophore synthesis might claim similar interest as mugineic acids for it is ubiquitous for plants and only one terminal NH_2 group is replaced by OH group in the molecule compared to the efficient chelator, mugineic acid. Nicotianamine synthase (NAS) was found in all plants examined, it reaches its highest activity at pH 9 referring to its presence in organelles. Its activity is remarkably elevated by iron deficiency but it is decreased to zero in normal Fe supply (Higuchi et al., 1995b). Another enzyme, nicotianamine aminotransferase (NAAT) is found only in graminaceous plants that turns nicotianamine to mugineic acid. The activity of NAS is induced directly by iron deficiency while that of NAAT depends on the concentration of nicotianamine. It is interesting that the release of mugineic acids has a diurnal rythm, while there is no periodicity in the activities of NAS and NAAT and the endogeneous mugineic acid level. This refers to a regulation mechanism in mugineic acid release (Kanazawa et al., 1995).

Several nicotianamine synthase genes (7) have been isolated which encode NAS and NAS-like proteins from iron deficient plants. *Escherichia coli* expressing *nas1* showed NAS activity, confirming that this gene encodes a functional NAS. NAS proteins are relatively hydrophobic, indicating that they may be membrane bound and localized in some organelle. It is assumed that *nas* genes belong to a multigene family and might have a role wider than mugineic acid synthesis in graminaceous plants (Higuchi et al., 1999). In Strategy I plants iron deficiency does not increase the activity of NAS (Higuchi et al., 1995) therefore *nas* gene is not regulated by Fe in these plants. Two distinct cDNA clones encoding nicotianamine aminotrasferase, *naat-A* and *naat-B*, were indentified. The expression of both *naat-A* and *naat-B* is increased in Fe-deficient roots, while *naat-B* has a low level of constitutive expression in Fe-sufficient barley roots (Takahashi et al., 1999). It was supported that NAS and NAAT enzymes are present in specific vesicles derived from the rough ER. Mugineic acids enclosed in vesicles were shown by electromicrographs by Nishizawa and Mori (1987). They also suspected that the synthesis of specific proteins increases in iron deficient root cells, then, after the release of mugineic acids it decreases. The evidence was provided 10 years after.

The "normalising factor" was discovered by Böhme and Scholz in 1960: the morphological changes and chlorotic mesophyll of the tomato mutant *chloronerva* could be cured by the extract of other plants. The mutant plant was normalised when grafted on the wild type so the substance is transported from the root to the shoot (Scholz and Rudolph, 1968). The normalising factor turned out to be the nicotianamine (Kristensen and Larsen, 1974; Noma and Noguchi, 1976). Since it is ubiquitous for plants it was hypothesized that it might act as a phytosiderophore in non-graminaceous plants (Scholz et al., 1985). It is able to complex not only Fe but also other divalent metals. the stability constant of the complexes (log K) are as follows: Mn(II) 8.8, Fe(II) 12.1, Zn 14.7, Co 14.8, Ni 16.1, Cu(II) 18.6 (Scholz et al., 1987). Based on the facts that it is found in the whole plants and it is transported in the xylem (Pich et al., 1995) and phloem (Stephan et al., 1994) but not cells surrounding the xylem control the transport to the shoot as a second step (De Boer and Wegner, 1997; De Boer, 1999). This finding is important because a certain part of ions translocated upwards are not used for the shoot growth and transported back to the root through the phloem. Root growth also makes use of a certain part of materials taken up from the soil or transported from the shoot and the excess moves upwards again. The bulk of the circulated and recirculated ions is made up of potassium but certainly a considerable amount comes from N, P, S and Mg. Concerning microelements Fe, Zn and Cu is released from the plants it was assumed to function within the plants in complexing Fe^{2+} protecting the cell against free radical formation induced by this metal (Scholz et al., 1988). Neither nicotianamine nor mugineic acid Fe-complex is autooxidable (Walker and Welch, 1986). After the discovery of nicotianamine the primary question remained about its role in plants. It was assumed to be the transporter for Fe, Mn, Zn, Cu in the phloem for the complexes are stabile in alkaline pH and these metals may be circulated (Stephan and Scholz, 1993; Stephan et al., 1994).

It might refer to its role in the translocation pathways that the accumulation of nicotianamine was localised in the vacuoles of stele parenchymatous cells in the 1 cm zone of the root tip by immunohistochemical methods. Can it be a detoxifying agent as phytochelatins (Pich et al., 1997)? Recent investigations do not support its role in translocation despite its presence in both the xylem and the phloem. It may bind to Fe and other heavy metals (Zn, Mn, Cu) during the loading and unloading of the phloem. In the phloem sap nicotianamine seems to be present in non-complexed form (Schmidke et al., 1999).

One of the most argued questions about the role of nicotianamine is that it can only bind Fe^{2+} which was believed recently. However, detailed studies by von Wirén et al. (1999) showed that Fe^{3+} complexes of nicotianamine and mugineic acid are structurally very similar and the Fe(III)-NA complex has a much greater stability constant than Fe(II)-NA. The Fe(III)-NA complex is unstable in acidic medium but stabile at alkaline pH (7-9) that is the pH 7-7.5 value of the cytoplasm. Fe(II)-NA is not autooxidable as opposed to Fe(II)-EDTA and Fe(II)-citrate, so it prevents Fenton-reaction, the OH free radical formation from H_2O_2 i.e. protects the cell against the oxidative stress caused by Fe(II).

6. THE CIRCULATION OF IONS IN THE PLANTS

It used to be a matter of argument that the regulation of ion uptake would only consist of controlling of ion movement to the symplast of roots by transporters and ion channels and then they would be transported to the shoot through the plasmodesmata along the symplast to the xylem with the transpiration stream. The ion discrimination would be regulated by the transport systems in the tonoplast. Recently it has been revealed that the parenchyma cells surrounding the xylem control the transport to the shoot as a second step (De Boer and Wegner, 1997; De Boer, 1999). This finding is important because a certain part of ions translocated upwards are not used for the shoot growth and are transported back to the root through the phloem. Root growth also makes use of a certain part of materials taken up from the soil or transported from the shoot and the excess moves upwards again. The bulk of the circulated and recirculated ions is made up of potassium but certainly a considerable amount comes from N, P, S and Mg. Concerning microelements Fe, Zn and Cu is recirculated through the phloem while the transport of Mn is ambiguous. Ca is not transported in the phloem and its concentration in the plasma is insignificantly low (Marschner et al., 1997). The parenchymatous cell of the stele play an important role in recirculation because the substances are likely to move from the phloem to the phloem parenchyma then to its apoplast and from them to the xylem parenchyma on the same way. These cells were specialized to uptake: they are rich in cytoplasm, ER and mitochondria and bear various channel systems. (According to this hypothesis the endodermis would keep the ions in the stele). There are four different channels described to function in the xylem parenchyma cells: the KIRC and the KORC channels ensure the uptake of K^+ into the parenchymatous cells and its release to the xylem elements. The KIRC is selective for cytosol. K^+ but the KORC transports Ca^{2+} besides K^+. It is regulated by the Ca concentration of the The NORC channels are likely to transport various ions because they are non-selective for cations and anions. Their activity is increased by cytosolic Ca. The lack of selectivity is unique for plant channels described so far and their physiological role is unknown. The fourth channel is an aquaporin making water transport possible (De Boer and Wegner, 1997; De Boer, 1999). Heavy metals should enter the xylem, too, but the specific transporters are unrevealed so far.

The transport of heavy metals from the root to the shoot is highly influenced by the chelating agent used to keep iron soluble in the nutrient solution. Cucumber plants grown until flowering in 10^{-5} M Pb-contaminated solution are as green as the control and root growth is stimulated up to a few percent. Investigations carried out with ^{59}Fe-, ^{64}Cu-, ^{51}Cr-EDTA showed that the EDTA complex of heavy metals bind to roots in very low amounts. Cucumber has a significant reducing capacity (Cseh and Fodor, 1997) and the stability constant for Fe(II)-EDTA is much lower than for Fe(III)-EDTA and Pb-EDTA so as a consequence of iron reduction Pb-EDTA is formed which mostly remains in the solution preventing the toxic effect of Pb. When the nutrient solution contained ^{59}Fe-EDTA and Pb the adsorption of Fe on the cucumber and wheat roots increased 3-fold and 4-5-fold, respectively, compared to the control while with $FeCl_3$ this is not the case (Fig. 3) which seems to prove the previous hypothesis. With Fe-citrate the Pb concentration of the exudation sap increases almost 10-fold compared to the plants grown with Fe-EDTA. With Fe-citrate Pb, Ni and Cd inhibit chlorophyll synthesis and growth. The amounts of all transported heavy metals increase due to the growth inhibition but it is obscure why does citrate increase the transport of heavy metals. The reason can be the increase in citrate concentration of the xylem sap as affected by heavy

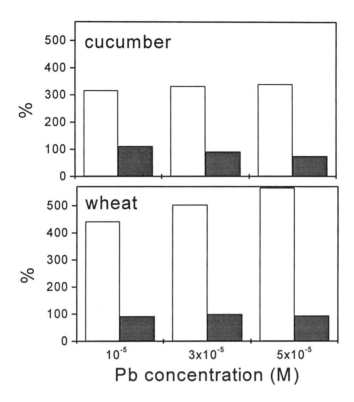

Figure 3. Uptake of ^{59}Fe-EDTA (empty) and ^{59}FeCl$_3$ (filled) in cucumber and wheat roots, respectively, at different external Pb concentrations in the percentage of the control (Pb-free solution).

metals (Tatár et al., 1998) or citrate prevents iron or other heavy metals from binding to the cell walls of the xylem (Senden et al., 1994).

7. THE EFFECTS OF HEAVY METALS ON THE EFFLUX: THE EFFECTS OF ALUMINUM

Free radical production due to the effect of heavy metals can lead to lipid peroxidation and the destruction of the membranes which could be measured by the increased ion efflux through the leaky membranes. However, plants possess numerous defensive mechanisms so the loss of semipermeability appears only at unreasonably high concentrations. It is much more interesting to monitor the change in natural ion efflux in the presence of metals. The best example for this is the alleviation of Al toxicity.

Aluminum exerts its damaging effect prevalently in acidic soils because Al is released from silicates and oxides in a phytotoxic form. It is a matter of dispute which is the toxic form of Al because it may occur as a free ion, as a complex with small molecules, in mononuclear or polynuclear form. At pH values lower than 5 Al speciates to a soluble octahedral hexahydrate form, commonly called Al^{3+}, which is believed to be the primary phytotoxic Al species. As the pH is increased the octahedral hexahydrate undergoes successive deprotonations to form relatively insoluble $Al(OH)_3$. Al^{3+} interacts most strongly with oxygen donor ligands such as carboxylate, phosphate, and sulphate groups and with number of macromolecules, including these groups (Kochian, 1995).

7.1. Sensitive and tolerant cultivars: efflux of malic, citric and oxalic acid

Plant species and cultivars within species vary widely in their resistance to Al toxicity between the exceptionally sensitive wheat varieties and tea plants containing Al in extreme concentrations (30 mg/g leaf dry weight), *Hydrangea* leaves containing 15,7 mmol Al/ fresh weight (Ma et al., 1997a). During the research of tolerance numerous valuable observations and data but also unanswered questions emerged. 2-3 mm part of the root tip was reported to be damaged in maize (Ryan et al., 1993; Kochian, 1995; Jorge and Arruda, 1997; Delhaize and Ryan, 1995) in wheat and barley (Delhaize et al., 1993a, b; Ma et al., 1997b; Samuels et al., 1997). In the root tip the cell division or more likely cell expansion is inhibited. The inhibition can be detected so quickly that no involvement of transcriptional or translational processes can be supposed. It is not easy to answer and more precisely there is no answer at present, why this inhibition appears within 30 minutes? This type of fast reaction occurs in the stomatal guard cells which is caused by the change in turgescence following water uptake. The change in the water potential is caused by the opening of the K^+_{in} channel and the consequent K^+ influx to guard cells due to the H^+ATPase activity. There are only a few studies dealing with ATPase activity of the plasmalemma and tonoplast *in vitro* and the rate of proton efflux in connection with resistance to Al. According to Matsumoto (1988) Al inhibits the proton transport in microsome membrane vesicles derived from barley roots. Lindberg and Griffiths (1993) observed similar ATPase inhibition in the plasma membrane of sugarbeet and Zhang et al. (1998) in tonoplast derived from wheat roots. Disturbances in the activity of H^+ATPases may alter the E_m values of membranes (Papernik and Kochian, 1997). The membrane can be hyperpolarized or depolarized which results in the opening or closure of voltage gated channels. Consequently the pH and ion composition especially the Ca^{2+} concentration of the cytoplasm is changed. Since the fastest process in the cell is the proton movement and its consequence the water and ion flow, this can be an acceptable explanation for the fast inhibitory effect of Al. Although Al may directly influence the activity of aquaporins.

From an economical point of view the existence of Al-tolerant and Al-sensitive varieties within a species is an important question to be solved. There is a general agreement that the site of the toxic effect is the root apex and the tolerant genotypes are capable of excluding Al from apical tissues. In tolerant wheat genotypes several authors observed malate efflux stimulated by Al which was minimal or absent in Al-sensitive lines (Delhaize et al., 1993a,b; Basu et al., 1994; Ryan et al. 1995a,b; Pellet et al., 1996). Malate release continued for 6 hours at a constant rate in the tolerant wheat genotype (Andrade et al., 1997). The released malate can compensate for the Al and prevents its uptake to the root. In other plants other organic acids may be used for the same purpose. Maize and bean releases citrate to the external medium (Pellet et al., 1995; Miyasaka et al., 1991). The Al-tolerant *Fagopyrum esculentum* releases oxalate (Zheng et al., 1998b). The stability constant of Al-oxalate depends on the Al:oxalate ratio (logK=12.4 if the ratio=1:3). LogK for Al-citrate with 1:1 ratio is 10.72 and for Al-malate is 6.0. The amount of organic acids involved in the metabolism increases in both the roots and shoots of both tolerant and sensitive maize plants, but the tolerant type has a considerably increased citrate and malate content (Pintro et al., 1997). The Al-resistant *Cassia tora L.* also releases citrate but the excretion has a long, 4-hour lag period followed by a linear increase. The excretion increases as Al concentration is increased. La^{3+} did not induce citrate excretion so that it is specific to Al (Ma et al., 1997c).

Besides the fast responses there are long-lasting ones that are observable for several days. The excretion of malate, citrate, oxalate (buckwheat, radish, oilseed rape, oat, wheat) depend on both plant species and organic acids and it changes considerably with time. The excretion of organic acids is continuous and lasts for several days at a high rate in some plants and low in others (Zheng et al., 1998a). A tolerant *Fagopyrum esculentum* accumulated a significant concentration in its leaves after 10 days of Al uptake (450 mg kg^{-1} dry-weight basis). This tolerant plants is therefore capable for transporting Al to the shoot without damage. It was assumed that it is the Al-oxalate that is taken up but more Al accumulates in the leaves from $AlCl_3$ so the metal must be complexed within the cells (Ma et al., 1998). Based on these results the question emerges: Is organic acid excretion enough for the alleviation of Al effects? It is against the „malate hypothesis" that this organic acid has the lowest stability constant compared to citrate and oxalate. The malate excretion is relatively small, thus other Al exclusion mechanisms can be supposed (Parker and Pedler, 1998). One of the possibilities for this is the excretion of PO_4^{3+} which besides precipitating with Al is protonated when released from the neutral (pH=7) cytoplasm so it alkalinize the external medium increasing the insolubility of Al (Pellet et al., 1997). Similar pH increasing effect was described in the *alr*-104 *Arabidopsis* mutant. Compared to the wild type the proton influx of the root tip increased in the presence of Al and consequently the pH of the root surface is shifted. This mutant does not give off organic acids.

Al complexes of the organic acids released from the root apex may prevent the uptake of Al but the retention can only be partial. In the evaluation of metal effects the most difficult problem is that they can be located in the cell walls or the cytosol which is hard to distinguish between. Since the different washing techniques are not able to totally remove the metal, a certain portion of Al must get into the cells. The cytosolic metal content can be directly measured only with large algae in which the cell wall and the cytosol can be separated. Al is transported through the plasmalemma of the alga, moreover some of the Al retained after desorption is taken up from the cell wall (Rengel and Reid, 1997). The same question was investigated with membrane vesicles of tolerant and sensitive wheat varieties which were isolated from the root tip and the parts above the tip (Yermiyahu et al., 1997). The adsorption of Al on plasmalemma vesicles was pH dependent: at pH 3.7 much more Al was adsorbed on the membranes from the sensitive line (Scout 66) compared to the tolerant line (Atlas 66). The membranes from Scout 66 have 26 % more negative charges than those from Atlas. Thus, there are differences in the characteristics of the plasmalemma from lines with different sensitivity to Al. The Al distribution between the symplast and apoplast can also be different in the 0-5 mm zone of the root apex. The apical cells of the resistant line take up half as much of Al to the symplast than those of the sensitive line.

The reason for the accumulation in the apoplast is the acidic groups of the substances constituting the matrix. The primary cell walls located in the elongation zone may act as ion exchangers due to their pectin content. Unusually significant is the cation exchange capacity of the non-graminaceous plants. When the pectin content was artificially increased by salt stress a positive correlation was found between the Al and pectin contents (Horst et al., 1999). Al bound to the pectin alters the physical properties of the cell wall e.g. decreases the expansibility. However it takes a lot more experiments to prove this hypothesis.

In the tolerant plants Al must get into the symplast because significant amounts are transported to their shoot. *Fagopyrum esculentum* releases oxalate to the external medium but it takes up Al in free ionic form in higher amounts than the Al-oxalate complex. *Fagopyrum* roots and leaves contain a lot of oxalate (Ma et al., 1998). In the root cells Al is complexed with oxalate which keeps the cellular Al concentration at a very low level so that Al uptake may occur along the Al concentration gradient. Probably this transport can be a passive process involving a transporter or channel. Since La^{3+} inhibits the uptake, Al may be transported through the Ca channel. (La^{3+} is a channel inhibitor, see: Huang et al., 1994). It is a matter of argument that in the tolerant plants Al is taken up at the root tip or above it before it reaches a 10-fold concentration in the xylem. The most interesting result of these investigations is that Al is present in the roots as Al-oxalate, it is transported in the xylem as Al-citrate but in the leaves there is again Al-oxalate. The xylem sap contains no Al-oxalate (Ma and Hiradate, 2000). According to the authors the complex is changed because oxalate would precipitate with Ca in the xylem sap. In other plants (e.g. sorrel) Ca-oxalate crystals are present in leaf vacuoles where they bind the excess of Ca. The remaining question is – and not only in case of Al but also heavy metals – where and how does the complex change and why does Al leave the stable oxalate complex? Where does it form the new complex with citrate? In the xylem? Let us recall a few thoughts mentioned before: the presence of channels seems to be proven and among them Ca-channels and non-selective channels were described. However Al^{3+} can act as a channel inhibitor. It inhibits the K^+_{in} and K^+_{out} channels but activates the anion channels making organic acid release possible. But this does not explain the difference between the tolerant and sensitive lines (Piñeros and Kochian, 2001), only that how can the large amount of organic acids be released from the cells. A new idea that has recently emerged concerning the inhibitory effect of Al suggests that Al activates the 1→3-β-D-glucan synthase i.e. callose synthesis at the plasmalemma–cell wall interface where the cells are connected by plasmodesmata. Apoplastic Al was proven to induce plasmodesmatal closure. The specific inhibitor of callose synthesis, 2-deoxy-D-glucose, significantly decrease the inhibitory effect of Al. The open or closed state of plasmodesmata determines the cell-cell translocation, consequently the movement of growth regulators and signals (Sivaguru et al., 2000). In pea roots, Al-induced callose synthesis is in close correlation with lipid peroxidation (Yamamoto et al., 2001). Callose synthesis is activated by the increase in the cellular concentration of Ca^{2+} which occurs when Ca^{2+} permeates the damaged membranes (Delmer and Amor, 1995).

7.2. Effect of combination of Al and Fe(II)

The combined effect of Al and Fe claims interest in tropical plants and others with considerable root reducing capacity. Yamamoto et al. (1997, 1998) and Ikegawa et al. (1998), investigating intensively growing cells (mature cells are not sensitive) in tobacco cell cultures, found that neither Al nor Fe(II) cause loss of cell viability. The two ions together induce lipid peroxidation in the membranes and cause almost complete cell destruction. Antoxidants protect the cells proving the effect of lipid peroxidation (Ikegawa et al., 1998). The same results were derived with the cells of soybean (Rath and Barz, 2000). It is interesting that in another set of experiments where Fe(III)-EDTA was used (Yamamoto et al., 1997) Al also inhibited growth when supplied with Fe-complex but to a smaller extent as compared to $FeSO_4$. The reason for this can be the iron reducing ability of tobacco cells producing Fe(II) which has a much lower stability constant with EDTA than Fe(III). Therefore this phenomenon should be investigated in dicotyledonous and monocotyledonous but not graminaceous plants belonging to Strategy I concerning Fe uptake, acidifying the external medium under iron deficiency and taking up Fe after reduction so the combined effect of Fe(II) and Al may occur in natural conditions in the presence of Al.

This connection between Al and Fe is worth investigating in graminaceous plants, too, because the mugineic acids transporting Fe(III) through the membranes in complexed form are able to bind other heavy metals. The connection between Al and Fe was suspected because Al also induced chlorosis in wheat. Al may enter mugineic acid metabolism at several points: it may inhibit its synthesis, its excretion and the reabsorption of its Fe-complex. Since the release of mugineic acids has a diurnal rythm the treatment of wheat with Al could have been carried out separately during the periods of synthesis and excretion. The results show that Al inhibits both processes but especially the synthesis (Chang et al., 1998). It was also found that Al does not form complex with mugineic acids and does not interfere substantially with the uptake of Fe-mugineic acid complex (Chang et al., 1999).

8. LEAD

The mechanism of uptake and translocation can be considered discovered if all data are available from the gene expression to the amino acid sequence of transport and channel proteins. In such details only the potassium uptake and circulation is known for a few plants. Concerning the toxic heavy metals the situation is completely different. It might be possible to describe the genetic background of defence mechanisms against them but a specific transporter protein facilitating their uptake, never. Why should ions or substances without any metabolic role have an uptake system. But the elements occurring in the soil surrounding the plants can be found also in the plants i.e. data show that they are able to get into the roots and transported to the shoots. They may use the transport pathways of similar elements or diffuse through the minor discontinuities due to damage on membranes? Some of the heavy metals have an unusually high stability constant with organic chelating agents (e.g. Pb, Cu, Fe) so they are not likely to stay in the cell in free ionic state. Concerning Pb, literature data is unambiguous that the majority of the metal can be found in the roots because the bulk is attached to the cell walls. 75 % of the primary cell wall is water which can be found in the interfibrillar and intermicellar spaces providing a large surface for cation adsorption on the carboxylic groups of matrix substances, the galacturonic acids and glucuronic acids. Ca and Pb may directly compete for these binding sites (Lane et al., 1978). This is proven by the role of Ca in the tolerance of Pb (Antosiewicz, 1995). In onion roots the cell wall remarkably thickened by the effect of Pb which was observed within one hour at a higher Pb concentration (Wierzbicka, 1998). This thickening is not uniform. After 24 hours tubular invaginations of the cell membrane (plasmatubules) were formed which bound the majority of Pb. The first step of Pb effect is the attachment to the cell wall.

Permeation through the membrane may occur in traces through Ca-ATPases or Ca^{2+}/H^+ antiport but there is no data available to support this hypothesis. Pb is more likely to enter through the Ca-channels. In *Helix* neurons Pb was proven to block the resting Na conductance (Osipenko et al., 1992). In animal cells it is known (Simons and Pocock, 1987) and in plant cells it is assumed (Huang and Cunningham, 1996) that the voltage gated Ca-channels are permeable to Pb, too.

There are plants in which despite the retention of Pb in the root a significant amount of it is transported to the shoot. Such example is the cucumber. Entering the xylem is evidenced by the significant amount of Pb measured in the xylem sap (Varga et al., 1999). Not only the growth of the plants but also the transported amount of Pb depend, on the supplied form of Fe. Comparing Pb concentrations in one g of fresh weight and one g of exudation sap the accumulation of Pb in the root is 2000-fold with Fe-EDTA and only 500-fold with $FeCl_3$ and Fe-citrate. However, the Pb and Fe concentrations in the root and the xylem sap, respectively, are not significantly different with the exception of the plants grown with Fe-EDTA and treated with 10^{-5} M Pb where Pb concentration in the roots was about 5-fold compared to that of Fe but only 1/6 of that in the exudation sap (unpublished). These data show that there are plants in which Pb transport to the shoot is significant. Movement through channels is also possible in this case because non-selective channels transporting both cations and anions were found in the xylem parenchyma cells (De Boer and Wegner, 1997; De Boer, 1999). The mechanism by which cucumber roots tolerate the presence of the high amount of Pb has not been revealed, yet. It was found that the concentration of citrate is significantly increased in the exudation sap (Tatár et al., 1998).

In rice it was found that one of the reasons for Pb tolerance is the excretion of oxalate from the roots. The adventitious roots of the sensitive plants are able to grow in the nutrient solution of the tolerant ones. A mixture of Pb and oxaláte (1:5) caused 80 % recovery in the growth of sensitive plants while without oxalate the inhibition was almost complete (Yang et al., 2000). A calmodulin binding protein (NtCBP4) that can modulate heavy metal tolerance was managed to be isolated from tobacco plants. This protein is structurally similar to the K^+ channel of vertebrates and invertebrates and the non-selective cation channel protein which was isolated in barley and *Arabidopsis*. This protein is responsible for the Pb-hypersensitive reaction of transgenic tobacco plants. NtCBP4 is an integral membrane protein, it can be one of the components of an ion-channel. It undoubtedly increases the Pb accumulation of tobacco plants (Arazi et al., 1999).

Unfortunately, the extremely heterogeneous Pb adsorption in plants and the inhibition of Pb transport to the shoot and the variability in this are unresolved questions at the current state of the research (Huang and Cunningham, 1996). However, it should be kept in mind that the atomic weight of Pb considerably exceeds that of other elements occurring in plants.

9. CADMIUM

Soils in general are contaminated with Cd to a certain extent where PO_4 fertilizers are applied (Cakmak et al., 2000) because phosphate fertilizers contain considerable amount of Cd: 70-150 mg Cd per kg P (Andersoon and Siman, 1991). Consequently, the cultured plants surely accumulate some Cd.

Just as in case of Pb there are no data available about the mechanism of Cd uptake. Cd^{2+} is a more mobile ion than Pb^{2+}, its complexes are less stable, it can occur in free ionic state so that it has a significantly toxic effect in both animals and plants. However, the evaluation of data on Cd uptake or its inhibitory effect is sometimes difficult because of high concentrations applied for the treatment (Welch and Norvell, 1999). Other difficulty in comparing the results is that the inhibitory effect of Cd is greatly dependent on the age of plants: the younger the plant the greater the damage (Shaw and Rout, 1998). Fully expanded leaves were not significantly damaged by Cd (Láng et al., 1998). At concentrations which induce growth inhibition significant changes in the metabolism are prevalent so the changes in the uptake systems may be of secondary importance, i.e. Cd did not damage or not only the uptake system.

Cadmium similarly to lead accumulates mostly in root cell walls, and it considerably inhibits the elongation growth of the roots. It has long been known that salt stress facilitates the suberisation of the endodermis and exodermis and the primary and tertiary state of endodermis is extended towards the root tip shortening the zone of high water uptake capacity. The structure of the primary cell walls also changes. Heavy metals induce the incorporation of lignin (Degenhardt and Gimmler, 2000). Lignification increases the rigidity of the walls and inhibits their expansion. These directly cause the decrease of water and solute uptake. In the primary cell walls, there are glycoproteins in various composition and amount which are characteristic in the formation of cell wall structure due to their net-like orientation. Besides the hydroxyproline-rich extensin and the glycine-rich glycoprotein proline-rich cell wall protein was isolated that was proven to be produced under varying stress conditions in the cell wall. Under heavy metal stress a 11 kDa protein is produced in green tissues (Chai et al., 1998) and a 14 kDa protein in roots (Choi et al., 1996). Cell wall glycoproteins increase the rigidity of the wall.

Cadmium may exert its inhibitory effect in the cells in two different ways: it may bind to specific groups on proteins and lipids inhibiting their normal function and it may induce free radical formation inducing oxidative stress (Stroiński, 1999). The former may occur at the transport and channel proteins of membranes disturbing the uptake of many other macro-and microelements while the latter is proved by the inactivation of antioxidant enzymes by Cd. Destruction of the cell membranes alters the ratio of essential elements and causes the decrease in their concentration e.g. Cd inhibits the transport of Fe into the shoot. The most apparent effect of Cd is the chlorosis of young leaves. In iron deficient cucumber plants 98-99 % of Fe taken up remained in the roots as affected by Cd (unpublished). Iron deficiency not only increases the rate of Fe uptake but also increase the accumulation of Cd (Rodecap et al., 1994). Cohen et al. (1998) came to the same conclusion with pea seedlings. It is assumed that a divalent cation transporter may be induced, possibly the Irt1 Fe(II) transporter isolated in *Arabidopsis*, which can speed up the uptake of Cd^{2+} and Zn^{2+}. Iron deficiency stimulates the high affinity system more than the low affinity system. However, the stimulation of uptake is only in the roots, more Cd remains in the root and less is transported to the shoot under iron deficiency. One of the reasons for the increased Cd accumulation can be the increase of the number of SH⁻ groups on the root surface and they can bind Cd covalently.

Cd may have antagonistic or synergistic effect on the macro- and microelements in 16 wheat lines examined by Zhang et al. (2000). Hart et al. (1998) measuring the uptake of Cd in two wheat varieties at a reasonably low Cd concentration determined the V_{max} and K_m of the saturating uptake period. The movement may occur through a transporter or channel. Cd is most likely transported through the substitution of other divalent cations limiting their uptake. First of all, one of this metal can be Zn for its chemical similarity to Cd, the transport mechanism of which is more and more established in both hyperaccumulator plants (*Thlaspi caerulescens*) and Zn deficient plants (Assunção et al., 2001). Zinc may be in competition with Cd in the soil and also in the nutrient solution. The excess of Zn may decrease the growth inhibition and the inhibition of CO_2 fixation by Cd (Ali et al., 2000) because Zn not only inhibits Cd uptake but also Cd translocation from the root to the shoot. It is assumed that Cd may substitute Ca in the basic processes and that is why it is so inhibitory. The lowering or increasing of Ca concentration may alter Cd distribution in bean plants. In case of low Ca concentration much more Cd is translocated to the shoot compared to high concentration when most of the Cd remained in the root (Skórzyńska-Polit et al., 1998). Despite Cd is far more mobile than Pb there are great differences between plant species and varieties concerning the root/shoot ratio of distribution (Guo-Yan et al., 1996). Differences between genotypes may be found more likely in the root/shoot distribution ratio than in the uptake (Florijn and Van Beusichem, 1993a,b). Cd is transported in the xylem sap towards the shoot. As there are SH-compoun ds, cysteine, glutathione in the xylem sap and Cd may bind to N and O ligands it can be transported as a complex. However, Cd-complexes are not very stable so it moves – at least in part – in ionic form.

Cadmium circulates in the plant i.e. it is transported in the phloem, too. Since phytochelatins are stable at alkaline pH values their Cd-complexes may be translocated in the phloem. There is no specific evidence for the existence *in vivo* complexes with mugineic acids and nicotianamine. Cd is capable of retranslocating from the leaves to the shoots – even if only up to a few % – which can be increased by Zn in essential concentration. However, 5 µM Zn inhibits Cd transport in the phloem (Cakmak et al., 2000).

10. SUMMARY

In order to understand the mode of action of heavy metals it is necessary to reveal their uptake and translocation. Considering either essential or toxic heavy metals, at the present state of our knowledge, it is not possible to determine, after administration to the root, how much of them stay in the cell walls and how much get into the symplast. We cannot accept the view that the heavy metals adsorbed in the cell walls do not exert any effect on metabolism. It is evidenced that the stress factors increase the activity of cell wall peroxidases and stimulate lignification. The cell wall becoming more and more rigid due to this process looses its ability to expand. The water uptake of the cells decreases or stops and only the exchange of solutes may occur. It is also known that the proline-rich glycoprotein is produced as a response to stresses. Expansins and the rest of the glycoproteins form the structural network of the cell wall which also increase rigidity. To date no one has examined the inhibition of production and function of proteins (α- and β-expansins) facilitating the expansion of the cell wall by heavy metals. The production of callose instead of cellulose next to plasmodesmata disrupting the cell-to-cell connection, in turn, was demonstrated under Al toxicity.

Microelements that are present in the xylem – because of the endodermis – should be assumed to have entered the symplast i.e. they were transported through the plasmalemma. This fact causes a problem for the metals which have high stability constants (logK) in their organic complexes, e.g. Pb, Cu and Fe(III). They form complexes with N, O, SH+, COO$^-$ ligands so they are not transported in ionic state. One of the most interesting questions is the chelating agent itself. In graminaceous plants a special compound, mugineic acid is produced which is first of all responsible for the transport of Fe but forms complexes with Cu, Zn and Cd, too. In barley that produces the highest amount of mugineic acid it may be detected to accumulate on contaminated soils. The uptake of Zn, Ni and Cd may be doubled since they permeate the membrane in complexed form. It is widely accepted that iron is transported to the shoot as Fe-citrate. However, the role of nicotianamine a compound which is widespread in the plant kingdom and which also forms complexes with heavy metals and it is not autooxidable, remains to be elucidated.

Most of the information on the genetic background of uptake is available on iron transporters and reductases, generally on the basis of the results achieved with yeast cells from which the *Fre1* and *Fre2* genes were described and the corresponding proteins were isolated from the membranes just as a multicopper oxidase that is responsible for iron transport together with Ftr1 protein.

It is generally accepted that not only macroelements but also microelements have both high and low affinity uptake systems. The low affinity systems are possibly not else than channel proteins. There were no specific transporters identified for toxic heavy metals. The merely plausible reason for it is that they are not essential and thus they have no uptake systems. Because of chemical similarity (same size) it is assumed that that Cd may utilize the transport systems of Zn but Cd may also substitute Ca. An important finding is the existance of Cu-ATPases which might transport other heavy metals, too.

Al induces organic acid efflux. Tolerant genotypes excrete malate, citrate or oxalate at the root apex. However, tolerance may have several other reasons: e.g. the extrusion of PO_4^{3-}, OH^-, and the difference between the surface charge of the plasmalemma. In graminaceous plants Al inhibits the synthesis and excretion of mugineic acids but does not form complex with them.

ACKNOWLEDGEMENTS

This work was supported by the Hungarian National Science Foundation T022913, T030837 and EEC IC15-CT98-0126.

REFERENCES

Albersheim, P., Darvill, A., Roberts, K., Staehelin, L.A., and Varner, J.E. (1997) Do the structures of cell wall polysaccaharides define their mode of synthesis?, *Plant Physiol.* **113**, 1–3.
Aldington, S. and Fry, S.C. (1993) Oligosaccharins, *Adv. Bot. Res.* **19**, 1–101.
Aldington, S. and Fry, S.C. (1994) Rhamnogalacturonan II – a biologically active fragment, *J. Exp. Bot.* **45**, 287–293.
Ali, G., Srivastava, P.E., and Iqbal, M. (2000) Influence of cadmium and zinc on growth and photosynthesis of *Bacopa monniera* cultivated *in vitro*, *Biol. Plant.* **43**, 599–601.
Andersoon, A. and Siman, G. (1991) Levels of Cd and some other trace elements in soils and crops as influenced by lime and fertilizer level, *Acta Agric. Scand.* **41**, 3–11.
Andrade, L.R.M., Ikeda, M., and Ishizuka, J. (1997) Excretion and metabolism of malic acid produced by dark carbon fixation in roots in relation to aluminium tolerance of wheat, in T. Anda et al. (eds.), *Plant Nutrition- for Sustainable Food Production and Environment*, Kluwer Acad. Pub., Japan, pp. 445–446.
Antosiewicz, D.M. (1993) Mineral status of dicotyledonous crop plants in relation to their constitutional tolerance to lead, *Environ. Exp. Bot.* **53**, 575–589.
Antosiewicz, D.M. (1995) The relationships between constitutional and inducible Pb-tolerance and tolerance to mineral deficits in *Biscutella laevigata* and *Silene inflata*, *Environ. Exp. Bot.* **35**, 55–69.
Arazi, T., Sunkar, R., Kaplan, B., and Fromm, H. (1999) A tobacco plasma membrane calmodulin-binding transporter confers Ni^{2+} tolerance and Pb^{2+} hypersensitivity in transgenic plants, *Plant J.* **20**, 171–182.
Askerlund, P. and Larsson, Ch. (1991) Transmembrane electron transport in plasma membrane vesicles loaded with an NADH-generating system or ascorbate, *Plant Physiol.* **96**, 1178–1184.
Assunção, A.G.L., Da Costa Martins, P., De Folter, S., Vooijs, R., Schat, H., and Aarts, M.G.M. (2001) Elevated expression of metal transporter genes in three accessions of the metal hyperaccumulator *Thlaspi caerulescens*, *Plant Cell Environ.* **24**, 217–226.
Bagnaresi, P., Basso B., and Pupillo, P. (1997) The NADH-dependent Fe^{3+}-chelate reductases of tomato roots, *Planta* **202**, 427–434.
Bagnaresi, P. and Pupillo, P. (1995) Characterization of NADH-dependent Fe^{3+}-chelate reductases of maize roots, *J. Exp. Bot.* **46**, 1497–1503.
Baker, A.J.M. (1978) Ecophysiological aspects of zinc tolerance in *Silene maritima* With., *New Phytol.* **80**, 635–642.
Baker, A.J.M. (1981) Accumulators and excluders – strategies in the response of plants to heavy metals, *J. Plant Nutr.* **3**, 643–654.
Barceló, J., Vázquez, M.D., and Poschenrieder, Ch. (1988) Structural and ultrastructural disorders in cadmium-treated bush bean plants (*Phaseolus vulgaris* L.), *New Phytol.* **108**, 37–49.
Basu, U., Godbold, D., and Taylor, G.J. (1994) Aluminium resistance in *Triticum aestivum* associated with enhanced exudation of malate, *J. Plant Physiol.* **144**, 747–753.
Baudouin, M. and Boutry, M. (1995). The plasma membrane H^+-ATPase, *Plant Physiol.* **108**, 1–6.
Bérczi, A. and Møller, I.M. (1998) NADH-monodehydroascorbate oxido-reductase is one of the redox enzymes in spinach leaf plasma membrane, *Plant Pysiol.* **116**, 1029–1036.
Bérczi, A. and Møller, I.M. (2000) Redox enzymes in the plant plasma membrane and their possible roles, *Plant Cell Environ.* **23**, 1287–1302.
Bienfait, H.F. (1985) Regulated redox processes at the plasmalemma of plant root cells and their function in iron uptake, *J. Bioenerg. Biomembranes* **17**, 73–83.

Bienfait, H.F. (1988) Mechanisms in Fe efficiency reactions of higher plants, *J. Plant Nutr.* **11**, 605–629.
Bienfait, H.F., Van den Briel, W., and Mesland-Mul, N.T. (1985) Free space iron pools in roots, generation and mobilization, *Plant Physiol.* **78**, 596–600.
Böhme, H. and Scholz, G. (1960) Versuche zur Normalisierung des Phänotyps der Mutante *chloronerva* von *Lycopersicon esculentum* Mill., *Kulturpflanze* **8**, 93–109.
Boutry, M., Michelet, B., and Goffeau, A. (1989) Molecular cloning of a family of plant genes encoding a protein homologous to plasma membrane H^+-translocating ATPases, *Biochem. Biophys. Res. Comm.* **162**, 567–574.
Briat, J.-F. and Lobréaux, S. (1997) Iron transport and storage in plants, *Trends Plant Sci.* **2**, 187–192.
Briggs, G.E. and Robertson, R.N. (1957) Apparent free space, *Annu. Rev. Plant Physiol.* **8**, 11–30.
Brown Jr, R.M. and Saxena, I.M. (2000) Cellulose biosynthesis: A model for understanding the assembly of biopolymers, *Plant Physiol. Biochem.* **38**, 57–67.
Brüggeman, W., Moog, P.R., Nakagawa, H., Janiesch, P., and Kuiper, P.J.C. (1990) Plasma membrane-bound NADH:Fe^{3+}-EDTA reductase and iron deficiency in tomato (*Lycopersicon esculetum*). Is there a turbo reductase?, *Physiol. Plant.* **79**, 339–346.
Cakmak, I., Erenoglu, B., Gülüt, K.Y., Derici, R., and Römheld, V., (1998) Light-mediated release of phytosiderophores in wheat and barley under iron or zinc deficiency, *Plant Soil* **202**, 309–315.
Cakmak, I., Sari, N., Marschner, H., Ekiz, H., Kalayci, M., Yilmaz, A., and Braun, H. J. (1996) Phytosiderophore release in bread and durum wheat genotypes differing in zinc efficincy, *Plant Soil* **180**, 183–189.
Cakmak, I., Welch, R.M., Erenoglu, B., Römheld, V., Norvell, W.A., and Kochian, L. V. (2000) Influence of varied zinc supply on re-translocation of cadmium (^{109}Cd) and rubidium (^{86}Rb) applied on mature leaf of durum wheat seedlings, *Plant Soil* **219**, 279–284.
Carpita, N.C. (1996) Structure and biogenesis of the cell walls of grasses, *Annu. Rev. Plant Physiol. Plant Mol. Biol.* **47**, 445–476.
Carpita, N.C. and Gibeaut, D.M. (1993) Structural models of the primary cell walls in flowering plants: consistency of molecular structure with the physical properties of the walls during growth, *Plant J.* **3**, 1–30.
Cassab, G.I. (1998) Plant cell wall proteins, *Annu. Rev. Plant Physiol. Plant Mol. Biol.* **49**, 281–309.
Cataldo, D.A., Garland, T.R., and Wildung, R.E. (1981) Cadmium distribution and chemical fate in soybean plants, *Plant Physiol.* **68**, 835–839.
Chai, T.Y., Didierjean, L., Burkard, G., and Genot, G. (1998) Expression of a green tissue-specific 11 kDa proline-rich protein gene in bean in response to heavy metals, *Plant Sci.* **133**, 47–56.
Chaney, R.L. Brown, J.C., and Tiffin, L.O. (1972) Obligatory reduction of ferric chelates in iron uptake by soybeans, *Plant Physiol.* **50**, 208–213.
Chang, Y-Ch., Ma, J.F., Iwashita, T., and Matsumoto, H. (1999) Effect of Al on the phytosiderophore-mediated solubilization of Fe and uptake of Fe-phytosiderophore complex in wheat (*Triticum aestivum*), *Physiol. Plant.* **106**, 62–68.
Chang, Y-Ch., Ma, J.F., and Matsumoto, H. (1998) Mechanisms of Al-induced iron chlorosis in wheat (*Triticum aestivum*). Al-inhibited biosynthesis and secretion of phytosiderophore, *Physiol. Plant.* **102**, 9–15.
Choi, D.W., Song, J.Y., Kwon, Y.M., and Kim, S.G. (1996) Characterization of a cDNA encoding a proline-rich 14 kDa protein in developing cortical cells of the roots of bean (*Phaseolus vulgaris*) seedlings, *Plant Mol. Biol.* **30**, 973–982.
Clarkson, D.T. and Lüttge, U. (1989) Divalent cations, transport and compartmentation, *Prog. Bot.* **51**, 93–112.
Clemens, S. (2001) Molecular mechanisms of plant metal tolerance and homeostasis, *Planta* **212**, 475–486.
Cohen, C.K., Fox, T.C., Garvin, D.F., and Kochian, L.V. (1998) The role of iron-deficiency stress responses in stimulating heavy-metal transport in plants, *Plant Physiol.* **116**, 1063–1072.
Cosgrove, D.J. (1993) How do plant cell walls extend?, *Plant Physiol.* **102**, 1–6.
Cosgrove, D.J. (1997) Relaxation in a high-stress environment: the molecular bases of extensible cell walls and cell enlargement, *Plant Cell* **9**, 1031–1041.
Cosgrove, D.J. (1999) Enzymes and other agents that enhance cell wall extensibility, *Annu. Rev. Plant Physiol. Plant Mol. Biol.* **50**, 391–417.
Cosgrove, D.J. (2000) Expansive growth of plant cell walls, *Plant Physiol. Biochem.* **38**, 109–124.
Crane, F.L., Sun, I.L., Barr, R., and Löw, H. (1991) Electron and proton transport across the plasma membrane, *J. Bioenerg. Biomembranes* **23**, 773–803.

Cutler, J.M. and Rains, D.W. (1974) Characterization of cadmium uptake by plant tissue, *Plant Physiol.* **54**, 67–71.

Cseh, E. and Fodor, F. (1997). Role of boron in the plasmalemma turbo reductase activity, in R.W. Bell and B. Rerkasem (eds.), *Boron in Soils and Plants.* Kluwer Acad. Publ., Nederlands, pp. 175–177.

Cseh, E., Fodor, F., Varga, A., and Záray, G. (2000) Effect of lead treatment on the distribution of essential elements in cucumber, *J. Plant Nutr.* **23**, 1095–1105.

Cseh, E., Váradi, Gy., and Fodor, F. (1994) Effect of Fe-complexes and N-forms on the Fe absorption, uptake and translocation of cucumber plants, *Bot. Közlem.* **81**, 47–55.

Czempinski, K., Gaedeke, N., Zimmermann, S., and Müller-Röber, B. (1999) Molecular mechanisms and regulation of plant ion channels, *J. Exp. Bot.* **50**, 955–966.

Davies, T.G.E. and Coleman, J.O.D. (2000) The *Arabidopsis thaliana* ATP-binding casette proteins: an emerging superfamily, *Plant Cell Environ.* **23**, 431–443.

De Boer, A.H. (1999) Potassium translocation into the root xylem, *Plant Biol.* **1**, 36–45.

De Boer, A.H. and Wegner, L.H. (1997) Regulatory mechanisms of ion channels in xylem parenchyma cells, *J. Exp. Bot.* **48**, 441–449.

De Cabo, R.C., Gonzáles-Reyes, J.A. Córdoba, F., and Navas, P. (1996) Rooting hastened in onions by ascorbate and ascorbate free radical, *J. Plant Growth Regul.* **15**, 53–56.

Degenhardt, B. and Gimmler, H. (2000) Cell wall adaptations to multiple environmental stresses in maize roots, *J. Exp. Bot.* **51**, 595–603.

Delhaize, E., Craig, S., Beaton, C.D., Bennet, R.J., Jagadish, V.C., and Randall, P.J. (1993a) Aluminium tolerance in wheat (*Triticum aestivum* L.) I. Uptake and distribution of aluminium in root apices, *Plant Physiol.* **103**, 685–693.

Delhaize, E. and Ryan, P.R. (1995) Update on environmental stress: aluminium toxicity and tolerance of plants, *Plant Physiol.* **107**, 315–321.

Delhaize, E., Ryan, P.R., and Randall, P.J. (1993b) Aluminium tolerance in wheat (*Triticum aestivum* L.) II. Aluminium-stimulated excretion of malic acid from root apices, *Plant Physiol.* **103**, 695–702.

Delmer, D.P. (1999) Cellulose biosynthesis: exciting times for a difficult field of study, *Annu. Rev. Plant Physiol. Plant Mol. Biol.* **50**, 245–276.

Delmer, D.P. and Amor, Y. (1995) Cellulose biosynthesis, *Plant Cell* **7**, 987–1000.

Dykema, Ph.E., Sipes, Ph.R., Anna Marie, Biermann, B.J., Crowell, D.N., and Randall, S.K. (1999) A new class of proteins capable of binding transition metals. *Plant Mol. Biol.* **41**, 139–150.

Eide, D.J. (1998) The molecular biology of metal ion transport in *Saccharomyces cerevisiae, Annu. Rev. Nutr.* **18**, 441–469.

Eide, D.J., Broderius, M.A., Fett, J., and Guerinot, M.L. (1996) A novel iron-regulated metal transporter from plants identified by functional expression in yeast, *Proc. Natl. Acad. Sci. USA* **93**, 5624–5628.

Epstein, E. (1972) *Mineral Nutrition of Plants: Principles and Perspectives*, Wiley, New York.

Ettinger, W.F., Clear, A.M., Fanning, K.J., and Peck, M.L. (1999) Idenfication of a Ca^{2+}/H^+ antiport in the plant chloroplast thylakoid membrane, *Plant Physiol.* **119**, 1379–1386.

Florijn, P.J. and van Beusichem, M.L. (1993a) Uptake and distribution of cadmium in maize inbred lines, *Plant Soil* **150**, 25–32.

Florijn, P.J. and van Beusichem, M.L. (1993b) Cadmium distribution in maize inbred lines. Effect of pH and level of Cd supply, *Plant Soil* **150**, 79–84.

Fodor, F., Cseh, E., Dao Thi Phuong, D., and Záray, Gy. (1996) Uptake and mobility of lead in *Urtica dioica* L., *J. Acta Phytogeogr. Suec.* **81**,106–108.

Fodor, F., Cseh, E., Varga, A., and Záray, Gy. (1998) Lead uptake, distribution, and remobilization in cucumber, *J. Plant Nutr.* **21**, 1363–1373.

Fry, S.C. (1982) Isodityrosine a new cross-linking amino acid from plant cell glycoprotein, *Biochem. J.* **204**, 388–397.

Fry, S.C., Aldington, S., Hetherington, P.R., and Aitken, J. (1993) Oligosaccharides as signals and substrates in the plant cell wall, *Plant Physiol.* **103**, 1–5.

Guo-Yan, L., Marschner, H., and Guo, Y.L. (1996) Genotypic differences in uptake and translocation of cadmium in bean and maize inbred lines, *Z. Pflanzenernähr. Düng. Bodenk.* **159**, 55–60.

Harper, J.F., Surowy, T.K., and Sussman, M.R. (1989) Molecular cloning and sequence of cDNA encoding the plasma membrane proton pump ($H^+ATPase$) of *Arabidopsis thaliana, Proc. Nat. Acad. Sci. USA* **86**, 1234–1238.

Harris, E.D. (2000) Cellular copper transport and metabolism, *Annu. Rev. Nutr.* **20**, 291–310.

Hart, J.J., Welch, R.M., Norvell, W.A., Sullivan, L.A., and Kochian, L.V. (1998) Characterization of cadmium binding, uptake, and translocation in intact seedlings of bread and durum wheat cultivars, *Plant Physiol.* **116**, 1413–1420.
He, Z.H., Cheeseman, I., He, D., and Kohorn, B.D. (1999) A cluster of five cell wall associated receptor kinase genes, Wakl-5, are expressed in specific organs of Arabidopsis, *Plant Mol. Biol.* **39**, 1189–1196.
He, Z.H., Fujiki, M., and Kohorn, B.D. (1996) A cell wall-associated receptor-like protein kinase, *J. Biol. Chem.* **271**, 19789–19793.
He, Z.H., He, D., and Kohorn, B.D. (1998) Requirement for the induced expression of a cell wall associated receptor kinase for survival during the pathogen response, *Plant J.* **14**, 55–63.
Higuchi, K., Kanazawa, K., Nishizawa, N.-K., Chino, M., and Mori, S. (1995a) Purification and characterization of nicotianamine synthase from Fe-deficient barley roots, in J. Abadia (ed.), *Iron Nutrition in Soils and Plants*, Kluwer Acad. Publ., Dordrecht, pp. 29–35.
Higuchi, K., Nishizawa, N-K., Yamaguchi, H., Römheld, V., Marschner, H., and Mori, S. (1995b) Response of nicotianamine synthase activity to Fe-deficiency in tobacco plants as compared with barley, *J. Exp. Bot.* **46**, 1061–1063.
Higuchi, K., Suzuki, K., Nakanishi, H., Yamaguchi, H., Nishizawa, N.-K., and Mori, S. (1999) Cloning of nicotianamine synthase genes, novel genes involved in biosynthesis of phytosiderophores, *Plant Physiol.* **119**, 471–479.
Hirschi, K.D., Zhen, R.G., Cunningham, K.W., Rea, P.A., and Fink, G.R. (1996) CAX1 and H^+/Ca^{2+} antiporter from Arabidopsis, *Proc. Natl. Acad. Sci. USA* **93**, 8782–8786.
Holden, M.J., Luster, D.G., Chaney, R.L., Buckhout, T.J., and Robinson, C. (1991) Fe^{3+}-chelate reductase activity of plasma membranes isolated from tomato (*Lycopersicon esculentum* Mill.) roots, *Plant Physiol.* **97**, 537–544.
Horemans, N., Asard, H., and Caubergs, R.J. (1994) The role of ascorbate free radicals as an electron acceptor to cytochrome b-mediated trans-plasma membrane electron transport in higher plants, *Plant Physiol.* **104**, 1455–1458.
Horemans, N., Asard, H., and Caubergs, R.J. (1997) The ascorbate carrier of higher plant plasma, *Plant Physiol.* **114**, 1247–1253.
Horemans, N., Asard, H., Van Gestelen, P., and Caubergs, R.J. (1998) Facilitated diffusion drives transport of oxidised ascorbate molecules into purified plasma membrane vesicles of *Phaseolus vulgaris*, *Physiol. Plant.* **104**, 783–789.
Horst, W.J., Schmohl, N., Kollmeier, M., Baluška, F., and Sivaguru, M. (1999) Does aluminium affect root growth of maize through interaction with the cell wall – plasma membrane – cytoskeleton continuum?, *Plant Soil* **215**, 163–174.
Huang, J.W., and Cunningham, S.D. (1996) Lead phytoextraction: species variation in lead uptake and translocation, *New Phytol.* **134**, 75–84.
Huang, J.W., Grunes, D.L., and Kochian, L.V. (1994) Voltage-dependent Ca^{2+} influx into right-side-out plasma membrane vesicles isolated from wheat roots characterization of a putative Ca^{2+} channel, *Proc. Natl. Acad. Sci.* **91**, 3473–3477.
Ikegawa, H., Yamamoto, Y., and Matsumoto, H. (1998) Cell death caused by a combination of aluminium and iron in cultured tobacco cells, *Physiol. Plant.* **104**, 474–478.
Ishii, T. (1997) Structure and functions of feruloylated polysaccharides, *Plant Sci.* **127**, 111–127.
Jarvis, S.C., Jones, H.P., and Hopper, M.C. (1976) Cadmium uptake from solution by plants and its transport from roots to shoots, *Plant Soil* **44**, 179–191.
Jorge, R.A. and Arruda, P. (1997) Aluminium induced organic acids exudation by roots of an aluminium tolerant tropical maize, *Phytochemistry* **45**, 675–681.
Kampfenkel, K., Kushnir, S., Babiychuk, E., Inz, É.D., and Van Montagu, M. (1995) Molecular characterization of a putative *Arabidopsis thaliana* copper transporter and its yeast homolog, *J. Biol. Chem.* **270**, 28479–28486.
Kanazawa, K., Higuchi, K., Fushiya, S., Nishizawa, N-K., Chino, M., and Mori, S. (1995) Inductions of two enzyme activities involved in the biosynthesis of mugineic acid, in Fe-deficient barley roots, in J. Abadia (ed.), *Iron Nutrition in Soils and Plants*, Kluwer Acad. Publ., Dordrecht, pp. 37–41.
Kawai, S., Takagi, S., and Sato, Y. (1988). Mugineic acid-family:phytosiderophores in root secretions of barley, corn and sorghum varieties, *J. Plant Nutr.* **11**, 633–642.
Keller, B. (1993) Structural cell wall proteins, *Plant Physiol.* **101**, 1127–1130.
Keller, P. and Deuel, H. (1957) Kationaustauschkapazität und Pektingehalt von Pflanzenwurzeln, *Z. Pflanzenernähr. Düng. Bodenk.* **79**, 119–131.

Kochian, L.V. (1995) Cellular mechanisms of aluminium toxicity and resistance in plants, *Annu. Rev. Plant Physiol. Plant Mol. Biol.* **46**, 237–260.
Kohorn, B.D. (2000) Plasma membrane-cell wall contacts, *Plant Physiol.* **124**, 31–38.
Kristensen, I. and Larsen, P.O. (1974) Azetidine-2-carboxylic acid derivates from seeds of *Fagus silvatica* L. and revised structure for nicotianeamine, *Phytochemistry* **13**, 2791–2798.
Kudlicka, K. and Brown R.M. Jr. (1997) Cellulose and callose biosynthesis in higher plants, *Plant Physiol.* **115**, 643–656.
Láng, F., Szigeti, Z., Fodor, F., Cseh, E., Zolla, L., and Sárvári, É. (1998) Influence of Cd and Pb on the ion content, growth and photosynthesis in cucumber, in G. Garab (ed.), *Photosynthesis: Mechanisms and Effects, Vol. IV,* Kluwer Acad. Publ., Dordrecht, pp. 2693–2696.
Lane, S.D., Martin, E.S., and Garrod, J.F. (1978) Lead toxicity effects on indole-3-Ylacetic acid induced cell elongation, *Planta* **144**, 79–84.
Liang, F., Cunningham, K.W., Harper, J.F., and Sze, H. (1997) ECA1 complements yeast mutants defective pumps and encodes an endoplasmatic reticulum-type Ca^{2+}-ATPase in *Arabidopsis thaliana, Proc. Nat. Acad. Sci. USA* **94**, 8579–8584.
Lindberg, S. and Griffiths, G. (1993) Aluminium effects on ATPase and lipid composition of plasma membranes in sugar beet roots, *J. Exp. Bot.* **44**, 1543–1550.
Luck, G., Liao, H., Murray, N.J., Grimmer, H.R., Warminski, E.E. (1994) Polyphenols astringency and prolinrich proteins, *Phytochemistry* **37**, 357–371.
Ma, J.F. and Hiradate, S. (2000) Form of aluminium for uptake and translocation in buckwheat (*Fagopyrum esculentum* Moench.), *Planta* **211**, 355–360.
Ma, J.F., Hiradate, S., and Matsumoto, H. (1998) High aluminium resistance in buckwheat. II. Oxalic acid detoxifies aluminium internally, *Plant Physiol.* **117**, 753–759.
Ma, J.F., Hiradate, S., Nomoto, K., Iwashita, T., and Matsumoto, H. (1997a) Internal detoxification mechanism of Al in *Hydrangea, Plant Physiol.* **113**, 1033–1039
Ma, J.F. and Nomoto, K. (1996) Effective regulation of iron acquisition in graminaceous plants. The role of mugineic acids as phytosiderophores, *Physiol. Plant.* **97**, 609–617.
Ma, J.F., Zheng, S.J., Li, X.F., Takeda, K., and Matsumoto, H. (1997b) A rapid hydroponic screening for aluminium tolerance in barley, *Plant Soil* **191**, 133–137.
Ma, J.F., Zheng, S.J., and Matsumoto, H. (1997c) Specific secretion of citric acid induced by Al stress in *Cassia cora* L., *Plant Cell Physiol.* **38**, 1019–1025.
Marschner, H. (1995) *Mineral Nutrition of Higher Plants,* Acad. Press, London, pp. 285–299.
Marschner, H., Kirkby, E.A., and Engels, C. (1997) Importance of cycling and recycling of mineral nutrients within plants for growth and development, *Bot. Acta* **110**, 265–273.
Marschner, H., Römheld, V., and Kissel, M. (1986) Different strategies in higher plants in mobilization and uptake of iron, *J. Plant Nutr.* **9**, 695–713.
Mathieu, Y., Guern, J., Spiro, M.D., O'Neill, M.A., Kates, K., Darvill, A.G., and Albersheim, P. (1998) The transient nature of the oligogalacturonide-induced ion fluxes of tobacco cells is not correlated with fragmentation of the oligogalacturonides, *Plant J.* **16**, 305–311.
Matsumoto, H. (1988) Inhibition of proton transport activity of microsomal membrane vesicles of barley roots by aluminium, *Soil Sci. Plant Nutr.* **34**, 499–506.
McQueen-Mason, S.J., Durachko, D.M., and Cosgrove, D.J. (1992) Two endogenous proteins that induce cell wall extension in plants, *Plant Cell* **4**, 1425–1433.
McQueen-Mason, S.J., Fry, S.C., Durachko, D.M., and Cosgrove, D.J. (1993) The relationship between xyloglucan endotransglycosylase and in-vitro cell wall extension in cucumber hypocotyls, *Planta* **190**, 327–331.
McQueen-Mason, S.J. and Rochange, F. (1999) Expansins in plant growth and development: an update on an emerging topic, *Plant. Biol.* **1**, 19–25.
Meharg, A.A. (1993) The role of the plasmalemma in metal tolerance in angiosperms, *Physiol. Plant.* **88**, 191–198.
Míyasaka, S.C., Bute, J.G., Howell, R.K., and Foy, C.D. (1991) Mechanism of aluminium tolerance in snapbean, root exudation of citric acid, *Plant Physiol.* **96**, 737–743.
Mok, M.C., Samuelsen, A.I., Martrin, R.C., and Mok, D.W.S. (2000) Fe(III) reductase, the FRE genes, and FRE-transformed tobacco, *J. Plant Nutr.* **23**, 1941–1951.
Mori, S. and Nishizawa, N. (1987) Methionine as a dominant precursor of phytosiderophores in Graminaceae plants, *Plant Cell Physiol.* **28**, 1081–1092.

Nishizawa, N. and Mori, S. (1987) The particular vesicles appearing in barley root cells and its relation to mugineic acid secretion, *J. Plant Nutr.* **10**, 1013–1020.
Noctor, G. and Foyer, Ch.H. (1998) Ascorbate and glutathione: Keeping active oxygen under control, *Annu. Rev. Plant Physiol. Plant Mol. Biol.* **49**, 249–279.
Noma, M. and Noguchi, M. (1976) Occurence of nicotianamine in higher plants, *Phytochemistry* **15**, 1701–1702.
Oki, H., Yamaguchi, H., Nakanishi, H., and Mori, S. (1999) Introduction of the reconstructed yeast ferric reductase gene, refre1 into tobacco, *Plant Soil* **215**, 211–220.
Osipenko, O.N., Győri, J., and Kiss, T. (1992) Lead ions close steady-state sodium channels in Helix neurons, *Neuroscience* **80**, 483–489.
Palmgren, M.G. and Harper, J.F. (1999) Pumping with plant P-type ATPases, *J. Exp. Bot.* **50**, 883–893.
Palmgren, M.G. and Axelsen, K. B. (1998) Evolution of P-type ATPases, *Biochim. Biophys. Acta* **1365**, 37–45.
Papernik, L.A. and Kochian, L.V. (1997) Possible involvement of Al-induced electric signals in Al-tolerance in wheat, *Plant Physiol.* **115**, 657–667.
Pardo, J. M. and Serrano, R. (1989) Structure of plasma membrane H^+-ATPase gene from the plant *Arabidopsis thaliana*, *J. Biol. Chem.* **264**, 8557–8562.
Parker, D.R. and Pedler, J.F. (1998) Probing the „malate hypothesis" of differencial aluminium tolerance in wheat by using other rhizotoxic ions as proxies for Al, *Planta* **205**, 389–396.
Pellet, D.M. Grunes, D.L., and Kochian, L.V. (1995) Organic acid exudation as an aluminium tolerance mechanism in maize (*Zea mays* L.), *Planta* **196**, 788–795.
Pellet, D.M., Papernik, L.A., Jones, D.L., Darrah, P.R., Grunes, D.L., and Kochian, L.V. (1997) Involvement of multiple aluminium exclusion mechanisms in aluminium tolerance in wheat, *Plant Soil* **192**, 63–68.
Pellet, D.M., Papernik, L.A., and Kochian, L.V. (1996) Multiple aluminium-resistance mechanisms in wheat: Roles of root apical phosphate and malate exudation, *Plant Physiol.* **112**, 591–597.
Pich, A., Hillmer, S., Manteuffel, R., and Scholz, G. (1997) First immunohistochemical localization of the endogenous Fe^{2+}-chelator nicotianamine, *J. Exp. Bot.* **308**, 759–767.
Pich, A., Scholz, G., and Stephan, U.W. (1995) Iron-dependent changes of heavy metals, nicotianamine, and citrate in different plant organs and in the xylem exudate of two tomato genotypes. Nicotianamine as possible copper translocator, in J. Abadia (ed.), *Iron Nutrition in Soils and Plants*. Kluwer Acad. Publ., Dordrecht, pp. 51–58.
Piñeros, M.A. and Kochian, L.V. (2001) A patch-clamp study on the physiology of aluminium toxicity and aluminium tolerance in maize. Identification and characterization of Al^{3+}-induced anion channels, *Plant Physiol.* **125**, 292–305.
Pintro, J., Barloy, J., and Fallavier, P. (1997) Effects of low aluminium activity in nutrient solutions on the organic acid concentrations in maize plants, *J. Plant Nutr.* **20**, 601–611.
Rautenkranz, A.A.F., Li, L., Mächler, F., Märtinoia, E., and Oertli, J.J. (1994) Transport of ascorbic and dehydroascorbic acids across protoplast and vacuole membranes isolated from barley (*Hordeum vulgare* L. cv Gerbel) leaves, *Plant Physiol.* **106**, 187–193.
Rath, I. and Barz, W. (2000) The role of lipid peroxidation in aluminium toxicity in soybean cell suspension cultures, *Z. Naturforsch.* **55c**, 957–964.
Rea, P.A. (1999) MRP subfamily ABC transporters from plants and yeasts, *J. Exp. Bot.* **50**, 895–913.
Reid, J.S.G. (1997) Carbohydrate metabolism. Structural Carbohydrates, in P. M. Dey and J. B. Harborne (eds.), *Plant Biochemistry,* Acad. Press, San Diego, pp. 205–236.
Reiter, W-D. (1998) The molecular analysis of cell wall components, *Trends Plant Sci.* **3**, 27–32.
Rengel, Z. and Reid, R. (1997) Uptake of Al across the plasma membrane of plant cells, *Plant Soil* **192**, 31–35.
Robinson, N.J., Procter, C.M., Connolly, E.L., and Guerinot, M.L. (1999) A ferric-chelate reductase for iron uptake from soils, *Nature* **397**, 694–697.
Rodecap, K. D., Tingey, D. T., and Lee, E. H. (1994) Iron nutrition influence on cadmium accumulation by *Arabidopsis thaliana* (L.) Heynh., *J. Environ. Qual.* **23**, 239–246.
Römheld, V. and Awad, F. (2000) Significance of root exudates in acquisition of heavy metals from a contaminated calcareous soil by graminaceous species, *J. Plant Nutr.* **23**, 1857–1866.
Römheld, V. and Marschner, H. (1986) Evidencefor a specific uptake system for iron phytosiderophores in roots of grasses, *Plant Physiol.* **80**,175–180.
Ryan, P.R., Delhaize, E., and Randall, P.J. (1995a) Characterization of Al-stimulated efflux of malate from the apices of Al-tolerant wheat roots, *Planta* **196**, 103–110.
Ryan, P.R., Delhaize, E., and Randall, P.J. (1995b) Malate efflux from root apices and of tolerance to aluminium are highly correlated in wheat roots, *Austr. J. Plant Physiol.* **22**, 531–536.

Ryan, P.R., Ditomaso, J.M., and Kochian, L.V. (1993) Aluminium toxicity in roots: an investigation of spatial sensitivity and the role of root cap, *J. Exp. Bot.* **44**, 437–446.

Sakaguchi, T., Nishizawa, N.K., Nakanishi, H., Yoshimura, E., and Mori, S. (1999) The role of potassium in the secretion of mugineic acid family phytosiderophores from iron-deficient barley roots, *Plant Soil* **215**, 221–227.

Samuels, T.D., Küçükakyüz, K., and Rincón-Zachary, M. (1997) Al partitioning patterns and root growth as related to Al sensitivity and Al tolerance in wheat, *Plant Physiol.* **113**, 527–534.

Sánchez, M., Queijeiro, E., Revilla, G., and Zarra, I.. (1997) Changes in ascorbic acid levels in apoplastic fluid during growth of pine hypocotyls. Effect on peroxidase activities associated with cell walls, *Physiol. Plant.* **101**, 815–820.

Schmidke, I., Krüger, C., Frömmichen, R., Scholz, G., and Stephan, U.W. (1999) Phloem loading and transport characteristics of iron in interaction with plant-endogenous ligands in castor bean seedlings, *Physiol. Plant.* **106**, 82–89.

Scholz, G., Becker, R., Stephan, U., Rudolph, A., and Pich, A. (1988) The regulation of iron uptake and possible functions of nicotianamine in higher plants, *Biochem. Physiol. Pflanzen* **183**, 257–269.

Scholz, G. and Rudolph, A. (1968) A biochemical mutant of *Lycopersicon esculentum* Mill. Isolation and properties of the ninhydrin-positive „normalizing factor" *Phytochemistry* **7**, 1759–1764.

Scholz, G., Schlesier, G., and Seifert, K. (1985) Effect of nicotianamine on iron uptake by the tomato mutant „chloronerva", *Physiol. Plant.* **63**, 99–104.

Scholz, G., Seifert, K., and Grün, M. (1987) The effect of nicotianamine on the uptake of Mn^{2+}, Zn^{2+}, Cu^{2+}, Rb^+ and PO^{3-}_4 by tomato mutant chloronerva, *Biochem. Physiol. Pflanzen* **182**, 189–194.

Senden, M.H.M.N., Van der Meer, A.J.G.M., Verburg, T.G., and Wolterbeek, H.Th. (1994) Effects of cadmium on the behaviour of citric acid in isolated tomato xylem cell walls, *J. Exp. Bot.* **45**, 597–606.

Serrano, R. (1989) Structure and function of plasma membrane ATPase, *Annu. Rev. Plant Physiol. Plant Mol. Biol.* **40**, 61–94.

Serrano, A., Córdoba, F., Gonzales-Reyes, J.A., Navas, P., and Villalba, J.M. (1994) Purification and characterization of two distict NAD(P)H dehydrogenazes from onion (*Allium cepa* L.) root plasma membrane, *Plant Physiol.* **106**, 87–96.

Shaw, B.P. and Rout, N.P. (1998) Age-dependent responses of *Phaseolus aureus* Roxb. to inorganic salts of mercury and cadmium, *Acta Physiol. Plant.* **20**, 85–90.

Showalter, A.M. (1993) Structure and function of plant cell wall proteins, *Plant Cell* **5**, 9–23.

Simons, T.J.B. and Pocock, G. (1987) Lead enters bovine adrenal medullary cells through calcium channells, *J. Neurochem.* **48**, 383–389.

Sivaguru, M., Fujiwara, T., Samaj, J., Baluška, F., Yang, Z., Osawa, H., MaedaT., Mori, T., Volkmann, D., and Matsumoto, H. (2000) Aluminium-induced 1→3-β-d-glucan inhibits cell-to-cell trafficking of molecules through plasmadesmata. A new mechanism of aluminium toxicity in plants, *Plant Physiol.* **124**, 991–1005.

Skórzyńska-Polit, E., Tukendorf, A., Selstam, E., and Baszyński, T. (1998) Calcium modifies Cd effect on runner bean plants, *Environ. Exp. Bot.* **40**, 275–286.

Sommer-Knudsen, J., Bacic, A., and Clarke, A.E. (1998) Hydroxyproline-rich plant glycoproteins, *Phytochemistry* **47**, 483–497.

Stephan, U.W., Schmidke, I., and Pich, A. (1994) Phloem translocation of Fe, Cu, Mn, and Zn in Ricinus seedlings in relation to the concentrations of nicotianamine, an endogenous chelator of divalent metal ions, in different seedling parts, *Plant Soil* **165**,181–188.

Stephan, U.W. and Scholz, G. (1993) Nicotianamine: mediator of transport of iron and heavy metals in the phloem?, *Physiol. Plant.* **88**, 522–529.

Stroiński, A. (1999) Some physiological and biochemical aspects of plant resistance to cadmium effect. I. Antioxidative system, *Acta Physiol. Plant.* **21**, 175–188.

Sugiura, Y. and Nomoto, K. (1984) Phytosiderophores. Structure and properties of mugineic acids and their metal complexes, in *Structure and Bonding 58*, Springer-Verlag, Berlin, Heidelberg, pp. 107–135.

Sussman, M.R. (1994) Molecular analysis of proteins in the plasma membrane, Annu. Rev. Plant Physiol. Plant Mol. Biol. **45**, 211–234.

Sze, H., Li, X., and Palmgren, M. G. (1999) Energization of plant cell membranes by H^+-pumping ATPases: reagulation and biosynthesis, *Plant Cell* **11**, 677–689.

Takahama, U. (1998) Ascorbic acid-dependent regulation of redox levels of chlorogenic acid and its isomers in the apoplast of leaves of *Nicotiana tabacum* L., *Plant Cell Physiol.* **39**, 681–689.

Takahashi. M., Yamaguchi, H., Nakanishi, H., Shioiri, T., Nishizawa, N-K., and Mori, S. (1999) Cloning two genes for nicotianamine aminotransferase, a critical enzyme in iron acquisition (Strategy II) in graminaceous plants, *Plant Physiol.* **121**, 947–956.

Takagi, S. (1976) Naturally occuring iron-chelating compounds in oat- and rice-root washings, *Soil Sci. Plant Nutr.* **22**, 423–433.

Takagi, S., Nomoto, K., and Takemoto, T. (1984) Physiological aspect of mugineic acid, a possible phytosiderophore of graminaceous plant, *J. Plant Nutr.* **7**, 469–477.

Takemoto, T., Nomoto, K., Fushiya, S., Ouchi, R., Kusano, G., Hikino, H., Takagi, S., Matsuura, Y., and Kakudo, M. (1978) Structure of mugineic acid, a new amino acid possessing an iron-chelating activity from root washings of water-cultured *Hordeum vulgare* L., *Proc. Japan Acad.* **54**, 469–473.

Tatár, E., Mihucz, V.G., Varga, A., Záray, Gy., and Fodor, F. (1998) Determination of organic acids in xylem sap of cucumber: Effect of lead contamination, *Microchem. J.* **58**, 306–314.

Ueoka-Nakanishi, H. and Maeshima, M. (2000) Quantification of Ca^{2+}/H^+ antiporter VCAX1p in vacuolar membranes and its absence in roots of mung bean, *Plant Cell Physiol.* **41**, 1067–1071.

Varga, A., Martinez, R.M.G., Záray, Gy., and Fodor, F. (1999) Investigation of effects of cadmium, lead, nickel and vanadium contamination on the uptake and transport processes in cucumber pőlants by TXRF spectrometry, *Spectrochim. Acta Part B* **54**, 1455–1462.

Von Wirén, N., Klair, S., Basal, S., Briat, J-F., Khodr, H., Shioiri, T., Leigh, R.A., and Hider, R.C. (1999) Nicotianamine chelates both Fe^{III} and Fe^{II}. Implications for metal transport in plants, *Plant Physiol.* **119**, 1107–1114.

Von Wirén, N., Marschner, H., and Römheld, V. (1996) Roots of iron-efficient maize also absorb phytosiderophore-chelated zinc, *Plant Physiol.* **111**, 1119–1125.

Walker, C.D. and Welch, R.M. (1986) Nicotianamine and related phytosideriphores: their physiological significance and advantages for plant metabolism, *J. Plant Nutr.* **9**, 523–534.

Welch, R.M. and Norvell, W.A. (1999) Mechanisms of cadmium uptake, translocation and deposition in plants, in M.J. McLaughlin and B.R. Singh (eds.), *Cadmium in Soils and Plants,* Kluwer Acad. Publ., Dordrecht.

White, P.J. (1998) Calcium channels in the plasma membrane of root cells, *Annu. Bot.* **81**, 173–183.

Wierzbicka, M. (1998) Lead in the apoplast of *Allium cepa* L. root tips – ultrastructural studies, *Plant Sci.* **133**, 105–119.

Yamamoto, Y., Hachiya, A., Hamada, H., and Matsumoto, H. (1998) Phenylpropanoids as a protectant of aluminium toxicity in cultured tobacco cells, *Plant Cell Physiol.* **39**, 950–957.

Yamamoto, Y., Hachiya, A., and Matsumoto, H. (1997) Oxidative damage to membranes by a combination of aluminium and iron in suspension-cultured tobacco cells, *Plant Cell Physiol.* **38**, 1333–1339.

Yamamoto, Y., Kobayashi, Y., and Matsumoto, H. (2001) Lipid peroxidation is a early symptom triggered by aluminium, but not the primary cause of elongation inhibition in pea roots, *Plant Physiol.* **125**, 199–208.

Yang, Y-Y., Jung, J-Y., Song, W-Y., Suh, H-S., and Lee, Y. (2000) Identification of rice varieties with high tolerance or sensitivity to lead and characterization of the mechanism of tolerance, *Plant Physiol.* **124**, 1019–1026.

Ye, Z-H. and Varner, J.E. (1991) Tissue-specific expression of cell wall proteins in developing soybean tissues, *Plant Cell* **3**, 23–37.

Yermiyahu, U., Brauer, K., and Kinraide, T.B. (1997) Sorption of aluminium to plasma membrane vesicles isolated from roots of Scout 66 and Atlas 66 cultivars of wheat, *Plant Physiol.* **115**, 1119–1125.

Zaharieva, T., Yamashita, K., and Matsumoto, H. (1999) Iron deficiency induced changes in ascorbate content and enzyme activities related to ascorbate metabolism in cucumber roots, *Plant Cell Physiol.* **40**, 273–280.

Zancani, M., and Nagy, G. (2000) Phenol-dependent H_2O_2 breakdown by soybean root plasma membrane-bound peroxidase is regulated by ascorbate and thiols, *J. Plant Physiol.* **156**, 295–299.

Zhang, F., Römheld, V., and Marschner, H. (1991a) Role of the root apoplasm for iron acquisition by wheat plants, *Plant Physiol.* **97**, 1302–1305.

Zhang, F.S., Treeby, M., Römheld, V., and Marschner, H. (1991b) Mobilization of iron by phytosiderophores as affected by other micronutrients, *Plant Soil* **130**, 173–178.

Zhang, G., Fukami, M., and Sekimoto, H. (2000) Genotypic differences in effects of cadmium on growth and nutrient compositions in wheat, *J. Plant Nutr.* **23**, 1337–1350.

Zhang, N. and Hasenstein, K.H. (2000) Distribution of expansins in graviresponding maize roots, *Plant Cell Physiol.* **41**, 1305–1312.

Zhang, W., Zhang, F., Shen, Z., and Liu, Y. (1998) Changes of H^+ pumps of tonoplast vesicle from wheat roots in vivo and in vitro under aluminium treatment and effect of calcium, *J. Plant Nutr.* **21**, 2515–2526.

Zheng, S.J., Ma, J.F., and Matsumoto, H. (1998a) Continuous secretion of organic acids is realted to aluminium resistance during relatively long-term exposure to aluminium stress, *Physiol. Plant.* **103**, 209–214.

Zheng, S.J., Ma, J.F., and Matsumoto, H. (1998b) High aliminium resistance in buckwheat. I. Al-induced specific secretion of oxalic acid from root tips, *Plant Physiol.* **117**, 745–751.

CHAPTER 2

GLUTATHIONE AND THIOL METABOLISM IN METAL EXPOSED PLANTS

BARBARA TOMASZEWSKA
Adam Mickiewicz University
Institute of Molecular Biology and Biotechnology
Department of Biochemistry
Fredry 10, 61–701 Poznań, Poland

1. INTRODUCTION

Glutathione (GSH) is a γ-glutamylcysteinylglycine tripeptide whose primary structure was discovered in 1935 (Meister, 1988) The interest in its physiological function is still very high, but its metabolism in response to stress is still only partly understood. Glutathione is a component of cells of different organisms – from bacteria to plants, fungi, and animals. It is a major low molecular weight thiol compound in plants. Its concentration in plant cells is determined by nutritional and environmental factors and it ranges between 3–10 mM, thus significantly surpassing concentrations of other thiol compounds e.g. cysteine. Under standard conditions it occurs mainly in the reduced form (GSH) and its oxidized form (GSSG) occurs only in small amounts. GSH is the main storage and transport form of reduced sulphur in the plant (Rennenberg and Brunold, 1994). Recently, S-methyl methionine was found to be a major transported form also (Leustek et al., 2000 and ref. cited therein). Due to the presence of SH group, glutathione preserves thiol groups of proteins in the reduced form which allows to maintain their biological activity and it likely plays a role in delivering cysteine to cells of whole plant.

2. GLUTATHIONE BIOSYNTHESIS

Glutathione represents the major pool of non-protein sulphur (Kunert and Foyer, 1993). GSH is synthesized in both the cytosol and chloroplast (Rennenberg and Brunold 1994; Noctor et al., 1998a). It is exported from leaves and redistributed through the phloem to

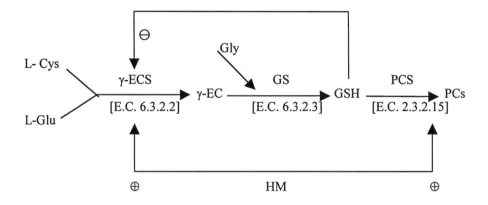

Figure 1. Regulation of biosynthesis of glutathione (GSH) and phytochelatins (PCs) from their constituent amino acids. HM (heavy metals), γ-ECS (γ-glutamylcysteinesynthetase), GS (glutathione synthetase), ⊕ activation, ⊖ inhibition.

$$L\text{-Glu} + L\text{-Cys} + ATP \xrightarrow{\gamma\text{-ECS}} L\text{-}\gamma\text{-Glu-L-Cys} + ADP + Pi.$$

fruits, seeds and roots (Leustek et al. 2000 and ref. cited therein). The level of glutathione in plant cells is a combined result of the processes of its synthesis and degradation. GSH is synthesized from glutamic acid, cysteine and glycine in two ATP-dependent reactions (Noctor et al., 1998a) (Fig. 1.). γ-glutamylcysteine synthetase (γ-ECS) (E.C. 6.3.2.2.) is the enzyme which catalyses the first reaction in which a peptide bond is created between the γ- carboxylic group of L-glutamic acid and the amine group of L-cysteine producing γ-glutamylcysteine. BSO – buthionine sulfoximine is the specific inhibitor of this enzyme. The second reaction in which glycine is bound to the C-terminus of γ-EC (Noctor et al., 1998) is catalysed by the ATP-dependent enzyme – glutathione synthetase (GS) (E.C. 6.3.2.3.).

$$L\text{-}\gamma\text{-Glu-L-Cys} + Gly + ATP \xrightarrow{GS} L\text{-}\gamma\text{-Glu-L-Cys-Gly} + ADP + Pi.$$

Apart of ATP the above two reactions are strictly dependent on Mg^{2+}. Both involved enzymes were found in cytosol and plastids of leaves and roots, but have not been found in mitochondria yet (May et al., 1998). The activity of γ-ECS undergoes inhibition by GSH, the end product of the both above-mentioned reactions at mM concentrations. In various organs of some plants, there are homologues of glutathione, differing in the C-terminal amino acid e.g.: γ-Glu–Cys–β-Ala (homoglutathione) which occurs in plants of the *Fabaceae* family, e.g. *Phaseolus vulgaris, Pisum sativum, Glycine max, Vigna radiata, Medicago sativa* in roots and leaves as well as in root nodules (Klapheck, 1988; Matamoros et al., 1999). The reaction of binding β-alanine to C-terminus of γ-EC is catalysed by an ATP-dependent enzyme – homoglutathione synthetase, hGS (Rauser, 1995; Klapheck, 1988). (Fig. 2).

The hGS enzyme from pea showed a six times higher affinity to β-alanine than to glycine, and in *Medicago trancatula* this enzyme was present only in roots (Frendo et al., 1999). These authors suggest that GS and hGS synthetase are encoded by two different genes. They isolated from *Medicago trancatula* full length cDNA encoding γ-ECS and two partial cDNAs *gsh1* and *gsh2*. *gsh1* gene had higher expression in leaves and flowers but *gsh2* was expressed only in roots and nodules. The authors hypothesise that *gsh1* and *gsh2* genes encode GS and hGS respectively. The proportions of GS/hGS varied in different legumes and in different organs of the same species (Frendo et al., 1999). There are also other GSH homologues in plants: γ-Glu–Cys–Ser (γ-glutamylcysteinylserine) and γ-Glu–Cys–Glu (γ-glutamylcysteinylglutamate). The biosynthesis of γ-Glu–Cys–Ser and Glu–Cys–Glu has not been explained yet (Rauser 1995, 1999).

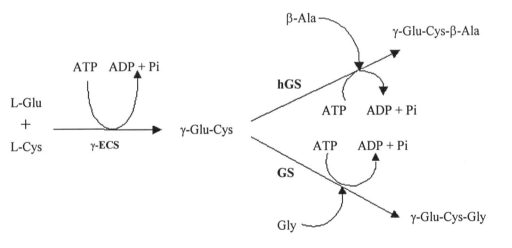

Figure 2. Pathway for glutathione (GSH) and homoglutathione (hGSH) biosynthesis in legumes.

The induction of glutathione biosynthesis is a fast mechanism protecting the plant against heavy metal ions which are toxic for them. This can be observed as an increase in the amount of foliar GSH, the substrate in the synthesis of PCs which bind heavy metals (HMs) with thiol groups. Zhu et al. (1999) demonstrated that *Brassica juncea* plants with an overexpression of *qsh2* gene encoding GS accumulated more Cd than non-transformed plants. Additionally, the transformed plants exhibited an increased tolerance to Cd both in seedlings and in mature plants. Cd accumulation and tolerance were correlated with the level of *qsh2* expression. Moreover, the transformated plants contained higher levels of GSH, PCs, thiol S and Ca than wild plants. The authors demonstrated in this way that, in the presence of Cd, GS is a rate-limiting enzyme for the biosynthesis of glutathione and phytochelatins. Transgenic plants with simultaneous γ-ECS and GS overexpression, which leads to an increased GSH production and accumulation, constitute a promising strategy of raising the phytoremediation capacity of heavy metals. Such plants would exhibit a specifically enhanced efficiency of HM phytoextraction from polluted soils and wastewater.

3. PHYTOCHELATINS AND THEIR STRUCTURE

A vast array of defence mechanisms such as: immobilization, exclusion, chelation and compartmentalization of ions of heavy metals, expression of stress proteins or activation of ethylene response to stress is activated in plants exposed to heavy metals stress (Cobbett, 2000a, b). The main mechanism of detoxificating heavy metals is chelating them into complexes and transporting these non-toxic complexes into vacuoles, thus separating metals from cell metabolism.

Phytochelatins are also called very often the class III metallothioneins (MT-IIIs). The primary structure of phytochelatins is known for five families (Rauser, 1995; Zenk, 1996). The common property of these compounds is that glutamate occupies the N-terminal position and cysteine is bound to its γ-COOH group, forming a γ-peptide bond instead of α-bond normally found in all proteins. C-terminal amino acid divides them into 5 families (Rauser 1995). Recently the amino acid sequence of the sixth family of phytochelatins isolated from horseradish (*Armoracia rusticana*) has been determined, in which glutamine is the C-terminal amino acid (Kubota et al., 2000).

Mehra et al. (1988) showed that *Candida glabrata* yeast exposed to Cu^{2+} ions produced both metallothioneins and phytochelatins. Phytochelatins are synthesized in plants treated with ions of heavy metals from glutathione (GSH) (Rüegsegger et al., 1990; Tukendorf and Rauser, 1990; Rauser, 1995; Zenk, 1996). PC-related peptides are also synthesized from proper (GSH) homologues. Such homologues: γ-Glu–Cys–β-Ala (Gekeler et al., 1989), γ-Glu–Cys–Ser (Klapheck et al., 1994) and γ-Glu–Cys–Glu (Meuwly et al., 1993) were detected in plants, whereas γ-Glu–Cys–Gln was determined only on the basis of the retention time after the separation on the HPLC column but identification of this peptide needs to be confirmed by determining the amino acid sequence (Kubota et al., 2000).

Zenk (1996) proposed to call PC-related peptides – isophytochelatins, and the substrates from which they are synthesized, i.e. GSH-related peptides – isotripeptides. According to this proposal, iso-phytochelatin, which has been recently detected in roots of horseradish by Kubota et al. (2000), should be called isoPC (Gln) = $(\gamma\text{-Glu-Cys})_n$ Gln. Such a naming pattern would leave some space for new isoPCs, which might be discovered and sequenced in the future. Classification of the known families of phytochelatins is shown in table 1.

4. COMPLEXES OF PHYTOCHELATINS WITH HEAVY METALS

The role of organic acids in heavy metal stress and function of some amino acid complexes containing heavy metals in detoxification mechanism has been recently reviewed by Rauser (1999). At present it is thought that the essential part in detoxification of heavy metal ions is played by phytochelatins – peptides containing large amounts of cysteine. Recently, some reviews have been published on the structure, biosynthesis and the function of phytochelatins in detoxification of heavy metals in plants (Rauser, 1995, 1999; Zenk, 1996); and quite recently a review was published by Cobbett (2000a) who discussed the latest data on the function of phytochelatins on the ground of research conducted with the use of molecular biology methods using genetically modified model plants. Phytochelatins (PCs) are low molecular thiol peptides. They were characterised for the first time in *Rauvolfia serpentina* cells treated with Cd ions (Grill et al., 1985). This discovery was the consequence of seeking plant proteins, called metallothioneins, which had been found earlier in animal organisms (Kägi and Nordberg, 1979) and which fulfilled the function of detoxificating heavy metals. At present it is well known that both metallothioneins and phytochelatins participate in detoxification of heavy metals in plants and fungi.

Table 1. First reports of the families of phytochelatins into 6 categories

1. Phytochelatins (PCs): $(\gamma\text{-Glu-Cys})_n$Gly in *Rauvolfia serpentina*	n = 2-11	Grill et al., 1985
2. Homophytochelatins (hPCs): $(\gamma\text{-Glu-Cys})_n\beta\text{-Ala}$ in family Fabaceae	n = 2-7	Gill et al., 1986
3. Hydroxymethylphytochelatins (isoPCS): $(\gamma\text{-Glu-Cys})_n$Ser in *Oryza sativa*	n = 2-4	Klapheck et al., 1994
4. IsoPCs (Glu): $(\gamma\text{-Glu-Cys})_n$Glu in *Zea mays*	n = 2-3	Meuwly, 1993
5. Desglycilpeptides: $(\gamma\text{-Glu-Cys})_n$ in *Candida glabrata* and *Schizosaccharomyces pombe* and in *Rubia tinctorum* L.	n = 2-5	Mehra et al., 1988 Kubota et al., 1995 Maitani et al., 1996
6. IsoPCs (Gln): $(\gamma\text{-Glu-Cys})_n$Gln in hairy roots of horseradish (*Armoracia rusticana*)	n = 3-1	Kubota et al., 2000

Mehra et al. (1988) estimated that the fungus *Candida glabrata* exposed to copper ions induced the production of both metallothioneins and phytochelatins. Later, Mehra et al. (1995) demonstrated that this resistance was based on the ability of the highly Cd-resistant strain of *Candida glabrata* to form huge amounts of PC-coated CdS crystallites. In 1999 an excellent review of II and III classes MT was published by Rauser. Phytochelatins are synthesized in plants from glutathione by means of cytosolic constitutive enzyme γ-glutamylcysteine transferase (E.C. 2.3.2.15) also called phytochelatin synthase (PCS) activated by a wide variety of heavy metals, e.g. Ag, Bi, Cd. Cu, Mg, Pb, Sn, Zn (Zenk, 1996). Additionally, multiatomic ions including SeO_4^{-2}, SeO_3^{-2} and AsO_4^{-3} also cause PCs synthesis.

Recently, Schmöger et al. (2000) demonstrated that both anions arsenite and arsenate induce the biosynthesis of PCs *in vivo* and *in vitro* in *Rauvolfia serpenina and Silene vulgaris* cell suspension culture. As-glutathione complexes are compatible with As-PCs complexes in which the ratio of PCs sulfhydryl groups to As is 3 to 1. Gel filtration studies and electrospray ionization mass spectroscopy indicate complexation and detoxification of As by the induced PCs.

Phytochelatins are able to chelate heavy metals (HMs) due to the high cysteine content in their molecules. Direct participation of PCs in the chelation of HM ions in plants has been verified for only few elements under *in vivo* or *in vitro* conditions. PCs have been shown to bind Cd and Cu directly and are believed to bind Pb and Hg in competition with Cd (Speiser et al. 1992). Grill et al. (1985) showed that the synthesis of the complexes which were isolated by means of gel filtration and which bound more than 90% of the metal, takes place during the exposition of *Rauvolfia serpentina* suspension cells to $CdSO_4$ at 200 μM concentration, well tolerated by these cells. The molecular weight of the complexes was 1040 Da and their peptide structure turned out to be $(NH_3)^+$–γ-Glu–Cys–γ-Glu–Cys–γ-Glu–Cys–GlyCOO$^-$ = (γ-Glu–Cys)$_3$Gly.

Other authors (Gekeler et al., 1989) studied more than 200 species of various plants and cell cultures of many taxonomic divisions treating them with ions of Cd^{2+} and other HM, and showed the presence of heavy metal-PCs complexes of varied length where n value ranged from 2 to 11. Complexes of phytochelatin molecules had variable length dependent on plant species, exposition time and the type of HM (Grill et al., 1987) and their molecular weight was situated between 2.5–3.6 kDa. Using the EXAFS spectroscopy method Strasdeit showed later (Strasdeit et al., 1991) that Cd-PCs complex shows Cd-thiolate co-ordination with Cd-S bond and excluded the participation of the carboxylate groups in the complexation (Zenk, 1996).

It is well known that associations of phytochelatins with heavy metals create complexes with S^{-2} ions, which enlarges their molecular weight. They are called LMW (low molecular weight) and HMW (high molecular weight) complexes. This also enlarges their stability and the capacity of toxic metals binding (Zenk, 1996). HMW-PCs complexes prevail in plants in general but there are differences between species. Moreover, the type of created complexes depends on the type of metal, its content in the growth medium and the separation method of gel filtration.

Recently, Rauser (2000) proposed that there are two types of Cd-building complexes: LMW and HMW, which occur in roots of maize cultivated hydroponically for 5–7 days, supplemented with 3 μmol Cd. These both complexes contained 3 families of cysteine-rich peptides (γ-Glu–Cys)$_n$Gly, (γ-Glu–Cys)$_n$ and (γ-Glu–Cys)$_n$Glu. The proportions of mono-, di-, and thiol peptides were higher in the LMW complex than in the HMW form. Vacuolar HMW complex contained more Cd per peptide thiol. LMW complexes were localized in cytosol and contained a considerable amount of acid labile sulphide. Together, both HMW and LMW complexes bound 75–88% of Cd ions in maize roots. Galli et al. (1996) showed in Cd and Cu treated maize the presence HMW complexes with relative molecular weight of 6200–7300 Da, whereas Kneer and Zenk (1997) demonstrated three types of PCs-Cd complexes in *Rauvolfia serpentina* roots: HMW, MMW (medium molecular weight) and LMW but only HMW in *Silene cucubalus* cells treated with Cd ions. In *Rauvolfia serpentina* MMW complexes have higher S:Cd ratio in comparison with the HMW complexes from *Silene cucubalus*.

Arabidopsis thaliana cad1 mutants had a characteristics of Cd-hypersensitivity and did not possess the ability to form a HMW complex, but only a slight amount of LMW (Howden et al., 1995a, b).

A LMW complex is transported into vacuole (Oritz et al., 1992, 1995) by an ATP-dependent transporter in fission yeast *Schizosaccharomyces pombe*. Salt and Rauser (1995) identified Mg ATP- dependent transport of phytochelatins in a form of LMW complexes across tonoplast of oat root cell vacuole which is identical with the one of the fission yeast. Cd^{2+} ions enter vacuoles via a $Cd^{2+}/2H^+$ antiport (Salt and Wagner, 1993) where they form with acid-labile sulphide and with apophytochelatin (LMW) stable complexes of higher molecular weight (HMW, PCs-Cd) (Speiser et al. 1992, Oritz et al., 1995). It seems that LMW complexes are not sufficient for Cd^{2+} detoxification. It is not clear yet whether sulphide incorporation takes place in the cytoplasm, on the way to the vacuole through the cytoplasm or in the vacuole.

The highest amounts of acid-labile sulphide are contained in HMW complexes from seedlings of *Brassica juncea* (Speiser et al., 1992), which is a plant well known as a good accumulator of heavy metals. Kneer and Zenk (1997) have recently isolated from cell cultures of *Rauvolfia serpentina* treated with Cd^{2+} ions, three groups of PCs-Cd complexes: LMW, MMW, HMW.

All these types of complexes contained labile sulphide, but its highest amounts were present in HMW fraction (Rauser, 1999 and ref. cit. therein). The same authors isolated some HMW complexes from cell cultures of *Silene vulgaris* treated with Cd^{2+}. Since 1988 the prevailing view has been that HMW are complexes resulting from Cd and sulphide binding with LMW complexes. Reese and Winge (1988) obtained HMW *in vitro* by addition of sulphide to LMW-Cd complexes isolated from *Schizosaccharomyces pombe*.

The comparison of LMW, MMW and HMW complexes from two plants – *Rauvolfia serpentina* and maize made by (Rauser, 1999 and ref. cit. therein) – proves that these complexes have different contents of phytochelatins *in vivo*. LMW from *R. serpentina* are made of PCs of the type (γ-Glu–Cys)$_{n=2-4}$, whereas MMW and HMW contain a majority of longer phytochelatins of the type (γ-Glu–Cys)$_{n=7}$. LMW and HMW from maize contain phytochelatins of the same length in which n=1–5. HMW complexes of both plants have the highest relation of Cd per peptide thiol, however only in *Rauvolfia serpentina* HMW complexes contain acid labile sulphide whereas in maize acid labile sulphide occurs both LMW and MMW. Obtaining Cd-sensitive mutants of fission yeast that are deficient in PC-Cd complexes, is the genetic evidence of the importance of phytochelatin bonds with sulphide. Speiser et al. (1992) showed that different single or double mutations in the adenine biosynthesis pathway led to the deficit of HMW complexes, which is related to sensitivity to Cd^{2+}.

Formation of phytochelatin complexes with other heavy metals was shown in *in vitro* research using $(\gamma\text{-Glu–Cys})_{n=2-4}$ which created complexes with the following metals: Cd^{2+}, Pb^{2+}, Ag^{1+}, Hg^{2+} (Mehra et al., 1995, 1996). The coordination of Pb^{2+} ions by PCs was dependent on the PCs length. Simultaneously, it was shown that Pb^{2+} and Hg^{2+} ions were inductors of phytochelatin synthesis (Grill et al.,1987; Tomaszewska et al., 1996; Maitani et al., 1996) and therefore it is very likely that complexes with these metals can be formed *in vivo* too.

The research conducted on plants tolerant to HM such as *Silene cucubalus, Minuarta verna* and *Armeria maritima* growing on the area of copper mine "Saugrund" Eisleben, Germany (Leopold et al., 1999) did not show the formation of HM-PCs complexes in these plants. HM ions in the soluble extract of these plants were bound with molecules of lower molecular weight than PCs. This research proves that PCs are not the main source of high tolerance of these plants to HMs under natural conditions although the same plants were able to create HM-PCs complexes in water cultures. Leopold et al. (1999) think that phytochelatins play only a transient role in heavy metal detoxification because PCs-Cu and Cd complexes disappeared from roots of *Silene vulgaris* grown in water between the 7th and 14th days of cultivation and metal ions were bound with molecules of lower molecular weight than phytochelatins. The transient formation of PCs-Pb complexes was also shown in roots of *Pisum sativum* treated with Pb^{2+} ions (Piechalak et al., in press).

The fact that sensitive plants synthesize sometimes more PCs than the tolerant plants under the influence of the same HM concentration (Tukendorf and Rauser, 1990; Verkleij et al., 1990; De Knecht et al., 1994; Harmens et al., 1993) is an additional argument against the essential role of phytochelatins in HM detoxification. The latter authors demonstrated lower synthesis of long-chain PCs in roots of Cd-tolerant *Silene vulgaris* than in sensitive plants and, moreover, stabile PCs-Cd complexes were formed much faster in sensitive plants than in tolerant ones. However, this fact can be caused by a faster transport of metal by tonoplast into vacuole and simultaneously, by a higher rate of peptide complexes breakdown in tolerant plants, which would explain the lower level of phytochelatins with simultaneous higher tolerance to the metal (De Knecht et al., 1992, 1994).

Not only phytochelatins but also glutathione can form complexes with HM. Most studies concern the formation of such complexes with Cd^{2+} ions but it can be assumed that other HM behave in the same way. The formation of Cd_1-$(GSH)_2$ complex and its transport across the vacuolar membrane were shown to be critical for Cd tolerance in the yeast *Saccharomyces cerevisiae* (Li et al., 1997). GSH-bound metal ions are transferred to shorter PCs which, in turn, transfer the metal ions to longer PCs. GSH-mediated transfer of Cd^{2+}, Pb^{2+}, Cu^{2+}, Hg^{2+} to PCs has been demonstrated by Mehra et al. (1995, 1996), as well as by Bae and Mehra (1997). Moreover, it was shown for the first time that PCs complexes with CdS crystallites were present in

Schizosaccharomyces pombe and *Candida glabrata* (Dameron et al., 1989) growing on 1 mM Cd.

5. BIOSYNTHESIS OF PHYTOCHELATINS

The primary structure of phytochelatins and iso-phytochelatins suggests that they are derivatives of glutathione and its homologues. γ-glutamyl bond between γ-carboxylic group of N-terminal glutamic acid and cysteine makes these peptides resistant to peptidases breaking down peptide bond of peptides which were formed in the translation process. Determination of the sequence of isoPC (β-Ala) peptide also points to the fact that it did not originate in the process of translation because tRNA adequate for β-alanine is not known yet (Zenk, 1996). In 1987 Grill et al. already demonstrated that the enzyme catalysing the reaction of phytochelatin synthesis does not have the inducible character since adding $Cd(NO_3)_2$ to cell suspension of *Rauvolfia serpentina* led to a large immediate decrease in endogenous glutathione concentration without a lag phase, characteristic for the induction of protein biosynthesis through transcription and translation (Zenk, 1996). The enzyme catalysing the synthesis of phytochelatins is the constitutive enzyme γ-glutamylcysteinedipeptidyl transpeptidase (E.C. 2.3.2.15), called phytochelatin synthase (PCS). This enzyme catalyses the transport of γ-Glu–Cys group from glutathione donor molecule to an acceptor molecule or to a short phytochelatin with n=2 (Grill et al., 1989; Loeffler et al., 1989). γ-Glu–Cys polymerisation from $(γ-EC)_n$ and GSH to $(γ-EC)_{n+1}$ followed by Gly addition with GSH synthetase (Hayashi et al., 1991) is an alternative pathway found in fission yeast. In reaction to heavy metal ions, higher plants, algae and some fungi synthesize phytochelatin peptides consisting of recurrent units of γ-glutamylcysteine followed by C-terminal glycine (Gekeler et al., 1988,1989; Steffens, 1990). The number of recurrent units ranges from 2 to 11 and varies with concentration and duration of heavy metal exposure (Tukendorf and Rauser, 1990).

The fact that glutathione is being used up is confirmed by a drastic decrease in GSH concentration in cytosol leading at the same time to the induction of enzyme synthesizing glutathione – γ-Glu–Cys synthetase (γ-ECS) and glutathione synthetase (GS). Consequently, this leads to a more intense synthesis of GSH, which is used in PC synthesis (Rüegseger et al., 1990; Rüegsegger and Brunold, 1992). PCs synthase is activated by heavy metal ions given in order of their decreasing induction ability (Grill et al., 1987): $Cd^{2+}>Pb^{2+}>Zn^{2+}>Sb^{3+}>Ag^{+1}>Hg^{2+}>As^{-5}> Cu^{+1} >Sn^{2+}> Au^{3+}> Bi^{3+}$. Each of these metals has a different concentration optimum for enzyme activation. Regulation of phytochelatin synthase activity is done in such a way that the end product of the reaction, phytochelatin, forming stabile complexes with metal ions, removes these ions from cytosol, which leads to the inhibition of the PCs synthesis. Phytochelatins containing longer chains with a characteristic higher relative complexing affinity than the shorter ones, have the ability of fast inhibition of the enzymatic reaction catalysed by PCs synthase (Zenk, 1996). Consistent with earlier findings, phytochelatin synthase is an enzyme built of four subunits of molecular weight 95 000 Da each; the optimum temperature is 35°C and pH 7.9; K_m GSH = 6,7 mM. PCS was isolated and characterized for different plants, and activated with ions of various metals in: *Silene cucubalus* cell cultures (Loeffler et al., 1989; Grill et al., 1989), *Pisum sativum* roots (Piechalak et al., in press; Klapheck et al., 1995) *Arabidopsis thaliana* (Howden, 1995b), *in vitro* carrot plants and in carrot cell suspension cultures (Sanitá di Toppi et al., 1999), roots and cell cultures of tomato (Chen and Goldsbrough, 1994; Chen et al., 1997), tobacco cell cultures (Nakazawa and Tokenaga, 1998), *Rauvolfia serpentina* cell suspension (Friederich et al., 1998). The enzyme was isolated and its activity was determined *in vitro* after activation with metal ions. HPLC method was used to measure the amount of phytochelatins produced after incubation. The purified PCS exhibited some activity only in the presence of metal ions. The best activator were Cd^{+2} ions. There is a limited knowledge about the tissue specificity of phytochelatin synthase. Chen et al. (1997) demonstrated its activity in roots and stems of tomato plants, but found no traces of activity in tomato leaves and fruits.

The role of glutathione in the synthesis of phytochelatins and in tolerance to heavy metals was proven in research using inhibitors and mutants. Buthionine sulfoximine (BSO) is a specific inhibitor of the first enzyme of glutathione synthesis, γ-EC synthetase (Griffith and Meister, 1979). Addition of this inhibitor to nutrient solution decreased both GSH and PCs levels and reduced tolerance to metal ions in many plants (Grill et al., 1987; Reese and Wagner, 1987; Tomaszewska et al., 1996). *Arabidopsis thaliana* mutants lacking either γ-EC synthetase or GSH synthetase did not synthesize PCs and were hypersensitive to Cd^{2+} ions (Howden et al., 1995a, b; Cobbett et al., 1998). Overexpression of tomato and *Arabidopsis* γ-ECS in cadmium-sensitive *Arabidopsis* mutant *cad2* restores cadmium tolerance (May et al., 1998). Moreover, *Arabidopsis* mutants (*cad1*) obtained as a result of an artificial mutagenesis, does not contain PC synthase, and exhibits unique sensitivity to Cd^{2+} ions (Howden et al., 1995b). Cultured tomato cells selected for increased Cd tolerance, had an increased γ-ECS activity (Chen and Goldsbrough, 1994). The level of glutathione decreased in roots of *Pisum sativum* treated with Cd^{2+} (Klapheck et al., 1995). Similarly, the content of glutathione was lower in roots and stems of maize treated with Cd^{2+} ions (Rauser et al., 1991) which was concomitant with an increase of γ-ECS activity in roots (Rüegsegger and Brunold, 1992). The activity of both enzymes of glutathione biosynthesis γ-ECS and GS, increased in tobacco cell cultures treated with Cd ions (Schneider and Bergman, 1995).

Transformed poplars overexpressing γ-ECS *qsh1* gene (from *E. coli*) in cytosol (Noctor et al., 1998a, b; Arisi et al., 2000) contained increased levels of GSH and γ-EC in leaves and 30 times higher activity of γ-ECS as compared with non-transformed plants. GS overexpression did not exert any influence on the level of GSH in leaves of transformed poplars (Noctor et al., 1998b). Therefore, the authors postulate that γ-ECS is the rate-limiting enzyme for glutathione biosynthesis in poplars in the absence of Cd. Exposure of transformed poplars to Cd^{2+} enhanced tissue cadmium accumulation that had only a marginal impact on tolerance (Arisi et al., 2000). In the case of *Arabidopsis thaliana*, treating plants with Cd^{2+} ions led to the induction of γ-ECS gene expression, which was demonstrated in the rise of transcript amount (Xiang and Oliver, 1998). Similarly, Cu^{2+} ions led to the increase in γ-ECS mRNA amount in roots and shoots of *Brassica juncea* (Schäfer et al., 1997). Moreover, Haag-Kerwer et al. (1999) showed that expression of γ-ECS genes was correlated with the synthesis of PCs under the influence of Cd^{2+} ions in *Brassica juncea* plants.

The biosynthesis of PCs in plants takes place in the presence of metal ions and leads to use of GSH, which is a product of the activity of two enzymes – γ-ECS and GS. Their activity increases in the presence of heavy metals ions. The use of GSH for the production of PCs during heavy metal ions treatment frees γ-ECS from the feedback inhibition by GSH (Fig. 1.).

6. PHYTOCHELATIN SYNTHASE GENES

The gene encoding phytochelatin synthase has been recently cloned simultaneously in three laboratories (Clemens et al., 1999; Ha et al., 1999; Vatamaniuk et al., 1999). Two reviews describing in detail the results obtained by the three research groups were published soon afterwards (Cobbett, 1998, 2000a, b). In 1995 Howden et al. (1995a, b) obtained *cad1* mutants of *Arabidopsis thaliana* sensitive to cadmium which were not able to synthesize phytochelatins and did not exhibit *in vitro* PCS activity despite the fact that they showed the same level of glutathione as the wild-type ones. The authors came to the conclusion that the *cad1* gene is a structural gene of PC synthase.

Ha et al. (1999) isolated genes of PC synthase from *Arabidopsis* and *Schizosaccharomyces pombe* yeast using the positional cloning strategy and called them *AtPCS1* for *Arabidopsis* (Vatamaniuk et al.1999), *SpPCS1* for yeast and other group of scientists isolated cDNA of *TaPCS1* gene from wheat (Clemens et al. 1999). Wheat cDNA encoding PC synthase was identified through its expression in yeast *S. cerevisiae* and resistance to Cd was obtained. Yeast mutants with *SpPCs1* gene deletion were sensitive to cadmium and were PCs-deficient. These features indicate the identical function of the products of these genes in plants and yeast. A gene similar to the PC synthase gene from *Arabidopsis thaliana* was identified in the nematode *Caenorhabditis elegans* [Fig. 3]. The N-terminal region of the predicted *CePCS1* gene product is similar to those found in plants and yeast. The C-terminal domain is much less conservative but it contains also multiple pairs of Cys residues. Till now there has been no evidence that these nematode genes encode the PC synthase enzyme but this is highly probable. The deduced product of the *CePCS1* gene was composed of 371 amino acid residues, while the product of the *AtPCS1* gene is a polypeptide with the molecular weight 55 kDa and it is composed of 485 amino acids residues. The product of the *SpPCS1* gene has 414 residues and the one from *Triticum aestivum TaPCs* is the longest one as it contains 500 amino acid residues. Homologous genes were found in *Schizosaccharomyces pombe* and in *Caenorhabditis elegans* (*CePCS1*) (Cobbett, 2000a, b). Identification of PC synthase genes in such unrelated organisms indicates that this enzyme plays an essential role in living organisms. Both *AtPCS1* and *TaPCS1* genes are transcribed in the absence of HM stress (Vatamaniuk et al., 1999; Clemens et al., 1999; Ha et al., 1999), which supports the earlier observation that PC synthase is expressed in plant tissues or cell cultures in a constitutive way. Glutathione, can replace these co-substrates and thus the need of heavy metals presence for PCs production is avoided. Therefore, it can be assumed that during the *in vitro* reaction of PCs synthase in the medium containing the

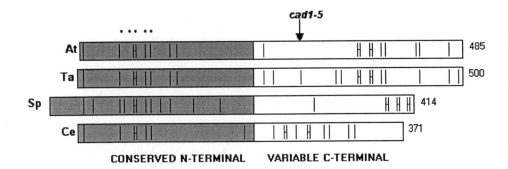

Figure 3. Comparison of PC synthase polypeptides from different organisms: Arabidopsis thaliana ATPCS1, Triticum aestivum Ta PCS1, Schizosaccharomyces pombe SpPCS1 and nematode Caenorhabditis elegans CePCS1. The numbers on the right correspond to the total number of amino acids. The vertical bars indicate the number of Cys residues; horizontal bars connect the adjacent Cys residues. Cys residues conserved across the three sequences are highlighted with an asterisk. The position of the Arabidopsis cad1–5 nonsense mutation is indicated by an arrow. (According to Cobbett, 2000a.)

optimum GSH concentration for PCs synthesis dependent on HM, most of the metal occurs in the form of a complex with GSH. Additionally, HMs do not activate catalysis in a medium containing free thiols through the direct interaction with the enzyme, as it has been thought so far, but through an interaction with the substrate.

The comparison of the deduced structures of phytochelatin synthases isolated from the above-mentioned organisms (Fig. 3) allows to draw the conclusion that these enzymes have the N-terminal highly conservative domain (40% identical amino acids) which shows catalytic activity and C-terminal variable domain which fulfils the function of a local sensor reacting to HM ions. The C-terminal domain of all four phytochelatin synthases has abundant amounts of cysteine residues, often occurring in pairs, which bind ions of metals and bring them in contact with the catalytic domain. The variable C-domain, is not necessary for catalysis or the activation by metal. This conclusion results from molecular characteristics of the *cad1-5* mutant from *Arabidopsis,* which has a nonsense mutation leading to a premature translation termination of the conservative N-terminal domain (Ha et al., 1999). It is predicted that this truncated polypeptide does not contain 9 out of 10 cysteine residues in the C-terminal domain. Nevertheless, this mutant had the same specific activity expressed through an identical amount of PCs synthesized *in vivo*. Consistent with the activation model, GSH-Cd complexes bind to the active center of PCs synthase, while the remaining free ions bind with the product of the PCS catalysed reaction.

Products similar to those of plant and yeast genes were identified in nematode *C. elegans* exhibiting 40%–50% of identical amino acids in the N-terminal region and C-terminal domain exhibiting also similar 10 Cys residues including two cysteine pairs

(Fig. 3). This fact allows to conclude that PCS synthesis also takes place in animal organisms although there is no evidence that animal genes encode the PC synthase protein. Similar genes were identified in slime mould (*Dictyostelium discoideum*) and in the aquatic midge *Chironomonas oppositus* (Cobbett, 2000a).

Besides the genes of PC synthase, a whole range of genes dealing with the functions of phytochelatins were identified and these included: genes of PCs complexes transport across the vacuolar membrane (*hmt1*), genes of metabolism of cysteine sulphinate in sulphide biosynthesis, genes of adenine biosynthesis (*ade2,6,7,8*), genes of detoxification of sulphide (*hmt2*) and genes of porphobilinogen synthase involved in sirohem biosynthesis the cofactor for sulphite reductase (*hem2*) (Fig. 4) (Cobbett 2000a and ref. cited therein). Molecular genetic approach allowing to identify a whole range of genes connected with the mechanism of heavy metals detoxification makes it possible to assume that PCs play an essential role not only in the detoxification process but also in essential heavy metal homeostasis. They are also linked with sulphur metabolism. It must be emphasized however, that detoxification of heavy metal ions by complexing them with phytochelatins and their transport to vacuoles are not the only mechanisms which enable plants to cope with the excess of these toxic ions.

7. PHYTOCHELATIN SYNTHASE ACTIVATION

The mechanism of PC synthase activation by heavy metals is not well-understood, although some metals are known to be more active than others. Regulation of PC synthase synthase activation by metal in: cell cultures of *Silene cucubalus* (Grill et al., 1989), tomato cell cultures (Chen and Goldsbrough, 1994), tomato plants (Chen et al., 1997), and *Arabidopsis thaliana* (Howden et al., 1995a, b), allowed to draw the conclusion that the activation occurs in a few minutes after the exposure to Cd^{2+} ions, which proves that it is independent of the *de novo* synthesis of proteins. The expression of PC synthase gene takes place regardless the exposure to ions of heavy metals. activity is most probably the main point in regulation of PCs synthesis. The study of PC

A recently published study has cast light on the mechanism of PC synthase activation by heavy metals (Vatamaniuk et al., 2000). The authors conducted an *in vitro* enzymatic reaction with Cd^{2+} and with Zn^{2+} and demonstrated through analyses of immunopurified recombinant PCS1 from *Arabidopsis thaliana* (*AtPCS1*) that the presence of free ions of heavy metals is not necessary for the reaction catalysed by PC synthase. Product of gene *AtPCS1* appears to be primarily activated postranslationally in plants. Although heavy metals are capable of direct binding the AtPCS1 enzyme, it is not the metal *per se* which is responsible for the catalysis but rather its complex with glutathione: heavy metal glutathione thiolate with the following structures: $Cd-GS_2$ or $Zn-GS_2$. Therefore, the enzyme catalyses bissubstrate transpeptidation in which free GSH and the proper thiolate of heavy metals fulfil the function of co-substrates. The authors showed that although both these co-substrates are necessary for reaching the maximum level of activation, other components such as, S-substituted derivatives of

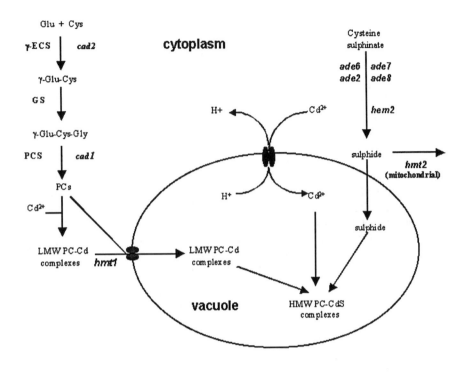

Figure 4. Genes and functions contributing to Cd detoxification in plants and fungi: *cad1, cad2* in *Arabidopsis thaliana; hmt, hmt2, ade2, ade6, ade7, ade8* in fission yeast; and *hem2* in *Candida glabrata*. Enzyme abbreviations are shown in bold. γ-ECS, γ-glutamylcysteine synthetase; GS, glutathione synthetase; PCS, phytochelatin synthase. Gene loci are shown in bold italics.
(According to Cobbett, 2000a.)

Vatamaniuk et al. (2000) demonstrated that PC synthase recognises GSH bonds with substituted thiol groups, e.g. S-alkyl-glutathiones, which can act both as γ-Glu–Cys donors and acceptors in the transpeptidation reaction. AtPCS1 synthase exhibited 100 times lower affinity to S-methylglutathione than metal thiolates, but its affinity to S-methylglutathione and GSH was almost identical. The authors proposed activation mechanism, which is presented in the last Vatamaniuk et al. 2000 paper.

In contrast with the previous activation model in which free heavy metal ions are essential for catalysis and directly bound to the PCS enzyme, the final stage of the reaction catalysed by PCS cannot last till the chelation of all free heavy metal ions because both GSH and PCs complexes containing these ions are also active substrate for PCS enzyme. Probably, free thiols: GSH and apoPCs, compete with heavy metal glutathione and PCs thiolate such as e.g. $Cd-GS_2$ or $Zn-GS_2$ for high affinity site of PCS and can act as γ-Glu–Cys acceptor and donor. Increasing the concentrations of heavy metal ions does not, even transiently, lead to increased level of PCs synthesis. Heavy metal glutathione thiolates are formed and transformed by PC synthase activity into PCs, which also bind free heavy metals but only those with much higher affinity.

8. THIOL METABOLISM AND N_2 FIXATION IN LEGUMES

Legumes contain also thiol tripeptides other than GSH, mainly homoglutathione (γ-Glu–Cys–β-Ala) which substitutes GSH (e.g. in *Phaseolus vulgaris*) or occurs together with GSH (e.g. in *Pisum sativum*). Synthesis of homoglutathione from γ-EC and β-Ala is catalysed by a specific enzyme – hGSH synthetase – of a high affinity to β-Ala and low affinity to glycine (Klapheck, 1988). Based on the analysis of thiol content in different organs of these plants, and especially in root nodules of eight legume species, one can draw the conclusion that thiol compounds play a key role in N_2 fixation (Matamoros et al., 1999). cDNA encoding enzymes of tripeptide synthesis: γ-ECS, GS and hGS – was isolated and the transcripts in nodules were quantified. It was found that, γ-ECS was localized in plastids, hGS in cytosol, and GS both in cytosol and in the mitochondria of root nodules (Matamoros et al., 1999; Moran et al., 2000).

Using the strategy combining PCR screening of cDNA libraries, 5'-RACE and RT-PCR of leaf and nodule DNA, the authors (Moran et al., 2000) obtained the sequence of all the above mentioned synthetases (γ-ECS, GS, hGS). It was shown that hGS is the only synthetase present in *Phaseous vulgaris* and *Glycine max* leaves and nodule host cells, and hGS isoenzyme is present in cytosol. On the other hand only γ-ECS was found in nodules of legumes (in plastids and bacteroids) producing only GSH, (e.g. in cowpea,) and GS activity was traced in cytosol and mitochondria (Matamoros et al. 1999). γ-ECS from nodules was different from the one of tobacco cell suspension. Nodule γ-ECS had molecular weight of 51 kDa, which was lower than that of the foliar one (Hell and Bergmann, 1990). Bacteroids contained high concentrations of thiols and exhibited high specific activity of γ-ECS and GS. Bean and cowpea nodule bacteroids contained 2.29 nmol Cys and 10 nmol GSH per mg of protein. Cowpea bacteroids had no hGSH but bean bacteroids contained 1.5 nmol hGSH per mg protein (Moran et al., 2000) Thus, nodule bacteroids play a key role in regulation of thiol compounds synthesis and in the antioxidative protection of the metabolism of N_2-fixing nodules. It has been thought so far, that the main function of plastids in root nodules is their participation in ammonia assimilation (Temple et al., 1998) and purine synthesis (Atkins et al., 1997). The key role of nodule plastids in antioxidative defence of root nodules was evidenced by demonstrating the localization of such enzymes of glutathione metabolism as: γ-ECS, GR (glutathione reductase) in nodule plastids (Matamoros et al., 1999). Moreover, in nodule plastids antioxidative proteins Fe-SOD (Moran et al., 2000 and ref. cited therein) and ferritin were present, whose expression is probably induced by ROS (reactive oxygen species) (Matamoros et al., 1999; May et al., 1998).

9. THE LINK OF THIOL COMPOUNDS WITH SULPHUR METABOLISM

It is well known that glutathione is the main storage pool of non-protein reduced sulphur in both prokaryotic and eukaryotic cells (Kunert and Foyer, 1993). The actual concentration of glutathione in plants is dependent, among others, on the supply of sulphates from the soil (Herschbach and Rennenberg, 1994). Most plants acquire sulphur from the soil in the form of sulphate. Although a significant progress has been made in explaining the assimilation process of sulphates many aspects remain unclear until now. Sulphate is reduced to sulphide, which is transferred to O-acetyl-L-serine (OAS) by the OAS (thiol) lyase enzyme (OAS-TL, E.C. 4.2.99.8). Finally most of the reduced sulphur is transformed into cysteine. O-acetyl-L-serine, an acceptor molecule for sulphide is produced from serine and acetyl-CoA in the reaction catalysed by serine acetyltransferase (SAT) enzyme (E.C. 2.3.1.30) (Leustek, 1996; Leustek and Saito, 1999).

The reaction of OAS synthesis is essential for sulphate metabolism because its availability may regulate the reaction of sulphation and cysteine biosynthesis. OAS does not occur in plants in the free form because it is immediately transformed into cysteine. The reaction of cysteine synthesis is catalysed by an enzymatic complex called cysteine synthase (CS, E.C. 4.2.99.8) composed of SAT and OAS-TL. The products of this reaction are cysteine and acetate (Takahashi and Saito, 1996).

Cysteine, one of the end products is an allosteric SAT inhibitor (Leustek et al., 2000). Both enzymes of SAT and OAS-TL complex are localized in plant cells in plastids, mitochondria and cytosol (Rennenberg and Brunold, 1994; Hell, 1997; Hesse et al., 1999). This proves that cysteine is necessary in all these cell compartments in which protein biosynthesis takes place. Cysteine occurs in plant tissues at low concentrations because of its high reactivity (Rennenberg and Brunold, 1994). It is transformed into less reactive glutathione or its homologues, in which SH group exhibits lower reactivity. Moreover, cysteine is a component of many regulatory proteins – defensins, metallothioneins and other low molecular proteins with high amounts of cysteine (Hell, 1997). It is also a substrate in methionine synthesis (Leustek 1996). It has been recently demonstrated (Harms et al., 2000) that the expression of SAT enzyme one of the enzymes of the bacterial cysteine synthesis in transgenic tomatoes, led to an increase in the level of cysteine and glutathione. Other authors (Błaszczyk et al., 1999) obtained transgenic tobacco with overexpression of the bacterial *cysE* gene encoding SAT enzyme, which led to increased tolerance to oxidative stress. The above experiments could be applied in obtaining transgenic plants with higher resistance to HMs. Such plants with increased tolerance to HM could be used in phytoremediation of soils polluted with HMs.

Heavy metals induce changes in the metabolism of sulphates in plants which synthesize phytochelatins (Tukendorf and Rauser, 1990). Phytochelatins are synthesized from glutathione and have high amounts of cysteine, thus high level of phytochelatins in plants treated with metals suggests that the synthesis pathway of glutathione precursors, such as glutamate, cysteine and glycine, must be upregulated under the influence of metals. Cysteine accessibility is the main factor limiting GSH synthesis (Noctor et al., 1998a, b). It was found that the rate of cysteine synthesis is correlated with the synthesis of GSH and other components which need this amino acid (Smith et al., 1985).

During heavy metal-induced PCs formation the cellular GSH level may transiently decline. Heavy metals induce the synthesis of PCs and an excess of metal ions increases the demand for cysteine. Heavy metal treatment induces the accumulation of 5-adanylylsulphate-sulphotransferase (APS-sulphotransferase) mRNA but only in roots (Heiss et al., 1999; Lee and Leustek, 1999). It is well known that APS-sulphotransferase is a key regulation point in SO_4^{-2} assimilation (Rennenberg and Brunold, 1994). It is localized only in plastids (Rotte, 1998). APS-sulphotransferase and APS-reductase (AR) they are the same enzyme (Leustek et al., 2000). Nussbaum et al. (1988) showed that both ATP-sulphurylase (AS) and APS-reductase are induced in Cd-treated maize seedlings. Later Rüegsegger et al. (1990) showed that AR activity is induced together with GS in Cd^{2+}-treated plants. Lee and Leustek (1999) reported that the expression of mRNA for both AS and AR is induced in *Brassica juncea* treated with Cd^{2+} ions or in oxidative stress (Leustek et al., 2000). When the same plants were treated with Zn, Pb, Cu, and Hg ions, the AR mRNA and enzyme activity increased synchronously (Lee and Leustek, 1999). Heiss et al. (1999) cloned cDNAs for three enzymes of sulphur assimilation: sulphate transporter (LAST), ATP-sulphurylase and APS-reductase. After the exposure to Cd^{2+}, the expression of ATP-sulphurylase and APS-reductase in roots and leaves of *Brassica juncea* plants increased. At the same time cysteine concentration increased by 81% in roots and 25% in leaves but GSH concentration decreased. The authors suggest that coordinated changes in activity of ATP sulphurylase, APS-reductase, and γ-ECS synthetase are necessary to increase GSH concentration during heavy metal treatment.

10. GSH AS AN ANTIOXIDANT

Heavy metal stress in plants results in the production of reactive oxygen species (ROS): O_2^-, H_2O_2 and OH^-, destructively affecting cell structure and metabolism (Weckx and Clijsters, 1996, 1997). This may result in a decreased activity of oxidation-reduction enzymes or electron transport system leading to the fast production of ROS in the cell. Oxidative stress occurs when the level of ROS in cells increases and the oxidative-antioxidative balance is shifted towards oxidation. An increase in ROS level in cells is extremely dangerous because these highly reactive molecules damage the main classes of cell components such as: proteins, polysaccharides, lipids and nucleic acids. The main sites of ROS formation in plant cells are chloroplasts, peroxisomes and mitochondria. The steady level of ROS in the plant cell is determined by the activity of the antioxidative system, which includes enzymes such as superoxide dismutase – SOD (E.C. 1.15.1.1), catalase – CAT (EC 1.11.1.6), ascorbate peroxidase – APOX (EC 1.11.1.11), glutathione peroxidase – GPX (E.C. 1.11.1.9) and glutathione reductase GR (E.C. 1.6.4.2). Glutathione and glutathione reductase are components of ascorbate-glutathione cycle. It is well-known that the ascorbate-glutathione cycle plays an essential role in detoxification of ROS (Gupta et al., 1999). Moreover, in plants there are non-enzymatic antioxidants such as: ascorbic acid, cysteine, glutathione, α-tocopherol, hydroquinone, carotenoids and polyamines. Information on the relationship between the heavy metal effect and oxidative stress in plants is rather scarce. It was described earlier that heavy metal treatment of plants activates the phytochelatin detoxification system that is connected with the decrease of glutathione concentration (Klapheck, 1995; Tukendorf and Rauser, 1990; Tomaszewska et al., 1996), and causes the occurrence of oxidative stress in plant cells. Both these processes are strictly connected and act destructively on plant cell metabolism.

Many authors observed an increased production of ROS in plants exposed to heavy metals e.g. in: leaves of sunflower plants (Gallego et al., 1996), Cd-stressed potato tubers (Stroiński and Kozłowska, 1997), Pb^{2+}-stressed pea roots (Malecka et al., in press), lupin roots exposed to Cu^{2+}, Cd^{2+} and Pb^{2+} ions (Rucińska et al., 1999), *Brassica juncea* treated with Zn^{2+} ions (Prasad et al., 1999) and *Phaseolus vulgaris* treated with Cu^{2+} ions (Gupta et al., 1999).

Recently, Nagalakshmi and Prasad (2001), observed a progressive depletion of GSH content in *Scenedesmus bijugatus* cells treated with increasing concentration of copper. Copper stress increased the activity of γ-ECS, the first enzyme of GSH synthesis. The authors suggest that Cu^{2+} alters equilibrium between synthesis and utilisation of GSH as the main antioxidant and precursor in PCs synthesis. They also suggest that ascorbate-glutathione cycle plays a key role in heavy metal detoxification by means of PCs synthesis. Manipulation of antioxidative enzymes activity in order to raise plants tolerance to oxidative stress would lead at the same time to an increase in amount of accumulated metal in plant and to a decrease in amount of metals in soil, thus, enhancing phytoremediation abilities.

11. CONCLUSIONS

Sulphur is an indispensable element in the life of plants. Plants are the first producer of organic sulphur through the reduction of sulphates taken up from the soil and their assimilation to cysteine and further to glutathione and many other thiol compounds. Our knowledge about the metabolism of thiol compounds in plants, their synthesis and degradation is rather superficial if compared with the knowledge on animal organisms. Thiol coumpounds in plants are connected with both the detoxification process of heavy metals and other xenobiotics such as herbicides. It is of great practical importance to explain their metabolism in plants. The application of molecular biology methods arises high expectations and broadens the possibilities of identification of genes responsible for high tolerance and resulting high accumulation of metals by plants called hyperaccumulators. Introduction of such genes into plants with high biomass production would make it possible to use such plants in phytoremediation of soils and water contaminated with heavy metals and herbicides.

REFERENCES

Arisi, A.C., Mocquot, B., Lagriffoul, A., Mench, M., Foyer, C.H., and Jouanin, L. (2000) Responses to cadmium in leaves of transformed poplars overexpressing γ-glutamylcysteine synthetase, *Physiol. Plant.* **109**, 143–149.

Atkins, C.A., Smith, P.M.C., and Storer, P.J. (1997) Reexamination of the intracellular localization of *de novo* purine synthesis in cowpea nodules, *Plant Physiol.* **113**, 127–135.

Bae, W. and Mehra, R.K. (1997) Metal-binding characteristics of a phytochelatin analog (Glu–Cys)$_2$Gly, *J. Inorg. Biochem.* **68**, 201–210.

Błaszczyk, A., Brodzik, R., Sirko, A. (1999) Increased resistance to oxidative stress in transgenic tobacco plants overexpressing bacterial serine acetyltransferase, *Plant J.* **20**, 237–243.

Chen, J. and Goldsbrough, P.B. (1994) Increased activity of γ-glutamylcysteine synthetase in tomato cells selected for cadmium tolerance, *Plant Physiol.* **106**, 233–239.

Chen, J., Zhou, J., and Goldsbrought, P.B. (1997) Characterization of phytochelatin synthase from tomato, *Physiol. Plant.* **101**, 165–172.

Clemens, S., Kim, E.J., Neumann, D., and Schroeder, J.I. (1999) Tolerance to toxic metals by a gene family of phytochelatin synthases from plants and yeast, *EMBO J.* **18**, 3325–3333.

Cobbett, Ch.S., May, M.J., Howden, R., Rolls, B. (1998) The glutathione-deficient, cadmium-sensitive mutant, *cad2-1*, of *Arabidopsis thaliana* is deficient in γ-glutamylcysteine synthetase, *Plant J.* **16**, 73–78.

Cobbett, Ch.S. (2000a) Phytochelatins and their roles in heavy metal detoxification, *Plant Physiol.* **123**, 825–832.

Cobbett, Ch.S. (2000b) Phytochelatins biosynthesis and function in heavy-metal detoxification, *Curr. Op. Plant Biol.* **3**, 211–216

Dameron, C.T., Smith, B.R., and Winge, D.R. (1989) Gluthatione coated cadmium sulfide crystallites in *Candida glabrata*, *J. Biol. Chem.* **264**, 17355–37360.

De Knecht, J.A., Koevoets, P.L.M., Verkleij, J.A.C., and Ernst, W.H.O. (1992). Evidence against a role for phytochelatins in naturally selected increased Cd tolerance in *Silene vulgaris* (Moench) Garcke, *New Phytol.* **122**, 681–688.

De Knecht, J.A., Van Dillen, M., Koevoets, P.L.M., Schot, H., Verkleij, A.C., and Ernst, W.H.O. (1994) Phytochelatins in cadmium sensitive and cadmium-tolerant *Silene vulgaris*, *Plant Physiol.* **104**, 255–261.

Frendo, P., Gallesi, D., Turnbull, R., Van de Sype, G., Herouart D., and Puppo, A. (1999) Localization of glutathione and homoglutathione in *Medicago truncatula* is correlated to a differential expression of genes involved in their synthesis, *Plant J.* **17**, 215–219.

Friederich, M., Kneer, R., and Zenk M. (1998) Enzymic synthesis of phytochelatins in gram quantities, *Phytochemistry* **49**, 2323–2329.

Gallego, S.M., Benavides, M.P., and Tomaro, M.L. (1996) Effect of heavy metal ion excess on sunflower leaves: evidence for involvement of oxidative stress, *Plant Sci.* **121**, 151–159.

Galli, U., Schüepp, H., and Brunold, C. (1996) Thiols in cadmium- and copper-treated maize (*Zea mays* L.), *Planta* **198**, 139–143.

Gekeler, W., Winnacker, E.-L., and Zenk, M.H. (1988) Algae sequester heavy metals via synthesis of phytochelatin complexes, *Arch. Microbiol.* **150**, 197–202.

Gekeler, W., Grill, E., Winnacker, E.L., and Zenk, M.H. (1989) Survey of the plant kingdom for the ability to bind heavy metals through phytochelatins, *Z. Naturforsch.* **44c**, 361–369.

Griffith, O.W. and Meister, A. (1979) Potent and specific inhibition of glutathione synthesis by buthionine sulfoximine (S-n-butyl homocysteine sulfoximine), *J. Biol. Chem.* **25**, 7558–7560.

Grill, E., Gekeler, W., Winnacker, E.L., Zenk, M.H. (1986) Homo-phytochelatins are heavy metal-binding peptides of homo-glutathione containing *Fabales*, *FEBS Lett.* **205**, 47–50.

Grill, E., Winnacker, E.L., and Zenk, M.H. (1987) Phytochelatins, a class of heavy-metal-binding peptides from plants, are functionally analogous to metallothioneins, *Proc. Natl. Acad. Sci. USA* **84**, 439–443.

Grill, E., Löffler, S., Winnacker, E.-L., and Zenk, M.H. (1989) Phytochelatins, the heavy metal-binding peptides of plants, are synthetized from glutathione by a specific γ-glutamylcysteine dipeptidyl transpeptidase (phytochelatin synthase), *Proc. Natl. Acad. Sci. USA* **86**, 6838–6842.

Grill, E., Winnacker, E.-L., and Zenk, M.H. (1985) Phytochelatins; The principal heavy-metal complexing peptides of higher plants, *Science* **230**, 674–676.

Gupta, M., Cuypers, A., Vangrosveld, and Clijsters, H. (1999) Copper affects the enzymes of ascorbate-glutathione cycle and its related metabolites in roots of *Phaseolus vulgaris*, *Physiol. Plant.* **106**, 262–267.

Ha, S.B, Smith, A.P., Howden, R., Dietrich, W.M., Bugg, S., O'Connel, M.J., Goldsbrough, P.B., and Cobbett, C.S. (1999) Phytochelatin synthase genes from *Arabidopsis* and the yeast *Schizosaccharomyces pombe*, *Plant Cell* **11**, 1153–1163.

Haag-Kerwer, A., Schäfer, H.J., Heiss, S., Walter, C., and Rausch, T. (1999) Cadmium exposure in *Brassica juncea* causes a decline in transpiration rate and leaf expansion without effect on photosynthesis, *J. Exp. Bot.* **341**, 1827–1835.

Harms, K., Ballmoos, P., Brunold, C., Höfgen, R., and Holger, H. (2000) Expression of a bacterial serine acetyltransferase in transgenic potato plants leads to increased levels of cysteine and glutathione, *Plant J.* **24**, 335–343.

Harmens, H., Den Hortog, A.R., Ten Boblaum, W.M., and Verkleij, J.A.C (1993) Increased zinc tolerance in *Silene vulgaris* (Moench) Garcke is not due to production of phytochelatins, *Plant Physiol.* **103**, 1305–1309.

Hayashi, Y., Nakagawa, C.W., Mutoh, N., Isobe, M., and Goto, T. (1991) Two pathways in the biosynthesis of cadystins $(\gamma EC)_n G$ in the cell free system of the fission yeast, *Biochem. Cell Biol.* **69**, 115–121.
Heiss, S., Holger, J., Schäfer, A., Haag-Kerwer, A., and Rausch, T. (1999) Cloning sulfur assimilation genes of *Brassica juncea* L.: cadmium differentially affects the expression of a putative low-affinity sulfate transporter and isoforms of ATP sulfurylase and APS reductase, *Plant Mol. Biol.* **39**, 847–857.
Hell, R. (1997) Molecular physiology of plant sulfur metabolism, *Planta* **202**, 138–148.
Hell, R. and Bergman, L. (1990) γGlutamylcysteine synthetase in higher plants; catalytic properties and subcellular localization, *Planta* **180**, 603–612.
Herschbach, C. and Rennenberg, H. (1994) Influence of glutathione (GSH) on net uptake of sulfate and sulfate transport in tabacco plants, *J. Exp. Bot.* **45**, 1069–1076.
Hesse, H., Lipke, J., Altmann, T., and Höfgen, R. (1999) Molecular cloning and expression analyses of mitochondrial and plastidic isoforms of cysteine synthase (O-acetylserine (thiol) lyase) from *Arabidopsis thaliana*, *Amino Acids* **16**, 113–131.
Howden, R., Andersen, C.R., Goldsbrough, P.B., and Cobbett, C.S. (1995a) A cadmium-sensitive, glutathione-deficient mutant of *Arabidopsis thaliana*, *Plant Physiol.* **107**, 1067–1073.
Howden, R., Goldsbrough, P.B., Andersen, C.R., and Cobbett, C.S. (1995b) Cadmium-sensitive, *cad1* mutants of *Arabidopsis thaliana* are phytochelatin deficient, *Plant Physiol.* **107**, 1059–1066.
Kägi, J.H.R. and Nordberg, M. (1979) *Metallothionein I*, Birkhäser Verlag, Basel, Switzerland.
Klapheck, S. (1988) Homoglutatione: isolation, quantification and occurence in legumes, *Physiol. Plant.* **74**, 727–32.
Klapheck, S., Fliegner, W., and Zimmer, I. (1994) Hydroxymethyl-phytochelatins [$(\gamma$-Glu–Cys$)_n$–Ser] are metal induced peptides of the *Poaceae*, *Plant Physiol.* **104**, 1325–1332.
Klapheck, S., Schlunz, S., and Bergmann, L. (1995) Synthesis of phytochelatins and homo-phytochelatins in *Pisum sativum* L., *Plant Physiol.* **107**, 515–521.
Kneer, R. and Zenk, M.H. (1997) The formation of Cd–phytochelatin complexes in plant cell cultures, *Phytochemistry* **44**, 69–74.
Kubota, H., Sato, K., Yamada, T., and Maitani, T. (1995) Phytochelatins (class III metallothioneins) and their desglycyl peptides induced by cadmium in normal root cultures of *Rubia tinctorum* L., *Plant Sci.* **106**, 157–166.
Kubota, H., Sato, K., Yamada, T., and Maitani, T. (2000) Phytochelatin homologs induced in hairy roots of horseradish, *Phytochemistry* **53**, 239–245.
Kunert, K.J. and Foyer, C.H. (1993) Thiol/disulphide exchange in plants., in De Kok L.J. (ed.), *Sulfur nutrition and assimilation in higher plants*, The Hague, The Netherlands: SPB Acad. Publishing, 139–151.
Lee, S. and Leustek, T. (1999) The effect of cadmium on sulfate assimilation enzymes in *Brassica juncea*, *Plant Sci.* **141**, 201–207.
Leopold, I., Günther, D., Schmidt, J., and Neumann, D. (1999) Phytochelatins and heavy metal tolerance, *Phytochemistry* **50**, 1323–1328.
Leustek, T. (1996) Molecular genetics of sulfate assimilation in plants, *Physiol. Plant.* **97**, 411–419.
Leustek, T., Martin, M.N., Bick, J.A., and Davies, J. (2000) Pathways and regulation of sulfur metabolism revealed through molecular and genetic studies, *Annu. Rev. Plant Physiol. Plant Mol. Biol.* **51**, 141–65.
Leustek,T. and Saito, K. (1999) Sulfate transport and assimilation in plants, *Plant Physiol.* **120**, 637–643.
Li, Z.-S., Lu, Y.-P., Zhen, R.-G., Szczypka, M., Thiele, D.J., and Rea, P.A. (1997) A new pathway for vacuolar cadmium sequestration in *Saccharomyces cerevisae;* YCF1-catalyzed transport of bis(glutathione) cadmium, *Proc. Natl. Acad. Sci. USA* **94**, 42–47.
Loeffler, S., Hochberger, A., Grill, E., Winnacker, E.-L., and Zenk, M.H. (1989) Termination of the phytochelatin synthase reaction through sequestration of heavy metals by the reaction product, *FEBS Lett.* **258**, 42–46.
Maitani, T., Kubota, H., Sato, K., and Yamada, T. (1996) The composition of metals bound to class III metallothionein (phytochelatin and its desglycyl peptide) induced by various metals in root cultures of *Rubia tinctorum*, *Plant Physiol.* **110**, 1145–1150.
Małecka, A., Jarmuszkiewicz, W., and Tomaszewska, B. (2001) Antioxidative defence to lead stress in subcellular compartments of pea root cells, *Acta Biochim. Polon.***3**, (in press).
May, M.J., Vernoux, T., Leaver, C., Van Montague, M., and Inze, D. (1998) Glutathione homeostasis in plants: implications for environmental sensing and plant development, *J. Exp. Bot.* **49**, 649–667.

Matamoros, M.A., Baird, L.M., Escuredo, P.R., Dalton, D.A., Minchin, F.R., Iturbe-Ormaetxe, I., Rubio, M.C., Moran, J.F., Gordon, A.J., and Becana, M. (1999) Stress-induced legume root nodule senescence: physiological, biochemical and structural alterations, *Plant Physiol.* **121**, 97-111.

Mehra, R.K., Kodati, R., and Abdullah, R. (1995) Chain length-dependent Pb(II)-coordination in phytochelatins, *Biochem. Biophys. Res. Commun.* **215**, 730-736.

Mehra, R.K., Miclat, J., Kodati, R., Abdullah, R., Hunter, T.C., and Mulchandani, P. (1996) Optical spectroscopic and reverse phase HPLC analyses of Hg (II) binding to phytochelatins, *Biochem. J.* **314**, 73-82.

Mehra, R.K., Tarbet, E.B., Gray, W.R., and Winge, D.R. (1988) Metalospecific synthesis of two metallothioneins and γ-glutamyl peptides in *Candida glabrata*, *Proc. Natl. Acad. Sci. USA*, **85**, 8815-8819.

Meister, A. (1988) Glutathione metabolism and its selective modification, *J. Biol. Chem.* **263**, 17205-17208.

Meuwly, P., Thibault,. P., and Rauser, W.E. (1993) γGlutamylcysteinyl-glutamic acid – a new homologue of glutathione in maize seedlings exposed to cadmium, *FEBS Lett.* **336**, 472-476.

Moran, J.F., Iturbe-Ormaetxe, I., Matamoros, M.A, Rubio, M.C, Clemente, M.R., Brewin, N.J., and Becana, M. (2000) Glutathione and homoglutathione synthetase of legume nodules, Cloning, expression, and subcellular localization, *Plant Physiol.* **124**, 1381-1392.

Nakazawa, R. and Takenaga, H. (1998). Interaction between cadmium and several heavy metals in the activation of the catalytic activity of phytochelatin synthase, *Soil Sci. Plant Nutr.* **44**, 265-268.

Nagalakshmi, N. and Prasad M.N.V. (2001) Responses of glutathione cycle enzymes and glutathione metabolism to copper stress in *Scenedesmus bijugatus*, *Plant Sci.* **160**, 291-299.

Noctor, G., Arisi, A. -C.M., Jouanin, L., Kunert, K.-J., Rennenberg, H., and Foyer, C.H. (1998a) Glutathione: biosynthesis, metabolism and relationship to stress tolerance explored in transformed plants, *J. Exp. Bot.* **49**, 623-647.

Noctor, G. and Mills, J.D. (1998b) Thiol-modulation of the thylakoid ATPase. Lack of oxidation of the enzyme in the presence of $\Delta\mu H^+$ *in vivo* and a possible explanation of the physiological requirement for the thiol regulation of the enzyme, *Biochim. Biophys. Acta* **935**, 53-60.

Noctor, G. and Foyer, C.H. (1998c) Ascorbate and glutathione: keeping active oxygen under control, *Annu. Rev. Plant Physiol. Plant Mol. Biol.* **49**, 249-79.

Nussbaum, S., Schmutz, D., and Brunold, C. (1988) Regulation of assimilatory sulfate reduction by cadmium in *Zea mays* L., *Plant Physiol.* **88**, 1407-1410.

Ortiz, D.F., Kreppel, L., Speiser, D.M., Scheel, G., McDonald, G., and Ow, D.W. (1992) Heavy-metal tolerance in the fission yeast requires an ATP-binding cassette-type vacuolar membrane transporter, *EMBO J.* **11**, 3491-3499.

Ortiz, D.F., Ruscitti, T., McCue, K.F., and Ow, D.W. (1995) Transport of metal-binding peptides by HMT1, a fission yeast ABC-type vacuolar membrane protein, *J. Biol. Chem.* **270**, 4721-4728.

Prasad, K.V.S.K., Paradha Saradhi, P., and Sharmila, P. (1999) Concerted action of antioxidant enzymes and curtailed growth under zinc toxicity in *Brassica juncea*, *Environ. Exp. Bot.* **42**, 1-10.

Piechalak, A., Tomaszewska, B., Barałkiewicz, D., and Małecka, A. (in press) Accumulation and detoxification of lead ions in legumes, *Phytochemistry.*

Rauser, W.E. (1995) Phytochelatins and related peptides. Structure, biosynthesis and function, *Plant Physiol.* **109**, 1141-1149.

Rauser, W.E. (1999) Structure and function of metals chelators produced by plants, *Cell Biochem. Biophys.* **31**, 19-48

Rauser, W.E. (2000) Roots of maize seedlings retain most of their cadmium through two complexes, *J. Plant Physiol.* **156**, 545-551.

Rauser, W.E., Schupp, R., and Rennenberg, H. (1991) Cysteine, γ-glutamyl-cysteine, and gluthatione levels in maize seedlings. Distribution and translocation in normal and cadmium-exposed plants, *Plant Physiol.* **97**, 128-138.

Reese, R.N. and Winge, D.R. (1988) Sulfide stabilization of the cadmium γ-glutamyl peptide complex of *Schizosaccharomyces pombe*, *J. Biol. Chem.* **263**, 12832-12835.

Reese, R.N, Wagner, G.J. (1987) Effects of buthionine sulfoximine on Cd-binding peptide levels in suspension-cultured tobacco cells treated with Cd, Zn, or Cu, *Plant Physiol.* **84**, 574-577.

Rennenberg, H. and Brunold C. (1994) Significance of glutathione metabolism in plants under stress, *Press Bot.* **55**, 142-153.

Rotte, C. (1998) Subcellular localization of sulfur assimilation enzymes in *Arabidopsis thaliana* (L.) Heynh., *Diplomarbeit thesis*, Carl von Ossietzky Univ., Oldenburg, Germany.

Rucińska, R., Waplak, S., and Gwóźdź, E.A. (1999) Free radical formation and activity of antioxidant enzymes in lupin roots exposed to lead, *Plant Physiol. Biochem.* **37**, 187–194.

Rüegsegger, A., Schmutz, D., and Brunold, C. (1990) Regulation of glutathione synthesis by cadmium in *Pisum sativum* L., *Plant Physiol.* **93**, 1579–1584.

Rüegsegger, A. and Brunold, C. (1992) Effect of cadmium on γ-glutamylcysteine synthesis in maize seedlings, *Plant Physiol.* **99**, 428–433.

Salt, D.E. and Rauser, W.E. (1995) MgATP-dependent transport of phytochelatins across the tonoplast of oat roots, *Plant Physiol.* **107**, 1293–1301.

Salt, D.E., Wagner, G.J. (1993) Cadmium transport across tonoplast of vesicles from oat roots: evidence for a $Cd^{2+}/H^{(+)}$ antiport activity, *J. Biol. Chem.* **268**, 1297–1302.

Sanitá di Toppi, L., Lambardi, M., Pecchioni, N., Pazzagli, L., Durante, M., and Gabbrielli, R. (1999) Effects of cadmium stress on hairy roots of *Daucus carota*, *J. Plant Physiol.* **154**, 385–391.

Schäfer, H.J., Greiner, S., Rausch, T., and Haag-Kerwer, A. (1997) In seedlings of heavy metal accumulator *Brassica juncea* Cu^{+2} differentially affects transcript amounts for γ-glutamycysteine synthetase (γ-ECS) and metallothionein (MT2), *FEBS Lett.* **404**, 216–220.

Schmöger, A., Marcus, E.V., Oven, M., and Grill, E. (2000) Detoxification of arsenic by phytochelatins in plants, *Plant Physiol.* **122**, 793–801.

Schneider, S. and Bergmann, L. (1995) Regulation of glutathione synthesis in suspension cultures of parsley and tobacco, *Bot. Acta* **108**, 34–40.

Smith, I.K. (1985) Stimulation of glutathione synthesis in photorespiring plants by catalase inhibitors, *Plant Physiol.* **79**, 1044–1047.

Speiser, J.L., Abrahamson, S.L., Banuelos, G., and Ow, D.W. (1992) *Brassica juncea* produces a phytochelatin cadmium sulfide complex, *Plant Physiol.* **99**, 817–821.

Steffens, J.C. (1990) The heavy metal-binding peptides of plants, *Annu. Rev. Plant Physiol. Plant Mol. Biol.* **41**, 553–75.

Strasdeit, H., Duhme, A.-K., Kneer, R., Zenk, M.H., Hermes, C., and Nolting, H.F. (1991) Evidence for discrete $Cd(SCys)_4$ units in cadmium complexes from EXAFS spectroscopy, *J. Chem. Soc. Chem. Commun.* **16**, 1129–1130.

Stroiński, A. and Kozłowska, M. (1997) Cadmium-induced oxidative stress in potato tuber, *Acta Soc. Bot. Polon.* **66**, 189–195.

Takahashi, H. and Saito, K. (1996) Subcellular localization of spinach cysteine synthase isoforms and regulation of their gene expression by nitrogen and sulfur, *Plant Physiol.* **112**, 273–280.

Temple, S.J., Vance, C.P., and Gantt, J.S. (1998) Glutamate synthase and nitrogen assimilation, *Trends Plant Sci.* **3**, 51–56.

Tomaszewska, B., Tukendorf, A., and Barałkiewicz, D. (1996) The synthesis of phytochelatins in lupin roots treated with lead ions, *Sci. Legumes* **3**, 206–217.

Tukendorf, A. and Rauser, W.E. (1990) Changes in glutathione and phytochelatins in roots of maize seedlings exposed to cadmium, *Plant Sci.* **70**, 155–166.

Vatamaniuk, O.K., Mari, S., Lu, Y.-P., and Rea, P.A. (1999) *AtPCS1*, a phytochelatin synthase from *Arabidopsis*: isolation and *in vitro* reconstitution, *Proc. Natl. Acad. Sci. USA* **96**, 7110–7115.

Vatamaniuk, O.K., Stephane, M., Lu, Y.-P., and Rea, P.A. (2000) Mechanism of heavy metal ion activation of phytochelatin (PC) synthase, *J. Biol. Chem.* **275**, 31451–31459.

Verkleij, J.A.C., Koevoets, P., Van't Reit, J., Bank, R., Nijdam, Y., and Ernst, W.H.O. (1990) Poly-γ-glutamylcysteinylglycines or phytochelatins and their role in cadmium tolerance of *Silene vulgaris*, *Plant Cell Environ.* **13**, 913–922.

Weckx, J.E.J. and Clijsters, H. (1996) Oxidative damage and defense mechanisms in primary leaves of *Phaseolus vulgaris* as a result of root assimilation of toxic amount of copper, *Physiol. Plant.* **96**, 506–512.

Weckx, J.E.J. and Clijsters, H. (1997) Zn phytotoxicity induces oxidative stress in primary leaves of *Phaseoulus vulgaris*, *Plant Physiol. Biochem.* **35**, 405–410.

Xiang, C. and Oliver, D.J. (1998) Glutathione metabolic genes coordinately respond to heavy metals and jasmonic acid in *Arabidopsis*, *Plant Cell* **10**, 1539–1550.

Zhu, Y.L., Pilon-Smith, E.A.H., Jouanin, L., and Terry, N. (1999) Overexpression of glutathione synthetase in Indian mustard enhances cadmium accumulation and tolerance, *Plant Physiol.* **119**, 73–79.

Zenk, M.H. (1996) Heavy metal detoxification in higher plants: a review, *Gene* **179**, 21–30.

CHAPTER 3

METAL CHELATING PEPTIDES AND PROTEINS IN PLANTS

L. SANITÀ DI TOPPI[a], M.N.V. PRASAD[b] AND S. OTTONELLO[c]

[a]*Department of Evolutionary and Functional Biology*
Section of Plant Biology, University of Parma, Parma, Italy
[b]*Department of Plant Sciences, School of Life Sciences, University of Hyderabad, Hyderabad, India*
[c]*Department of Biochemistry and Molecular Biology, University of Parma, Parma, Italy*

1. INTRODUCTION

A group of metals having a density higher than 5 g cm^{-3} are generally termed as heavy metals (HM) and are toxic to plants when available in excess. However, a subset of them, at low concentrations, are essential micronutrients. Wild-type plants cope with HM stress (and homeostasis) by synthesizing a variety of chemically distinct metal-chelating ligands (Figure 1), including organic acids, and by exploting a broad range of different response mechanisms, which may act in an additive and/or in a synergistic manner (Prasad, 2001). The main defence mechanisms involved in HM detoxification are revolved around the previously discussed "fan-shaped" model (Figure 2) (Sanità di Toppi and Gabbrielli, 1999), which is based on the "General Adaptation Syndrome" (GAS) hypothesis (Leshem and Kuiper, 1996; Selye, 1936). In this chapter, we focus on HM chelating peptides and proteins, with particular emphasis on phytochelatins. Updates on plant metallothioneins, ferritins and nicotianamine are given as well. Nonprotein metal chelators, in particular organic acids, single amino acids and phytin, not covered in this chapter, were recently reviewed by Rauser (1999).

2. PHYTOCHELATINS

2.1 Generalities

Once entered the plant cell, several chemical elements – in particular HMs, metalloids, and a few multiatomic anions (see below) – rapidly induce the cytosolic synthesis of metal-binding ligands, termed PCs, which may contribute decisively towards HM detoxification

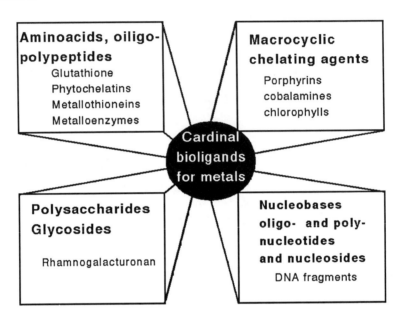

Figure1. Cardinal metal binding bioligands

in algae, fungi, mosses and higher plants (Figure 3) (for recent reviews see: Clemens, 2001; Cobbett, 2000a; 2000b; Kotrba et al., 1999; Rauser, 1999; Sanità di Toppi and Gabbrielli, 1999). Here, we shall use the most widely accepted term "phytochelatins" (PCs), since the terms "class-III MTs", "MT-IIIs", "γ-glutamyl peptides", "poly(γ-glutamylcysteinyl)$_n$ gly", "(γ-EC)$_n$ G", "cadystins", "Cd-binding peptides", etc., are in general all synonyms less widespread and proper than the term PCs. PCs were isolated and characterized in higher plants for the first time by Grill et al. (1985), although their presence in Hg-treated tobacco leaves (where they were designated as "organo-mercury complexes") had been supposed since 1973 (Anelli et al., 1973; Pelosi and Galoppini, 1973). PCs are peptides derived from the tripeptide glutathione (GSH = γ-Glu-Cys-Gly or γ-ECG) and, thus, have the general structure (γ-Glu-Cys)$_n$-Gly, with n = number of repetitions of the γ-Glu-Cys unit, which can vary from 2 to 11, more commonly from 2 to 5. PCs are characterized by the presence of the γ-Glu-Cys carboxylamide bond in place of the α-carboxylamide bond typical of ribosome-synthesized peptides. Due to the presence of the metal chelating sulfhydryl groups of Cys, PCs can form complexes with several HMs, thus preventing their free circulation inside the cytosol (Grill et al., 1985). In the absence of largely over-saturating metal ion concentrations, only the sulfhydryl group of Cys is involved in such complexes, with no direct participation of either the nitrogen or oxygen atoms of the peptides (Pickering et al., 1999) (Figure 4). The isopeptide γ-EG linkage increases the structural flexibility of PCs, but is not absolutely required for metal complexation, and synthetic α-PC analogues are also competent for metal binding (Bae and Mehra, 1997). Besides its

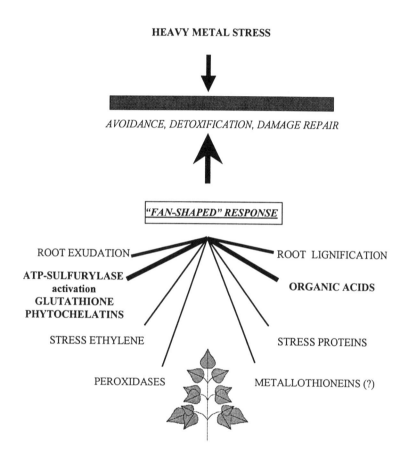

Figure 2. "Fan-shaped" response to heavy metal stress in higher plants. In this multicomponent model, plants are proposed to cope with heavy metal stress by modulating the "expression" of each ray of the fan. (after Sanità di Toppi and Gabbrielli, 1999, with permission from Elsevier Science)

obvious biotechnological interest as to the feasibility of overproducing gene-encoded PC-like γα-peptides for HM phytoremediation purposes (Bae et al., 2000), this observation adds functional significance to the presence of identical C-x-C sequence motifs in class II metallothioneins (MTs) and MT-like proteins (see below) as well as in PCs. In keeping with this view, the UV and CD spectra of PCs and mammalian MTs, which exhibit all the characteristic features of metal-thiolate bonds, are quite similar (Rauser, 1990). Because of its toxicity to humans, environmental diffusion, marked thiophilicity, and consequent high affinity binding to PCs (see below), Cd has been the metal of choice for most PC-metal binding studies (Kotrba et al., 1999; Rauser, 1999; Sanità di Toppi and Gabbrielli, 1999). The molecular masses of metal-PC complexes formed with Cd ions range from about 1800 to 4000 Da, but may be as high as 8000 Da, depending on the ionic strength of the solvent employed for gel-filtration chromatography (Grill et al., 1987; Murasugi et al., 1981; Rauser, 1999). A Cd-S bond length of 2.52±0.02Å for the Cd-PC complex was determined by extended X-ray absorption fine structure (EXAFS) spectroscopy (Strasdeit et al., 1991). Moreover, X-ray absorption spectroscopy of Cd-PC complexes extracted from maize seedlings exposed to low levels of Cd ions evidenced a polynuclear Cd cluster similar to those formed by vertebrate MTs (Pickering et al., 1999). Some Cd-PC complexes also include acid-labile sulfur (S^{2-}), which gives them an increased stability and a higher Cd sequestration capacity (see below).

Capillary electrophoresis has been successfully employed for the characterization of Cd-induced peptides, probably PCs (Mori and Leita, 1998). Moreover, very powerful and sensitive analyses of PC-Cd complexes from *Datura innoxia* tissue cultures have been developed by nano-electrospray ionization tandem mass spectrometry (nano-ESI-MS/MS) and by capillary liquid chromatrography/ electrospray ionization tandem mass spectrometry (LC/ESI-MS/MS) (Yen et al., 1999). The most effective metal ions (and multiatomic anions) in the induction of PC biosynthesis were are Ag^+, AsO_2^-, AsO_4^{3-}, Cd^{2+}, Cu^{2+}, Hg^{2+}, Pb^{2+} and Zn^{2+}. Moderate PC biosynthesis was detected in response to Au^+, Bi^{3+}, Ga^{3+}, In^{3-}, Sb^{3+}, SeO_3^{2-}, SeO_4^{2-} and Sn^{2+}. On the contrary, other metals, such as Al^{2+}, Co^{2+}, Cr^{3+}, CrO_4^{2-}, Fe^{2+}, MoO_4^{2-}, Mn^{2+}, Ni^{2+}, Te^{4+} and W^{6+}, did not appear to be effective stimulators of PC synthesis (Cobbett, 2000a; Kotrba et al., 1999; Zenk, 1996). Cd ions are probably the most powerful inducers of PC synthesis.

Although PCs ending with a C-terminal Gly residue are by far the most widespread (Gekeler et al., 1989), several C-terminal variants, collectively termed *iso*-PCs, have been identified in different plant species (Zenk, 1996) (Figure 5). For instance, homo-PCs, derived from homo-GSH (hGSH), are typically present in a few members of the Fabaceae, especially the tribe Phaseoleae. Homo-PCs are characterized by the presence of γ-Ala instead of Gly as the C-terminal amino acid (Grill et al., 1986). Moreover, in some Poaceae (oat, rye, rice and wheat) were discovered hydroxymethyl-PCs, which end with a Ser rather than a Gly residue (Klapheck et al., 1994), while in maize roots were isolated iso-PCs with a C-terminal Glu instead of a Gly residue (Meuwly et al., 1993). Another PC homologue, ending with a Gln instead of a Gly residue, has recently been identified in hairy roots of horseradish (Kubota et al., 2000). In plants producing both PCs and *iso*-PCs, also desGly-PCs, i.e. PCs lacking the C-terminal Gly residue, can be present in varying percentages (Bernhard and Kägi, 1987).

Two additional variants, namely PCs and desGly-PCs lacking the N-terminal γ-Glu residue (designated as PCs-γGlu and desGly-PCs-γGlu) have been identified very recently in maize roots (Chassaigne et al., 2001). In all *iso*-PCs and desGly-PCs, n values ranging from 2 to 7, more often from 2 to 4, are the most common. The mechanism(s) of formation and the functional significance of this largely species-specific C-terminal and N-terminal heterogeneity are not understood. They may reflect either simply the abundance or the biosynthetic preference for pre-existing GSH derivatives (e.g.: hGSH) in the case of *iso*-PCs, or the action of PC-selective, but as yet unidentified, carboxypeptidases in the case of desGly-PCs (Kubota et al., 1995), or γ-glutamyl transpeptidases in the case of PC_n-γGlu peptides (Chassaigne et al., 2001). Interestingly, Imai et al. (1996) reported the production of PCs in a *Schizosaccharomyces pombe* extract in the complete absence of Cd (and or other HMs), but in the presence of a carboxypeptidase and of high GSH concentrations.

2.2 PC biosynthesis

Two features of PCs, namely the presence of a non-translational γ-carboxamide bond as in GSH, and the presence in *iso*-PC-(β-Ala) of an unusual C-terminal amino acid for which no tRNA is available, early suggested a non-ribosomal origin of these peptides. Other important hints pointing to the non-translational formation of PCs through the extension of the GSH tripeptide were: i) the structural similarity between PCs and GSH, particularly the isopeptide bond; ii) the existence of an inverse relationship between HM-induced (especially Cd) PC accumulation and GSH levels (Delhaize et al., 1989a; Grill et al., 1987; Scheller et al., 1987); iii) the blockage of PC synthesis by buthionine sulfoximine (BSO), a specific inhibitor of γ-EC synthetase (Reese and Wagner, 1987; Scheller et al., 1987; Steffens et al., 1986). All these strong circumstantial evidenceswith respect to the origin of PCs were substantiated and extended by the isolation of a homogenously purified enzyme preparation from *Silene vulgaris*, capable of catalyzing the *in vitro* synthesis of PC_{2-4} in the presence of GSH and micromolar Cd^{2+} ion concentrations (Grill et al., 1989). The *in vitro* reaction catalyzed by this enzyme involved the transpeptidation of the γ-EC moiety of GSH onto another GSH molecule forming PC_2, or, at later stages, onto a growing PC_n unit to yield a PC_{n+1} oligomer. The earliest transpeptidation product (formed within a few minutes after Cd addition) was PC_2, followed at 10-20 minute intervals by the appearance of higher order polymers (PC_3, PC_4).

The enzyme responsible for PC synthesis, which can also utilize preformed PCs as its sole substrate, but neither γ-EC nor GSSG, is thus a γ-EC dipeptidyl-transpeptidase (E.C.2.3.2.15), and is commonly referred to as PC synthase (PCS) (Grill et al., 1989). Under denaturing SDS-PAGE conditions, *Silene vulgaris* PCS migrated as a single polypeptide species with an apparent molecular mass of 25 kDa, while apparent molecular weight estimates of 50,000 and 95,000 were derived from gel filtration analysis under non-denaturing conditions (Grill et al., 1989). This was interpreted as an indication that active PCS can be present in either a dimeric or a tetrameric form (see

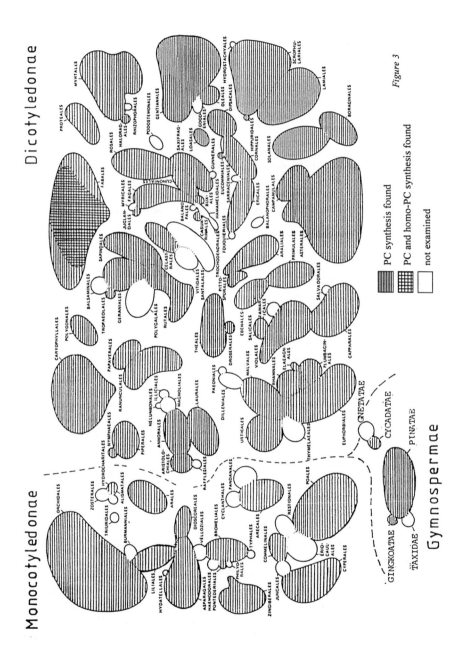

Figure 3. Phytochelatin production in various higher plant orders (courtesy of Prof. Dr. M.H. Zenk).

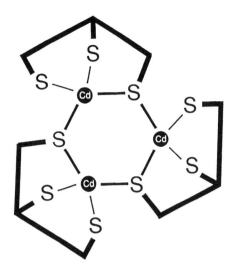

Figure 4. Structural model of a Cd(SCys)$_4$ complex (Courtesy Prof. Dr.M.H.Zenk)

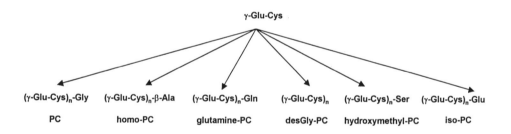

Figure 5. Phytochelatins, iso-phytochelatins and desGly phytochelatins.

below). The enzyme from *Silene vulgaris* has an isoelectric point of 4.8, an apparent K_m for GSH of 6.7 mM, and, consistent with a cytosolic rather than a vacuolar localization, has a pH optimum for activity of 7.9. Many of the basic properties of *Silene vulgaris* PCs have subsequently been confirmed with partially purified PCs preparations from other plants (Chen et al., 1997; Howden et al., 1995a; Klapheck et al., 1995; Yan et al., 2000). As indicated by the relatively fast appearance of PC$_2$ products upon Cd addition to *in vitro* reconstituted reaction mixtures (Grill et al., 1989), and by the insensitivity of *in vivo* PC synthesis to the translation inhibitor cycloheximide (Robinson et al., 1988; Scheller et al., 1987), PCS is a constitutive enzyme expressed at basal levels even in the absence of HM exposure. In Cd-treated tomato plants, PCS activity was detected in roots and stems, but not in leaves or fruits (Chen et al., 1997). This constitutive, root preferential expression of PCS activity is consistent with the postulated role of PCs as a rapid first line of defense against HM poisoning. Further to this point, PCS-supported, *in vitro* synthesis of PCs has an absolute requirement for HM ions. Unlike MTs, for which a number of inductive stimuli and various instances of developmental regulation have been described (Kägi, 1991; Steffens, 1990; Clemens, 2001), HM-dependent postranscriptional activation appears to be the main regulatory strategy operating in the case of PCS. Only one case of ethylene-related positive regulation of PCS activity in carrot (Sanità di Toppi et al., 1998), and one instance of metal-induced PCS mRNA upregulation in wheat (Clemens et al., 1999; see below) have recently been reported. As expected, the availability of free (or GSH-complexed, see below) HM ions (the activators of an enzyme reaction that leads to the production of metal-chelating peptides) controls the termination of such reaction (Loeffler et al., 1989). This has been demonstrated in the case of Cd-activated PC synthesis by *Silene vulgaris* PCS, where transpeptidation autonomously terminates at a critical PC:Cd ratio (corresponding to an exceedingly low $[Cd^{2+}]_{free}$), but can also be blocked artificially by the addition of exogenous metal-free PCs or EDTA (Grill et al., 1989; Loeffler et al., 1989). This postranslational autoregulatory loop, as opposed to a relatively slow and persistent transcriptional activation, likely evolved as a very effective means to control, in a demand-driven manner, the investment of GSH units into PC synthesis.

Species-specific variations on the above described biosynthetic theme have been reported. One of them regards *Schizosaccharomyces pombe*, where PC synthesis has been reported to proceed either through a PCS-indepedent pathway relying on γ-EC polymerization followed by C-terminal Gly addition by GSH synthetase (GS), or through a standard biosynthetic pathway, supported by a PCS-like enzyme with some peculiar functional features as compared to its plant counterpart (Hayashi et al., 1991). The ability of fission yeast GS (encoded by the *gsh2* gene) to act as a bifunctional enzyme capable of adding a C-terminal Gly residue to either the GSH precursor γ-EC, or to a $(\gamma\text{-EC})_n$ PC precursor has recently been confirmed (Al-Lahham et al., 1999).

Another important PCS-related variation regards *iso*-PC synthesis in plants, such as pea, that produce Gly-terminated PCs along with *iso*-(βAla) and *iso*-(Ser) PCs, as well as *iso*-(Glu) and desGly-PC formation in maize. When supplied with GSH as the sole PC precursor, a crude PCS preparation from pea roots produced $(\gamma\text{-EC})_n$-G PCs at high levels, yet when the only *in vitro* PC precursors were either γ-EC-βA (hGSH) or γ-GC-S, synthesis of the corresponding *iso*-PC derivatives proceeded at a much lower rate. The use of a mixed, GSH and γ-GC-βA or γ-GC-S substrate mixture greatly enhanced *iso*-PC synthesis with a corresponding reduction of $(\gamma\text{-EC})_n$-G formation (Klapheck et al., 1995). Based on this finding, it has been proposed that pea PCS may have a γ-EC donor binding site highly specific for GSH and a less selective γ-EC acceptor site that can utilize other tripeptides, besides GSH, as substrates. Since *Silene vulgaris* PCS does not utilize hGSH as substrate (Zenk, 1996), the behaviour of the pea enzyme likely reflects either a species-specific active site difference, or the existence of hGSH-specific, *iso*-PC synthase forms. Less is known about the formation of desGly-PCs and *iso*-PC_n(Glu) peptides, which accumulate at late stages of the Cd response in maize (Meuwly et al., 1995; Rauser and Meuwly, 1995). They may be formed either through a *Schizosaccharomyces pombe*-like pathway without the addition of the C-terminal Gly residue or through the standard pathway followed by carboxypeptidase-mediated removal of the terminal Gly in the case of desGly-PCs, or, in the case of *iso*-PC$_n$(Glu), through the action of an as yet unidentified γ-glutamyl (trans) peptidase, cleaving intramolecular isopeptide (γ-Glu) bonds within preformed PCs. Given this extra biosynthetic complexity and the seemingly equal chelating capacities of *iso*-, des- and standard-PCs, an important, as yet unanswered question is whether these modified PCs confer any selective advantage to the organisms that produce them. For *Silene vulgaris* PCS, the decreasing rank order of *in vitro* enzyme activation efficiency by different HM ions was the following: Cd^{2+}, Ag^+, Bi^{3+}, Pb^{2+}, Zn^{2+}, Cu^{2+}, Hg^{2+} and Au^+ (L. Sanità di Toppi, unpublished; Zenk, 1996). As further reported by Zenk (1996), other ions that activate *Silene vulgaris* PCS to varying extents are AsO_4^{3-}, Ga^{3+}, Ge^{4+}, In^{3-}, Sn^{2+} and Tl^{3+}.

Albeit with some differences, mainly regarding the relative efficiencies of Pb^{2+}, Zn^{2+} and Cu^{2+}, and the activating effects of Co^{2+} (Yan et al., 2000) and Fe^{2+} (Chen et al., 1997) for the rice and tomato enzymes, respectively, similar results have been obtained with partially purified PCS from other plant sources. Importantly, there also is a reasonably good agreement between the above *in vitro* data and the relative ability of various metals to induce PC synthesis in cell suspension cultures. A number of other elements (Al, Co, Mn, K, Na, Mg, Ca, Cr, Cs, MoO_4^{2-} and V ions) are devoid of any detectable inducing activity. Contrasting (often, poorly reproducible) results have been reported for SeO_4^{2-} and Ni^{2+}. The concentration of the free metal species, rather than the total metal concentration, appears to be critical for induction. Moreover, even though various metals are able to activate PCS and induce PC synthesis, only a subset of them - most notably, Cd, Pb, As, Ag and Hg - have been shown to be actually bound by PCs, either *in vitro* or *in vivo* (Maitani et al., 1996b; Rauser, 1999, and references therein).

2.3 PC synthase genes

A major recent advance in the PC field has been the isolation of three distinct cDNAs and one genomic DNA clone coding for PCS. Positional cloning within the previously identified *CAD1* locus (Howden et al., 1995a) was employed for the isolation of the *Arabidopsis* PCS gene (Ha et al., 1999), whereas a functional complementation strategy, exploiting the ability of PCS to confer increased Cd resistance to a *Saccharomyces cerevisiae* oxidative stress (and Cd) hypersensitive mutant or to wild-type yeast cells, was utilized for the isolation of *Arabidopsis* (*AtPCS1*) (Vatamaniuk et al., 1999) and wheat (*TaPCS1*) (Clemens et al., 1999) PCS cDNAs, respectively. An intronless, single copy gene encoding *Schizosaccharomyces pombe* PCS, named *SpPCS*, was identified based on its homology with the two plant PCS cDNAs (Clemens et al., 1999; Ha et al., 1999). The *Arabidopsis CAD1* gene contains eight introns and is located on chromosome 5. A largely incomplete, likely non-functional sequence resembling *CAD1* is also present on chromosome 5, while a full-length *CAD1* paralog, that is very poorly expressed under all the conditions and tissues examined so far, is present on chromosome 1 (Ha et al., 1999). The conceptual translation product of the *AtPCS1*

(*CAD1*) cDNA is a 485 aminoacid-long polypeptide of 55 kDa. Recombinant AtPCS1 (CAD1), expressed in *Escherichia coli* or *Saccharomyces cerevisiae*, catalyzes the GSH/Cd-dependent synthesis of mainly PC_{2-3} products. On SDS-polyacrylamide gels it migrates as a single polypeptide species with an apparent molecular mass of ~55 kDa (Vatamaniuk et al., 1999). It thus appears that the 25 kDa subunit molecular mass previously reported for the *Silene vulgaris* enzyme (Grill et al., 1989) is either due to a rather large species-specific size difference or, more likely, to some purification or gel migration artifact. In fact, native *Silene vulgaris* PCS species with apparent molecular masses of 50 kDa and 95 kDa were identified at different purification stages, suggesting that the enzyme is a monomer under at least some conditions, but it either dimerizes, or strongly associates with a distinct ~40 kDa protein component under some other, as yet unidentified experimental conditions. As revealed by sequence alignment of the three predicted polypeptides, PCS is made up by a conserved N-terminal region (~220 amino acids; 76%-45% identity) containing 6 aligned Cys residues (two of which adjacent), and a highly divergent (41%-6% sequence identity) C-terminal region, whose most characteristic feature is the presence of 7 to 14 Cys residues, at least four of which occur as contiguous pairs (Figure 6). Based on this different, region-specific conservation of the PCS polypeptide sequence, it has been proposed that the transpeptidase active site is located within the highly conserved N-terminal region, while a less critical functional role is associated to the much less conserved C-terminal region (Cobbett 2000a). In keeping with this view, three out of four *cad1* mutants with a strongly impaired PC biosynthetic capacity map within the N-terminal region, while the fourth, less severe mutant (*cad1-5*) has a nonsense mutation downstream of the conserved N-terminal region (Ha et al., 1999; Howden et al., 1995a). A most unexpected result produced by database searches using the above predicted polypeptide sequences as queries, has been the identification of putative PCS homologs in organisms such as the nematode *Caenorhabditis elegans* (Clemens et al., 1999; Ha et al., 1999; Vatamaniuk et al., 1999) and the slime mold *Dictyostelium discoideum* (Cobbett, 2000a). Recently, it has been shown that the *C. elegans* PCS homologue does indeed code for a functional, Cd-dependent γ-EC dipeptidyl-transpeptidase, whose activity is critical for HM tolerance in the intact organism (Vatamaniuk et al., 2001). Contrary to what was commonly thought, a PC-based HM detoxification strategy thus appears to operate in organisms other than plants and fungi. As revealed by a more recent database search (R. Percudani and S. ottonello, unpublished), a full-length genomic clone 90% identical to the putative *Caenorhabditis elegans* PCS and with the same three-intron structure is present also in the nematode *Caenorhabditis briggsae* (accession no. AC084525) (Figure 6). In addition, consistent with the expected widespread occurrence of PCS in plants and fungi, a BLAST analysis using AtPCS1 as a query identified a number of PCS-homologous Expressed Sequence-Tags, that have recently been sequenced in a variety of land and aquatic plants (*Glycine max*; *Medicago truncatula*; *Sorghum propinquum*; *Sorghum bicolor*; *Lycopersicon esculentum*; *Oryza sativa*; *Oryza sativa* subsp. *japonica*; *Typha latifolia*) and in the oomycete *Phytophthora sojae*.

The availability of PCS cDNA clones, along with sizeable amounts of highly purified, catalytically competent recombinant enzyme (rPCS) has led to new insights as well as to important validations of previously known or expected PCS features. These include: i) the complete dependence of PCS-mediated Cd detoxification on an adequate GSH supply; ii) the constitutive expression of the *AtPCS1* mRNA under HM-free conditions; iii) the fact that PCS-mediated Cd protection is well-distinct from other HM detoxification mechanisms, such as the uptake of Cd-GSH complexes by the yeast YCF1 transporter or MT (*CUP1*)-dependent protection; iv) the cytosolic localization of PCS and its ability to function in a vacuole-independent manner. As predicted by its relatively high Cys content, PCS itself is a HM binding protein. Recombinant AtPCS1 binds 7.09 ± 0.94 Cd ions/ molecule, with an aggregate dissociation constant of 0.5 ± 0.2 µM (Vatamaniuk et al., 1999). As expected, *in vitro* PC synthesis supported by either rAtPCS1 or rSpPCS was strongly enhanced by a variety of HM ions. Cd, Cu, Hg and Ag were all very effective positive regulators of AtPCS1 activity, with either Cd^{2+} (Vatamaniuk et al., 1999) or Cu^{2+} (Ha et al., 1999) as the most powerful activators. Arsenic, both as AsI_3 or in the form of the AsO_2^- oxyanion, was also a strong activator, while Zn^{2+}, Pb^{2+} and the AsO_4^{3-} oxyanion were all rather poor activators, and nearly background activity was observed with either Mg^{2+}, Ni^{2+} or Co^{2+}. There is also a generally good agreement between the HM sensitivity displayed by the *cad 1-3 Arabidopsis* mutant and the extent of HM tolerance conferred by the *AtPCS1* cDNA to *Saccharomyces cerevisiae* (Ha et al., 1999). At the top of the HM rank order thus obtained there is Cd^{2+}, then either Cu^{2+} (Vatamaniuk et al., 1999) or AsO_4^{3-} (Ha et al., 1999) and Ag^+, followed by AsO_2^- and Hg^{2+}. No significant activation was observed with either Zn^{2+}, Ni^{2+}, or the SeO_3^{2-} oxyanion. Considerably different metal response results have been reported for the *Schizosaccharomyces pombe* recombinant enzyme (rSpPCS; Ha et al., 1999), which exhibited: i) a relatively flat HM response, with only a ~20-fold difference between Cd-activated and basal PC synthesis (as opposed to a >1000-fold difference in the case of rAtPCS1); ii) an only 1.4-fold higher activation by Cd^{2+} as compared to Cu^{2+}; iii) a less than two-fold difference between Cd^{2+}- and Zn^{2+}-induced activation. Although consistent with earlier data obtained with *Schizosaccharomyces pombe* cell free extracts (Hayashi et al., 1991), these results may also reflect some *in vitro* artifacts caused by the use of rSpPCS, which was over 40-fold less active than the corresponding *Arabidopsis* enzyme. Indeed, much more conventional results were produced by *in vivo* experiments measuring either metal-induced PC synthesis or HM sensitivity, which revealed a strong (~100-fold) Cd-dependent activation of PC_{2-3} synthesis (Clemens et al., 1999; Ha et al., 1999) and a marked sensitivity of *SpPCS* deletion mutants to Cd^{2+} and AsO_4^{3-}, little or none towards Cu^{2+} (Clemens et al., 1999), and no sensitivity towards Zn^{2+}, Hg^{2+}, SeO_3^{2-}, Ag^+ and Ni^{2+} ions (Ha et al., 1999). As indicated by the marked HM sensitivity and the extremely low levels of PC synthesis exhibited by fission yeast *SpPCS* disruptants, PC formation in this organism also mainly relies on PCS-catalyzed transpeptidation. The previously reported PCS-independent pathway (Hayashi et al., 1991), which may operate under as yet unidentified growth conditions, thus appears to be of minor physiological significance; it may explain, however, PC synthesis in organisms, such as *Saccharomyces cerevisiae*, which do not contain any PCS homolog in their genome (Mewes et al., 1997).

2.4 Catalytic activation of PC synthase by HM ions

Interaction of either free HM ions or GSH-HM complexes with the PCS active site have been considered as equally likely, not mutually exclusive mechanisms underlying PCS activation by HMs (Cobbett, 2000a). As revealed by the results of a very elegant mechanistic study recently conducted with yeast expressed rAtPCS1 (Vatamaniuk et al., 2000), the latter model (i.e interaction of the enzyme with a metal-thiolate substrate) explains most of the activating effect of metal ions. In fact, based on the stability constants of the Cd-GS and Cd-GS$_2$ complexes, it was calculated that the free Cd ion concentration in a metal-limited, yet fully productive reaction mixture (typically containing 0.025 mM Cd^{2+} and 3.3 mM GSH) is about six orders of magnitude lower than the Cd binding constant of AtPCS1 (Kd=0.54 ± 0.21 µM). Under these conditions, moreover, a further increase of the free GSH concentration enhanced, rather than depressed PC synthesis. In keeping with the above observations, a detailed kinetic analysis showed that PCS-supported PC synthesis proceeds via a substituted enzyme ("ping-pong") mechanism, in which both free GSH and its corresponding metal-thiolate complex serve as co-substrates with apparent K_m values of 13.6 mM and 9.2 µM, respectively. In agreement with the previously unexplained observation that S-bimane GSH can also serve as a PCS substrate (Grill et al., 1989), it was further shown that various S-alkyl GSH derivatives efficiently support the synthesis of S-substituted PCs in the absence of metal ions. Thus, the minimum requirement for basal PC synthesis is that the thiol group of at least one of the co-substrates is blocked either through HM-thiolate formation or S-alkylation (Figure 7). Although S-substituted GSH is by itself sufficient for PC synthesis, an about two-fold increase of the biosynthetic rate was elicited by Cd addition to S-methyl GSH-supported reactions (Vatamaniuk et al., 2000). This HM activating effect, which required a Cd ion concentration compatible with the metal binding affinity of apo-PCS (see above), but about five orders of magnitude higher than the free metal ion concentrations typically found in reaction mixtures containing unsusbstituted GSH, is likely due to a direct Cd-enzyme interaction. However, if one considers that the intracellular GSH concentration is typically comprised between 1 and 10 mM (Inzé and van Montagu, 1995; Noctor et al., 1998) and that Cd ion concentrations ranging from 0.04 to 0.3 µM and from 0.3 to 1 µM have been reported for non-polluted and polluted soil solutions, respectively, (Sanità di Toppi and Gabbrielli, 1999; Wagner, 1993), it follows that the enhancement of PC synthesis resulting from direct Cd-enzyme interaction is likely of little physiological significance under most environmental conditions.Direct HM activation may, perhaps, be more significant in the presence of relatively high concentrations of metals such as Cu and Zn that form less stable complexes with GSH (thus leading to a higher free metal ion concentration), or under the conditions of mixed soil pollution (e.g. Cd plus Zn) commonly found in natural environments (Baker et al., 1990). Contrary to the once postulated, high intrinsic HM resistance of PCS, a significant inhibition of PCS activity

METAL CHELATING PEPTIDES AND PROTEINS 71

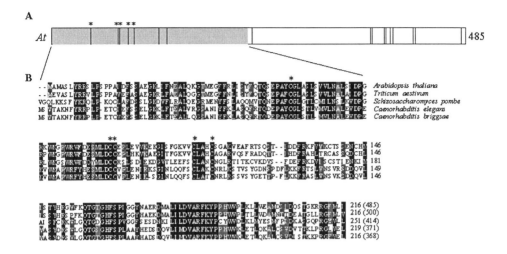

Figure 6. Alignment of phytochelatin synthase genes from different organisms. **A**. Outline of the predicted AtPCS1/CAD1 polypeptide sequence highlighting the conserved N-terminal region (positions 1-216; gray) and the highly divergent C-terminal region (positions 217-485; white). Vertical bars indicate Cys residues; cysteines that are aligned in the five sequences are marked with asterisks. **B**. Alignment of the deduced polypeptide sequences of plant and fission yeast phytochelatin synthases and of the putative enzymes from the nematodes C. elegans and C. briggsae. Amino acid residues that are identical or similar in all of the five sequences, or in at least four of them, are shown on a black or a gray background, respectively; Cys residues that are aligned in all sequences are marked with asterisks.

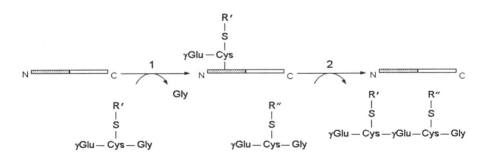

Figure 7. Schematic representation of dipeptidyl transpeptidation by phytochelatin synthase.

at Cd ion concentrations higher than 5 μM was observed in reactions supported by the S-substituted GSH derivative S-methyl GSH. According to the kinetic data of Vatamaniuk et al. (2000), GSH (or GS_2-HM) is predicted to form a covalent acyl-enzyme intermediate (γ-Glu-Cys-enzyme) in step 1; this is followed by the transfer of the γ-Glu-Cys moiety from such intermediate to a second GSH (or GS_2-HM) molecule to yield PC_2. The minimum requirement for enzyme-catalyzed transpeptidation is that either one of the two substrate molecules (R' or R'') is present as a heavy metal (e.g. Cd^{2+})-GS_2 complex.

In the "low heavy-metal/high GSH concentration" scenario delineated by the study conducted by Vatamaniuk et al. (2000), two other important aspects of the mode of action of PCS pertain to: i) the mechanism of PC synthesis termination, which does not primarily rely on HM chelation by GSH and PCs, the actual transpeptidation co-substrates, but rather on competitive enzyme inhibition by free GSH and apo-PCs upon depletion of the HM pool (e.g. through LMW complex formation or vacuolar internalization; see below); ii) the fact that, contrary to a previous proposal envisaging PC-mediated HM sequestration as a defense mechanism mainly operating in the presence of relatively high Cd amounts (Wagner, 1993), even extremely low Cd ion concentrations (as long as they are compatible with the stability constants of Cd-GSH complexes) are sufficient to form the metal-thiolate co-substrate, thus immediately triggering PC synthesis by the constitutively expressed PCS enzyme.

2.5 Vacuolar compartmentalization of HM-PC complexes

An essential role in HM detoxification is played by vacuolar compartmentalization (Figure 8). This mechanism prevents the accumulation of HMs, in particular Cd ions, in the cytosol, where they can exert toxic effects. It has been demonstrated that exposure to Cd ions stimulates the biosynthesis of PCs, which are rapidly converted into a "low molecular weight" (LMW) complex (mainly Cd bound to PC_3) (Abrahamson et al., 1992; Vögeli-Lange and Wagner, 1990; 1996), and, in *Rauvolfia serpentina* cell cultures, into a "medium molecular weight" (MMW) complex (mainly Cd bound to PCs with a higher polymerization level) (Kneer and Zenk, 1997). These complexes enter the vacuole, where they gain acid-labile sulfur (S^{2-}) and form a "high molecular weight" (HMW) complex (Speiser et al., 1992a) with a higher affinity towards Cd. The formation of the HMW complex, which is highly stabilized by S^{2-}, is crucial for Cd detoxification. Acid-labile S^{2-} is probably also involved in the detoxification of Se in *Brassica juncea* (Speiser et al., 1992a). In fact, at variance with Se isologues, Se toxicity is counteracted by the incorporation of S-containing amino acids into proteins (Zayed and Terry, 1992; Terry and Zayed, 1994). Besides molecular weight differences among Cd-PC complexes - which in the case of fungal complexes are typically accounted for by the presence of 2-3 PC_{2-5} molecules in LMW species, as opposed to 6-8 molecules in HMW species - the main differences between LMW and HMW complexes are: i) the amount of associated sulfide ions, with typical S^{2-}:Cd^{2+} molar ratios of 0.01 and 0.1-1, respectively; ii) the generally higher intracellular abundance of the HMW species; iii) the higher degree of PC polymerization, metal binding capacity, and acid stability of the HMW complexes, with typical half-dissociation pH values of 3.8-4 as compared to 5-5.4 for the LMW complexes; iv) the different subcellular distribution and biogenesis of LMW (mainly cytosolic) and HMW (mainly vacuolar) species. There also seems to be a

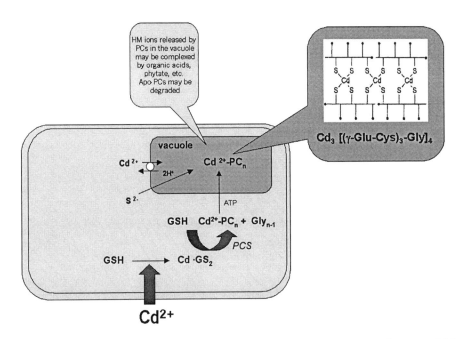

Figure 8. Phytochelatin biosynthesis, cadmium complexation, and subsequent translocation of Cd (and other metals) into the vacuole. See the text (paragraph 2.5) for details. (after Sanità di Toppi and Gabbrielli, 1999, with permission from Elsevier Science)

different tendency of different PCs to be recruited into HMW complexes. In Cd-challenged maize seedlings, for instance, n3 and n4, but not n2 (γ-EC)$_n$ peptides were found to be most represented in HMW complexes (Rauser and Meuwly, 1995). The peptide-sulfhydryl:Cd^{2+} molar ratio is also quite variable, ranging from a value close to the theoretical one (3.78 vs. 4.00) in the case of soluble, low aggregation complexes, down to values around 1, with an involvement of peptide O and N atoms, under *in vitro* over-saturation conditions or in natural HMW complexes (Rauser and Meuwly, 1995; Strasdeit et al., 1991). Although PC determination has successfully been employed to estimate environmental Cd poisoning (Knauer et al., 1998, and references therein), the above mentioned variability complicates the use of bulk PC measurements as quantitative indicators of Cd accumulation.

Little is known about the HMW/S^{2-} transition in plants, but genetic and biochemical data from fungi clearly point to the importance of this step for the PC-dependent detoxification of Cd, and perhaps other HMs. Sulfide is a toxic intermediate product of the sulfur assimilation pathway, which is generated through the ferredoxin-dependent reduction of sulfite (catalyzed by the siroheme containing enzyme sulfite reductase) and is then incorporated into *O*-acetyl serine to produce Cys. Various plant cDNAs coding for sulfite reductase as well as a cDNA encoding the first committed enzyme of siroheme biosynthesis from *Arabidopsis* have been isolated (Leustek et al., 2000), but there are as yet no reports either about Cd hypersensitivity phenotypes arising from mutations in these genes, or their response to Cd exposure. However, pointing to the importance of an adequate S^{2-} production for HM detoxification, a Cd hypersensitive mutant of *Candida glabrata* (*hem2*) has been shown to be deficient in an enzyme, porphobilinogen synthase, involved in siroheme biosynthesis (Hunter and Mehra, 1998), while a fission yeast mutant (*hmt2*), lacking a mitochondrial sulfide/quinone oxidoreductase involved in the detoxification of endogenous sulfide, accumulates high levels of S^{2-} regardless of Cd exposure (Vande Weghe and Ow, 1999). Other fission yeast genes indirectly related to Cd tolerance code for adenine biosynthetic enzymes (Speiser et al., 1992b). In fact, biochemical analysis has shown that besides catalyzing the formation of aspartic acid-derived adenine precursors, these enzymes also act on Cys sulfinate (an aspartate analog) to produce sulfur-containing compounds that are thought to be involved in sulfide incorporation into HMW complexes.

Genetic and biochemical studies in *Schizosaccharomyces pombe* led to the identification of the *HMT1* gene, which codes for a membrane transporter that translocates Cd-PC complexes into the vacuole (Ortiz et al., 1992; 1995). The internalization of both apo-PCs and LMW PC-Cd complexes, but not free Cd or Cd-GS thiolates, is impaired in *hmt1* mutants, and yeast cells overexpressing the HMT1 transporter exhibit an enhanced HM tolerance along with higher intracellular levels of Cd (Ortiz et al., 1992). HMT1-mediated PC-Cd uptake requires ATP, but is functionally independent from the trans-vacuolar proton gradient (Ortiz et al., 1995). Although no HMT1-related, vacuolar ABC transporter has thus far been reported in plants, HMT1-like activities have been identified in mesophyll tobacco protoplasts (Vögeli-Lange and Wagner, 1990) and in tonoplast vesicles from oat roots (Salt and Rauser, 1995). In keeping with the recently documented existence PC-based detoxification mechanisms outside plants and fungi (Vatamaniuk et al., 2001), ABC (MRP) transporter mutants of *Caenorhabditis elegans* are hypersensitive to Cd and arsenite (Broeks et al., 1996). In the *Schizosaccharomyces pombe* mutant JS237, cAMP and Ca ions appear to be involved in the vacuolar accumulation of Cd (see Ow, 1996, and references therein). Moreover, in the presence of Mg-ATP, Cd-PC complexes (as well as apo-PCs) are transported against the concentration gradient across the tonoplast by means of specific carriers. They accumulate inside tonoplast vesicles from oat roots at levels that are up to 38 times higher than those of PC-Cd complexes present in the outer solution (Salt and Rauser, 1995). A distinct ABC transporter (YCF1), that translocates into the vacuole GSH conjugates and Cd-GS$_2$ complexes, but not Cd-PC complexes, is present in *Saccharomyces cerevisiae* (Li et al., 1997). Related Cd transport activities, distinct from the PC-ABC transporter, have been identified in both fission yeast and plants (reviewed in Rea et al., 1998).

Free Cd ions, if present, may enter the vacuole by means of a $Cd^{2+}/2H^+$ antiport (Salt and Wagner, 1993; Gries and Wagner, 1998). In the acidic environment of the vacuole the HMW complex dissociates (Krotz et al., 1989). At pH 5.4, typical of the vacuolar lumen, Cd-PC complexes dissociate and free Cd ions are released in amounts ranging from 62 % to 100 % of the total vacuolar Cd, depending on whether the total Cd concentration is either 50 or 5 µM, respectively (Johanning and Strasdeit, 1998). Cd can then be complexed by vacuolar organic acids (citrate, oxalate, malate) (Krotz et al., 1989) and, possibly, by phytate or amino acids. Apo-PCs present in the vacuole may be degraded by vacuolar hydrolases and/or return to the cytosol, where they can continue to carry out their shuttle role. Interestingly, X-ray microanalysis of water hyacinth leaves clearly indicated the incorporation of Cd, Pb and Sr ions in Ca oxalate crystals (Mazen and El Maghraby, 1997/98) and similar results on non-peptide HM ligands were obtained by Lichtenberger and Neumann (1997). For a comprehensive review on this topic see also Rauser (1999) and references therein.

As to metals other than Cd, *in vitro* PC (n = 2-4)-metal complexes with Ag, Cu, Hg, Pb, and Zn have been identified (Kneer and Zenk, 1992; Mehra et al., 1995; 1996a; 1996b; Mehra and Mulchandani, 1995). In the case of Pb, metal-PC complexes exhibited apparent peptide-sulfhydryl:Pb stoichiometries ranging from 1 to 0.4. Similar data have been reported for Cu and Zn complexes. To the best of our knowledge, however, no *in vivo* HM-PC complexes, other than those with Cu, have been characterized thus far by means of HPLC-ICP-MS analysis (Leopold and Günther, 1997). No Pb- or Zn-PC complexes were evidenced by these authors. Interestingly, a peptide-sulfhydryl:metal molar ratio ranging from 1 to 2, similar to that found in mammalian and yeast MTs, has been reported for a natural Cu-PC complex (Grill et al., 1989). Experiments performed with barley and garlic plants treated with Cd ions, evidenced an improved, "indirect" protection against Hg ions (Panda et al., 1997; Subhadra and Panda, 1994), but no clear and reproducible results were obtained thereafter. An inductively coupled plasma-atomic emission spectroscopic analysis of *Rubia tinctorum* root cultures exposed to various metals (including the metalloid As), revealed predominant Cd-, Ag-, and Cu-PC complexes, but no As-PC complex (Maitani et al., 1996). This has later been explained by the ease, compared to other metals such as Cd, with which As (especially under alkaline pH conditions) undergoes redox reactions leading to PC sulfhydryl oxidation and complex disruption (Schmöger et al., 2000). The results of experiments on the reduction and coordination of As in *Brassica juncea* (Pickering et al., 2000) and the detoxification of As compounds by PCs (Schmöger et al., 2000) have been reported recently. Both AsO_2^- and AsO_4^{3-} markedly induced the *in vivo* synthesis of PCs in cell suspensions of *Rauvolfia serpentina* and in seedlings of *Arabidopsis thaliana*. In addition, both oxyanions enhanced PCS activity in cell-free extracts of *Silene vulgaris*. Based on ESI-MS analyses, probably three SH groups (provided by two PC_2 molecules) chelate one As (Schmöger et al., 2000). Interestingly, a fern (*Pteris vittata*) that hyperaccumulates AsO_4^{3-} into its above-ground biomass has recently been discovered by Ma et al. (2001).

Although different PC-metal complexes have been identified both *in vitro* and in metal-challenged cell cultures, the actual physiological significance of the complexation by PCs of metals other than Cd remains to be established.

2.6 A likely dual role for PCs: metal detoxification and essential metal homeostasis

No doubt, PCs play a central role in the detoxification of several HMs, in particular Cd, the metal that has been addressed more extensively till now. The use of

Cd-sensitive, PC-deficient mutants of *Arabidopsis thaliana* (*cad1*) was especially instrumental for demonstrating the crucial role of PCs in Cd detoxification (Howden and Cobbett, 1992; Howden et al., 1995a; 1995b). More recently, suspension cultured cells from a plant species that is naturally impaired in PC synthesis (Azuki bean; *Vigna angularis*) have been shown to be hypersensitive to Cd (Inouhe et al., 2000).

The amount of PCs induced by Cd seems to be roughly related to the intensity and duration of Cd exposure. In addition, the existence of a positive correlation between the duration and the level of Cd exposure and the number of γ-Glu-Cys repeat units incorporated into PCs has been reported (Grill et al., 1987; Jemal et al., 1998; Vögeli-Lange and Wagner, 1996). In some experiments (Salt et al., 1995; 1997), about 60% of the intracellular Cd was chelated by sulfur ligands, presumably PCs, and the remaining 40% by oxygen ligands, probably organic acids. Thus, Rauser (1999) correctly asserted that PCs "function to bind some of the intracellular Cd from small to large proportions, depending on the duration of exposure to Cd, and that other ligands participate to chelate some of the remaining Cd". Consequently, PCs should be viewed as effective HM detoxification systems, but especially as sinks for excess Cd and other metals with affinity towards sulfur ligands. Kneer and Zenk (1992) demonstrated that several metal-sensitive enzymes tolerate 10 to 1000-fold greater amounts of PC-chelated Cd as compared to free Cd ions, and that PCs are able to reactivate Cd-poisoned enzymes. However, especially for HMs other than Cd, there are a few studies (see above) evidencing the formation of HM-PC complexes *in vivo*. Following the "fan-shaped" model proposed previously (Sanità di Toppi and Gabbrielli, 1999) and schematically presented in Figure 2, other effective, PC-independent detoxification mechanisms can be concurrently put into action depending on the type of metal supplied, its concentration, exposure duration, and plant species (or ecotype), etc. In this sense, with regard to Cd, Cu, Pb and Zn ions, Leopold at al. (1999) concluded that "the synthesis of HM-PC complexes is a fast and probably transient answer of the cells and not necessarily important for the HM tolerance of plants, which are able to grow on HM-polluted soils under natural conditions".

As discussed by Ow (1996), other factors, besides intracellular PC levels, may be crucial for efficient Cd detoxification. For instance, in the *Schizosaccharomyces pombe* mutant JS563, a FAD/NAD-linked disulphide reductase (perhaps a PC reductase) was discovered (see Ow, 1996, and references therein). This enzyme seems to be essential to guarantee sufficient reducing power to prevent the oxidation, and consequent inactivation of Cd-induced PCs. Vacuolar accumulation of Cd and Zn has also been shown to play a key role in the prompt detoxification and long-term tolerance of these HMs (Chardonnens et al., 1998; 1999). Additionally, the rapid formation of HMW complexes, highly stabilized by S^{2-} groups, seems to be particularly important for Cd detoxification (Kneer and Zenk, 1997; Ortiz et al., 1992; Speiser et al., 1992a; Verkleij et al., 1990). As noted by Vande Weghe and Ow (1997), fission yeast *hmt2* mutants are defective in S^{2-} production, are hypersensitive to cadmium, and unable to accumulate PCs. Other *Schizosaccharomyces pombe* mutants defective in sulfur-reduction are also unable to form Cd-PC HMW complexes (Mehra and Mulchandani, 1995). In some cases, GSH itself might have a direct role in HM detoxification, especially in the presence of *low* Cd ion concentrations (Inouhe et al., 2000; Vögeli-Lange and Wagner,

1996) (see below). Thus, it is conceivable that also in plants high levels of PCs may not be sufficient *per se* for complete Cd detoxification.

More generally, it appears that reductive sulfate assimilation and an adequate Cys and GSH supply play crucial roles in PC-mediated HM tolerance. PCS itself is a Cys-rich protein. In addition, as predicted by the relatively high K_m of free GSH (6.7-13.6 mM), GSH availability can be a critical factor limiting PC formation. Consistent with this view, both the activity (Chen et al., 1997; De Knecht et al., 1995; Nagalakshmi and Prasad, 2001; Noctor et al., 1998; Nussbaum et al., 1988; Pilon-Smits et al., 1999a Rüegsegger et al., 1990; Rüegsegger and Brunold, 1992) and the mRNA levels (Dominguez-Solis et al., 2001; Heiss et al., 1999; Lee and Leustek, 1999; Schäfer et al., 1998) of enzymes involved in sulfur assimilation and GSH biosynthesis are upregulated following HM (usually, Cd and Cu) exposure. In particular, the mRNAs encoding γ-EC synthetase and GSH synthetase (GS) as well as the GSH reductase mRNA are coordinately upregulated in Cd or Cu (but not Li, Zn, Mg, Ca) exposed *Arabidopsis* seedlings, concomitantly with a substantial increase of PC levels (Xiang and Oliver, 1998). Interestingly, the signal molecule jasmonate upregulates these three genes even in the absence of Cd ions (Xiang and Oliver, 1998), thus pointing to its possible involvement as a mediator of this HM response.

In the presence of Cd, GS is the rate limiting enzyme for the synthesis of GSH and PCs in *Brassica juncea*, and PC formation as well as Cd tolerance are both increased in Indian mustard transgenic plants overexpressing GSH biosynthetic enzymes (Zhu et al., 1999a; 1999b). On the same note, studies conducted on transgenic *Brassica juncea*, *Arabidopsis thaliana*, and poplar, overexpressing O-acetylserine(thiol)lyase (OAS-TL), γ-glutamylcysteine synthetase (γ-ECS) and GS, respectively, have shown that increased Cd tolerance and PC production in such plants may both be explained by the increased accumulation of Cys and GSH (Arisi et al., 2000; Dominguez-Solis et al., 2001; Pilon-Smits et al., 1999b; Zhu et al., 1999b). Further to this point, a mitochondrial γ-ECS isoform has been found to be induced by Cd^{2+} ions in the roots of *Brassica juncea* (Schäfer et al., 1998).

Altogether, the above data suggest that sulfur metabolism and HM detoxification are closely related processes (Figure 9). In addition, the fact that: i) there are low, but detectable levels of PCs even in the absence of HM exposure (Kneer and Zenk, 1992); ii) PC levels increase concomitantly with Cu and Zn depletion upon transfer of cell suspension cultures to a minimal micronutrient medium (Grill et al., 1988), clearly points to a possible homeostatic role of PCs towards essential HMs. In fact, it has been hypothesized that PCs are not simply a HM-detoxification system *sensu stricto*, especially in the presence of low concentrations of (essential) metals. Under these conditions, instead, PCs may primarily act as key components of metal homeostasis (Thumann et al., 1991). The constitutive expression of PCS(Cobbett, 2000b; Grill et al., 1989) might also be considered as an indication of a more general role of PCs, not exclusively related to HM detoxification. Further supporting this view is the strong protective effect of PCs against Cd-mediated inactivation of metal-sensitive enzymes (Kneer and Zenk, 1992) and the ability of Zn- and Cu-PC complexes, mainly of the PC_2/PC_3 type, to reactivate metal - depleted or metal- poisoned metalloenzymes

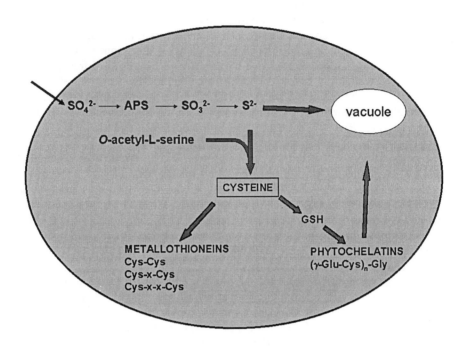

Figure 9. Multiple connections between sulfur metabolism and heavy metal detoxification and homeostasis in plants.

(albeit not more efficiently than the corresponding free salts) (Thumann et al., 1991). Stronger evidence in favour of a metal homeostatic role has been obtained for mammalian MTs (Jacob et al., 1998, and references therein). In plants, it is thus possible that PCs and MT-like proteins cooperatively act in the homeostasis of essential HMs. PCs have also been proposed as activated sulfate acceptors in the formation of a thiosulfate intermediate leading to sulfite formation upon reduction by thiosulfonate reductase (Steffens, 1990, and references therein). This hypothesis, however, has been challenged by the fact that no plant thiosulfonate reductase has been identified so far, and by the recent demonstration that the main sulfite forming pathway in plants relies on an enzyme (Adenosine 5'-Phospho Sulfate reductase) that directly reduces activated sulfate (APS) using an intramolecular glutaredoxin domain (reviewed by Leustek et al., 2000). Although the metal detoxification and homeostatic roles of PCs are not mutually exclusive and may coexist at the whole plant level, the fact that *Arabidopsis* PC-deficient mutants (*cad1*) grow well in the presence of Cu and Zn micronutient concentrations (Howden et al., 1995a) suggests that the latter role, if real, is either dispensable or easily replaceable by other metal binding components such as MTs. On the other hand, the idea of an exclusive metal detoxification function of PCs is somewhat weakened by the lack of correlation between the PC content and the HM sensitivity of metal-tolerant and nontolerant ecotypes of *Silene vulgaris* (de Knecht et al., 1992; Harmens et al., 1993) and *Datura innoxia* (Reese and Wagner, 1987), and by similar findings recently reported for the HM hyperaccumulator *Thlaspi caerulescens* and the closely related, non accumulator species *Thlaspi arvense* (Ebbs et al., 2000; Sanità di Toppi et al., 2001, and references therein). Neither PCS activity, nor PC turnover upon transfer to a Cd-less medium differed between wild-type and HM tolerant *Silene vulgaris* (de Knecht et al., 1995). In addition, while PC_2 was the most abundant PC peptide in metal-tolerant plants, the more effective metal chelator, PC_3, prevailed in the non-tolerant ecotype (de Knecht et al., 1994). Further to this point, a stronger correlation between Cu tolerance and the accumulation of MT mRNAs (r values ranging from 0.89 to 0.998), than with the total amount of intracellular non-protein thiols, including PCs, ($r = 0.77$) has been reported in *Arabidopsis* (Murphy and Taiz, 1995). Therefore, it cannot be excluded that other systems, either autonomously or in combination with PCs, may regulate HM homeostasis in the plant cell. For instance, Cu-MTs and Cu chaperones (termed Atx1, Lys7 and Cox 17) seem to be involved in Cu ion traffic in yeast cells (Valentine and Gralla, 1997) and analogous mechanisms might operate in higher plants.

2.7 PC Synthase as a Biotechnological Tool for HM Remediation:
Promises and Limitations

PCS genes, either overexpressed in a PCS-positive background or transplanted into organisms that lack an endogenous PCS homolog, hold great promise as potential tools for HM bioremediation. However, an integrated view of PC biosynthesis and action is required to understand the biotechnological potentialities and limitations of a PCS-based metal detoxification strategy. Although several cases of variably improved metal tolerance have been reported for transgenic plants overexpressing single sulfur assimilation or GSH biosynthetic genes (Arisi et al., 2000; Dominguez-Solis et al., 2001; Pilon-Smits et al., 1999b; Zhu et al., 1999a; 1999b), it is conceivable to imagine that the balanced and concerted action of most of the above mentioned pathways (i.e. reductive sulfate assimilation and GSH synthesis, LMW complex formation and vacuolar internalization, S^{2-} production and incorporation into HMW complexes) is required for an optimal exploitation of PC-mediated HM detoxification. An adequate sulfur nutrition (either natural or with fertilizer supplementation) and a high capacity for sulfur assimilation through the multienzyme pathways leading from sulfate to sulfide and then to Cys and GSH appear to be particularly important. In fact, natural HM accumulator plants such as *Brassica juncea* have an especially strong sulfur assimilation capacity and an *Arabidopsis* mutant, *man1*, with the characteristics of a natural polymetallic accumulator has a sulfur content in leaves about 3-fold higher than wild-type (Delhaize, 1996). On the other hand, the multiple connections between sulfur metabolism, PC biosynthesis, and downstream processing of PC-metal complexes, leave room to a number of potentially negative, cellular and environmental interferences that may impair the effectiveness of the PC-mediated response. For example, reductive sulfate assimilation in fungi is blocked by micromolar concentrations of Li ions (Murguia et al., 1996) and Cd resistance as well as the total PC content (especially PC_3

and PC_4) have been reported to be lowered in tobacco cells simultaneously exposed to Cd and arsenic (AsO_2^- or AsO_4^{3-}) (Nakazawa et al., 2000). On the same note, a reduced, rather than an increased *in vivo* PC synthesis has been observed in Cu-tolerant *Silene vulgaris* (Schat and Kalff, 1992). This response has been interpreted as a way to limit the GSH cost of HM-induced PC synthesis, thus preventing drastic alterations of intracellular GSH homeostasis and consequent oxidative stress (de Vos et al., 1992). Clearly, founding the entire HM adaptation capacity on a single, metabolically expensive response such as PC synthesis would be quite risky. In fact, other types of HM ligands (e.g. organic acids such as malate and citrate, amino acids, and phytate) and responses (e.g. avoidance and damage repair ; Figure 2) are known to markedly contribute to HM detoxification in plants (Rauser, 1999, and references therein; Sanità di Toppi and Gabbrielli, 1999). For example, neither wild-type nor Cu-tolerant variants of spanish clover (*Lotus purshianus*) accumulated PCs upon Cu challenge and most of the metal was found associated with an acidic, Asp- and Glu-rich, but Cys-poor protein (Lin and Wu, 1994). Further supporting the functional plasticity of the HM response, both PCs and MTs are produced in Cd-challenged *Saccharomyces cerevisiae* cells, but in a different yeast (*Saccharomyces exiguus*), a comparable Cd tolerance (with no intracellular Cd accumulation) is achieved through cell wall immobilization of Cd ions (i.e. avoidance) (Inouhe et al., 1996). Similarly, *Candida glabrata* exclusively produced PCs in response to Cd, but MTs in response to Cu (Mehra et al., 1988), and PC synthesis was inhibited in cells of the same organism constitutively expressing the *Saccharomyces cerevisiae* MT (*CUP1*) gene (Yu et al., 1994).

Other concerns regard: i) the possible requirement for a Cd pre-treatment in order to induce the synthesis of adequate PC levels and subsequent detoxification of a different HM ion, as suggested by the Cd-induced, Hg tolerance observed in both barley (*Hordeum vulgare*) (Subhadra and Panda, 1994) and garlic (*Allium cepa*) (Panda et al., 1997); ii) the potentially detrimental effects, in terms of GSH loss, that may occur in PCS-overproducing cells exposed to "gratuitous" PCS activators (i.e. HM ions that can promote GSH transpeptidation, but form PC complexes of low stability and/or inefficiently translocated into the vacuole); iii) the seemingly rather limited range of HMs, besides Cd, that may be effectively detoxified by PCs under natural field conditions. Consistent with the latter point, BSO addition rendered tobacco cells more sensitive to Cd without any effect on either Zn or Cu tolerance (Reese and Wagner, 1987), and Cd-tolerant tomato cells were only slightly tolerant to Cu and as sensitive as wild-type cells to either Ag, Hg, Zn, and Pb (Huang et al., 1987).

Additional aspects of HM trafficking that may limit the effectiveness of PCS as a bioremediation tool are the mobilization and uptake of metal ions from the soil, the fate of vacuole-accumulated metals, and their translocation from the roots, which are the primary accumulation sites, to the shoots, which are much easier to harvest. Cd ions cross the root plasma membrane by exploiting surface transporters normally utilized for the uptake of essential metals. Three well documented examples of such "Cd-abused" permeases are the *Arabidopsis* Fe(II) transporter IRT1 (Eide et al., 1996), the wheat Ca transporter LCT1 (Clemens et al., 1998), and the *Thlaspi caerulescens* Zn transporter ZNT1 (Lasat et al., 2000; Pence et al., 2000). A gene (*PAA1*) encoding a putative metal-transporting, P-type ATPase that may be involved in Cd extrusion has been identified in

Arabidopsis (Tabata et al., 1997). Polycarboxylic acids and phytate have been proposed as potential Cd acceptors following HMW complex dissociation and HM mobilization inside (and outside) the vacuole (Briat and Lebrun, 1999; Rauser 1999; and references therein). Xylem loading and long-distance transport of metal ions through the transpiration stream is also mainly mediated by organic acids (e.g. citrate), nicotianamine, and amino acids (especially, histidine) (Cataldo et al., 1988; Kramer et al., 1996; Stephan et al., 1996). As the main form of reduced sulfur translocated from leaves to roots, reduced GSH is fairly abundant in the phloem, but not in the xylem sap. An increased translocation of Cd ions from roots to shoots was observed in Cd-challenged, wild-type *Silene vulgaris* plants (de Knecht et al., 1992). In xylem samples from *Brassica juncea*, Cd ions have been found coordinated to O and N atoms, indicating that PCs are probably not involved in long-distance Cd transport in this plant (Salt et al., 1995b). PC complexes, however, have recently been identified as a possible transport form of copper from roots to leaves in Cu-exposed Creosote bush (*Larrea tridentata*) (Polette et al., 2000).

As suggested by the above considerations, although PC overproduction can certainly improve HM (especially Cd) tolerance, it may also suffer from many limitations. Several aspects of the overall HM response, besides PC synthesis, need to be taken into account in the construction of HM remediating organisms with optimal performance under the variable (and often poorly controllable) conditions found in metal-contaminated environments.

3. OTHER HM CHELATING POLYPEPTIDES

3.1 Metallothioneins

In animals, cyanobacteria and fungi, various HMs can be complexed and detoxified by metallothioneins (MTs) (Figure 11), a group of gene-encoded Cys-rich peptides generally lacking aromatic amino acid residues (Kägi, 1991; Robinson et al., 1993), and with molecular masses ranging from 6 to 8 kDa. Cys residues are present in MTs as Cys-x-Cys, Cys-x-x-Cys, or Cys-Cys clusters (where x is any amino acid other than Cys). The first MTs were discovered in equine kidney cortex and termed class I MTs (Margoshes and Vallee, 1957).

The identification of MT-like genes in a number of (higher) plants clearly points to the existence of MT-like gene products also in these organisms, but only a few MT proteins have been characterized up to now in the plant kingdom. In this paragraph we will not deal with MT coding *genes* or with the *expression* of MT transcripts, two topics that are exhaustively covered in another chapter of this book, but only with those few MT-like *proteins* hitherto recognized in (higher) plants. To the best of our knowledge, in fact, there are no broad and repeatable indications of the existence in higher plants of MTs induced by HMs, in particular by Cd. At the moment, the role of plant MTs in HM detoxification seems to be of secondary importance compared to PCs and other non-thiol HM ligands (see below).

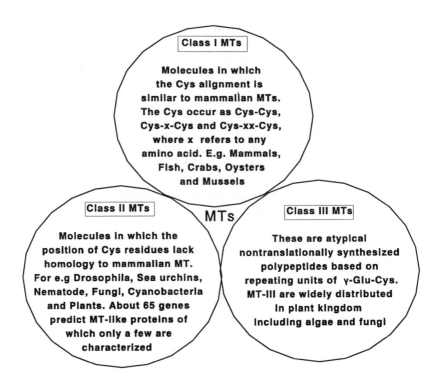

Figure 11. Classification of metallothioneins

In various angiosperms there are indications pointing to the existence of "MT-like proteins" (for reviews, see Kotrba et al, 1999; Prasad, 1999, Rauser, 1999; and references therein), but, with rare exceptions, no MT *sensu stricto* has been purified and sequenced till now. As a matter of fact, the first "certain" MT isolated from higher plants was a wheat germ E_c-MT, probably involved in the regulation of Zn homeostasis during early seed germination (Lane et al., 1987). Because of its homology with mammalian MTs (MTs of class I = MT-I), albeit with a different distribution of Cys residues, it was designated as a class II MT (MT-II). Afterwards, transgenic plants overexpressing animal MT genes were produced by several authors (see Ow, 1996, and references therein). The first protocol for the determination of Cd partitioning in transgenic plants overexpressing a chimeric mouse MT gene was set up by Wagner (1995), and subsequently utilized and improved by others.

To our knowledge, however, as of 1997 only one group has reported the amino acid sequences of *Arabidopsis* MT gene products (Murphy et al., 1997). In fact, in *Arabidopsis thaliana* root apex, two MTs with molecular masses of 4.5 and 8.0 kDa (termed MT1 and MT2) were isolated with the use of polyclonal antibodies raised against plant MT fusion proteins produced in *Escherichia coli* (Murphy et al., 1997).

MT1 appeared to be constitutively expressed in the root apex and inducible in leaves, whereas the opposite behaviour was exhibited by MT2. A conceivable hypothesis, supported by experiments with Cd ions (De Miranda et al., 1990), is that MT1 has a homeostatic function toward Cu, whilst MT2 has a protecting function against *excess* Cu. MT 1 and 2 expression patterns were also determined in leaf mesophyll, trichomes, and vascular tissue of developing fruits and seeds derived from Cu-exposed *Arabidopsis* seedlings (Garcia-Hernandez et al., 1998).

More generally, it cannot be ruled out that MTs may have specialized functions depending on the particular tissue in which they are expressed, and, similar to PCs, they might allow the transfer of essential metals to metal-requiring enzymes. One of the problems encountered with the isolation of MTs from natural sources is their low abundance as compared, for example, to PCs. In this regard, it is possible that plant MTs, which have an interdomain linker usually much longer than that present in animal MTs (up to 30-45 amino acids as compared to 2-4 amino acids in animal MT-I), are intrinsically more susceptible to proteolytic degradation and thus have a much shorter *in vivo* half-life compared to their animal counterparts (Kille et al., 1991; Nelson and Winge, 1983; Robinson et al., 1993). No doubt, further research on plant MT purification and sequencing, and on their contribution, if any, to HM homeostasis and detoxification is urgently needed.

3.2 Ferritins

In plants and other organisms, the specific sequestration of Fe ions is carried out by ubiquitous proteins, termed ferritins (FTs) FTs are typically present in plastids (Briat and Lobréaux, 1998), both in non-photosynthetic tissues (i.e. roots, seeds, legume nodules) and in photosynthetic tissues ((Briat and Lebrun, 1999, and references therein; Harrison and Arosio, 1996).

FTs form complexes with up to 4500 Fe atoms per FT molecule (Guerinot and Salt, 2001). Fe sequestration into such complexes protects the cell against Fe-induced oxidative stress and damage - especially membrane damage and genotoxic effects caused by activated oxygen species produced through the Fenton reaction. FTs are coded by a small gene family (Briat and Lobréaux, 1998), and their expression is activated by many environmental stimuli, particularly Fe ions. Some plant FT genes have been found to be activated by abscisic acid, others by antioxidants and serine/threonine phosphatase inhibitors (Briat and Lobréaux, 1997). Interestingly, transgenic rice plants overexpressing soybean FT genes had a 3-fold higher Fe content in seeds compared to untransformed plants (Goto et al., 1999). FT is an essential protein for aerobic metabolism. The atmospheric transition and the evolution of an aerobic atmosphere on earth placed several demands on Fe metabolism, especially in view of the extreme insolubility and potential toxicity of free Fe ions. Therefore, organisms have developed a wide variety of strategies to deal with this challenge, including extracellular Fe^{3+} chelators (siderophores), Fe^{3+} transport systems (transferrins), and iron storage proteins (ferritins).

Variations in the iron content occur among phytoferritins. Instead, the iron content of plant FTs (seeds, undifferentiated cells) was found to be relatively insensitive to Fe overload. The phosphorus content was considered to be a constant component (Fe:P= 8:1) of FT- Fe cores (Table 1).

3.3 Metal-binding properties of ferritin - role in heavy metal detoxification

Fe is the predominant metal stored in FT. However, FT can bind also non-ferrous metal ions and this phenomenon has been demonstrated *in vitro* and *in vivo* (Sczekan and Joshi, 1989). Other divalent and trivalent cations may also be bound albeit in lesser amounts in animal FTs (Price and Joshi, 1982). In plants, investigations of Sczekan and Joshi (1989) have shown that FT from *Glycine max* is capable of binding HMs *in vitro*. It was also demonstrated that metal-binding by animal FTs results in the appearance of peaks with maximum absorbance at 242 and 295 nm in UV difference spectra. These difference spectra are generally considered to be characteristic of deprotonated tyrosine. Similar results were obtained with *Vigna mungo* seed FT, which after Cd binding exhibited absorbance peaks at 242 and 290 nm in UV difference spectra. The ratio of protein to Cd varied from 1:2 to 1:5 (Rama Kumar and Prasad, 2001) and absorbance peaks at 240 and 275 nm were observed with Cu, thus indicating the possibile involvement of tyrosine residues in metal-binding (Harris and Madisen, 1988; Tan and Woodworth, 1969). It was reported by Price and Joshi (1983) that Be^{2+} is bound to the carboxyl groups of aspartic or glutamic acid residues or to the hydroxyls of tyrosine on the protein shell.

In vitro binding assay using ^{109}Cd as a ligand also evidenced a radioactively labeled polypeptide band corresponding to FT, thus further confirming the Cd binding ability of FT (Rama Kumar and Prasad, 2001). Moreover, fluorescence spectra, as depicted by the quenching of fluorescence upon heavy metal binding to FT (Rama Kumar and Prasad, 2000), clearly point to the fact that besides its iron storage function, FT also chelates various other HMs.

Table 1. Fe content (atoms/molecule) of FTs from some legume seeds

Taxon	Fe per molecule	Reference
Pea (*Pisum sativum*)	1800-2100	Crichton et al., 1978
Soybean (*Glycine max*)	2500	Laulhere et al., 1988
Lentil (*Lens esculenta*)	2100	Laulhere et al., 1988
Jackbean (*Canavalia ensiformis*)	900	Briat et al., 1990
Black gram (*Vigna mungo*)	1100	Kumar and Prasad, 2000

3.4 Nicotianamine

Nicotianamine (NA) is a plant amino acid-derivative generated by the condensation of three S-adenosyl-L-methionine units (SAM) (Briat and Lebrun, 1999, and references therein; Shojima et al., 1990). Fe and other HM ions can be bound by three carboxylic groups of NA (Stephan and Scholz, 1993). NA can form complexes with Co^{2+}, Cu^{2+}, Fe^{2+}, Mn^{2+}, Zn^{2+} and other divalent transition metals (Pich et al., 1997; Stephan et al., 1996). NA may contribute to the vacuolar accumulation of some of the above metals as well as to their xylematic and/or phloematic relocation (Pich and Scholz, 1996). The tomato mutant "chloronerva", which lacks NA, displays a manifest intercostal chlorosis of the leaves due to Fe deficiency. The stability constants of various metal-NA complexes were found to range from 9 to 18.5 (log K) (Beneš et al., 1983).

3.5 Stress proteins

Plants exhibit multiple stress resistance mechanisms. *Lycopersicon peruvianum* cell cultures when stressed with $CdSO_4$ (10^{-3} M) for 4h after a heat-shock for 15 min at 40°C acquired tolerance to HMs by preventing the membrane damage. Cells showed normal ultrastructure. The protective function of heat stress preceeding Cd stress was not observed with cyclohexamide, a potent inhibitor of protein synthesis.

Conversely, barley seedlings (*Hordeum vulgare* cv. Fatran) pretreated with various concentrations of Cu, Fe, Mn, Mg and Zn for 3 days exhibited thermoresistance to sub-lethal (40 °C) and lethal (45 °C) temperature upon treatment with heat shock subsequently. Each metal had a different degree of effectiveness and Cu being the most effective one amongst the tested. Thermoprotection was evident in the shoots though the HM stress was supplied through roots. HMs induce `hsc' synthesis which ultimately prevent damage of the membranes against sub-lethal and lethal temperature. In plants, thermoprotection by heavy metals *via* heat-shock cognates and the role of heat-shock proteins in protecting membranes from toxic heavy metals is a manifestation of a state of co-stress, which evokes coping mechanisms for these reciprocal stressors resulting in multifaceted adaptive responses at both the cellular and the molecular level (Prasad, 1997). (Figure 12)

Pretreatment with

⤺ ⤻

➢ Heavy metals   ➢ Heat shock

Induced tolerance for
Sublethal and lethal
temperatures
e.g. barley, sorghum
(Prasad, 1997)

Induced heat shock
proteins (hsps) act as
molecular chaperones
and protect membranes
against heavy metal
Toxicity. e.g. Tomato
cell cultures, wheat
Leaves (Prasad, 1997)

Figure 12. Thermoprotection by heavy metals via heat shock cognates and the role of heat shock proteins in protecting membrane damage by functioning as molecular chaperones is a manifestation of co-stress

REFERENCES

Abrahamson, S.L., Speiser, D.M., Ow, D.W. (1992). A gel electrophoresis assay for phytochelatin. Anal. Biochem. **200**, 239-243.
Aisen, P., Listowsky, I. (1980) Iron transport and storage proteins. Ann Rev Bio Chem **4**, 357-393
Al-Lahham, A., Rohde, V., Heim, P., Leuchter, R., Veeck, J., Wunderlich, C., Wolf, K., Zimmermann, M. (1999). Biosynthesis of phytochelatins in the fission yeast. Phytochelatin synthesis: a second role for the glutathione synthetase gene of *Schizosaccharomyces pombe*. Yeast **15**, 385-396.
Anelli, G., Pelosi, P., Galoppini, C. (1973). Influence of mercury on the amino acidic composition of tobaccoleaves. Agr. Biol. Chem. **37**, 1579-1582.
Arisi, A.-C. M., Mocquot, B., Lagriffoul, A., Mench, M., Foyer, C.H., Jouanin, L. (2000). Responses to cadmium in leaves of transformed poplars overexpressing γ-glutamylcysteine synthetase. Physiol. Plant. **109**, 143-149.
Bae, W., Chen, W., Mulchandani, A., Mehra, R. K. (2000). Enhanced bioaccumulation of heavy metals by bacterial cells displaying synthetic phytochelatins. Biotech. Bioeng. **70**, 518-524.
Bae, W., Mehra, R.K. (1997). Metal binding characteristics of a phytochelatin analog (Glu-Cys)$_2$-Gly. J. Inorg. Biochem. **68**, 201-210.

Baker A.J.M., Ewart, K., Hendry, G.A.F., Thorpe, P.C., Walker, P.L. (1990). The evolutionary basis of cadmium tolerance in higher plants. In: 4th International Conference on Environmental Contamination. October 1990, Barcelona, Spain, pp. 23-29.

Beneš, I., Schreiber, K., Ripperger, H., Kircheiss, A. (1983). Metal complex formation by nicotianamine, a possible phytosirophore. Experientia **39**, 261-262.

Bernhard, W.R., Kägi, J.H.R. (1987). Purification and characterization of atypical cadmium-binding polypeptides from *Zea mays*. Experientia **52** (suppl.), 309-315.

Briat, J.-F., Lebrun, M. (1999). Plant responses to metal toxicity. C. R. Acad. Sci. Paris/Life Sci. **322**, 43-54.

Briat, J.-F., Lobréaux, S. (1998). Iron storage and ferritin in plants. In: H. Sigel, A. Sigel (Eds.), Iron transport and storage in microorganisms, plants and animals, vol. 35 of Metal ions in Biological Systems. Marcel Dekker, New York, pp. 563-584.

Briat, J.-F. and Lobréaux, S. (1997). Iron transport and storage in plants. Trends Plant Sci. **2**, 187-193.

Broeks, A., Gerrard, B., Allikmets, R., Dean, M., Plasterk, R.H. (1996). Homologues of the human multidrug resistance genes MRP and MDR contribute to heavy metal resistance in the soil nematode *Caenorhabditis elegans*. EMBO J. **15**, 6132-6143.

Cataldo, D.A., McFadden, K.M., Garland, T.R., Wildung, R.E. (1988). Organic costituents and complexation of nickel, iron, cadmium and plutonium in soybean xylem exudates. Plant Physiol. **86**, 734-739.

Chardonnens, A.N., Koevoets, P.L.M., van Zanten, A., Schat, H., Verkleij, J. (1999). Properties of enhanced tonoplast zinc transport in naturally selected zinc-tolerant *Silene vulgaris*. Plant Physiol. **120**, 779-785.

Chardonnens, A.N., ten Bookum, W.M., Kuijper, L.D.J., Verkleij, J.A.C., Ernst, W.H.O. (1998). Distribution of cadmium in leaves of cadmium tolerant and sensitive ecotypes of *Silene vulgaris*. Physiol. Plant. **104**, 75-80.

Chassaigne, H., Vacchina, V., Kutchan, T.M, Zenk, M.H. (2001). Identification of phytochelatin-related peptides in maize seedlings exposed to cadmium and obtained enzymatically in vitro. Phytochemistry **56**, 657-668.

Chen, J., Zhou, J., Goldsbrough, P.B. (1997). Characterization of phytochelatin synthase from tomato. Physiol. Plant. **101**, 165-172.

Clemens, S., Antosiewicz, D.M., Ward, J.M., Schachtman, D.P., Schroeder, J.I. (1998). The plant cDNA LCT1 mediates the uptake of calcium and cadmium in yeast. Proc. Natl. Acad. Sci. USA **95**, 12043-12048.

Clemens, S., Kim, E.J., Neumann, D., Schroeder, J.I. (1999). Tolerance to toxic metals by a gene family of phytochelatin synthases from plants and yeast. EMBO J. **18**, 3325-3333.

Clemens, S. (2001) Molecular mechanisms of plant metal tolerance and homeostasis. Planta **212**, 475-486.

Cobbet, C.S. (2000a). Phytochelatins and their roles in heavy metal detoxification. Plant Physiol. **123**, 825-832.

Cobbet, C.S. (2000b). Phytochelatins biosynthesis and function in heavy-metal detoxification. Curr. Opin. Plant Biol. **3**, 211-216.

de Knecht, J.A., Koevoets, P.L.M., Verkleij, J.A.C., Ernst, W.H.O. (1992). Evidence against a role for phytochelatins in naturally selected increased cadmium tolerance in *Silene vulgaris* (Moench) Garcke. New Phytol. **122**, 681-688.

de Knecht, J.A., van Baren, N., Ten Bookum, W.T., Wong Fong Sang, H.W., Koevoets, P.L.M., Schat, H., Verkleij, J.A.C. (1995). Synthesis and degradation of phytochelatins in cadmium-sensitive and cadmium-tolerant *Silene vulgaris*. Plant Sci. **106**, 9-18.

de Knecht, J.A., van Dillen, M., Koevoets, P.L.M., Schat, H., Verkleij, J.A.C., Ernst, W.H.O. (1994). Phytochelatins in cadmium-sensitive and cadmium-tolerant *Silene vulgaris*. Plant Physiol. **104**, 255-261.

Delhaize, E. (1996). A metal-accumulator mutant of *Arabidopsis thaliana*. Plant Physiol. **111**, 849-855.

Delhaize, E., Jackson, P.J., Lujan, L.D., Robinson, N.J. (1989). Poly(γ-glutamylcysteinyl)glycine synthesis in *Datura innoxia* and binding with cadmium. Plant Physiol. **89**, 700-706.

De Miranda, J.R., Thomas, M.A., Thurman, D.A., Tomsett, A.B. (1990). Metallothionein genes from the flowering plant *Mimulus guttatus*. FEBS Lett. **260**, 277-280.

de Vos, C.H.R., Vonk, M.J., Vooijs, R., Schat, H. (1992). Glutathione depletion due to copper-induced phytochelatin synthesis causes oxidative stress in *Silene cucubalus*. Plant Physiol. **98**, 853-858.

Dominguez-Solis, J.R., Gutierrez-Alcala, G., Romero, L.C., Gotor, C. (2001). Cytosolic O-acetylserine(thiol) lyase gene is regulated by heavy metals and can function in metal tolerance. J. Biol. Chem. **276**, in press (manuscript M009574200).

Ebbs, S.D., Lau, I., Ahner, B.A., Kochian, L.V. (2000). Phytochelatins synthesis is not responsible for Cd tolerance in the Zn/Cd hyperaccumulator *Thlaspi caerulescens*. Plant Biology 2000 (The American Society of Plant Physiologists), Abstract no. 21004.

Eide, D., Broderius, M., Fett, J., Guerinot, M.L. (1996). A novel iron-regulated metal transporter from plants identified by functional expression in yeast. Proc. Natl. Acad. Sci. USA **93**, 5624-5628.

Garcia-Hernandez, M., Murphy, A., Taiz, L. (1998). Metallothioneins 1 and 2 have distinct but overlapping expression patterns in *Arabidopsis*. Plant Physiol. **118**, 387-397.

Gekeler, W., Grill, E., Winnacker, E.-L., Zenk, M.H. (1989). Survey of the plant kingdom for the ability to bind heavy metals through phytochelatins. Z. Naturforsch. **44c**, 361-369.

Goto, F., Yoshihara, T., Shigemoto, N., Toki, S., Takaiwa, F. (1999). Nat. Biotechnol. **17**, 282-286.

Gries, G.E., Wagner, G.J. (1998). Association of nickel versus transport of cadmium and calcium in tonoplast vesicles of oat roots. Planta **204**, 390-396.

Grill, E., Gekeler, W., Winnacker, E.-L., Zenk, M.H. (1986). Homo-phytochelatins are heavy metal-binding peptides of homo-glutathione containing Fabales. FEBS Lett. **205**, 47-50.

Grill, E., Loeffler, S., Winnacker, E.-L., Zenk, M.H. (1989). Phytochelatins, the heavy-metal-binding peptides of plants, are synthesized from glutathione by a specific γ-glutamylcysteine dipeptidyl transpeptidase (phytochelatin synthase). Proc. Natl. Acad. Sci. USA **86**, 6838-6842.

Grill, E., Thumann, J., Winnacker, E.-L., Zenk, M.H. (1988). Induction of heavy metal binding phytochelatins by inoculation of cell cultures in standard media. Plant cell Rep. **7**, 375-378.

Grill, E., Winnacker, E.-L., Zenk, M.H. (1985). Phytochelatins: the principal heavy-metal complexing peptides of higher plants. Science **230**, 674-676.

Grill, E., Winnacker, E.-L., Zenk, M.H. (1987). Phytochelatins, a class of heavy-metal-binding peptides from plants are functionally analogous to metallothioneins. Proc. Natl. Acad. Sci. USA **84**, 439-443.

Guerinot, M.L., Salt, D.E. (2001). Fortified foods and phytoremediation. Two sides of the same coin. Plant Physiol. **125**, 164-167.

Ha, S.B., Smith, A.P., Howden, R., Dietrich, W.M., Bugg, S., O'Connell, M.J., Goldsbrough, P.B., Cobbett, C.S. (1999). Phytochelatin synthase from Arabidopsis and the yeast *Schizosaccharomyces pombe*. Plant Cell **11**, 1153-1163.

Harmens, H., den Hartog, P.R., Ten Bokum, W.M., Verkleij, J.A.C. (1993). Increased zinc tolerance in *Silene vulgaris* (Moench) Garcke is not due to increased production of phytochelatins. Plant Physiol. **103**, 1305-1309.

Harris W.R., Madisen L.J. (1988) Equilibrium studies on the binding of cadmium(II) human transferrin. Biochem J 27:284-289

Harrison, P.M., Arosio, P. (1996). The ferritins: molecular properties, iron storage function and cellular regulation. Biochem. Biophys. Acta **1275**, 161-203.

Hayashi, Y., Nakagawa, C.W., Mutoh, N., Isobe, M., Goto, T. (1991). Two pathways in the biosynthesis of cadystins (γEC)$_n$G in the cell-free system of the fission yeast. Biochem. Cell Biol. **69**, 115-121.

Heiss, S., Schäfer, H.J., Haag-Kerwer, A., Rausch, T. (1999). Cloning sulfur assimilation genes of *Brassica juncea* L.: cadmium differentially affects the expression of a putative low-affinity sulfate transporter and isoforms of ATP sulfurylase and APS reductase. Plant Mol. Biol. **39**, 847-857.

Howden, R., Andersen, C.S., Goldsbrough, P.B., Cobbett, C.S. (1995a). A cadmium-sensitive, glutathione-deficient mutant of *Arabidopsis thaliana*. Plant Physiol. **107**, 1067-1073.

Howden, R., Cobbett, C.S. (1992). Cadmium-sensitive mutants of *Arabidopsis thaliana*. Plant Physiol. **99**, 100-107.

Howden, R., Goldsbrough, P.B., Andersen, C.S., Cobbett, C.S. (1995b). Cadmium-sensitive, *cad1* mutants of *Arabidopsis thaliana* are phytochelatin deficient. Plant Physiol. **107**, 1059-1066.

Huang, B., Hatch, E., Goldsbrough, P.B. (1987). Selection and characterization of cadmium tolerant cells in tomato. Plant Sci. **52**, 211-221.

Hunter, T.C., Mehra, R.K. (1998). A role for *HEM2* in cadmium tolerance. J. Inorg. Biochem. **69**, 293-303.

Imai, K., Obata, H., Shimizu, K., Komiya, T. (1996). Conversion of glutathione into cadystins and their analogs catalyzed by carboxypeptidase Y. Biosci. Biotech. Biochem. **60**, 1193-1194.

Inouhe, M., Ito, R., Ito, S., Sasada, N., Tohoyama, H., Joho, M. (2000). Azuki bean cells are hypersensistive to cadmium and do not synthesisze phytochelatins. Plant Physiol. **123**, 1029-1036.

Inouhe, M., Sumiyoshi, M., Tohoyama, H., Joho, M. (1996). Resistance to cadmium ions and formation of a cadmium-binding complex in various wild-type yeasts. Plant Cell Physiol. **37**, 341-346.

Inzé, D., van Montagu, M. (1995). Oxidative stress in plants. Curr. Opin. Biotechnol. **6**, 153-158
Jacob, C., Maret, W., Vallee, B.L. (1998). Control of zinc transfer between thionein, metallothionein, and zinc proteins. Proc. Natl. Acad. Sci. USA. **95**, 3489-3494.
Jemal, F., Didierjean, L., Ghrir, R., Ghorbal, M.H., Burkard, G. (1998). Characterization of cadmium binding peptides from pepper (*Capsicum annuum*). Plant Sci. **137**, 143-154.
Johanning, J., Strasdeit, H. (1998). A coordination-chemical basis for the biological function of the phytochelatins. Angew. Chem. Int. Ed. **37**, 2464-2466.
Kägi, J.H.R. (1991). Overview of metallothionein. Methods Enzymol. **205**, 613-626.
Kille, P., Winge, D.R., Harwood, J.L., Kay, J. (1991). A Plant metallothionein produced in *E. coli*. FEBS Lett. **295**, 171-175.
Klapheck, S., Fliegner, W., Zimmer, I. (1994). Hydroxymethyl-phytochelatins are metal-induced peptides of the Poaceae. Plant Physiol. **104**, 1325-1332.
Klapheck, S., Schlunz, S., Bergmann, L. (1995). Synthesis of phytochelatins and homo-phytochelatins in *Pisum sativum* L. Plant Physiol. **107**, 515-521.
Kneer, R., Zenk, M.H. (1992). Phytochelatins protect plant enzymes from heavy metal poisoning. Phytochemistry, **31**, 2663-2667.
Knauer, K.B.A., Ahner, H., Sigg, L. (1998). Phytochelatin and metal content in phytoplankton from freshwater lakes with different metal concentrations. Environ. Toxicol. Chem. **17**, 2444-2452.
Kotrba, P., Macek, T., Ruml, T. (1999). Heavy metal-binding peptides and proteins in plants. A review. Collection. Czec. Chem. Commun. **64**, 1057-1086.
Kneer, R., Zenk, M.H. (1992). Phytochelatins protect plant enzymes from heavy metal poisoning. Phytochemistry, **31**, 2663-2667.
Kneer, R., Zenk, M.H. (1997). The formation of Cd-phytochelatins complexes in plant cell cultures. Phytochemistry **44**, 69-74.
Kramer, U., Cotter-Howells, J.D., Baker, A.J.M., Smith, J.A.C. (1996). Free histidine as a metal chelator in plants that accumulate nickel. Nature **379**, 635-638.
Krotz, R.M., Evangelou, B.P., Wagner, G.J. (1989). Relationships between cadmium, zinc, Cd-binding peptide, and organic acid in tobacco suspension cells. Plant Physiol. **91**, 780-787.
Kubota, H, Sato, K., Yamada, T., Maitani, T. (1995). Phytochelatins (class III metallothionein) and their desglycyl peptides induced by cadmium in normal root cultures of *Rubia tinctorum* L. Plant Sci. **106**, 157-166.
Kubota, H, Sato, K., Yamada, T., Maitani, T. (2000). Phytochelatin homologs induced in hairy roots of horseradish. Phytochem. **53**, 239-245.
Lane, B., Kajoika, R., Kennedy, R. (1987). The wheat-germ E_c protein is a zinc-containing metallothionein. Biochem. Cell Biol. **65**, 1001-1005.
Lasat, M.M., Pence, N.S., Garvin, D.F., Ebbs, S.D., Kochian, L.V. (2000). Molecular physiology of zinc transport in the Zn hyperaccumulator *Thlaspi caerulescens*. J. Exp. Bot. **51**, 71-79.
Lee, S., Leustek, T. (1999). The affect of cadmium on sulfate assimilation enzymes in *Brassica juncea*. Plant Sci. **141**, 201-207.
Leopold, I., Günther, D. (1997). Investigation of the binding properties of heavy-metal-peptide complexes in plant cell cultures using HPLC-ICP-MS. Fresenius J. Anal. Chem. **359**, 364-370.
Leopold, I., Günther, D., Schmidt, J., Neumann, D. (1999). Phytochelatins and heavy metal tolerance. Phytochem. **50**, 1323-1328.
Leshem, Y.Y., Kuiper, P.J.C. (1996). Is there a GAS (general adaptation syndrome) response to various types of environmental stress? Biol. Plant. **38**, 1-18.
Leustek, T., Martin, M.L., Bick, J.A., Davies, J.P. (2000). Pathways and regulation of sulfur metabolism revealed through molecular and genetic studies. Annu. Rev. Plant Physiol. Mol. Biol. **51**, 141-165.
Li, Z.S., Lu, Y.P., Zhen, R.G., Szczypka, M., Thiele, D.J., Rea, P.A. (1997). A new pathway for vacuolar sequestration in *Saccharomyces cerevisiae*: YCF1-catalyzed transport of bis(glutathionato) cadmium. Proc. Natl. Acad. Sci. USA **94**, 42-47.
Lichtenberger, O., Neumann, D. (1997). Analytical electron microscopy as a powerful tool in plant cell biology: examples using electron energy loss spectroscopy and X-ray microanalysis. Eur. J., Cell Biol. **73**, 378-386.
Lin, S.L., Wu, L. (1994). Effects of copper concentration on mineral nutrient uptake and copper accumulation in protein of copper-tolerant and nontolerant *Lotus purshianus* L. Ecotoxicol. Environ. Saf. **29**, 214-228.

Loeffler, S., Hochberger, A., Grill, E., Gekeler, W., Winnacker, E.-L., Zenk, M.H. (1989). Termination of the phytochelatin synthase reaction through sequestration of heavy metals by the reaction product. FEBS Lett. **258**, 42-46.

Ma, L.Q., Komar, K.M., Tu, C., Zhang, W., Cai, Y., Kennelley, E.D. (2001). A fern that hyperaccumulates arsenic. Nature **409**, 579.

Maitani, T., Kubota, H., Sato, K., Yamada, I. (1996). The composition of metals bound to class III metallothionein (phytochelatin and its desglycyl peptide) induced by various metals in root cultures of *Rubia tinctorum*. Plant Physiol. **110**, 1145-1150.

Margoshes, M., Vallee, B.L. (1957). A cadmium protein from equine kidney cortex. J. Am. Chem. Soc. **79**, 4813-4814.

Mazen, A.M.A., El Maghraby, O.M.O. (1997/98). Accumulation of cadmium, lead and strontium, and a role of calcium oxalate in water hyacinth tolerance. Biol. Plant. **40**, 411-417.

Mehra, R.K., Kodati, V.R., Abdullah, R. (1995). Chain length-dependent Pb(II)-coordination in phytochelatins. Biochem. Biophys. Res: Commun. **215**, 730-736.

Mehra, R.K., Miclat, J., Kodati, V.R., Abdullah, R., Hunter, T.C., Mulchandani, P. (1996a). Optical spectroscopic and reverse-phase HPLC analyses of Hg(II) binding to phytochelatins. Biochem. J. **314**, 73-82.

Mehra, R.K., Tarbet, E.B., Gray, W.R., Winge, D.R. (1988). Metal-specific synthesis of two metallothioneins and gamma-glutamyl peptides in *Candida glabrata*. Proc. Natl. Acad. Sci. USA **85**, 8815-8819.

Mehra, R.K., Tran, K., Scott, G.W., Mulchandani, P., Saini, S.S. (1996b). J. Ag(I)-binding to phytochelatins. J. Inorg. Biochem. **61**, 125-142.

Mehra, R.K., Mulchandani, P. (1995). Glutathione-mediated transfer of Cu(I) into phytochelatins. Biochem. J. **307**, 697-705.

Meuwly, P., Thibault, P., Rauser, W.E. (1993). γ-glutamylcysteinyl-glutamic acid - a new homologue of glutathione in maize seedlings exposed to cadmium. FEBS Lett. **336**, 472-476.

Meuwly, P., Thibault, P., Schwan, A.L., Rauser, W.E. (1995). Three families of thiol peptides are induced by cadmium in maize. Plant J. **7**, 391-400.

Mewes, H.W., Albermann, K., Bahr, M., Frishman, D., Gleissner, A., Hani, J., Heumann, K., Kleine, K., Maierl, A., Oliver, S.G., Pfeiffer, F., Zollner, A. (1997). Overview of the yeast genome. Nature **387** (6632 Suppl), 7-65.

Mori, A., Leita, L. (1998). Application of capillary electrophoresis on the characterization of cadmium-induced polypeptides in roots of pea plants. J. Plant Nutr. **21**, 2335-2341.

Murasugi, A., Wada, C., Hayashi, Y. (1981). Cadmium-binding peptide induced in fission yeast, *Schizosaccharomyces pombe*. J. Biochem. **90**, 1561-1564.

Murguia, J.R., Belles, J.M., Serrano, R. (1996). The yeast HAL2 nucleotidase is an *in vivo* target of salt toxicity. J. Biol. Chem. **271**, 29029-29033.

Murphy, A., Taiz, L. (1995). Comparison of metallothionein gene expression and nonprotein thiols in ten *Arabidopsis* ecotypes. Correlation with copper tolerance. Plant Physiol. **109**, 945-954.

Murphy, A., Zhou, J., Goldsbrough, P.B., Taiz, L. (1997). Purification and immunological identification of metallothioneins 1 and 2 from *Arabidopsis thaliana*. Plant Physiol. **113**, 1293-1301.

Nagalakshmi, N., Prasad, M.N.V. (2001). Responses of glutathione cycle enzymes and glutathione metabolism to copper stress in *Scendesmus bijugatus*. Plant Sci. **160**, 291-299.

Nakazawa, R., Ikawa, M., Yasuda, K., Takenaga, H. (2000). Synergistic inhibition of the growth of suspension cultured tobacco cells by simultaneous treatment with cadmium and arsenic in relation to phytochelatin synthesis. Soil Sci. Plant Nutrition **46**, 271-275.

Nelson, K.B., Winge, D.R. (1983). Order of metal binding in metallothionein. J. Biol. Chem. **258**, 13063-13069.

Noctor, G., Arisi, A.-C. M., Jouanin, L., Kunert, K.-J., Rennenberg, H., Foyer, C.H. (1998). Glutathione: biosynthesis, metabolism and relationship to stress tolerance explored in transformed plants. J. Exp. Bot. **49**, 623-647.

Nussbaum, S., Schmutz, D., Brunold, C. (1988). Regulation of assimilatory sulfate reduction by cadmium in *Zea mays* L. Plant Physiol. **88**, 1407-1410.

Ortiz, D.F., Kreppel, L., Speiser, D.M., Scheel, G., McDonald, G., Ow, D.W. (1992). Heavy metal tolerance in the fission yeast requires an ATP-binding cassette-type vacuolar membrane transporter. EMBO J. **11**, 3491-3499.

Ortiz, D.F., Ruscitti, T., McCue, K., Ow, D.W. (1995). Transport of metal-binding peptides by HMT1, a fission yeast ABC-type vacuolar membrane protein. J. Biol. Chem. **270**, 4721-4728.

Ow, D.W. (1996). Heavy metal tolerance genes: prospective tools for bioremediations. Res. Conserv. Recycl. **18**, 135-149.
Panda, K. K., Patra, J., Panda, B. B. (1997). Persistence of cadmium-induced adaptive response against genotoxicity of maleic hydrazide and methyl mercuric chloride in root meristem cells of *Allium cepa* L. Differential inhibition by cycloheximide and buthionine sulfoximine. Mutat. Res. **389**, 129-139.
Pelosi, P., Galoppini, C. (1973). Sulla natura dei composti mercurio-organici nelle foglie di tabacco. Atti Soc. Tosc. Sci. Nat. Mem. **80**, 215-220.
Pence, N.S., Larsen, P.B., Ebbs, S.D., Letham, D.L., Lasat, M.M., Garvin, D.F., Eide, D., Kochian, L.V., (2000). The molecular physiology of heavy metal transport in the Zn/Cd hyperaccumulator *Thlaspi caerulescens*. Proc. Natl. Acad. Sci. USA **97**, 4956-4960.
Pich, A., Hillmer, S., Manteuffel, R., Scholz, G. (1997). First immunohistochemical localization of the endogenous Fe^{2+}-chelator nicotianamine. J. Exp. Bot. **48**, 759-767.
Pich, A., Scholz, G. (1996). Translocation of copper and other micronutrients in tomato plants (*Lycopersicon esculentum* Mill.): nicotianamine-stimulated copper transport in the xylem. J. Exp. Bot. **47**, 41-47.
Pickering, I.J., Prince, R.C., George, G.N., Rauser, W.E., Wickramasinge, W.A., Watson, A.A., Dameron, C.T., Dance, I.G., Firlie, D.P., Salt, D.E. (1999). X-ray absorption spectroscopy of cadmium phytochelatin and model systems. Biochim. Biophys. Acta **1429**, 351-364.
Pickering, I.J., Prince, R.C., George, M.J., Smith, R.D., George, G.M., Salt, D.E. (2000). Reduction and coordination of arsenic in Indian mustard. Plant Physiol. **122**, 1171-1177.
Pilon-Smits, E.A.H., Hwang, S., Lytle, C.M., Zhu, Y.L., Tay, J.C., Bravo, R.C., Chen, Y., Leustek, T., Terry, N. (1999a). Overexpression of ATP sulfurylase in Indian mustard leads to increased selenate uptake, reduction, and tolerance. Plant Physiol. **119**, 123-132.
Pilon-Smits, E.A.H., Zhu, Y.L., Pilon, M., Terry, N. (1999b). Overexpression of glutathione synthesizing enzymes enhances cadmium accumulation in *Brassica juncea*. Proceedings of the 5th International Conference on the Biogeochemistry of Trace Elements, Vienna, July 11-15, pp 890-891.
Polette, L.A., Gardea-Torresdey, J.L., Chianelli, R.R., George, G.N., Pickering, I.J., Arenas, J. (2000). XAS and microscopy studies of the uptake and bio-transformation of copper in *Larrea tridentata* (creosote bush). Microchem. J. **65**, 227-236.
Price DJ, Joshi JG (1982) Ferritin: a zinc detoxicant and a zinc ion donor. Proc Natl Acad Sci (USA) 79:3116-3119
Price DJ, Joshi JG (1983) Ferritin: Binding of beryllium and other divalent metal ions. J Biol Chem 258: 10873-10880
Prasad, M.N.V. (1997). Trace metals. In: M.N.V.Prasad (Ed.), Plant ecophysiology. John Wiley and Sons. Inc., New York, pp. 207-249.
Prasad, M.N.V. (1999). Metallothioneins and metal binding complexes in plants. In: M.N.V. Prasad, J. Hagemeyer (Eds.), Heavy metal stress in plants: from molecules to ecosystems. Springer-Verlag, Berlin Heidelberg, pp. 51-72.
Prasad, M.N.V. (ed) (2001) *Metals in the Environment - Analysis by biodiversity*. Marcel Dekker Inc. New York. pp. 504
Rama Kumar, T and M.N.V. Prasad (2001) Possible involvement of calcium/ calmodulin dependent kinase(s) in phosphorylation of ferritin in *Vigna mungo* (L.) Hepper (Black Gram). *J Plant Biology* 28:1-5
Rama Kumar, T and M.N.V. Prasad (2000) Partial purification and characterization of Ferritin from *Vigna mungo* (Black gram) seeds. *J. Plant Biology* 27: 241-246
Rauser, W.E. (1990). Phytochelatins. Annu. Rev. Biochem. **59**, 61-86.
Rauser, W.E. (1999). Structure and function of metal chelators produced by plants. Cell Biochem. Biophys. **31**, 19-48.
Rauser, W.E., Meuwly, P. (1995). Retention of cadmium in roots of maize seedlings. Plant Physiol. **109**, 195-202.
Rea, P.A., Li, Z.S., Lu, Y.P., Drozdowicz, Y., Martinoia, E. (1998). From vascular GS-X pumps to multispecific ABC transporters. Annu. Rev. Plant Physiol. Mol. Biol. **49**, 727-760.
Reese, R.N., Wagner, G.J. (1987). Effects of buthionine sulfoximine on Cd-binding peptide levels in suspension-cultured tobacco cells treated with Cd, Zn, or Cu. Plant Physiol. **84**, 574-577.
Robinson, N.J., Ratliff, R.L., Anderson, P.J., Delhaize, E., Berger, J.M., Jackson, P.J. (1988). Biosynthesis of poly(γ-glutamylcysteinyl)glycines in cadmium-tolerant *Datura innoxia* (Mill.) cells. Plant Sci. **56**, 197-204.

Robinson, N.J., Tommey, A.M., Kuske, C., Jackson, P.J. (1993). Plant metallothioneins. Biochem. J. **295**, 1-10.
Rüegsegger, A., Brunold, C. (1992). Effects of cadmium on γ-glumilcysteinyl-glycine synthesis in maize seedlings. Plant Physiol. **99**, 428-433.
Rüegsegger, A., Schmutz, D., Brunold, C. (1990). Regulation of glutathione synthesis by cadmium in *Pisum sativum* L. Plant Physiol. **93**, 1579-1584.
Salt, D.E., Pickering, I.J., Prince, R.C., Gleba, D., Dushenkov, S., Smith, R.D., Raskin, I. (1997). Metal accumulation by aquacultured deedlings of Indian mustard. Environ. Sci. Technol. **31**, 1636-1644.
Salt, D.E., Prince, R.C., Pickering, I.J., Raskin, I. (1995). Mechanisms of cadmium mobility and accumulation in indian mustard. Plant Physiol. **109**, 1427-1433.
Salt, D.E., Rauser, W.E. (1995). MgATP-dependent transport of phytochelatins across the tonoplast of oat roots. Plant Physiol. **107**, 1293-1301.
Salt, D.E., Wagner, G.J. (1993). Cadmium transport across tonoplast of vescicles from oat roots. J. Biol. Chem. **268**, 12297-12302.
Sanità di Toppi, L., Favali, M.A., Gabbrielli, R., Gremigni, P. (2001). Brassicaceae. In: M.N.V. Prasad (Ed.), Metals in the environment: analysis by biodiversity. Marcel Dekker, New York, pp. 219-258.
Sanità di Toppi, L., Gabbrielli, R. (1999). Response to cadmium in higher plants. Environ. Exp. Bot. **41**, 105-130.
Sanità di Toppi, L., Lambardi, M., Pazzagli, L., Cappugi, G., Durante, M., Gabbrielli, R. (1998). Response to cadmium in carrot *in vitro* plants and cell suspension cultures. Plant Sci. **137**, 119-129.
Sanità di Toppi, L., Lambardi, M., Pecchioni, N., Pazzagli, L., Durante, M., Gabbrielli, R. (1999). Effects of cadmium stress on hairy roots of *Daucus carota*. J. Plant Physiol. **154**, 385-391.
Schäfer, H.J., Haag-Kerwer, A., Rausch, T. (1998). cDNA cloning and expression analysis of genes encoding GSH synthesis in roots of the heavy-metal accumulator *Brassica juncea* L.: evidence for Cd-induction of a putative mitochondrial γ-glutamylcysteine synthetase isoform. Plant Mol. Biol. **37**, 87-97.
Schat, H., Kalff, M.M. A. (1992). Are phytochelatins involved in differential metal tolerance or do they merely reflect metal-imposed strain? Plant Physiol. **99**, 1475-1480.
Scheller, H.V., Huang, B., Hatch, E., Goldsbrough, P.B. (1987). Phytochelatin synthesis and glutathione levels n response to heavy metals in tomato cells. Plant Physiol. **85**, 1031-1035.
Schmöger, M.E.V., Oven, M., Grill, E. (2000). Detoxification of arsenic by phytochelatins in plants. Plant Physiol. **122**, 793-801.
Sczekan SR, Joshi JG (1989) Metal binding properties of phytoferritin and synthetic iron cores. Biochim Biophys Acta 990:8-14
Selye, H. (1936). A syndrome produced by various noxious agents. Nature **138**, 32-34.
Shojima, S., Nishizawa, N.K., Fushiya, S., Nozoe, S., Irifune, T., Mori, S. (1990). Biosynthesis of phytosiderophores. In vitro biosynthesis of 2'-deoxymugineic acid from L-methionine and nicotianamine. Plant Physiol. **93**, 1497-1503.
Speiser, D.M., Abrahamson, S.L., Banuelos, G., Ow, D.W. (1992a). *Brassica juncea* produces a phytochelatin-cadmium-sulfide complex. Plant Physiol. **99**, 817-821.
Speiser, D.M., Ortiz, D.F., Kreppel, L., Ow, D.W. (1992b). Purine biosynthetic genes are required for cadmium tolerance in *Schizosaccharomyces pombe*. Mol. Cell. Biol. **12**, 5301-5310.
Steffens, J.C. (1990). The heavy-metal binding peptides of plants. Annu. Rev. Plant Physiol. Plant Mol. Biol. **41**, 553-575.
Steffens, J.C., Hunt, D.F., Williams, B.G. (1986). Accumulation of non-protein metal-binding polypeptides (γ-glutamyl-cysteinyl $_n$-glycine in selected cadmium-resistant tomato cells. J. Biol. Chem. **261**, 13879-13882.
Stephan, U.W., Schmidke, I., Stephan, V.W., Scholz, G. (1996). The nicotianamine molecule is made-to-measure for complexation of metal micronutrients in plants. Biometals **9**, 84-90.
Stephan, U.W., Scholz, G (1993). Nicotianamine: mediator of transport of iron and heavy metals in the phloem? Physiol. Plant. **88**, 522-529.
Strasdeit, H., Duhme, A.-K., Kneer, R., Zenk, M.H., Hermes, C., Nolting, H.-F. (1991). Evidence for discrete $Cd(SCys)_4$ units in cadmium phytochelatin complexes from EXAFS spectroscopy. J. Chem. Soc. Chem. Commun. **16**, 1129-1130.
Subhadra, A. V., Panda, B. B. (1994). Metal-induced genotoxic adaptation in barley (*Hordeum vulgare* L.) to maleic hydrazide and methyl mercuric chloride. Mutat. Res. **321**, 93-102.
Tabata, K., Kashiwagi, S., Mori, H., Ueguchi, C., Mizuno, T. (1997). Cloning of a cDNA encoding a putative metal-transporting P-type ATPase from *Arabidopsis thaliana*. Biochim. Biophys. Acta **1326**, 1-6.

Terry N., Zayed, A.M. (1994). Selenium volatilisation by plants. In: W.T. Frankenberger, S. Benson (Eds.), Selenium in the Environment. Marcel Dekker, New York.
Tan AT, Woodworth RC (1969) Ultraviolet difference spectral studies of conalbumin complexes with transition metal ions. Biochemistry 8:3711-3716
Thumann, J., Grill, E., Winnacker, E.-L., Zenk, M.H. (1991). Reactivation of metal-requiring apoenzymes by phytochelatin-metal complexes. FEBS Lett. **284**, 66-69.
Valentine, J.S., Gralla, E.B. (1997). Delivering copper inside yeast and human cells. Science **278**, 817-818.
Vande Weghe, J.G., Ow, D.W. (1997). A novel mitochondrial oxidoreductase required for phytochelatin accumulation and cadmium tolerance in fission yeast. J. Exp. Bot. **48** (suppl.), 96.
Vande Weghe, J.G., Ow, D.W. (1999). A fission yeast gene for mitochondrial sulfide oxidation. J. Biol. Chem. **274**, 13250-13257.
Vatamaniuk, O.K., Mari, S., Lu, Y.P., Rea, P.A. (1999). AtPCS1, a phytochelatin synthase from *Arabidopsis*: isolation and in vitro reconstitution. Proc. Natl. Acad. Sci. USA **96**, 7110-7115.
Vatamaniuk, O.K., Mari, S., Lu, Y. P., Rea, P.A. (2000). Mechanism of heavy metal activation of phytochelatin (PC) synthase. J. Biol. Chem. **275**, 31451-31459.
Vatamaniuk, O.K., Bucher, E.A., Ward, J.T., Rea, P.A. (2001). A new pathway for heavy metal detoxification in animals. J. Biol. Chem. **276**, 20817-20820.
Verkleij, J.A.C., Koevoets, P., van't Riet, J., Bark, R., Mijdam, Y., Ernst, W.H.O. (1990). Poly(γ-glutamylcysteinyl)-glycines or phytochelatins and their role in cadmium tolerance of *Silene vulgaris*. Plant Cell Environ. **13**, 913-921.
Vögeli-Lange, R., Wagner, G.J. (1990). Subcellular localization of cadmium and cadmium-binding peptides in tobacco leaves. Plant Physiol. **92**, 1086-1093.
Vögeli-Lange, R., Wagner, G.J. (1996). Relationship between cadmium, glutathione and cadmium-binding peptides (phytochelatins) in leaves of intact tobacco seedlings. Plant Sci. **114**, 11-18.
Wagner, G.J. (1993). Accumulation of cadmium in crop plants and its consequences to human health. Adv. Agron. **51**, 173-212.
Wagner, G.J. (1995). Assessing cadmium partitioning in transgenic plants. In: Methods in molecular biology, vol. 44: *Agrobacterium* protocols, K.M.A. Gartland, M.R. Davey (Eds.). Humana Press Inc., Totowa, pp. 295-308.
Xiang, C., Oliver, D.J. (1998). Glutathione metabolic genes co-ordinately respond to heavy metals and jasmonic acid in *Arabidopsis*. Plant Cell **10**, 1539-1550.
Yan, S.L., Tsay, C.C., Chen, Y.R. (2000). Isolation and characterization of phytochelatin synthase in rice seedlings. Proc. Natl. Sci. Counc. Repub. China **24**, 202-207.
Yen, T.-Y., Villa, J.A., DeWitt, J.G. (1999). Analysis of phytochelatin-cadmium complexes from plant tissue culture using nano-electrospray ionization tandem mass spectrometry and capillary liquid chromatography/electrophoresis ionization tandem mass spectrometry. J. Mass Spectrom. **34**, 930-941.
Yu, W., Santhanagopalan, V., Sewell, A.K., Jensen, L.T., Winge, D.R. (1994). Dominance of metallothionein in metal ion buffering in yeast capable of synthesis of (γEC)$_n$G isopeptides. J. Biol. Chem. **269**, 21010-21015.
Zayed, A.M., Terry, N. (1992). Selenium volatilisation in broccoli as influenced by sulphate supply. J. Plant Physiol. **140**, 646-652.
Zenk, M.H. (1996). Heavy metal detoxification in higher plants - a review. Gene **179**, 21-30.
Zhu, Y.L., Pilon-Smits, E.A.H., Jouanin, L., Terry, N. (1999a). Overexpression of glutathione synthetase in Indian mustard enhances cadmium accumulation and tolerance. Plant Physiol. **119**, 73-79.
Zhu, Y.L., Pilon-Smits, E.A.H., Tarun, A.S., Weber, S.U., Jouanin, L, Terry, N. (1999b). Cadmium tolerance and accumulation in Indian mustard is enhanced by overexpressing γ-glutamylcysteine synthase. Plant Physiol. **121**, 1169-1177.

CHAPTER 4

METABOLISM OF ORGANIC ACIDS AND METAL TOLERANCE IN PLANTS EXPOSED TO ALUMINUM

H. MATSUMOTO
Research Institute for Bioresources, Okayama University, Chuo, Kurashiki, Okayama, 710-0046, Japan

1. INTRODUCTION

Aluminum (Al) is one of the most potent toxic metals for plant growth in acidic soils. Root elongation is inhibited even at a µmol level of Al^{3+} and the inhibition can be observed within hours.

Thus many researches have been conducted to unaderstand the mechanism of Al tolerance for the improvement of crop production in acid soil. Al-sensitive plants absorb more Al than Al-tolerant plants, and thus the Al exclusion mechanism is the major idea for Al tolerance (Kochian, 1995; Matsumoto, 2000). The most important exclusion mechanism is based on the chelation of toxic Al with organic acids resulting in the detoxification of Al intracellularly and extracellularly. Generally speaking, Al-tolerant plants excrete more organic acids than the Al-sensitive plants when they are exposed to Al. The major organic acids are citric, malic and oxalic acids, depending on the plant species. Excretion of organic acids can be classified into two patterns according to the time required for this process to occur. The excretion of organic acids may be carried out through the anion channel.

However, an important problem remaining to be solved is the understanding of the regulatory mechanism of the synthesis and excretion of organic acids upon the detection of the Al signal.

2. ALUMINUM TOXICITY IN THE ACID SOIL

Acidity is a major degradative factor of soils and covers an extensive area of both tropical and temperate regions. It is estimated that approximately 30 - 40% of arable land in the world is acid soils. The factors involved in producing acid soil are the prolonged leaching by rainwater, soil-forming processes and climatic conditions. Furthermore, the acidity of the soil is gradually increasing as a result of agricultural farming processes and management practices or environmental problems including acid rains.

Al is one of the most abundant minerals in the soil, comprising approximately 7%. The chemistry of Al in water and soil is very complex as it exists in various forms depending on the pH. At neutral or weakly acidic pH, Al exists in the form of insoluble aluminosilicate or oxide which are non-toxic for plants. However, as the soil becomes more acidic, Al is solubilized into a phytotoxic form. Soluble Al can be classified into several groups, free or mononuclear forms of Al^{3+}, polynuclear Al, and Al as a low-molecular-weight complex (Kochian, 1995). As the pH increases, $Al(H_2O)_6^{3+}$ changes to $Al(OH)^{2+}$ and $Al(OH)_2^{+}$. At a neutral pH, insoluble $Al(OH)_3$ (gibbsite) is formed and aluminate anion $Al(OH)_4^-$ dominates at an alkaline pH.

Furthermore, the toxicity of Al differs markedly with the chemical form, and the study of Al-related processes is complicated by the complex chemistry of Al. $Al(H_2O)_6^{3+}$, which by convention is usually called Al^{3+}, is dominant in acid soil below pH 4.5 and is believed to be the most toxic form.

3. ALUMINUM TOXICITY IN PLANTS

The toxic effects of Al include stunted root growth, poor root hair development, swollen root apices, stubby and brittle roots, resulting in the decrease in absorption and translocation of water and nutrient elements.

Inhibition of root elongation is the first visible symptom of Al stress (Matsumoto, 2000, 2002; Matsumoto et al., 2001). In most plant species, root elongation is markedly inhibited by Al at the μmol level in a simple solution containing Ca^{2+} (Matsumoto, 2000). The Al-induced inhibition of root elongation of Al-sensitive maize occurred within 30 min (Llugany et al., 1995). It is also known that inhibition of root elongation by Al is different among plant species and cultivars.

The root apex (root cap, meristem and elongation zone) accumulates more Al and plays a major role in the Al-perception mechanism. Indeed, only the apical 2 - 3 mm of maize and pea roots need to be exposed to Al for the growth to be inhibited (Delhaize and Ryan, 1995; Matsumoto et al., 1996). In near-isogenic wheat (*Triticum aestivum* L.) lines differing in Al tolerance, root apices of Al-sensitive genotypes were stained with hematoxylin for the detection of Al after a short exposure to Al (10 min to 1 hour), whereas apices of Al-tolerant seedlings showed less intense staining (Delhaize et al., 1993). The results indicate that: (a) differential Al accumulation in the growing root tissue is related to differential Al sensitivity, (b) inhibition of root growth is related to the Al content in the root tissue as well as cultured cells, and (c) tolerant cultivars encode a mechanism that exclude Al from root apices (Samuels et al., 1997; Yamamoto et al., 1994; Rincón and Gonzales, 1992).

In the research line directing to (c), the role of organic acids as Al-chelating compounds resulting in the reduced toxicity of Al has been examined in detail in the past decade. Therefore, the information obtained on organic acids participating in the mechanism of Al tolerance is discussed in this chapter.

4. MECHANISM OF ALUMINUM TOLERANCE

Higher plants are usually kept stationary with roots in the soil and can not move away from a stress. Therefore, unlike an animal, plants have specific strategies to avoid the stress in the soil by themselves. Al^{3+} is very toxic to plants through the binding with cell wall, DNA, membrane, etc (Matsumoto, 2000). Therefore, an economic and sustainable approach to improve crop production in acid soils is the selection and breeding of Al-resistant plants. For this approach, it is necessary to gain an understanding of the Al-resistance mechanism.

Two strategies have been proposed, 1) exclusion of Al from the root apex and 2) internal tolerance once Al enters the plant symplasm (Kochain, 1995; Delhaize and Ryan, 1995). The internal tolerance mechanisms may include Al-chelating in the cytosol, compartmentation in the vacuole, Al-binding by protein, and elevated enzyme activity, while the exclusion mechanisms may involve secretion of Al-chelating ligands, binding of Al with the cell wall and mucilage, plant-induced pH barrier in the rhizosphere or root apoplasm, selective permeability of the plasma membrane and Al efflux (Taylor, 1991).

Recent experimental evidence supports the view that the major Al resistance mechanism in some species or cultivars of plants is the secretion of Al-chelating exudates from the root upon Al-stress signals.

5. EXCRETION OF ORGANIC ACIDS AS Al-CHELATING SUBSTANCES

Organic acids are the most powerful natural chelators of Al. Al-tolerant plants excrete organic acids into the rhizosphere and act as Al chelator (Ryan et al., 2001). The first reported chelating compound was the citrate from snapbean under Al stress (Miyasaka et al., 1991). The root of the Al-resistant cultivar excreted 70 times more citrate in the presence of Al than in its absence, and it excreted 10 times more citrate than Al-sensitive cultivars.

So far several organic acids have been found to be excreted from different plant species and cultivars under Al stress. When compared with an equal concentration of each organic acid, the efficiency of Al-tolerance activity is dependent on the chelating ability expressed by stability constant, log Keq of the Al-chelate complex. For example, the molecular ratio of citrate to Al (1 : 1) completely protects the root from the inhibition by Al (log Keq of Al-citrate; 12.26). On the other hand, 5 to 8 times more malate is required for the complete suppression of Al toxicity for wheat (log Keq of Al-malate; 6.0). Figure 1 shows the effect of different molar ratios of Al to citrate or malate on the root elongation of Al-sensitive wheat, Scout 66, grown with 50 µM Al (pH 4.5) (Li, 2000).

6. DIFFERENT PATTERNS OF THE EXCRETION OF ORGANIC ACIDS

The pattern of Al-induced secretion of organic acids differs with the plant. Excretion of malate from wheat, citrate from maize, *Cassia tora*, soybean and rye, and oxalate from buckwheat and taro has been reported. Some plants excrete both malate and citrate under Al stress (Li et al., 2000). It is generally observed that malate and oxalate are excreted instantly upon Al stress but citrate is excreted only after a lag phase (Ma, 2000). These

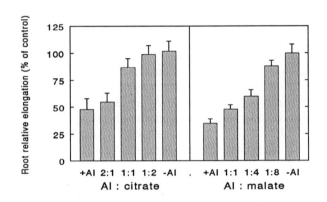

Figure 1. Effect of different molar ratios of Al to citric or malic acid on the root elongation of wheat, Scout 66. The Al concentration was 50µM in 0.1 mM $CaCl_2$ solution (pH 4.5). Seedlings were exposed to different treatment solution for 9 h. Error bars represent ±SD (n=3). Data from Li (2000).

results suggest that stored malate and oxalate are excreted while citrate is synthesized by a gene-regulated system.

Although the types of organic acids differ with the plant species, it has been demonstrated that Al-resistant plant species/cultivars can release larger amounts of organic acid than Al-sensitive ones. Ryan et al. (1995a) screened 36 wheat cultivars differing in Al tolerance and showed that Al-induced malate release correlated with Al resistance.

7. CHARACTERISTICS OF THE EXCRETION OF ORGANIC ACIDS

Malate is excreted from wheat and citrate from soybean specifically in response to Al (Yang et al., 2000). No other cations including In^{3+}, La^{3+}, Yb^{3+}, Ga^{3+} and Cd^{2+} caused excretion of citrate.

The site of excretion of the organic acid has been investigated. The root apex (0 - 3 mm of root) is the primary source of malic acid excretion. Similarly the region 0 - 5 mm from the root tip is the site of excretion of oxalate in buckwheat (Zheng et al., 1998). These facts also suggest that exposure of only the root tip to Al is enough for the excretion of organic acid, suggesting that the plant root has an effective potential for the survival under Al stress, because the site of the root injured by Al is similar to that of excretion of organic acids.

The temperature treatment experiment showed that excretion of malate in Al-tolerant wheat (Atlas 66) root and citrate in rye depended on the temperature. Al treatment at 4°C extremely repressed the excretion of both organic acids (Li et al., 2000; Osawa and Matsumoto, 2001), suggesting that the initiation of Al-induced excretion of organic acids is dependent on temperature.

When we consider the long-term excretion of organic acids, the source of carbon for organic acids must be supplied by either dark CO_2 fixation in the root or flow from the fixed carbon by photosynthesis in the leaf. Citrate excretion from Al-tolerant soybean (*Glycine max* L. cv Suzunari) increased with increasing light-exposure time, and only low citrate efflux was monitored under 24-h continuous dark (Figure 2) (Yang et al., 2001). However, the photosynthetic rate during a 6, 12 or 24 h period of exposure to 50 μM Al produced no differences compared to the controls without Al treatment. The carbons transported from the leaves are metabolized into organic acids which are extruded from the roots. In accordance with this idea, the Al-dependent triphenyl-tetrazolium chloride (TTC) reduction which represents the mitochondrial activity was apparently higher in an Al-tolerant bean cultivar (*Phaseolus vulgaris* L. cv. Rosecocco) than an Al-sensitive line (*Phaseolus vulgaris* L. vig. French bean) (Mugai et al., 2000).

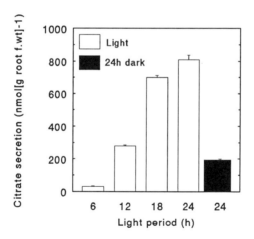

Figure 2. Effect of light-exposure time on the citrate secretion in an Al tolerant soybean cultivar. Seedlings were grown in complete nutrient solution for ten days, pre-treated in 0.5 mM $CaCl_2$ (pH 4.5) solution for 12 h, and then subjected to Al in the same solution containing 50 μM Al (pH 4.5). Root exudates were collected at 6, 12, 18, 24 h light (□) and after 24 h continuous dark (■) and the secreted citric acid was analysed. Values are mean of three replicates ±SE and representative of at least two independent experiments. Data from Yang et al. (2001).

The loss of a certain chromosome arm resulted in a decrease in Al resistance in ditelosomic wheat lines and decreased the rate of root apical malate release. With Al-resistant (alr) mutants of *Arabidopsis thaliana*, the mutants mapped on chromosome 1 released greater amounts of citrate or malate compared with the wild type concomitant with decreased Al exclusion (Larsen et al., 1998). Recently, it was shown that the short arm of chromosome 3R carries genes necessary for Al tolerance in triticale (x *Triticosecale Wittmark* cv. Currency) (Ma et al., 2000)

8. MECHANISM OF EXCRETION OF ORGANIC ACIDS UNDER Al STRESS

8.1 Anion channel

Schroeder (1995) speculated about the function of anion channels in roots in Al tolerance. Activation of the anion channel by Al is a possible mechanism in malate release, because several antagonists of the anion channel, anthracene-9-carboxylic acid (A9C) (Papernik and Kochian, 1997) and niflumic acid inhibited the malate excretion (Ryan et al., 1995b).

The mode of malate efflux mediated by the channel in plasma membrane proposed by Delhaize and Ryan (1995) is as follows : (1) Al interacts directly with the channel protein changing its conformation and increasing its mean open time or conductance, (2) Al interacts with a specific receptor on the membrane surface or with the membrane itself which, through a series of secondary messengers in the cytoplasm, changes channel activity, and (3) Al enters into the cytoplasm and alters channel activity either directly by binding with the channel or indirectly through the signal transduction pathway.

Ryan et al. (1997) reported that, in the whole cell configurate, 20-50 µM $AlCl_3$ depolarized the plasma membrane of protoplasts and activated an inward current that could remain active for more than 60 min. This channel is more selective for anions than for cations. Outside patch recordings revealed a multistate channel with a single-channel conductance between 27 and 66 pS. Papernik and Kochian (1997) reported that exposure to Al induced depolarization of the membrane potential (Em) in an Al-tolerant wheat cultivar but not in an Al-sensitive cultivar. The depolarization was specific to Al. All results suggest the possible involvement of voltage dependent anion channel in the malate efflux through the plasma membrane.

Recently Zhang et al. (2001) found the Al-activated anion channel in wheat root. Using whole-cell patch clamp they found that Al^{3+} stimulated an electrical current carried by anion efflux across the plasma membrane in the Al^{3+}-tolerant (ET8) and Al^{3+} sensitive (ES8) genotypes. Addition of 50 µM $AlCl_3$ activated an inward current that was reversed between the equilibrium potentials of $malate^{2-}$ and Cl^-. The permeability of the channel to malate anions is about 2.6 times greater than to chloride and was inhibited by anion channel antagonists. The Al-activated anion current was compared in ET8 and ES8 genotypes of wheat. These genotypes represent isogenic lines that differ in Al sensitivity at a single gene/locus. The frequency of responses measured in ET8 (39% of cells, n=109) was greater than in ES8 (11%, n=44) and the current density of the inward current was greater in ET8 (-68 ± 8 mA m^{-2}) than in ES8 (-27 ± 6 mA m^{-2}). Furthermore, the time between the addition of Al and the increase of inward current was shorter in ET8 (9 ± 1 min) than in ES8 (36 ± 7 min). They also showed that the addition of

0.5 mM cAMP reversed the inhibition of outward-rectifying potassium channel in wheat root cells with approximately 25 µM Al. This response occurrs in the Al-tolerant ET8 only and suggests that a nucleotide-gated potassium channel might be involved in the potassium efflux that occurs concurrently with malate efflux.

Piñeros and Kochian (2001) also found the existence of Al-dependent anion channel with excised membrane patches from Al-tolerant maize (*Zea mays*) hybrid South American 3. The channels were highly selective for anions over cations. Activation of this channel with extracellular Al^{3+} in membrane patches excised prior to any Al^{3+} exposure indicates that the machinery required for Al^{3+} activation of this channel, and consequently the whole root Al^{3+} response, is located in the root-cell plasma membrane. This Al-activated anion channel may also be permeable to organic acids, thus mediating the Al tolerance response (i.e. Al-induced organic acid exudation) observed in intact maize root apices.

Citrate and oxalic acid are also important for the detoxification of Al. However, the mechanism of their release through the plasma membrane are scarcely known except that anion channel blockers inhibit the excretion of citrate and oxalate. Phenylglyoxal (PG) inhibited the excretion of oxalate in buckwheat (Zheng et al., 1998).

Citric acid synthesized in the mitochondria is transported via citrate-carrier on the membrane under Al stress. Using an Al-tolerant rye, Li et al. (2000) found that pyridoxal 5'-P and phenylisothiocyanate, an inhibitor of citrate carrier on the mitochondrial membrane, inhibited the Al-induced secretion of citrate in the rye.

8.2 Enzyme activities related to the Al-induced excretion of organic acids

So far Al-induced enzyme activities in terms of malate metabolism have been examined in several plants including rye, wheat and bean. The activities of phosphoenolpyruvate carboxylase (PEPC) and NADP-malate dehydrogenase scarcely changed with Al-stress. Furthermore, there were no differences between tolerant and sensitive varieties (Li et al., 2000; Ryan et al., 1995b; Andrade, 1997; Mugai et al., 2000). On the contrary, citrate excretion needs a lag period of several hours and the activities of citrate synthase (CS) and NADP-isocitrate dehydrogenase (NADP-ICDH) which regulate the citric acid content through the degradation of isocitrate to 2-oxoglutarate have been investigated in detail. CS activity generally increased in Al-tolerant rye (Li et al., 2000) and *Phaseolus vulgaris* (Mugai et al., 2000). These results support the increased expression of CS gene in Al-tolerant plants exposed to Al stress. In transgenic tobacco and papaya harboring the CS gene from *Pseudomonas aeruginosa* with 35S promoter of cauliflower mosaic virus increased CS activity was apparently followed by a higher level of citrate in vivo and excreted citrate. Furthermore, these transgenic plants showed tolerance against Al (de la Fuente et al., 1997).

Similarly Koyama et al. (1999) reported that the mitochondrial CS gene improved the growth of carrot cells selected as an Al-tolerant mutant line in Al-phosphate medium due to increased synthesis of citrate which led to the release of phosphate from the Al-phosphate complex by Al-citrate chelation.

8.3 Al signal and the excretion of organic acids

Since there are marked differences in the lag time required for the excretion of organic acids between plant species, the regulatory mechanism of the excretion of organic acids in response to Al stress needs to be elucidated. From the agricultural point of view, the regulation of excretion of organic acids by Al signal is important. Continuous excretion of citrate due to overexpression of the CS gene in the transgenic plant may consume an extraordinary amount of energy and carbon. In this regard, we must elucidate the detailed mechanism specifying how the efflux of organic is switched on/or off by the Al^{3+} signal in the rhizosphere.

With Al-tolerant wheat (*Triticum aestivum* L. cv. Atlas 66), it was found that the malate efflux from the root apex (0 - 2 mm from the tip) occurred within 5 min after the addition of Al (Figure 3).

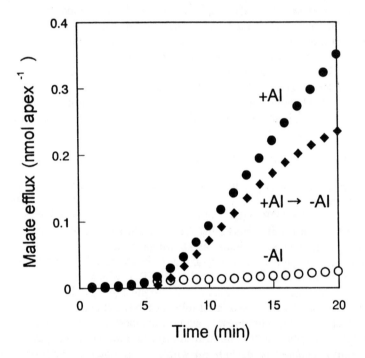

Figure 3. Time course of malate efflux from root apexes of Al-resistant (cv Atlas) wheat after exposure to 200 µM $AlCl_3$. Excised root apexes (2 mm in length from root apices) were exposed to 200 µM $CaCl_2$ (○ = Ca solution) or the Ca solution containing 200 µM Al (pH 4.2) (● = Ca+Al solution) passing through at a flow rate of 0.5 mL min^{-1}. Root apexes were exposed to the Ca+Al solution for 1 min (♦) followed by three rinses with the Ca solution, and subsequent exposure to the Ca solution at a flow rate of 0.5 mL min^{-1} from 5 min after addition of the Ca+Al solution. Eluents passed through root apexes were collected every 1 min and assayed for malate concentration. Treatments replicated three times gave similar results. Data from a representative experiment are shown. Data from Osawa and Matsumoto (2001).

This response was sensitive to temperature and a signal transduction pathway triggered by Al signal might be involved. The possible involvement of protein phosphorylation was investigated because of the high binding nature of Al with phosphate compounds (Jones and Kochian, 1997; Matsumoto, 2000). K-252a, a wide range inhibitor of protein kinase, effectively blocked the Al-induced malate efflux accompanied with an increased accumulation of Al and intensified Al-induced inhibition of root growth (Osawa and Matsumoto, 2001) (Figure 4). A transient activation of a 48 kD protein kinase was observed preceding the initiation of malate efflux, and this activation was cancelled by K-252a (Figure 5). The malate efflux was accompanied with a rapid decrease in the contents of organic anions in the root apex, such as citrate, succinate and malate, but with no change in the contents of inorganic anions such as chloride, nitrate and phosphate (Osawa and Matsumoto, 2001). These results suggest that protein phosphorylation is involved in the Al-responsive malate efflux in the wheat root apex and that organic acid-specific channels might be a terminal target which responds to Al-signaling mediated by phosphorylation

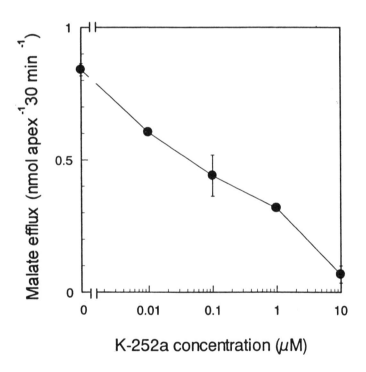

Figure 4. Dose-response curve for Al-induced malate efflux in response to a protein kinase inhibitor, K-252a. Values are means $\pm SE$ (n=3). Data from Osawa and Matsumoto (2001).

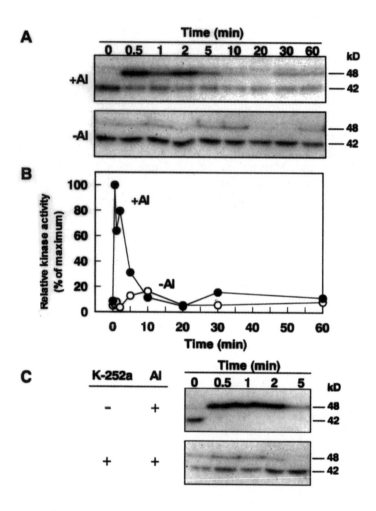

*Figure 5. Activation of a 48 kD kinase in response to Al. **A**, Excised root apexes of Al-resitant (cv Atlas) wheat were exposed to the Ca solution (-Al) or the Ca+Al solution (+Al; pH4.2) at indicated times. Crude protein extracts from the root apexes were assayed for in-gel kinase activity using myelin basic protein as a substrate. **B**, 48 kD kinase activity (○ = -Al; ● = +Al) was quantified from the intensity of bands on the X ray-film digitized with an LAS 1000 image analyzer (Fuji-Film, Tokyo, Japan). The relative kinase activity was normalized according to a maximum band intensity of the Ca+Al solution at 0.5 min. **C**, K-252a blocked Al-induced activation of 48 kD kinase and Al-induced inactivation of 42kD kinase. Exised root apexes of Al-resistant (cv. Atlas) wheat were pretreated with the Ca solution without (-K-252a) or with 10 μM K-252a (+K-252a) for 30 min. After the treatment, root apexes were rinsed three times with the Ca solution to remove excess K-252a and then exposed to the Ca + Al solution at indicated time. Data from Osawa and Matsumoto (2001).*

8.4 Gene(s) possibly involved in the excretion of malate in isogenic wheat lines

Al-tolerant (ET8) and sensitive (ES8) wheat cultivars were used to detect specific genes participating in malate excretion. The cultivars are nearly pure isogenic lines whose difference is considered to be an *alt 1* gene coding the protein(s) for malate release in ET8. The subtractive cloning method was used to detect the specific gene(s) which is expressed only in ET8 under Al stress. The specific gene was expressed only in ET8 (our unpublished result). The full cDNA sequence of the gene was about 1500 bp and the deduced amino acid sequence had 459 residues. The hydrophobility of this gene showed that the coded protein has transmembrane domains. This gene is constitutively expressed in only ET8 and expression of the gene was observed in the root apexes specifically where malate is excreted. Therefore, this gene might be related to the excretion of malate in the root of Al-tolerant wheat ET8 (Sasaki et al., unpublished results).

9. INTRACELLULAR MECHANISM OF Al RESISTANCE

Chelation of Al with an organic acid occurrs not only extracellularly but also intracellularly rendering in the formation of less toxic Al.

Hydrangea is characterized by a high Al content, especially the blue color of flowers is formed by the pigment chelated with Al. The leaves of *Hydrangea* contain as much as 15.66 mmol Al g^{-1} fresh weight and 77% of the total Al exists in the cell sap of leaves in soluble form. This Al concentration is harmful to most plants. Therefore, *Hydrangea* must have some resistance mechanism against a high concentration of Al. The chelate form of Al in the cell sap has been investigated and the ligand in the Al complex was determined to be citrate, and the molecular ratio of Al to citrate was 1 : 1, according to the measurement of ^{27}Al-NMR. Furthermore, the purified Al : citrate complex (1 : 1) from Hydrangea leaves did not inhibit the root elongation of corn and did not decrease the cell viability although both parameters were strongly inhibited by the same $AlCl_3$ concentration as used in the Al citrate complex (Ma et al., 1997a).

Buckwheat is also an Al-tolerant plant and excretes oxalate immediately after the exposure to Al (Ma et al., 1997b). The excreted oxalate can easily chelate with Al and Al-chelated with oxalate lost its toxicity due to the marked reduction of its absorption into roots (Ma et al., 1998; Zheng et al., 1998).

In spite of this defence mechanism upon the excreted oxalate promptly in the rhizosphere, the content of Al in both root and leaves of buckwheat exposed to Al is high. However, the buckwheat can grow under Al stress suggesting that buckwheat has intracellular resistance mechanism against Al.

In both leaves and roots Al exists in the form of Al-oxalate chelated form (1 : 3) (Figure 6). This form is the most stable with a stability constant of 12.4 (Nordstrom and May, 1996) compared to other Al-oxalate complexes as 1 : 1 or 1 : 2. Therefore, the formation of Al : oxalate (1 : 3) hinders the formation of Al with cellular compounds resulting in Al toxicity. The high content of oxalate may not be induced by Al because there is almost no difference in the concentration of oxalate between buckwheat treated with or without Al.

Taken together, the results suggest that backwheat seems to have a duplicate resistant mechanism functioning in intracellularly and extracellularly through the formation of Al-oxalate chelate complex.

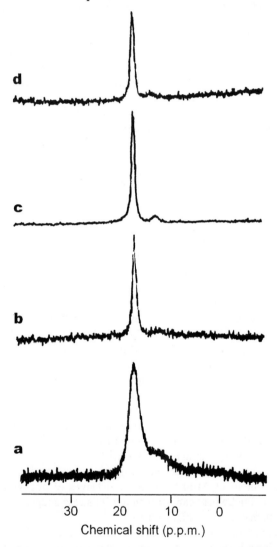

Figure 6. ^{27}*Al-NMR spectra.* ***a****, Intact buckwheat leaves exposed to AlCl$_3$ solution.* ***b****, Crude cell sap from the leaves.* ***c****, Purified sap.* ***d****, 1 : 3 Al-oxalate complex. The aluminium complex in the sap was purified by Sephadex G-10. Spectra were obtained at 156.3 MHz using a JNM-α600 spectrometer. Data from Ma et al. (1997b).*

10. FUTURE PROSPECTS

Al toxicity is one of the most deleterious factors for plant growth in acidic soils. Much effort has been paid to improve crop production in acidic soils. The role of organic acids in the Al-tolerance mechanism, both intracellularly and extracellularly, has been examined intensively. The mechanism is based on the chelation of Al by the organic acids rendering the Al^{3+} less toxic.

In the future the problem of the complicated behaviour of the excreted organic acids in the soil should be elucidated. Organic acids become adsorbed onto the soil's exchange complex (> 80% within 10 min) (Jones and Brassington, 1998) and the degradation of organic acids by bacteria occurs. The breakdown of malate is rapid in soil with a half life of approximately 1.7 h, Km of 1.7 mM and V_{max} of 70 nmol g^{-1} soil h^{-1} (Jones et al., 1996). Therefore, the organic acids in the soil may not be effective in protecting the plant from Al stress.

Furthermore, Parker and Pedler (1998) reported that the organic acid excreted in the rhizosphere may have only a limited role and a more integrative, multifaceted model of tolerance is needed.

From the economic and energetics point of view in plant metabolism, we must elucidate the regulatory system on the mechanism of synthesis and release of organic acids upon perception of the Al signal.

ACKNOWLEDGEMENT

Part of the presented work was conducted by the financial support of PROBRAIN (Program for Promotion of Basic Research Activities for Innovative Bioresources). My thanks are due to Mrs. S. Rikiishi for technical help in the preparation of the manuscript.

REFERENCES

Andrade, L.R.M., Ikeda, M., and Ishizuka, J. (1997) Excretion and metabolism of malic acid produced by dark carbon fixation in roots in relation to aluminium tolerance of wheat, in T. Ando, K. Fujita, T. Mae, H. Matsumoto, S. Mori and J. Sekiya (eds), Plant Nutrition for Sustainable Food Production and Environment, Kluwer Academic Publishers, Dordrecht, pp.445-446.

Delhaize, E., Craig, S., Beaton, C.D., Bennet, R.J., Jagadish, V.C., and Randall, P. (1993) Aluminum tolerance in wheat (*Triticum aestivum* L.), Plant Physiol **103**, 685-693.

Delhaize, E. and Ryan, R.P. (1995) Aluminum toxicity and tolerance in plants, Plant Physiol **107**, 315-321.

de la Fuente, J.M., Ramírez-Rodríguez, V., Cabrera-Ponce, J.L., and Herrera-Estrella, L. (1997) Aluminum tolerance in transgenic plants by alteration of citrate synthesis, Science **276**, 1566-1568.

Jones, D.L. and Brassington, D.S. (1998) Sorption of organic acids in acid soils and its implications in the rhizosphere, Eur J Soil Sci **49**, 447-455.

Jones, D.L. and Kochian, L.V. (1997) Aluminum interaction with plasma membrane lipids and enzyme metal binding sites and its potential role in Al cytotoxicity, FEBS Letters **400**, 51-57.

Jones, D.L., Prabowo, A.M., and Kochian, L.V. (1996) Kinetics of malate transport and decomposition in acid soils and isolated bacterial populations: The effect of microorganisms on root exudation of malate under Al stress, Plant Soil **182**, 239-247.

Kochian, L.V. (1995) Cellular mechanisms of aluminum toxicity and resistance in plants, Ann Rev Plant Physiol Plant Mol Biol **46**, 237-260.

Koyama, H., Takita, E., Kawamura, A., Hara, T., and Shibata, D. (1999) Over expression of mitochondrial citrate synthase gene improves the growth of carrot cells in Al-phosphate medium, Plant Cell Physiol **40**, 482-488.

Larsen, P.B., Degenhardt, J., Tai, C-Y., Stenzler, L.M., Howell, S.H., and Kochian, L.V. (1998) Aluminum-resistant Arabidopsis mutants that exhibit altered patterns of aluminum accumulation and organic acid release from roots, Plant Physiol **117**, 9-18.

Li, X.F. (2000) Aluminum detoxification mechanism by organic acids and mucilage in higher plants, PhD dissertation, Okayama University, Kurashiki, Okayama.

Li, X.F., Ma, J.F., and Matsumoto, H. (2000) Pattern of aluminum-induced secretion of organic acids differs between rye and wheat, Plant Physiol **123**, 1537-1543.

Llugany, M., Poschenrieder, C., and Barceló, J. (1995) Monitoring of aluminium-induced inhibition of root elongation in four maize cultivars differing in tolerance to aluminium and proton toxicity, Physiol Plant **93**, 265-271.

Ma, J.F. (2000) Role of organic acids in detoxification of Al in higher plant, Plant Cell Physiol **113**, 1033-1039.

Ma, J.F., Hiradate, S., and Matsumoto, H. (1998) High aluminum resistance in buckwheat. II. Oxalic acid detoxifies aluminum internally, Plant Physiol **117**, 753-759.

Ma, J.F., Hiradate, S., Nomoto, K., Iwashita, T., and Matsumoto, H. (1997a) Internal detoxification mechanism of Al in Hydrangea. Identification of Al form in the leaves, Plant Physiol **113**, :1033-1039.

Ma, J.F., Taketa, S., and Yang, Z.M. (2000) Aluminum tolerance genes on the short arm of chromosome 3R are linked to organic acid release in Triticale, Plant Physiol **122**,1-8.

Ma, J.F., Zheng, S.J., Matsumoto, H., and Hiradate, S. (1997b) Detoxifying aluminium with buckwheat, Nature **390**, 569-570.

Matsumoto, H. (2000) Cell biology of aluminum toxicity and tolerance in higher plants, Int Rev Cytol **200**, 1-46.

Matsumoto, H. (2002) Plant roots under aluminum stress: Toxicity and tolerance, in Y. Waisel, A. Eshel and U. Kafkafi (eds), Plant Roots: The Hidden Half, Third Edition. Marcel Dekker, Inc. pp. ISBN 0-8247-0631-5,in press

Matsumoto, H., Senoo, Y., Kasai, M., and Maeshima, M. (1996) Response of the plant root to aluminum stress: Analysis of the inhibition of the root elongation and changes in membrane function, J Plant Res **109**, 99-105.

Matsumoto, H., Yamamoto, Y., and Devi, S.R. (2001) Aluminum toxicity in acid soils: Plant response to aluminum, in M.N.V.Prasad (ed), Metals in the Environment: Analysis by Biodiversity, Marcel Dekker, Inc. pp. 289-320. ISBN 0-82470523-8,

Miyasaka, S.C., Buta, J.G., Howell, R.K., and Foy, C.D. (1991) Mechanism of aluminum tolerance in snapbeans. Root exudation of citric acid, Plant Physiol **96**, 737-743.

Mugai, E.N., Agong, S.G., and Matsumoto, H. (2000) Aluminium tolerance mechanisms in *Phaseolus vulgaris* L.: Citrate synthase activity and TTC reduction are well correlated with citrate secretion, Soil Sci Plant Nutr **46**, 939-950.

Nordstrom, D.K. and May, H.M. (1996) Aqueous equilibrium data for mononuclear aluminum species. in G. Sposito (ed), Environment Chemistry of Aluminum, CRC Press, Florida, pp. 39-80.

Osawa, H. and Matsumoto, H. (2001) Possible involvement of protein phosphorylation in aluminum-responsive malate efflux from wheat root apex, Plant Physiol **126**, 411-420.

Papernik, L.A. and Kochian, L.V. (1997) Possible involvement of Al-induced electrical signals in Al tolerance in wheat, Plant Physiol **115**, 657-667.

Parker, D.R. and Pedler, J.F. (1998) Probing the "malate hypothesis" of differential aluminum tolerance in wheat by using other rhizotoxic ions as proxies for Al, Planta **205**, 389-396.

Piñeros, M.A. and Kochian, L.V. (2001) A patch clamp study on the physiology of aluminum toxicity and tolerance in *Zea mays*: Identification and characterization of Al^{3+}-induced anion channels, Plant Physiol **125**, 292-305.

Rincón, M. and Gonzales, A. (1992) Aluminum partitioning in intact roots of aluminum-tolerant and aluminum-sensitive wheat (*Triticum aestivum* L.) cultivars, Plant Physiol 99, 1021-1028.

Ryan, P.R., Delhaize, E., and Jones, D.L. (2001) Function and mechanism of organic anion exudation from plant roots, Annu Rev Plant Physiol Plant Mol Biol **52**, 527-560.

Ryan, P.R., Delhaize, E., and Randall, P.J. (1995a) Malate efflux from root apices and tolerance to aluminium are highly correlated in wheat, Aust J Plant Physiol **22**, 531-536.

Ryan, P.R., Delhaize, E., and Randall, P.J. (1995b) Characterization of Al-stimulated efflux of malate from the apices of Al-tolerant wheat roots, Planta **196**, 103-110.

Ryan, P.R., Skerrett, M., Findlay, G.P., Delhaize, E., and Tyerman, S.D. (1997) Aluminum activates an anion channel in the apical cells of wheat roots, Proc Natl Acad Sci USA **94**, 6547-6552.

Samuels, T.D., Kücükakyüz, K., and Rincón-Zachary, M. (1997) Al partitioning patterns and root growth as related to Al sensitivity and Al tolerance in wheat, Plant Physiol **113**, 527-534.

Schroeder, J.I. (1995) Anion channels as central mechanisms for signal transduction in guard cells and putative functions in roots for plant-soil interactions, Plant Mol Biol **28**, 353-361.

Taylor, G.J. (1991) Current views of the aluminum stress response; The physiological basis of tolerance, Curr Topics in Plant Biochem and Physiol **10**, 57-93.

Yamamoto, Y., Rikiishi, S., Chang, Y-C., Ono, K., Kasai, M., and Matsumoto, H. (1994) Quantitative estimation of aluminium toxicity in cultured tobacco cells: Correlation between aluminium uptake and growth inhibition, Plant Cell Physiol **35**, 575-583.

Yang, Z.M., Nian, H., Sivaguru, M., Tanakamaru, S., and Matsumoto, H. (2001) Characterization of aluminum induced citrate secretion in aluminum tolerant soybean (*Glycine max* L.) plants, Physiol Plant, in press

Yang, Z.M., Sivaguru, M., Horst, W.J., and Matsumoto, H. (2000) Aluminum tolerance is achieved by exdudation of citric acid from roots of soybean (*Glycine max*), Physiol Plant **110**, 72-77.

Zhang, W.H., Ryan, P.R., and Tyerman, S.D. (2001) Malate-permeable channels and cation channels activated by aluminum in the apical cells of wheat roots, Plant Physiol **125**, 1459-1472.

Zheng, S.J., Ma, J.F., and Matsumoto, H. (1998) High aluminum resistance in buckwheat. I. Al-induced specific secretion of oxalic acid from root tips, Plant Physiol **117**, 745-751.

CHAPTER 5

PHYSIOLOGICAL RESPONSES OF NON-VASCULAR
PLANTS TO HEAVY METALS

N. MALLICK[a] AND L.C.RAI[b]
[a]*Agricultural and Food Engineering Department
Indian Institue of Technology, Kharagpur, India*
[b]*Department of Botany, Banaras Hindu University
Varanasi 221005,India*

1. INTRODUCTION

Metals have been in the service of human beings for thousands of years, but many of the most useful metals are highly toxic at elevated concentrations. In the second half of the twentieth century, under the ever-increasing impacts of the exponentially growing population coupled with industrialisation, metal contamination of air, water and soil has become a threat to the very existence of many plant and animal communities spread over contaminated ecosystems. Perhaps no other legacy of man's early ignorance of the price of releasing toxic compounds into the environment is better known than that of the catastrophe at the Minamata Bay, a small bay on the south-western shore of Kyushu Island, Japan. Understandably, many scientific, industrial and governmental agencies have diverted large chunk of money and manpower to combat this problem and provide effective control measures.

Generally speaking the term 'heavy metal' is often given to metals which are environmental pollutants (Tiller, 1989). Though this term has been redefined over the years (Rai et al. 1981; Gadd, 1992) but no satisfactory definition has emerged. From a chemical stand point, it is generally used for elements having density greater than 5g cm^{-3}.

Metals continue to pose a serious threat to biota due to their acute toxicity, non-biodegradable nature and gradual build up of high concentrations in the environment all over the world (Prasad 2001; Prasad and Hagemeyer 1999). The non-vascular plants, due to their lack of particular uptake and transport systems, absorb metals through the whole body, thereby increasing their body burden and thus, are more prone to the toxicants. Moreover, the extremely high capacity of certain algae, bryophytes and lichens to sorb and accumulate metals has led to their use as biosorbents as well as biomonitors of aquatic as well as atmospheric metal burdens. The unique lichen symbiosis presents a system in which basic effects of metals on biological systems can be monitored and probed (Branquinho 2001; Colpaert and Vandenkoornhuyse 2001; Lepp 2001; Pawlik- skowronska and Skowronski 2001; Van der Lelie and Tibazarwa 2001).

This chapter covers some aspects dealing with the sources and extent of metal pollution, essentiality, and bioavailability of metals as affected by physico-chemical conditions and various mechanism(s) which affect the metabolism of non-vascular plants. Metal tolerance a phenomenon in which organisms use their physiological, biochemical and molecular machinery to survive and proliferate in stressed environments shall also be discussed. In the aftermost part of this chapter a brief account of the mechanism(s) of survival of organism(s) in acid and metal-stressed environments is also included, since the co-existence of both stresses in such environments, as acid-mine drainage and tannery wastes, has aroused considerable interest because at acid pH metals tend to exist in their highly toxic free cationic forms. Since the physiological responses of organism to the combined insult of metal toxicity and acidity are extremely fascinating, their brief discussion should be hopefully befitting.

2. SOURCES AND EXTENT OF METAL POLLUTION

The entry of metal into the environment could be basically of two origins: (i) natural (geological weathering processes) and (ii) anthropogenic. Natural sources such as surface mineralization, volcanic outgassings, spontaneous combustions or forest fires are the main sources of metal release into the environment. Varrica et al. (2000) reported that particulate Cd, Hg, Se, Cu and Zn emitted by the Mt. Etna volcano is comparable to the amount of these elements released in the Mediterranean area by anthropogenic sources.

The major sources of metal enrichment in aquatic ecosystems including the oceans are coal-burning power plants (As, Hg, and Se in particular), the non-ferrous metal industry (Pb, As, Cd, Cu and Zn), iron and steel plants (Cr, Sb and Zn), domestic wastewater effluents (especially As, Cr, Cu and Ni) and dumping of sewage sludge (As, Mn and Pb). Likewise the two principal sources of metals in the soil, however, are the disposal of ash residues from coal combustion and the wastes of commercial products into the land. Incineration of solid wastes also significantly contributes to the emission of metals, such as Hg, Cd and Pb into the environment (Nriagu and Pacyna, 1988).

Very high concentrations of metal ions are found in industrial and mining effluents (Kelly, 1988). Pasternak (1973) found high levels of Zn (1800 mg L^{-1}) and Pb (300 mg L^{-1}) in mine drainage at Bolesaw, Poland. Iron level of 200 mg L^{-1} has been reported by Hill (1973). Klein et al. (1975) reported a high concentration of Cu in the wastewater discharges from the fur dressing and dying industries. High concentrations of Zn in metal processing wastes (40-1463 mg L^{-1}), wire mill pickle (36-374 mg L^{-1}) and in rayon wastes (250-1000 mg L^{-1}) were also detected (Sittig, 1975). Nickel concentrations ranging from 2 to 900 mg L^{-1} in the rinse waters of plating plants was reported by Ramaswamy and Somashekhar (1982). This electroplating effluent was also found to contain 56 mg L^{-1} of Cr. In industrial effluents, Cd concentrations ranging from 0.005-0.50 g L^{-1} were detected (De Filippis and Pallaghy, 1994). Xu et al. (2000) reported high concentrations of Cr, Pb, Zn, Cu and Ni in the Yangtse river, the largest river in China (Table 1). In central India, the area lying between, 18^0 to 23^0 N latitude and 80^0 to 84^0 E longitude is most prone to metal pollution due to the occurrence of minerals and coals and their utilization. Several steel, cement and aluminium industries, the thermal power plants mainly using coal as a source of energy are at work in this region. A recent report shows (Patel et al. 1999) that the pond and river sediments of this region are highly contaminated with metals. Lead contamination in aquatic ecosystems is unusual in that it arises mainly from atmospheric transport (De Filippis and Pallaghy, 1994). Al-Chalabi and Hawker (2000) detected a very high concentration of Pb in road side soil samples from Brisbane, Australia. However, studies by Monaci et al. (2000) reported a new tracer (Zn) in addition to Pb from vehicle emission. It is well established that the atmosphere is the primary transport path for lead found in the surface waters of the North Atlantic and North Pacific oceans (Schaule and Patterson, 1983).

High concentrations of metals are found in algae and phytoplankton in rivers contaminated with metals. Algae and phytoplankton acquire high concentrations of metal elements from their surrounding milieu. High concentrations of inorganic elements in algae and freshwater phytoplankton are reported by Crompton (1998); these

Table 1. Range of heavy metal contamination in some study area of different parts of the world

Location	Metal concentration (ppm)						
	Pb	Zn	Cu	Cr	Ni	Fe	Mn
Gulf of Venice, Italy (Donazzolo et al., 1981)	5-84	48-870	3-44	10-254	5-41	7,000-67,000	Nd
Weser Estuary, Germany (Shoer et al., 1982)	25-142	59-142	23-28	Nd	22-24	29,500-42,500	307-1,402
Ganges Estuary, India (Subramanian et al., 1988)	12-115	12-611	4-43	21-100	8-57	12,000-46,000	254-800
Belfast Inner Lough, UK (Smith and Orford, 1989)	52-207	83-798	11-54	150-335	Nd	Nd	Nd
Jurujuba Sound, Brazil (Neto et al., 2000)	5-123	15-337	5-213	10-223	15-79	1,000-21,250	10-414
Yangtse River, China (Xu et al., 2000)	10-40	50-130	15-50	90-120	20-50	Nd	330-820

Nd : not determined

range from 20,000µg kg^{-1} of arsenic to as high as 660,000 µg kg^{-1} copper in algae as compared to, respectively, 490 and 200 µg L^{-1} in water. Depending on the proximity of industrial sites, metal concentration in algae (µg g^{-1} dry weight) can be very high (e.g. 33 and 189 for As in phytoplankton and microalgae, 660 for Cu in unattached filamentous freshwater species, 1200 for Pb in *Enteromorpha* sp., 700 for Ni in algae and 800 for Zn in *Fucus vesiculosus* (Moore and Ramamurthy, 1984) and 342 for Cd in *Lamanea fluviatilis*, (Harding and Whitton, 1976). Denton and Burdon-Jones (1986) reported metal contents in contaminated marine algae from around the world (e.g. 25.6 µg g^{-1} Cd in *Fucus vesiculosus*, and 25.5 µg g^{-1} Hg in *Ulva lactuca* and 14 µg g^{-1} in *Ulva pertussa* from the site of the Minamata bay).

Very high levels of metals were also detected from the tissues of various fungi, lichens and bryophytes. Nash (1975) reported 334 and 3500 ppm (µg. g^{-1} dry wt.) of Cd and Zn in the tissues of lichen growing near a Zn factory. Schutte (1977) found 70 ppm Cr (µg. g^{-1} dry wt) in two corticolus lichens from Ohio and West Virginia. Lawrey and Hale (1979) observed 1131 ppm Pb (µg. g^{-1} dry wt) in the body tissue of roadside lichen. Likewise very high concentrations (ppm) of Zn (7166), Cd (148), Pb (11670), Ni (146), Hg (15), Cu (211) and Cr (937) were reported in various bryophytes by Goodman and Roberts (1971), Burkitt et al. (1972), Shimwell and Laurie (1972), Roberts and Goodman (1974), Rejment-Grochowska (1976) and Lee et al. (1977) respectively. High concentration of Zn was also reported in fungal mycelium grown in Zn waste in southern Poland. Elevated levels of Ag, Cd, Pb were also detected in some edible mushrooms (Agaricales) from the Chicago region (Aruguete et al. 1998). Studies by Sanchez et al. (1998) showed an accumulation of 1.1 mg Pb and 6.8 mg Zn kg^{-1} in the moss *Brachythecium rivulare* growing in the Urumea river valley, close to an old lead-zinc mine in Spain. Mosses, collected from the Antarctic region also showed high concentrations of Hg and Cd, though Pb concentration was remarkably lower (Bargagli et al., 1998). Recently, a metal level of 6.8, 6.87, 16.8 and 66.4 mg of Hg, Pb, Cd and Cu respectively was reported on per kg basis in four edible mushrooms (*Hydnum repandum, Russula delica, Agaricus silvicolla* and *Tricholoma terreum*) of Turkis origin (Demirbas, 2000).

3. ESSENTIALITY

Essential metal ions have many functions in the cells, not all of which are fully characterized. Such functions may reflect their fundamental chemical properties and rate of ligand exchange. One important area of interaction of metals in the cell is metalloenzymes. It has been estimated that up to one third of all known enzymes contain a metal ion as a functional participant (Dedyukhina and Eroshin, 1991). The second important process is the redox reaction, where metals are involved in electron transfer processes. Other important roles for metal ions include the formation of charge and concentration gradients across membranes (Hughes and Poole, 1991). At supraoptimal concentration micronutrients, however, become phytotoxic. Deprivation of an essential metal ion, by definition, will ultimately result in cell death. However, broader definitions of essentiality may only refer to impairment of growth and other functions in the absence of the essential metal(s) and the fact that beneficial effects cannot be completely replaced by any other element (Dedyukhina and Eroshin, 1991). Recent study by Kriedemann and Anderson (1988) showed trace element deficiency-induced photosynthetic disfunction resulted in lowered O_2 evolution and slowed down leaf gas exchange. Both initial slope and CO_2 saturated phases of A(pi) were affected under either Cu and Mn deficiency in wheat and barley.

However, under certain conditions the metabolic processes of some organisms may actually show better performance with elements which are not essential for those processes. For example, stimulation of growth, heterocyst differentiation, photosynthesis and nitrogenase activity in *Nostoc muscorum* by Ni (Rai and Raizada, 1986), increased carbon assimilation in cyanobacteria by low concentrations of Cd and Cr (Azeez and Banerjee, 1986, 1988), increased algal biomass and metabolic processes by Zn (Mohanty, 1989; Stauber and Florence, 1990), replacement of Zn by Cd in a marine diatom (Price and Morel, 1990a) and K by Cs in yeast (Jones and Gadd, 1990) and cyanobacteria (Avery et al., 1991). Cd at 10 µM has been shown to stimulate calmodulin-dependent phosphoesterase activation in *Chlamydomonas* extracts by altering the concentration of calcium ions (Behra, 1993). At low concentrations of Cd, some enhancement of growth and development was also noticed for *Funaria* and *Marchantia* (Lepp and Roberts, 1977). More interestingly DMS production was found to be stimulated by 10 fold in the marine alga, *Amphidinium hoefleri,* in the presence of Cd (Lin et al., 1999), a result which needs further explanation.

4. BIOAVAILABILITY

A host of factors which regulate the availability of metals in the ecosystem are: pH, redox potential (E_h), presence of anions and cations, aminoacids, sulphur containing compounds, complexing ligands, organic carbon, salinity etc. The sequential addition of metals also affect the toxicity. All the above factors adversely affect the metal availability thus toxicity. Some such cases are briefly documented.

The bioavailability and toxicity of metals for the living organisms has been well studied and extensively documented. The importance of metal speciation and the concentration of free metal ions in controlling metal toxicity has also been well recognised. Many reports claim that only the free metal ions are toxic, while others have suggested that metal complexes of hydroxides, carbonate or some organic molecules are also toxic. Nevertheless, some interesting patterns have already emerged (Table 2). Metal toxicity is strongly correlated with pH, either positively or negatively depending on the range of pH. At acidic pH, metals tend to exist as more toxic free hydrated ions, whereas at alkaline pH they may be precipitated as insoluble complexes. However, the response of pH is complex and low pH does not always lead to increased uptake and toxicity (Rai et al. 1990). A recent study showed that concentration of Al was higher in *Scapania undulata*, an aquatic liverwort, growing in neutral habitat than those at acidic pH (Yoshimura et al. 1998).

Table 2. Effect of physico-chemical factors on metal toxicity

Parameter	Effects on metal toxicity
Temperature	Increases at higher temperature
Light	Toxicity is light dependent in some cases
pH	Low pH increases toxicity
E_h	Negative E_h decreases toxicity
Monovalent cations (Na and K)	Decrease in toxicity with increasing concentration.
Divalent cations (Ca, Mg, Mn and Fe)	Decrease in toxicity with increasing concentration.
Heavy metals	Synergistic /antagonistic depending on metal combination, concentration and sequence of addition
Anions (Acetate, phosphate, nitrate and sulphate)	Reduce toxicity
Extracellular products (organic acids, polyphenols, polysaccharides, polypeptides)	Reduce toxicity
Sulphur containing aminoacids	Reduce toxicity
Sediment fractions (suspended solids and colloids)	Reduce toxicity

Adapted from De Filippis and Pallaghy (1994) with modification

Therefore, the pH specific response observed is probably a trade-off between metal speciation and pH optima of a particular organism (Kelly, 1988). In this regard the lichen *Byoria fuscescens* responded interestingly to toxic metals at lower pH. The algal partner was found to be more sensitive to metals at low pH. Damage was apparent in chloroplasts and mitochondria, where thylakoid and mitochondrial cristae were swollen (Tarhanen, 1998). The fungal partner was, however, found to be more sensitive to metals at higher pH, suggesting an ameliorative role of low pH on metal toxicity.

E_h (redox potential), a major property regulating the availability of electrons, with negative values being indicative of a reducing environment. Reducing conditions, such as those encountered in anaerobic environments, lead to the conversion of sulphate to sulphide with the subsequent precipitation of metals as metal sulphides, thereby reducing their availability and toxicity. Moreover, the E_h of the environment also determines the valency of some metals. For example, in the oxygenated part of the Fjord of Saanich Inlet, British Columbia, Cr occurs as Cr(VI), whereas in the anoxic zone it occurs as Cr(III). Differentially charged forms of the same element exert different toxicities, as Cr(VI) is generally more toxic than Cr(III) (Babich and Stotzky, 1983).

Metal toxicity was normally found to decrease in the presence of cations, either mono- or divalents (Table 2). The sensitivity to metals was generally higher at low salinity (Haglund et al, 1996). A reduction in Pb toxicity was reflected in the photosynthetic characteristics of *Fucus vesiculosus* at a salinity level of 4.5 and 8% – at a higher salinity (20%) a reverse trend was however, evident (Nygard and Ekelund, 1999). Increased toxicity of metals for *Scenedesmus armatus* and *Oocystis submarina* with increasing salinity was also reported by Adam and Waldemar (1998). A reduction in toxicity was also noticed in the presence of various anions, e.g. hydroxyl, acetate, sulphate, phosphate, nitrate and/or selenite. Water hardness, i.e. calcium, magnesium and carbonate ions, reduce the toxicity of metal ions. Gagnon et al. (1998) reported a reduced uptake of Cd with increasing water hardness in aquatic mosses. The presence of suspended solids, sulphur-containing amino acids and organic materials influence metal toxicity negatively. The addition of peptone or yeast extract was also found to reduce toxicity (EI-Hissy et al. 1993). Temperature and hydrostatic pressure were, however, found to influence metal toxicity in an additive way, i.e., with increasing temperature and hydrostatic pressure a progressive increase in the toxicity of metals was observed. A four-fold rise in Pb uptake and toxicity was observed in *Saccharomyces cerevisiae* with an increase in temperature from 20 to 50^0C (Jung-Ho, 1998).

Pollutant emissions seldom, if ever, consist only of one toxicant. Such emissions usually contain multiple metals and, consequently, deposition into a common environment will simultaneously expose the indigenous biota to several metals. Depending on the metal present, the interaction could be of an antagonistic or synergistic type. The specific metal-metal interaction was found to be dependent on (i) the test organism, (ii) the relative concentrations of metals and (iii) the sequence of metal exposure. For example, the effect of the combination of Cd + Pb on the photosynthesis of a brackish water phytoplankton community was antagonistic when the concentration of Pb was greater than that of Cd, whereas it was synergistic when the concentration of Cd was greater than that of Pb (Pietilanien, 1975). The effect of a combination of Hg + Ni on the growth rate of the cyanobactrerium *Anabaena inequalis* was synergistic when both Hg and Ni were added simultaneously or when Hg was added first, but it was antagonistic when Ni was added before Hg (Stratton and Corke, 1979). Pre-addition of Fe was also found to protect (antagonistic effect) the cyanobacterim *Anabaena doliolum* against Cu and Ni toxicity (Mallick and Rai, 1989).

5. METAL TOXICITY

Effects of toxic metals on algal cell structure and function have been extensively documented by Rai et al. (1981) and De Filippis and Pallaghy (1994). Sub-lethal doses of metals cause retardation of growth, and thus a loss of biomass. They can impose a longer lag phase and also depress the rates of cell division. Since some species are found to be more tolerant than others, metals would ensue a change in the species composition and thereby the community structure. Josef et al. (1999) studied the cyanobacteria and eukayotic algal population in 18 plots selected across Sverdrup Pass Valley of Central Ellesmere Island, 79^0N, Canada. A high species diversity totalling 136 taxa (cyanobacteria accounted for 52 and eukayotic algal species for 84) was reported. These microautotrophs occupied the soil profile upto a depth of 7 cm. Their highest density was not at the surface but at a depth of 3-4 cm. Analysis showed that the plot contaminated by wind-blown copper-rich dust from a nearby outcrop had the poorest

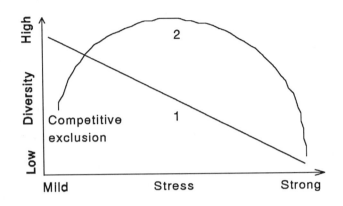

Figure 1. Two hypothetical models showing the effects of heavy metal on diversity of a community. In model 1 the effect of increasingly severe stress results in a decrease in diversity in the microbial community. In model 2 when the stress is mild, it is hypothesized that competitive species can predominate resulting in a lack of diversity, whereas as stress increases these individuals lose their competitive advantage and more types can proliferate. At higher level of the stress, however, progressive extinction of organisms leads to a loss of diversity as in model 1 (after Giller et al., 1998)..

content of photosynthetic pigments and low diversity. Figure 1 presents two hypothetical models of the effect of heavy metals on the diversity of a community (Giller et al., 1998).

Metals are also found to have strong effects on the morphometry and the ultrastucture of algae (Sicko-Goad et al., 1986; Rai et al., 1990). Changes in the volume of chloroplasts, mitochondria and vacuoles, dilation of membranes, tubuli, grana and thylakoid, multinucleation, disturbances in membrane microfibrils and changes in storage products have been observed in many test algae following metal treatment. Inhibitory effects of metals on motility, with decreasing the percentages of motile cells, change from flagellar to ameboid movement and later to a complete loss of motility was also observed with increasing metal concentration (De Filippis et al., 1981). Cell lysis, filament elongation and formation of abnormal cells are some of the other important morphological changes induced by metals.

Toxic effects of metals are also evident on pigments, protein, DNA and RNA contents of the cells. Changes in the relative abundance of fatty acids in marine macroalgae due to Cu has been observed by Jones and Harwood (1993). Xyländer and Braune (1994) reported a decrease in protein and carbohydrate content in the marine alga *Haematococcus* in the presence of Ni. Metal-induced depurination of DNA has also been reported (Schaaper et al., 1987). The uptake of nutrients and their assimilatory enzymes are also found to be severely disrupted by metal ions (Mallick and Rai, 1994).

The processes of photosynthesis, both O_2 evolution and carbon fixation, are known to be severely inhibited by metals. Inhibitory effects of metals are also reported on Hill reaction (Rai et al., 1991). Inactivation of enzymes of photosynthetic and respiratory pathways were reported in a number of studies (see De Filippis and Pallaghy, 1994). Metals are also found to repress the enzymes of nitrogen cycles, such as nitrogenase, nitrate reductase and glutamine synthetase. The activity of ATPase, the enzyme responsible for the movement of H^+ and other vital ions across the cell membrane, was found to be inhibited by metals (Rai et al., 1994, Husaini et al., 1996a). An alteration (decrease) in cytoplasmic pH and increase in the cytoplasmic viscosity in metal-stressed cells are also observed (Grabowski et al. 1998).

Data from both glass-house and field studies provide strong evidence of reduced percentages of fungal community as a result of pollution by Cd, Cu, Ni, Pb and Zn (see Cairney and Meharg, 1999). Even where percentages of colonization are unaffected, metals could influence ectomycorrhizal fungi symbiosis by inhibiting the growth of extramatrical mycelial systems (Colpaert and Van Assche, 1993). Mycelial growth of *Achlya racemosa* and *Alatospora acuminata* decreased significantly in the presence of Ni, Cd, Pb and Cu (EI-Hissy et al., 1993). Studies showed that arbuscular and ericoid mycorrhizal fungal colonization and spore formation were significantly reduced in the presence of toxic metals. Arbuscular mycorrhizal spore density and infectivity were lower in the vicinity of a zinc smelter where soils were rich in available Zn and Cd (Leyval et al. 1995). In contrast, Wiessenhorn et al. (1995) found that long-term (5-18 years) application of sewage sludge, contaminated with Zn, Cd and Ni, to agricultural soil had no significant effect on either spore density or soil infectivity. Nutrient utilization was found to be reduced significantly in the mycorrhizal fungus *Hebeloma mesophacus* in the presence of Al (Kong et al., 1997). Mycelial growth and development of fruit bodies in *Ganoderma lucidum* were also affected by metals (Xuan, 1999); the metal toxicity was found to decrease in the order Hg>Cd>Cu>U>Pb>Mn>Zn. Cu and Ni were also found to inhibit the growth of *Saccharomyces cerevisiae*. Thermochemical measurement based on the measurement of metabolic heat evolution showed that Cd, Cu, Hg and Pb exert toxic effect on the growth performance of *Rhizopus nigricans* (Yan et al., 1999). This thermochemical measurement by using a LKB2277 BioActivity Monitor (heat conduction microcalorimeter) is found to be quantitative, inexpensive and versatile method for obtaining toxicological information from species of interest. Jentschke et al. (1999) studied the effects of ectomycorrhizal colonisation by *Laccaria bicolor* and *Paxillus involutus* in Norway spruce seedlings (*Picea abies* L.Karst) in presence of Cd. Cd treatment decreased colonisation by *L. bicolor* but not by *P. involutus*. Fifty percent reduction in growth of *Suillus luteus* was recorded under 16µM Cu exposure (Van Tichlen et al. 1999).

Hartley et al. (1999a) studied the effects of single and multiple metal contamination (Cd, Pb, Zn, Sb and Cu) on Scots pine seedlings colonised by ectomycorrhizal (ECM) fungi from natural soil inoculum. Of the five metals tested in amended soils, Cd was the most toxic. When compared to the soil amended with only one metal, soils amended with a combination of all five metals tested had lower relative toxicity. This indicates that the toxicity of multiple metal contamination cannot be predicted from the toxicity of individual metal investigated. The effects of Cd and Zn on cross-colonisation by *Paxillus involutus* of Scots pine seedlings was examined by using pairs of ECM and non-mycorrhizal (NM) seedlings grown in the same vessel (Hartley et al., 1999b). This was done to assess i) the ability of *P. involutus* to colonise NM Scots pine seedlings by growth from colonised roots and other Scots pine seedlings in the presence of Cd or Zn, and ii) whether ECM colonisation of Scots pine by *P. involutus* provided a competitive advantage over NM seedlings. Ectomycorhizal colonisation of Scots pine was shown to be more sensitive than Scots pine itself to Cd and Zn, but prior colonisation did provide a competitive advantage with respect to biomass production. Cross-colonisation from an ECM to NM was reduced but not prevented by Cd and Zn. Cd had a more negative effect on cross-colonisation than on initial colonisation of seedlings, whereas Zn had an equally inhibitory effect on both the parameters. These results have important implications for plant establishment on metal contaminated sites. If cross-colonisation between plants is reduced by toxic metals, plant establishment on contaminated sites might be retarded.

Metals are also found to exert strong toxic effects on bryophytes (Shaw, 1990). Mosses don't possess true roots and derive water and dissolved inorganic nutrients through direct absorption by the whole plant body. There is no selective barrier to control metal absorption, such as that represented by the endodermis in vascular plants. In addition, bryophytes possess no cuticle which increases the potential of soluble metals to directly access the living cytoplasm (Richardson, 1981). Spore germination of *Graphis* and *Lecidea* species was considerably reduced by treatment with Cu and Hg solutions (Pyatt, 1979). Likewise spore germination and protonemal development of *Funaria hygrometrica* were severely reduced by Cu and atypical capsule cells were formed at the site of copper contamination. Zinc was considerably less toxic than copper for both spore germination of *Funaria* and the development of *Marchantia polymorpha* gemmae (Lepp and Roberts, 1977). Simola (1977) demonstrated toxic effects of Pb, Cd, Cu and Hg on *Sphagnum* growth. Investigations concerned with effects of metals on growth of *Salvinia natans* showed that Cd in concentrations greater than 0.1 ppm caused severe repression of growth. Reduction in size as well as in numbers of fronds was also observed for *Salvinia* in zinc solutions of 1 ppm and higher concentrations (Hutchinson and Czyrska, 1975). Lepp and Salmon (1999) found that plerocarpous species are more sensitive to Cu than apocarpous.

6. MECHANISM(S) OF METAL TOXICITY

Virtually all metals, essential or non-essential, can exhibit toxicity above threshold concentrations, which may be extremely low for highly toxic metal species. Toxicity mechanisms include the blocking of functional groups of important molecules, e.g. enzymes, polynucleotides, transport systems for essential nutrients and ions, displacement and/or substitution of essential ions from cellular sites, denaturation and inactivation of enzymes, and disruption of cell and organellar membrane integrity. In a biochemical context, every index of living cell's integrity may be affected under conditions of metal pollution (Nagalakshmi and Prasad 1998, 2001; Prasad 1999, Prasad et al. 1998, Reddy and Prasad 1992, a,b; 1994). These inhibitory mechanisms are outlined below.

6.1. Membrane damage

Plasmamembrane is the first target of metals. Thus only after passing the membrane metal can interact with other cellular components. Jensen and co-workers have extensively studied the effect of metals on membrane structure with the help of electron microscopy and X-ray energy dispersive approaches. Formation of extraneous membrane whorls in *Plectonema boryanum* following treatment with Zn, Cu, Hg, Cd and Ni has been demonstrated by Rachlin et al. (1982). Rachlin et al. (1984) using *Anabaena flos-aquae* showed that many internal components of the cell were abolished and plasma membrane was shrunken away from the cell wall resulting in an increase in the relative volume of cell wall number 1 following Cd exposure. A reduction in the volume of wall layers 2, 3 and 4 may indicate a depolymerization of wall mucopolymers and mobilisation of this material into wall layer 1. Rai et al. (1990) also observed a degradation of peptidoglycan layer of the cells of *Anabaena flos-aquae* follwing Cd treatment, degradation being more prominent at extreme pHs. The degradation of peptidoglycan layer was complete at $1.18\mu M$ Cd at pH 4.0.

Another important harmful effect of metals at the membrane level is the alteration of permeability, leading to leakage of ions like potassium and other solutes (De Filippis, 1979). Yeast cells exposed to mercuric chloride suffered irreversible damage of the membrane, resulting in a loss of K^+ and cellular anions into the medium. The maximal loss of K^+ was related to the concentration of mercuric chloride (Passow and Rothstein, 1960). Metals are known to alter membrane permeability in lichen and bryophytes too. Brown and Slingsby (1972) showed that low concentrations of nickel induced only slight losses of potassium from *Cladonia rangiformis* but there was an abrupt increase in K^+ loss at higher Ni concentrations. In contrast, large potassium loss occurred in low concentrations of Cu, Hg and Ag. Overnell (1975) studied K^+ release from algal cells and suggested that increased permeability of the plasma membrane may be considered to constitute the primary toxic effect of Cu. Quick release of amino acids, especially glutamate and aspartate following metal treatment has been reported earlier.

Two generally held mechanisms for metal-induced changes in the plama membrane function are: direct effects on sulphydryl groups of the membrane constituents, and free- radical-mediated lipid peroxidation. Metals are known to possess relatively high affinity for sulphydryl and carboxyl groups, depending on the physicochemical properties of the cations. In the cells of the green alga *Chlorella*, Zn, Hg and Cu induced potassium leakage was strongly correlated to the strength of the metal-sulphydryl bond (De Filippis, 1979). In addition, the ATPase of the plasma membrane, responsible for electrogenic pumping and transport of nutrients was found to be strongly affected by metals (Rai et al., 1994; Husaini et al., 1996a). The H^+-ATPase is a phosphate-linked enzyme in both prokaryotes and eukaryotes, and plays a significant role in regulating the H^+ movement across the cell membrane, thereby maintaining the cytoplasmic pH to neutrality. Therefore, inhibition of H^+-ATPase activity will lead to acidification of the cytoplasm and disruption of the H^+ gradient across the plasma as well as the thylakoid membrane leading to a failure of the photosynthetic machinery and other cellular processes. With the help of a fluorogenic compound, fluoroescein diacetate, Grabowski et al. (1997) observed a decrease in cytoplasmic pH and an increase in cytoplasmic viscosity following Co and Cd treatment. A decrease in cytoplamic pH would also decrease the effectiveness of intracellular enzymes, whereas an increase in cytoplasmic viscosity would decrease the diffusion rate thereby reducing the effectiveness of the reactions. Moreover, Husaini et al. (1996a) demonstrated severe inhibition of Ca and Mg-ATPase activity of *Nostoc linckia* in the presence of Al, which may trigger the action potentials thereby disrupting the intracellular signalling and regulation of different physiological processes, leading finally to the death of the organism (Trebacz et al. 1994).

At the biomembrane level lipid peroxidation is a very important destructive reaction. Lipid peroxidation initiated by the formation of reactive free radicals may severely affect the functioning of biological membranes. This process induces lipid phase transition and increases membrane permeability, thereby leading to cell decompartmentation as well as loss of cellular constituents. De Vos et al. (1989) suggested that copper-induced damage of the permeability barrier is caused by direct action on both the membrane lipids and thiols. They proposed that the first damaging effect of copper ions is the oxidation and cross-linking of membrane protein sulphydryls. However, they also adjudged an important role to copper-induced membrane lipid peroxidation, possibly due to direct free radical formation. Peroxidation of membrane lipid by Cu in the cyanobacterium *Anabaena* was also reported by Mallick and Rai (1999). However, in the case of Pb it was not as much evident, thus linking it to the specific ability of transition group metals only.

6.2. Enzyme inactivation

High metal concentrations produce significant effects on enzyme systems that control biochemical and physiological processes like photosynthesis, respiration and synthesis of biomolecules. Metals may inactivate enzymes by oxidising the -SH groups necessary for catalytic activity or by substitution for other divalent cations. Often the co-ordination sites of the enzymes are normally occupied by metals; inhibition of function occurs when an essential trace metal is displaced by another metal that does not possess the necessary chemical attributes to support biochemical activity. Competitive interaction can occur at a number of different cellular sites and among different metals. In this way, many metalloenzymes can be inhibited. Several enzymes of the Calvin cycle are directly inhibited by metals, e.g., glyceraldehyde-3-phosphate kinase (Weigel, 1985). In the green alga *Chlorella vulgaris* the enzyme protochlorophyllide reductase was found to be inhibited in the presence of sublethal concentration of Hg, which resulted in the reduction of chlorophyll biosynthesis and accumulation of protochlorophyll (De Filippis and Pallaghy, 1976). Zinc had a similar but less pronounced effect. In *Euglena gracilis* a decrease in chlorophyll and a marked enhancement in protochloropyllide level in the presence of sublethal concentrations of Zn, Cd and Hg was evident (De Filippis et al., 1981). RubisCo is also inhibited by heavy metals by one of the following mechanisms: SH group interaction and/or substitution of Mg by another toxic metal (Van Assche and Clijisters, 1990). In the brown alga *Laminaria saccharina* RubisCo is inhibited by cadmium (Kremer and Markham, 1982). No direct interaction of cadmium with the enzyme was found *in vivo*, but *de novo* protein synthesis was inhibited. It was concluded that cadmium in the culture medium inhibits one or several steps in the protein biosynthesis and thus leads to enzyme deficiency.

Rai and co-workers have extensively studied the effect of various metals (Cu, Ni, Cd, Hg, Ag, Cr, Fe, Al etc.) on several enzymes of the nitrogen and phosphorus metabolism of green algae as well as cyanobacteria. A general reduction in the activities of nitrate reductase, glutamine synthetase, urease, alkaline and acid phosphatase, ATPase (both Ca^{2+}- and Mg^{2+}- dependent) of green algae and cyanobacteria, and nitrogenase activity of the nitrogen-fixing cyanobacteria following supplementation of metals was observed (Raizada and Rai, 1985; Dubey and Rai, 1987; Mallick and Rai, 1990; Rai et al., 1996a). Mallick and Rai (1994) demonstrated a competitive inhibition of glutamine synthatase, urease and APase of the nitrogen fixing cyanobacterium *Anabaena doliolum*, whereas the NO_3^- reducing enzyme, nitrate reductase, was inhibited in a non-competitive manner by metals. This suggests a direct competition of metals for the substrate binding sites in the case of the former, whereas the latter was inhibited in an irreversible manner. The former type of inhibition can be overcome by increasing the substrate concentration, whereas metals were found to cause irreparable damage of nitrate reductase the most important enzyme of the nitrogen assimilatory pathway.

Ectoenzymes, enzymes attached to the cell surface with extracellularly orientated catalytic sites, are likely to be particularly sensitive to trace metals. The importance of these enzymes in plankton physiology and aquatic chemistry has been recently discussed (Price and Morel, 1990b). One such enzyme, alkaline phosphtase, has been shown to be inhibited by Cu at a concentration that normally does not have any inhibitory effect on the growth of the organism (Rueter, 1983).

Lichen thalli have been shown to exhibit extracellular enzyme activity. Boissiere (1973) demonstrated that some phosphatase activity was localised in the plasmalemma of the phycobiont (*Nostoc*) of *Peltigera canina*. Their extracellular location made them particularly vulnerable to changes in the immediate environment. Indeed, phosphatase activity has been used previously to measure the effects on lichens both in the field and under laboratory conditions. In the field surveys, the multiplicity of sources precluded the identification of any trace elements as agents responsible for the observed reductions in activity. Laboratory studies have, however, shown that several metals reduce phosphatase activity in whole thallus of *Cladonia rangiferina*.

6.3 Damage to energy yielding metabolism

All the metals studied were found to be potential inhibitors of photosystem II; photosystem I was reported to be less sensitive. From the *in vitro* experiments, at least two metal- sensitive sites in the photosynthetic electron transport chain can be identified: the water splitting or the oxidising side and the direct inactivation of the PS II reaction centre. In the green alga *Chlorella vulgaris*, PS II was found to be sensitive to Cd, Cu and Zn, and the water splitting site was found to be the site of action of all these metals. PS II of *Chlorella vulgaris* (Hg, Zn, Cu and Ni; Rai et al. 1991, 1994), *Anabaena doliolum* (Pb, Cu, Ni and Fe, Rai et al., 1991; Mallick and Rai, 1992), *Nostoc linckia* (Cd and Al, Husaini and Rai, 1992; Rai et al., 1996a) and *Nostoc muscorum* (Pb, Ag and Cr, Prasad et al., 1991) also displayed high sensitivity to metals. *In vitro* experiments with a supply of artificial electron donors suggest that metals exert their toxic action at the water splitting site of the PS II reaction centre. Further Rai et al. (1995) with *Anabaena doliolum* showed that Cu reacts at multiple sites of the photosynthetic electron transport chain. The reduction and blue shifting of absorption peaks, suppression of fluorescence intensity and the excitation spectra also indicate the involvement of phycobiliproteins in the inhibition of PS II activity.

It has recently been established that chlorophyll fluorescence competes with electron transport processes in the photosynthetic apparatus. Parameters based on fluorescence measurements do not directly measure the photosynthetic electron transport, but give a good estimation of the photochemical efficiency of photosystem II. At the same time measurements of *in vivo* chlorophyll *a* fluorescence are one of the most powerful tools for the detection of compounds that exhibit harmful effects on PS II (Fig. 2). Chlorophyll *a* fluorescence not only provides toxicity thresholds, but also sheds light on the mode of action of given xenobiotics (Salvetat et al., 1998). The standard laboratory techniques for studying fluorescence kinetics in plants employ different types of fluorometers. The advantage of pulse-amplitude-modulated (PAM) fluorometer is that only one measurement is sufficient to determine numerous important photosynthetic parameters. The minimum fluorescence value F_0 is estimated with a high accuracy, minimizing bias errors in the determination of the other photosynthetic parameters (F_m, F_v/F_m, Φ_{PSIIR}, CA, qN and qP). In photosynthetic studies CA, qN, and qP are related respectively to PS II working capacity, non-photochemical energy dissipation and also photochemical energy dissipation. Investigations employing F_v/F_m and Φ_{PSIIR} are indicators of maximum PS II photochemical efficiency.

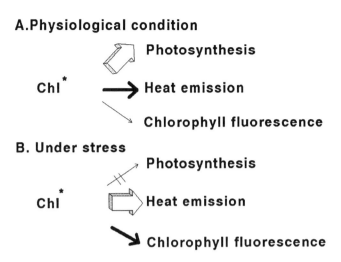

Figure 2. De-excitation of the excited states of chlorophyll a by photosynthesis, heat emission and chlorophyll fluorescence A) under physiological and B) under stress conditions. The thickness of the arrow indicates the relative proportions of the three de-excitation processes. (after Lichtenthaler, 1996).

However, information on the effect of metals on the chlorophyll fluorescence characteristics of lower plants are relatively meagre. Lu and Zhang (1999) demonstrated that Cu inhibition of PS II photochemistry in the cyanobacterium *Spirulina platensis* is linked to the reduction of Q_A. Cu also decreased the proportion of the active PS II reaction centre, but showed no effect on the Q_B-non-reducing sites. Cu was also found to induce an increase in non-photochemical quenching as well as a decrease in the photochemical quenching and the efficiency of energy excitation capture by open PS II reaction centres, both of which resulted in a decrease in the quantum yield of PS II electron transport. Mallick and Mohn (2001) demonstrated a general reduction in quantum yield of the green microalga *Scenedesmus obliquus* under of stress Cu, Ni, Pb, Cr and Cd. Thus an impaired photosystem efficiency in the presence of metals resulted in low photophosporylation and thus low ATP biosynthesis. Metals are also reported to exert a severe detrimental effect on oxidative phosphorylation (Kesseler et al., 1994a, b).

Profound toxic effects of metals on the level of adenylate nucleotides were also reported. Levels of AMP were affected very slightly, while ADP levels remain static or increased, but ATP decreased significantly (De Filippis, 1989). The energy charge, defined as the ratio of (ATP+0.5 ADP)/(ATP+ADP+AMP), decreases sharply with metal treatment. The decrease in the ATP pool is a characteristic response to heavy metals (De Filippis and Ziegler, 1993), which is responsible for the reduced metabolism of living cells.

6.4. Free radical formation

Aerobic organisms enjoy significant energetic advantages by using molecular oxygen as a terminal oxidant in respiration. However, the presence of oxygen in the cellular environment poses a constant oxidative threat to cellular structure and processes (Alscher et al., 1997). Although O_2 itself is a totally harmless molecule, it can potentially reduce to form toxic reactive oxygen species. In the plant system, reactive oxygen species (ROS) are always formed by the inevitable leakage of electrons onto molecular oxygen from the electron transport activities of chloroplasts, mitochondria and the plasma membrane (Fridovich, 1995; Foyer, 1997). In addition, ROS production is known to be stimulated by various toxic metals belonging to the group of transition elements.

All the metals in the first row of the d-block in the periodic table contain unpaired electrons and can thus qualify as radicals, with the sole exception of zinc. Copper does not really fit the definition of a transition element since its 3d-orbitals are full, but it readily forms Cu^{2+} ion by the loss of two electrons. The transition elements are of great biological importance. The most important feature from a radical point of view is their variable valence, which allows them to undergo changes in oxidation state involving one electron thereby stimulating the formation of free radicals. Reactive forms of oxygen include the superoxide radicals (O_2^-), the hydroxyl radical ($^\bullet OH$) and hydrogen peroxide (H_2O_2). All these can be highly toxic since they attack several cell constituents, such as amino acids, proteins, carbohydrates, nucleic acids and lipids. At the biomembrane level, lipid peroxidation is a very important destructive reaction. Additionally, several metal ions such as zinc, cadmium, lead, nickel and mercury form very stable complexes with glutathione and disturb the interconversion of oxidised and reduced glutathione thereby lowering the level of available antioxidants in the cells (Christie and Costa, 1984)

6.5 Metals and nucleic acids

In general metals can interact directly or indirectly with nucleic acids. Ernst (1980) reported an increase in the number of structural chromosomal aberrations after cadmium treatment in *Crepis capillaris* seed. Information from non-vascular plants is rather poor. Most of the reports on the interactions of metals with nucleic acids are based on studies with bacteria and animals. In general metals can interact directly or indirectly with nucleic acids.

The numerous neuclephilic centres in nucleic acids are favourite binding sites for metal ions. Induction of cross linking between both the strands of DNA, single-strand DNA breaks and the formation of complexes between DNA and metals are also reported (Gebhart and Rossman, 1991). Nevertheless, generation of free radicals by metals can indirectly induce breaks in the DNA strand. Exposure of DNA to free oxygen radicals was found to induce extensive strand breakage and degradation of DNA (Halliwell and Gutteridge, 1984). Similar reactions can also be expected for RNA, thereby altering their functions.

Denaturation of DNA microfibrils was observed in *Plectnema boryanum* cells following Cd treatment (Rachlin et al., 1982). Reduction in the total content of DNA and RNA of the nitrogen fixing cyanobacterium *Nostoc muscorum* was observed following exposure to Cr, Pb, Ni and Ag (Rai et al., 1990). Reduction in nucleic acid content of some fungi (*Aspergillus niger, A. ocharaceus and Penicillium digitatum*) and the green alga *Scenedesmus quadricauda* was observed under metal stress (Metwally and Abou-Zeid, 1996; Angadi and Mathad, 1998). Damage in DNA strand was also detected in the yeast *Saccharomyces cerevisiae* following exposure to Hg, Cd and Ni (Kungolos and Aoyama, 1992). Studies on the structure of DNA fragments by RNA slot blot hybridisation showed changes at the DNA transcription level following exposure of wheat plants to Cd, Pb and Zn (Meng et al., 1998). DNA fingerprinting analysis by a PCR-based method showed polymorphic DNA bands in plants exposed to metals, these were not detectable in unexposed plants (Conte et al., 1998). Chromosomal breaks and formation of bridges were observed immediately after Cd treatment. Cd was also found to alter the protein- DNA interaction (Thomine et al., 2000).

6.6. Transport interference

Metals have been reported to interfere with the transport of major as well as minor nutrients. In the diatom *Thalassiosira wiessflogii* Fe and Si transport was interrupted by Cd and Cu respectively (Rueter and Morel, 1981; Harrison and Morel, 1983). At low ferric ion concentrations, Fe uptake rates are competitively inhibited by Cd, whereas Si uptake was linearly dependent on the ratio of the cupric and zinc ion activities. These results demonstrate that Cu inhibits the transport of Si by binding to the zinc binding sites. Production of a low concentration of Fe-containing compounds concomitant with a reduction in Fe-dependent biochemical steps has been observed in Cd-stressed cells.

Cu has also been found to interfere with Mn uptake and metabolism (Sunda et al. 1981). With increasing concentration of Cu a decrease in cellular Mn quota and a reduction in growth rate was evident in oceanic phytoplankton. This effect was reversed by increasing Mn concentration. Husaini and Rai (1992) observed a competitive inhibition for nitrate uptake and non–competitive for phosphate uptake in *Nostoc linckia* in the presence of Cd. In contrast to the above, *Chlorella vulgaris* showed competitive inhibition of ammonium and non-competitive for nitrate uptake following Cu and Ni treatment. Competitive inhibition of major nutrients such as ammonium, urea and phosphate uptake by Cd, Cu, Ni, Fe and Al was also observed in *Anabaena doliolum*. However, a non-competitive uptake of nitrate with above metals was also observed (Mallick and Rai, 1994).

6.7. Major nutrient mimicry

The word 'mimicry' as it means 'to copy the way in which a particular organism/ thing works'. As described by Autor Hoffmann 'it is same, but not the same'. It is a condition existing among certain animals, especially insects, in which they have the appearance of other animals or insects thereby gain some protection. In the strict sense it means imitation.

In metal pollution studies some interesting mimicry has also been observed, particularly with inorganic metal complexes like AsO_4^{3-} and AlF_4^-. These complexes interfere with the metabolism of PO_4^{3-} by acting as an structural analogue of the latter. Competitive interference of PO_4^{3-} uptake and transport by AsO_4^{3-} was reported in a marine yeast as well as in phytoplankton (Button et al., 1973; Planas and Healey, 1978). In contrast, AlF_4^- was found to inhibit PO_4^{3-} metabolism in bacteria non-competitively (Missiaen et al., 1988). Rai and co-workers have done extensive work on the formation and role of AlF_4^- complex on the laboratory–grown cyanobacterium *Nostoc linckia* and the green microalga *Chlorella vulgaris*. The toxic effects of NaF, $AlCl_3$, AlF_3 and the combination of ($AlCl_3$ + NaF) were compared on various processes like growth, photosynthesis (O_2 evolution and carbon fixation), photosynthetic electron transport chain, uptake of nutrients and their kinetics, and activities of various enzymes of nitrogen and phosphorus metabolism at different pH values (Husaini et al., 1996b). The combination of ($AlCl_3$ + NaF) and AlF_3 was more inhibitory to all tested processes as compared to $AlCl_3$ and NaF separately. Toxicity of all these test chemicals increased with acidity. However, interaction of ($AlCl_3$ + NaF) was additive at pHs 7.5 and 6.8 whereas synergistic at pHs 6.0 and 4.5. The formation of fluoroaluminate complexes in the presence of aluminium chloride and sodium fluoride at low acidic pHs (pH 6.0 and 4.5) was demonstrated by Rai et al. (1996a) using IR spectroscopy. Further studies on the inhibition kinetics of Mg^{2+}- and Ca^{2+}-ATPase and characteristics of SDS-PAGE demonstrated that AlF_4^- inhibits ATPase activity of both the test algae by acting as a functional barrier without affecting the structure of the enzyme (Husaini et al., 1996a). In presence of high concentration of PO_4^{3-}, NaF (in the $AlCl_3$ + NaF) was required to produce the same degree of inhibition in ATP synthesis and ATPase activity, thus demonstrating a competition between PO_4^{3-} and AlF_4^- for the binding sites. However, except for beryllium to some extent, combinations of Cd, Co, Fe, Sn, and Zn with F were not as effective as Al in inhibiting ATPase activity. Thus, it is predicted that in acidified lakes containing high concentrations of dissolved Al, AlF_4^- complexes may exist and inhibit extracellular phosphatases and phosphate transporters.

7. MECHANISMS OF METAL TOLERANCE

Tolerance can be broadly defined as the ability of individuals to cope up with the stresses that are inhibitory or lethal to nontolerant individuals to which they are exposed. As described earlier high concentrations of metals occur naturally in many abused environments. Organisms in these environments, often characterised as endemic metalliferous forms, might have taken long periods of time to evolve tolerance strategies. For example populations occurring in mining and industrial effluents may have had scores of years to evolve tolerance.

The acquisition of tolerance is not a predictable consequence of exposure to metal stress. A comparison of species composition, number, and/or diversity reveals that polluted sites are often biologically less diverse. Many species present in the noncontaminated areas are absent from nearby contaminated sites. This is true even of species capable of movement or dispersal into the polluted sites. Presumably these species are not present because they do not have or cannot achieve the appropriate tolerance. From this perspective, the evolution of tolerance would be uncommon and most species are excluded from the contaminated sites.

The study of tolerance and genetic mechanisms involved in tolerance acquisition can be most effectively done when the tolerance or tolerance-related character is objectively defined and readily assayed. In the case of higher plants, the Tolerance Index (ratio of root growth in toxic solution relative to growth in control solution) has been widely used. Assays for aquatic animals include measures of body burdens, fin generation, larval abnormality, growth rate, longevity and fecundity. In lower plants, including algae, growth rate, species composition, diversity, different physiological and biochemical processes etc., are the common measures of tolerance. These attributes are assumed to be the components of or related to individual fitness. Under appropriate circumstances, the difference in these attributes between individuals and/or populations exposed to metals relative to their nonexposed counterparts can be used as indicators.

Metal tolerance may be defined therefore as the ability of an organism to survive metal toxicity by means of intrinsic properties, which may include possession of impermeable cell membrane, change in metal uptake or elimination rates, the ability to bind or sequester metals, and the differences in enzyme sensitivity to inhibition by metals. Such mechanisms have been termed 'gratuitous' mechanisms of tolerance/ resistance. However, distinctions are still difficult in many cases because of the frequent involvement of several direct and indirect mechanisms, both physico-chemical and biological, for survival of the organism in the field as well as under laboratory conditions. A given organism may directly and /or indirectly rely on several survival strategies, though this is not always appreciated in the literature. For example, metallothionein synthesis is a mechanism of copper resistance in yeast, yet copper binding or precipitation around the cell wall and transport across the cell membrane must be the component of the total cellular response. General mechanisms of tolerance to metals in the non-vascular plants fall into at least six categories (Table 3), which are discussed below.

7.1 Metal efflux or exclusion

The Cu exclusion mechanism was first reported by Foster (1977) in the green microalga *Chlorella*. In his study two strains of *Chlorella vulgaris* Beijerinck were isolated from the river Hayle, which drains disused copper mines in Cornwall, England. The non-tolerant strain came from a site upstream of mining influence where copper was not detectable in the water (< 0.002 mg l^{-1}). The tolerant strain came from a polluted site with a total copper concentration in the water of 0.12 l^{-1} (mean of four samples taken

Table.3. Mechanisms Of Resistance /Tolerance To Heavy Metals

Mechanism	Reference
Metal efflux or exclusion	Foster (1977)
Extracellular binding or external complexation	Shimwell & Laurie (1972)
Binding and precipitation within the cytoplasm or vacuole	Skarr et al. (1973)
Special detoxification by biotransformation of metals	De Filippis (1978)
Binding onto metal binding polypeptides and proteins	Gekeler et al. (1988)
Activation of antioxidant	Mallick and Rai (1999)

during 1975-76). The effect of Cu on the growth rates of the tolerant and non-tolerant strains was measured.

Although copper reduced the growth rates of both the strains, the tolerant strain was much less sensitive. A concentration of 0.3 mg l^{-1} was found completely inhibitory for the sensitive strain, whereas the tolerant strain was found to survive and grow even at a Cu concentration of 1.0 mg l^{-1}. The Cu uptake results showed that the non-tolerant strain accumulated 5-10 times more metal at comparable Cu concentration. Analysis of growth rate of each culture as a function of the amount of Cu accumulation by the algae showed lack of differences. This means the two strains respond identically to the same amount of cellular Cu. Other growth parameters, for example, lag period and yield, are also identical in the two strains at equivalent cellular Cu concentration. Thus Cu exclusion is likely to be the only mechanism of tolerance in this alga.

There is good evidence for some exclusion mechanisms operating at the cell membrane level in resistant organisms (Hall, 1981). Cadmium resistance of *Euglena gracilis* strain was suggested to be the result of inhibition of Cd transport across the membrane or the restriction of entry of Cd by the membrane (Bariaud et al., 1985). In *Nostoc calcicola*, an energy- dependent Cu efflux system was demonstrated in a resistant mutant (Verma and Singh, 1991). Rai et al. (1991) isolated a Cu-tolerant strain of the cyanobacterium *Anabaena doliolum* in laboratory by repeated subculturing in Cu-enriched medium and Cu tolerance was studied by comparing the physiological properties of its wild type with the tolerant strain. A concentration dependent reduction in growth, pigments, protein, sugar, lipid and ATP pools, photosynthetic electron transport chain, O_2 evolution, carbon fixation, uptake of nutrients and activities of enzymes like nitrate reductase, nitrogenase and glutamine synthetase was noticed in both the strains following Cu supplementation. The reduction in all the parameters was higher in wild type than the tolerant strain. The later was found to produce a larger (19.5%) lipid fraction than the wild type (10.3%) and also exhibited low Cu uptake and insignificant loss of K^+ and Na^+. The data on lipid production, loss of K^+ and Na^+, uptake of copper collectively indicate change in plasma membrane properties e.g. reduced permeability is responsible for reduced uptake of Cu. Yoshimura et al. (1999) studied Al exclusion mechanism in an acidophilic thermophilic alga, *Cyanidium caldarium*. The test alga was cultured in a medium containing various metal ions (Al, Cd, Cr, Cu, Mn, Ni, Zn, etc.). Of all these metals tested the alga was found to survive in especially high concentrations of Al (200 mM). However, the intracellular Al concentration was found to be very low, despite its high concentration in the external medium. Treatment of the alga with CCCP resulted in increased Al uptake. This suggested the operation of an energy-dependent Al exclusion mechanism in the acidophilic alga, *Cyanidium caldarium*.

7.2. Extracellular complexation

Shimwell and Laurie (1972) observed crystalline metal deposits on the carpets of the moss *Dicranella varica*. In the bryophyte *Grimmia diniana* grown in a lead mine area, Pb was found to bind to the extracellular sites. Zinc in aquatic bryophytes has also been reported to bind chiefly with cell wall. Electron microscopy has revealed the presence of Pb precipitates in the cell wall of *Hylocomium splendens* (Skaar et al., 1973).

Extracellular organic materials and metabolites have been involved in reducing toxicity of metals mainly in blue-green algae and in some green algae (Butler et al., 1980). Sheaths isolated from cultures of *Gloeothece* ATCC 27152 showed substantial biosorption of Cd (Tease and Walker, 1987). Cr precipitation in the mucilage of *Nostoc muscorum* and precipitation of Cu, Ni and Zn in the sheath of *Calothrix parietina* and *C. scopulorum* has also been reported (Weckesser et al., 1988). Parker et al. (1998) demonstrated that the bloom-forming cyanobacterium *Microcystis* slime or capsule has tremendous potential for interaction with cations.

Sarret et al. (1998) studied the Pb and Zn hyperaccumulation and tolerance in lichen *Xanthoria parietina* and *Diploschistes muscorum*. The speciation of Zn and Pb has been investigated by powder X-ray diffraction (XRD) and extended X-ray absorption fine structure (EXAFS) spectroscopy using the advanced third generation synchrotron radiation source. Their study revealed that both the lichen were protected from toxicity by the complexation of metals, but the strategies differed. In *D. muscorum* Pb and Zn accumulated through higher synthesis of oxalate, which precipitates toxic elements as insoluble salts, whereas in *X. parietina* Pb is complexed to carboxylic group of fungal cell walls.

7.3. Binding or precipitation within the cytoplasm or vacuole

Internal precipitation of lead within the cytoplasm was reported in *Rhytidiadelphus squarrous* (Skaar et al., 1973). Mosses from an urban polluted area showed electron-dense particles of lead enclosed in vesicles and in plasma membrane invaginations into the cytoplasm. Silverberg et al. (1976) reported intracellular inclusions in copper-tolerant *Scenedesmus,* and concluded that the formation of inclusion bodies leads to a lower concentration of copper in the cytoplasm. Internally accumulated metals, such as

Cd, Cu, Hg, Ni and Zn, have been shown by *in situ* X-ray dispersive analysis to be localized in polyphosphate bodies of *Plectonema boryanum, Anabaena cylindrica, Anacystis nidulans* and *Anabaena flos-aquae* (Crang and Jensen, 1975; Jensen et al., 1982; Rachlin et al., 1984; Pettersson et al, 1985). Intact vacuoles isolated from higher plants (tobacco and barley) exposed to Zn have been clearly shown to accumulate metals. Vacuolar Zn accumulation has also been confirmed in both roots and shoots of *Thalaspi caerulescens* (Vazquez et al., 1994). The cyanobacterium *Synechococcus* sp. synthesises large quantities of intracellular polymer that binds nickel, leaving the cell interior highly granular. X-ray analysis in algae showed that the electron-dense granules are complex co-precipitation sites for metals, Ca, phosphorus and sulphur (Jensen et al. 1982).

The responses of *Saccharomyces cerevisiae* towards the oxyanions (tellurite and chromate) were investigated in order to establish the three mutants of *S. cerevisiae* with defective vacuolar morphology and function (Gharieb and Gadd, 1998). All the mutant strains showed increased sensitivity to both the oxyanions. Their results indicate that accumulation of both tellurium and chromium occurred in the cytosolic compartment of the cell, with detoxification influenced by the presence of a functionally active vacuole, which plays a role in compartmentation as well as the regulation of the cytosolic compartment for optimal expression of a detoxification mechanism.

7.4 Biotransformation

Biotransformation, i.e, alteration of the form of metals by methylation and other reactions, thereby changing the volatility, solubility and toxicity, is another kind of detoxification mechanism for metals. Biomethylation has been reported in *Chlorella* (De Filippis, 1978), where ethylene and vitamin B_{12} are the proposed methylating agents. Detoxification by biotransformation is clearly evident for mercurial compounds where inorganic (Hg^{2+} or phenylmercuric ($Ph.Hg^{2+}$) ions are biologically transformed to metallic mercury (Hg^0) with the help of a reducing enzyme, which is then volatilized from the cell.

Volatilisation of selenium as dimethyl selenide from animals has a long history. In the plant system Lewis et al. (1966) was the first to report the above phenomenon. The volatile selenium compound released from the selenium accumulator species *Astragalus racemosus* was identified as dimethyl diselenide, whereas selenium released from alfalfa, a selenium non-accumulator, was identified as dimethyl selenide (Lewis et al., 1974). Unfortunately no such studies have been conducted in lower plant cells. However, volatilization of arsenic as dimethylarsenic has also been reported from marine algae (Francesconi and Edmonds, 1994). Chromium is also toxic to plants, but there is limited evidence that plants like certain bacteria and animals can reduce Cr (VI) to Cr (III) as part of their detoxification mechanism (Dushenkov et al., 1995).

7.5. Metal binding polypeptides or proteins

Metal tolerance in animals and plants (vascular and non-vascular) is known to be conferred by production of a special class of polypeptides or proteins, called metallothioneins (MTs, three classed MT I, MT II and MT III) or phytochelatins (PCs). MT I was also reported in some genera of fungi e.g. *Agaricus bisporus* and *Neurospora crassa,* whereas MT II was reported in the yeast *Saccharomyces cerevisiae.* Nevertheless, class II MTs have also been isolated from a marine cyanobacterium *Synechococcus* RRIMP NI and the freshwater strains *Synechococcus* UTEX-625 and *Synechococcus* TX-20 (Olafson et al., 1980). The MT from *Synechococcus* TX-20 was subsequently sequenced (Olafson et al., 1988). A metal-thiolate cluster, similar to that of eukaryotic MT II but with a single domain, is found to be regulated at transcription level. Class III metallothioneins, or phytochelatins, are reported to range from the yeast *(Schizosaccharomyces pombe),* to eukaryotic algae and also to higher plants. These are typical in the sense that they are non-translationally synthesised metal thiolate polypeptides and secondary metabolites. The presence of several γ-caboxymide linkages in this polypeptide was the turning point for classifying them into a different class of MT (Robinson, 1989). Purified proteins of this polypeptide are composed of three amino acids, namely L-glutamic acid, L-cystein and glycine. However, biosynthesis of these peptides revealed that they are not primary gene products. Their synthesis is catalysed by a specific γ–glutamyl cystein dipeptidyl transpeptidase, called phytochelatin synthase, which is activated in the presence of metal ions and uses glutathione as the substrate for the post-translationally activated metal-dependent enzymatic pathway (De Vos et al., 1992). Purified peptide from *Euglena* showed a ratio of Glu : Cys : Gly as 2 : 2 : 1 (De Filippis, 1989), whereas in the case of *Chlorella* it is 5 : 5 : 1 (Gekeler et al., 1988). In the nitrogen fixing cyanobacterium *Anabaena dolioum* the Cd-induced low molecular wt. protein resembled most of the characteristics of phytochelatins such as molecular wt. of only 3.3 kDa, richness in Cd and -SH contents, sensitivity to buthionine sulfoximine (BSO) and loss of absorbance at A_{254} following acidification which could not be resumed even after neutralisation. Synthesis of this protein was blocked in the presence of transcriptional and translational inhibitors, but resumed on supplementation of glutathione, thus suggesting that its synthesis is independent of genetic control (Mallick and Rai, 1998). Synthesis of this protein was not only found to confer co-tolerance to other metals in the cyanobacterium, but also provides multiple tolerance to a number of environmental stresses like anaerobiosis, heat and cold shocks, X-rays and ultraviolet–B radiation. Rijstenbil et al. (1998) showed that in the planktonic diatom *Thalassiosira pseudonana* the synthesis of phytochelatin is dependent on the cellular C:N ratios, i.e. the appearance of phytochelatin was noticed at cellular atomic C:N ratios lower than ~ 15. Synthesis of new protein bands under Cd stress was also reported in two cyanobacterial isolates of *Chroococcus* spp. (Santos and Gilda, 1999). However, further investigation is needed for full characterisation of this protein band.

Table 4. Classes of metal binding polypeptides covering a wide range of phenotypically related metallothiolate proteins (Robinson, 1989).

Class	Description	Species
Metallothionein MT I	Proteins (polypeptides) with locations of cysteine closely related to those in equine renal metallothionein	*Agaricus bisporus*, *Neurospora crassa*, vertebrate animals
Metallothionein MT II	Proteins (polypeptides) with locations of cysteine only distantly related to those in equine renal metallothionein	*Synechococcus* TX-20, *Saccharomyces cerevisiae*, *Anacystis nidulans*
Metallothionein MT III Phytochelatins (PC)	Atypical, non-translationally synthesized metal thiolate polypeptides (secondary metabolites)	*Shizosaccharomyces pombe*, eukaryotic algae, higher plants

Polymerase chain reaction products corresponding to part of an MT gene were generated using template DNA from *Synechococcus* PCC-6301, the products were sequenced and the gene called *Smt A* was identified (Morby et al., 1993). Amplification of the *Smt A* gene and development of unique *Smt A* restriction fragments were detected in Cd-tolerant lines of *Synechococcus* PCC-6301 (Gupta et al., 1992). These fragments were subsequently used as probes to isolate a MT divergon, *Smt*, which includes *Smt A* and a divergently transcribed gene, *Smt B*. According to the sequence homology, *Smt B* is considered to be a member of the *ArsR* family of metalloregulatory proteins. It has been proposed that, in addition to a predicted helix-turn-helix DNA binding domain, each member of the *ArsR* family contains a metal binding box having a consensus sequence of ELCVCDL. Binding of metal within this region is thought to induce a conformational change in the adjacent helix-turn-helix domain resulting in dissociation of the repressor from the DNA.

Smt B is also capable of direct interaction with metals as evidenced by ^{65}Zn binding to the *Smt B* protein as well as the inhibition of repressor DNA complex formation in the presence of various metal ions. Methylation interference analysis of such complexes identifies four protein contact points within the *Smt* operator/promoter DNA. The points of contact appear to represent two pairs of binding sites, one pair in each of two inverted repeats. The *Smt B* was found to bind to the *Smt* operator/promoter in a multimeric fashion (Erbe et al. 1995).

Recent developments in genetic research showed that the transfer of mouse MT-I gene into the transgenic cyanobacterium *Anabaena* sp. PCC7120 confers Cd resistance on the latter (Ren et al., 1998). A 300-bp fragment containing the entire mMT-I gene coding sequence was excised from the plasmid pBX-MT by digestion with *Bgl*II and *Bam*HI. The excised DNA fragment containing translation initiation and termination sites was inserted into the *Bam*HI site of the intermediary transformation vector pRL-439, downstream of the *psb*A-promoter derived from chloroplast DNA of *Amaranthus hybris*. Positive clones containing the plasmid pRL-MT were selected by rapid plasmid digestion of the recombinant DNA. The recombinant intermediary plasmids were digested with *Eco*RI and then ligated to the shuttle vector pKT-210. The shuttle expression plasmodia pKT-MT was then introduced to *Anabaena* by triparental conjugative transfer. Similar acquired resistance to metals was also observed by transferring metallothonein genes from higher plant cells to yeast and *vice versa* (Hasegawa et al., 1997; Ezaki et al., 1999).

7.6. Metals and antioxidants

Metal ions initiate the formation of free radicals inside the cell. Living cells have evolved, antioxidant defence mechanisms to combat the danger posed by the presence of reactive oxygen species (ROS). These include several enzymatic and non-enzymatic mechanisms such as superoxide dismutase (SOD), peroxidase (POD), catalase (CAT), glutathione reductase (GR), GSH (reduced glutathione), ascorbic acid, tocopherols, carotenoids, flavonoids, hydroquinones, phycocyanin, etc. Glutathione, ascorbic acid, α-tocopherol, hydroquinones, β-carotene and flavonoids act as reductant and free radical scavengers. Vitamins C and E, glutathione and hydroquinones react directly or via enzyme catalysis with $^{\bullet}OH$, H_2O_2 or O_2^-, while carotenes and flavonoids directly interact with singlet oxygen. Peroxidase, SOD and catalase function as effective quenchers of reactive intermediary forms of oxygen.

Increased activity of SOD in the marine alga *Dictylum brightwelli* under Cu supplementation was reported by Rijstenbil et al. (1994). Similar stimulation was reported in *Tetraselmis gracilis* in the presence of Cd (Okamoto et al., 1996). Supplementation of a metal mixture (Hg, Cd, Pb and Cu) was also found to stimulate SOD activity in *Gonyaulax polyedra* (Okamoto and Colepicolo, 1998). Rijstenbil et al. (1998) further reported a reduction in glutathione redox ratio, GSH : (GSH + 0.5 GSSG), in *Enteromorpha prolifera* under Cu stress. Mallick and Rai (1999) showed an increased activity of SOD and catalase in *Anabaena doliolum* following Cu treatment. However, a reduction in activities of ascorbate peroxidase and glutathione reductase demonstrated the operation of an antioxidant system though not enough to withstand the copper stress (Mallick and Rai, 1999).

Of several metals tested, only the addition of Fe^{2+} to the growth medium resulted in an increase in carotenoid levels in *Hematococcus* cells on a unit culture volume basis. However, the addition of EDTA-chelated $FeCl_3$ hence no significant difference in astaxanthin formation as reported by Borowitzka et al. (1991) was observed. As the ferrous form of iron is known to give rise to free radical formation *via* the Fenton reaction, it may play a role in astaxanthin formation. Kobayashi et al. (1993) reported that increased astaxanthin formation caused by Fe^{2+} was due to the formation of hydroxyl radicals and the addition of potassium iodide, scavenger of hydroxyl radical, inhibited astaxanthin biosynthesis. Other active oxygen species including superoxide, hydrogen peroxide, peroxyl etc. also stimulate the formation of astaxanthin (Tjahjono et al., 1994). Enhanced formation of carotenoids by free radicals was also reported in *Dunaliella* (Ben-Amotz and Avron, 1983) and *Phaffia* (Schroeder and Johnson, 1995). However, when the yeast cells are exposed to exogenous $^1O_2$, not peroxy radicals or H_2O_2, carotenoid synthesis is stimulated.

Park et al. (1998) isolated a Cd-induced gene, CIP_2, that specifically hybridizes to a mRNA of approximately 950 nucleotides. The CIP_2 mRNA was barely present in normal *Candida* sp., but synthesis was enhanced in Cd-treated cells. CIP_2 contains an open reading frame encoding a protein of 203 amino acids. This gene was found to play a crucial role in the specific cellular response to oxidative stress evolved by the cadmium treatment.

8. ACIDIFICATION AND METAL TOLERANCE

Atmospheric emissions of acidifying substances such as sulphur dioxide (SO_2) and nitrogen oxides (NOx), mainly from the burning of fossil fuels, can persist in the air for upto a few days and thus be transported over thousands of kilometres, where they are chemically converted into acids (sulphuric and nitric). After their deposition the primary pollutants sulphur dioxide, nitrogen dioxide and ammonia (NH_3), together with their reaction products, lead to changes in the chemical composition of the soil and surface water. The decline in forests in Central and Eastern Europe and the many dead lakes in Scandinavia and Canada are examples of damage which, in part, is due to acidification.

Anthropogenic sulphur dioxide emissions are due largely to the combustion of sulphur-containing fuels (oil and coal) used in power stations, other stationary combustion activities and processing industries (refineries). Nitrogen oxides are emitted by combustion processes; transport, power generation and heating. Most of the ammonia in the atmosphere is due to the production and spreading of animal manure. Although ammonia is a base gas, it may lead to acidification after it reaches the soil and is nitrified.

Emissions of compounds leading to acidification increased considerably in Europe after the industrial revolution and especially after World War II. In Scandinavia and North America, lakes provided the first warning signals for acidification: the water became peculiarly clear, beds are gradually covered with an ever thicker growth of *Sphagnum* mosses, and there was an increase in fish death and a decline in the number of other aquatic animals. Nowadays with increasing acidity, the rain water has been found to have a pH between 4 and 4.5, and even individual values as low as 3 have been reported (Stanners and Bourdeau, 1995).

Metal toxicity can be expected to produce important effects on the species of the acidifying environment. In natural acidic environments the availability of toxic metals always increases - at low pH metals tend to stay in the free ionic state. However, Yan (1999) showed that the phytoplankton in several acidic lakes close to smelters, which have high levels of metals such as Cu and Ni, did not differ from lakes of comparable pH without these additional metals. This suggests that species occurring in these acidic situations may be co-tolerant to metals. Gimmler et al. (1989, 1991) have verified the above hypothesis with the help of the unicellular acid resistant green alga *Dunaliella acidophila*. This alga exhibits optimal growth and photosynthesis at a pH of 1.0. The cytoplasmic pH of this non-vacuolate alga was found to be maintained at pH 7.0. Thus the cytoplasmic H^+ concentration is lower by a factor of 10^6 than that of the medium and the resulting H^+ concentration gradient across the plasma membrane was far greater than that of the cells growing at neutral pH. At least two adaptations are necessary in order to maintain a large trans-membrane H^+ gradient: (1) H^+ influx into the cell must be minimised by a low permeability coefficient of plasma membrane for H^+; and (2) protons which have entered the cells must be pumped out by efficient mechanism. It is worthwhile to indicate that the membrane potential of the test alga *D. acidophilla* was found to be positive (Gimmler et al. 1989); normally the plasma membrane of the plant cells have a net negative charge originating from phosphate and caboxylate groups of the membrane constituents. The zeta potential, which is a measure of surface potential, was also calculated from the electrophoretic mobility of the cells as measured by means of free-flow electrophoresis (Gimmler et al., 1991). *Dunaliella acidophila* cells exhibited a positive zeta potential (+5 to +20mV) at acidic pH, which enables the alga to maintain the cytoplasmic pH at neutrality. Besides, the photosynthesis and growth of this alga also showed resistance against Al, La, Cu, Hg and Cd (Gimmler et al, 1991). These cations also caused an increase of positive zeta potentials. Toxic polyvalent anions decrease positive zeta potentials, but increase negative zeta potentials. This result suggested that the zeta potential plays an important role in regulating cation and anion toxicity.

Rai and co-workers have also done considerable amount of work on the mechanism of survival of *Chlorella* in acid and metal stressed environment (Rai et al., 1996b). In their study an acid-tolerant strain of *Chlorella vulgaris* was isolated in laboratory by successive subculturing of the wild type alga at low acidic pH. The growth rate, uptake of NH_4^+, Na^+, K^+, Ca^{2+}, NO_3^-, PO_4^{3-}, efflux of Na^+ and K^+, activities of nitrate reductase, acid phosphatase and ATPase and internal pH of the acid-tolerant and the wild type strain of acid-sensitive *Chlorella vulgaris* exposed to Cu and Ni at different pHs were compared. A general reduction in physiological variables was noticed with decreasing pH; however, the acid-tolerant strain was metabolically more active than the sensitive one. Reduction in the uptake of cations coupled with an increased uptake of anions suggested the possible development of positive membrane /zeta potentials in the acid-tolerant strain. ATPase activity of the acid-tolerant strain increased approximately 2.5-9 folds at pH 5.0 and pH 3.5, respectively, relative to that of pH 6.8. A reverse trend was observed for the sensitive strain. Moreover, a relatively low loss of Na^+ and K^+ ions was observed in the tolerant strain following metal treatment at acidic pH. However, the internal pH of the acid-sensitive *Chlorella* decreased with decreasing external pH, whereas the internal pH of the tolerant strain was less responsive to external pH. Based on their results Rai et al. (1996b) proposed a mechanistic model to demonstrate the survival of organisms in acid- and metal- stressed environments (Fig. 3). This includes 4 mechanisms such as (i) accelerated uptake of anions for neutralizing internal pH; (ii) development of superactive ATPase for

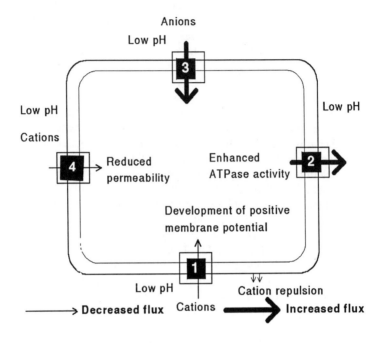

Figure 3. Proposed mechanistic model for acid and metal tolerance by Chlorella vulgaris. (after Rai et al., 1996).

extrusion of excess H^+ ions; (iii) repulsion of cations including metals and (iv) reduced membrane permeability coupled with increased lipid production. This indicates that the development of a superactive ATPase and a change in membrane potential and permeability not only offers protection against acidity by keeping internal pH neutral but also offers co-tolerance to metals.

ACKNOWLEDGEMENTS

Nirupama Mallick is thankful to the Agricultural and Food Engineering Department, IIT Kharagpur, West Bengal for facilities. L.C. Rai is thankful to the Ministry of Environment and Forests, New Delhi for Pitamber Pant National Environment Fellowship Award and National Science Foundation, USA for Indo-US Project.

REFERENCES

Adam, L. and Waldemar, S. (1998) The effect of salinity on toxic influence of heavy metals towards planktonic green algae, *Biologia* **53**, 547-555.

Al-Chalabi, A.S. and Hawker, D. (2000) Distribution of vehicular lead in roadside soils of major roads of Brisbane, Australia, *Water Air Soil Pollut.* **118**, 299-310.

Alscher, R.G., Donahue J.L. and Cramer, C.L. (1997) Reactive oxygen species and antioxidants: relationships in green cells. *Physiol. Plant.* **100**, 224-233.

Angadi, S.B. and Mathad, P. (1998) Effect of copper, cadmium and mercury on the morphological, physiological and biochemical characteristics of *Scenedesmus quadricauda* (Turp). De Breb, *J. Environ. Biol.* **19**, 119-124.

Aruguete, D.M., Aldstadt, J.H. and Mueller, G.M. (1998) accumulation of several heavy metals and lanthanides in mushrooms (agaricales) from chicago region, *Sci. Total Environ.* 224, 43-56.

Avery, S.V., Codd, G.A. and Gadd, G.M. (1991) Caesium accumulation and interactions with other monovalent cations in the cyanobacterium *Synechocystis* PCC 6803, *J. Gen. Microbiol.* **137**, 405-413.

Azeez, P.A. and Banerjee, D.K. (1986) Effect of copper and cadmium on carbon assimilation and uptake of metals by algae, *Toxicol. Environ. Chem.* **12**, 77-86.

Azeez, P.A. and Banerjee, D.K. (1988) Effect of chromium on cyanobacteria and its accumulation, *Toxicol. Environ. Chem.* **16**, 229-240.

Babich, H. and Stotzky, G.C. (1983) Influence of chemical speciation on the toxicity of heavy metals to the microbiota, in J.O. Nriagu (ed.), *Aquatic Toxicology*, John Wiley & Sons. Inc. New York, pp. 1-46.

Bargagli, R., Sanchez-Hernandez, J.C., Martella, L. and Monaci. F. (1998) Mercury, cadmium and lead accumulation in Antarctic mosses growing along nutrient and moisture gradients. *Polar Biol.* **19**, 316-322.

Bariaud, A., Bury, M. and Mestre, J.C. (1985) Mechanism of Cd^{2+} resistance in *Euglena gracilis*, *Physiol. Plant.* **63**, 382-386.

Behra, R. (1993) In vitro effects of cadmium, zinc and lead on calmodulin-dependent actions in *Oncorhynchus mykiss*, *Mytilus* sp. and *Chlamydomonas reindardtii*, *Arch. Environ. Contam. Toxicol.* **24**, 21-27.

Ben-Amotz, A. and Avron, M. (1983) Accumulation of metabolites by halotolerant algae and its industrial potential, *Ann. Rev. Microbiol.* **37**, 95-119.

Boissiere, M.C. (1973) Activi phosphatasique neutre chez le phycobionte de *Peltigera canina* comparie celled'un *Nostoc* libre, C.R. Acad. Sci. Paris. 277(Sirie D), pp. 1649-1651.

Borowitzka, M.A., Huisman, J.M. and Osborn, A. (1991) Cultures of the astaxanthin-producing green algae *Haematococcus pluvialis*. I. Effect of nutrient on growth and cell type, *J. Appl. Phycol.* **3**, 295-304.

Branquinho, C. (2001) Lichens, In: M.N.V Prasad (ed) *Metals in the environment analysis by biodiversity*. Marcel Dekker, New York, pp. 117-158.

Brown, D.H. and Slingsby, D.R. (1972) The cellular location of lead and potassium in the lichen *Cladonia rangiformis* (L.) Hoffm, *New Phytol.* **71**, 297-305.

Burkitt, A., Lester, P. and Nickless, G. (1972) Distribution of heavy metals in the vicinity of an industrial complex, *Nature* **238**, 327-328.

Butler, M., Haskew, A.E.J. and Young, M.M. (1980) Copper tolerance in the green alga, *Chlorella vulgaris*, *Plant Cell Environ.* **3**, 119-126.

Button, D.K., Dunker, S.S. and Morse, M.L. (1973) Continuous culture of *Rhodotorula rubra*: kinetics of phosphate-arsenate uptake, inhibition, and phosphate-limited growth, *J. Bacteriol.* **113**, 599-611.

Cairney, J.W. G. and Meharg, A.A. (1999) Influences of anthropogenic pollution on mycorrhizal fungal communities, *Environ. Pollution* **106**, 169-182.

Christie, N.T. and Costa, M. (1984) *In vitro* assessment of the toxicity of metal compounds. IV. Deposition of metals in cells: Interactions with membranes, glutathione, metallothionein and DNA, *Biol. Trace Elem. Res.* **6**, 139-158.

Colpaert, J.V. and Vandenkoornhuyse , P. (2001) Mycorrhizal Fungi, in M.N.V Prasad (ed), *Metals in the environment analysis by biodiversity*, Marcel Dekker, New York, pp. 37-58

Colpaert, J.V., and Van Assche, J.A. (1993) The effects of cadmium on ectomycorrhizal *Pinus sylvestris*. L, *New Phytol.* **123**, 325-333.

Conte, C., Mutti, I., Puglisi, P., Ferrarini, A., Regina, G., Maestri, E. and Marmiroli, N. (1998) DNA fingerprinting analysis by PCR-based method for monitoring of genotoxic effects of heavy metal pollution, *Chemosphere* **37**, 2739-2749.
Crang, R.E., and Jensen, T.E. (1975) Incorporation of titanium in polyphosphate bodies of *Anacystis nidulans*, *J. Cell Biol.* **67**, 80a.
Crompton, T.R. (1998) *Occurrence and Analysis of Organometallic Compounds in the Environment*, John Wiley and Sons. Chichester, England, pp. 237.
De Filippis, L.F. (1978) The effect of sub-lethal concentrations of mercury and zinc on *Chlorella*. IV characteristics of a general enzyme system for metallic ions, *Z. Pflanzenphysiol.* **86**, 339-352.
De Filippis, L.F. (1979) The effect of heavy metal compounds on the permeability of *Chlorella* cells, *Z. Pflanzenphysiol.* **92**, 39-49.
De Filippis, L.F. (1989) Toxicity and resistance of algae to heavy metals, in M.A. Ozturk (ed.), *Plants and Pollutants in Developed and Developing Countries*, Ege University Press, Izmir, pp. 19-35.
De Filippis, L.F. and Pallaghy, C.K. (1976) The effects of sub-lethal concentrations of mercury and zinc on *Chlorella*. II. Photosynthesis and pigment composition, *Z. Pflanzenphysiol.* **78**, 314-322.
De Filippis, L.F. and Pallaghy, C.K. (1994) Heavy metals: sources and biological effects, in L.C. Rai, J.P. Gaur and C.J. Soeder (eds.), *Algae and Water Pollution*, E. Schweizerbart'sche Varlagsbuchhandlung, Stuttgart, pp. 31-77.
De Filippis, L.F. and Ziegler, R. (1993) Effect of sublethal concentrations of zinc, cadmium and mercury on the photosynthetic carbon reduction cycle of *Euglena*, *J. Plant Physiol.* **142**, 167-172.
De Filippis, L.F., Hampp, R. and Ziegler, H. (1981) The effects of sub-lethal concentrations of zinc, cadmium and mercury on *Euglena*. Growth and pigments, *Z. Pflanzenphysiol.* **101**, 37-47.
De Vos, C.H.R., Vooijs, H., Schat, H. and Ernst, W.H.O. (1989) Copper-induced damage to the permeability barrier in roots of *Saline cucubalus*, *J. Plant Physiol.* **135**, 165-169.
De Vos, R.C.H., Marjolein, J., Vonk, R., Voojis, R. and Schat, H. (1992) Glutathione depletion due to copper-induced phytochelatin synthesis causes oxidative stress in *Saline cucubalus*, *Plant Physiol.* **98**, 853-858.
Dedyukhina, E.G. and Eroshin, V.K. (1991) Essential metal ions in the control of microbial metabolism, *Process Biochem.* **26**, 31-37.
Demirbas, A. (2000) Accumulation of heavy metals in some edible mushrooms from Turkey, *Food Chem.* **68**, 415- 419.
Denton, G.R.W. and Burdon-Jones, C. (1986) Trace metals in algae from the Great Barrier Reef, *Mar. Pollut. Bull.* **17**, 98-107.
Donazzolo, R., Merlin, O.H., Vitturi, L.M., Orio, A.A., Pavoni, B., Perin, G. and Rabitti, S. (1981) Heavy metal contamination in surface sediments from the Gulf of Venice, Italy, *Marine Pollution Bulletin* **12**, 417-425.
Dubey, S.K. and Rai, L.C. (1987) Effect of chromium and tin on survival, growth, carbon fixation, heterocyst differentiation, nitrogenase, nitrate reductase and glutamine synthetase activities of *Anabaena doliolum*, *J. Plant Physiol.* **130**,165-172.
Dushenkov, V., Kumar, P.B.A.N., Motto, H. and Raskin, I. (1995) Rhizofiltration-the use of plants to remove heavy metals from aqueous streams, *Environ. Sci. Technol.* **29**, 1239-1245.
El-Hissy, F.T., Khallil, A.M. and Abdel-Raheem, A.M. (1993) Effect of some heavy metals on the mycelial growth of *Achlya racemosa* and *Alatospora accuminata*, *Zent. Fuer Microbiol.* **148**, 535-542.
Erbe, J.L., Taylor, K.B. and Hall, L.M. (1995) Metalloregulation of the cyanobacterial *Smt* locus: identification of *Smt B* binding sites and direct interaction with metals, *Nucleic Acids Res.* **23**, 2472-2478.
Ernst, W.H.O. (1980) Biochemical aspects of cadmium in plants, in J.O.Nriagu (ed.),*Cadmium in the Environment, Part I, Ecological Cycling*, John Wiley and Sons, New York, pp. 639-653.
Ezaki, B., Sivaguru, M., Ezaki, Y., Matsumoto, H. and Gardner, R.C. (1999) Acquisition of aluminum tolerance in *Saccharomyces cerevisiae* by expression of the BCB or Nt GDII gene derived from plants, *FEMS Microbiol. Lett.* **171**, 81-87.
Foster, P.L. (1977) Copper exclusion as a mechanism of heavy metal tolerance in a green alga, *Nature* **269**, 322-323.
Foyer, C.H. (1997) Oxygen metabolism and electron transport in photosynthesis, in J. Scandalios (ed.), *Molecular Biology of Free Radical Scavenging Systems*, Cold Spring Harbor Laboratory Press, New York, pp. 587-621.
Francesconi, K.A. and Edmonds, J.S. (1994) Biotransformation of arsenic in the marine environment, in J.O. Nriagu, (ed.), *Arsenic in the Environment, Part I. Cycling and Characterization*, Wiley, New York, pp. 221-261.

Fridovich, I. (1995) Superoxide radical and superoxide dismutases, *Ann. Rev. Biochem.* **64**, 97-112.
Gadd, G.M. (1992) Metals and microorganisms a problem of defination, *FEMS Microbiol. Lett.* **100**, 197-204.
Gagnon, C., Vaillancourt, G. and Pazdernik, L. (1998) Influence of water hardness on accumulation and elimination of cadmium in two aquatic mosses under laboratory conditions, *Arch. Environ. Contam. Toxicol.* **34**, 12-20.
Gebhart, E. and Rossman, T.G. (1991) Mutagenicity, carcinogenicity and tetratogenicity, in E. Merian (ed.), *Metals and Their Compounds in the Environment. Occurrence, Analysis and Biological Relevance*, Weinheim: VCH Verlagsgesellschaft, pp. 617-640.
Gekeler, W., Grill, E., Winnacker, E.L. and Zenk, M.H. (1988) Algae sequester heavy metals via synthesis of phytochelatin complexes, *Arch. Microbiol.* **150**, 197-202.
Gharieb, M.M. and Gadd, G.M. (1998) Evidence for the involvement of vacuolar activity in metal(loid) tolerance: Vacuoler-lacking and defective mutants in *Saccharomyces cerevisiae* display lesser sensitivity to chromate, tellurite and selenite, *Biometals* **11**, 101-106.
Giller, K.E., Witter, E. and Mcgrath, S.P. (1998) Toxicity of heavy metals to microorganisms and microbial processes in agricultural soils. A review. *Soil. Biol. Biochem.* **30**, 1389-1414.
Gimmler, H., Treffny, B., Kowalski, M. and Zimmermann, U. (1991). The resistance of *Dunaliella acidophila* against heavy metals: The importance of the zeta potential, *J. Pl. Physiol.* **138**, 708-716.
Gimmler, H., Weiss, U., Weiss, C., Kugel, H. and Treffny, B. (1989) *Dunaliella acidophila* (Kalina) Masyuk-An alga with a positive membrane potential, *New Phytol.* **113**, 175-184.
Goodman, G.T. and Roberts, T.M. (1971) Plants and soils as indicators of metals in the air, *Nature* **231**, 287-292.
Grabowski, J., Ke-Cheng, H., Baker, P.R. and Bornman, C.H. (1997) Fluorogenic compound hydrolysis as a measure of toxicity-induced cytoplasmic viscosity and pH, *Environ. Pollut.* **98**, 1-5.
Gupta, A., Whitton, B.A. and Morby, A.P. (1992) Amplification and rearrangement of a prokaryotic metallothionein locus *smt* in *Synechococcus PCG 6301* selected for tolerance to cadmium, *Proc. R. Soc. Lond.* **B 248**, 273-281.
Haglund, K., Bjorklund, M., Gunnare, S., Sandberg, A., Olander, V. and Pedersen, M. (1996) New method for toxicity assessment in marine and brakish environments using the macroalga *Gracilaria tenuistipitata* (Glacilariales, Rhodophyta), *Hydrobiol.* **326**, 317-325.
Hall, A. (1981) Copper accumulation in copper-tolerant and non-tolerant populations of the marine fouling alga, *Ectocarpus siliculous* (Dillw.) Lyngbye, *Bot. Mar.* **29**, 223-228.
Halliwell, B. and Gutteridge, J.M.C. (1984) Oxygen toxicity, oxygen radicals, transition metals and disease, *Biochem. J.* **219**, 1-14.
Harding, J.P.C. and Whitton, B.A. (1976) Response to zinc of *Stigeoclonium tenue*, *Br. Phycol. J.* **11**, 417-426.
Harrison, G.I. and Morel, F.M.M. (1983) Antagonism between cadmium and iron in the marine diatom *Thalassiosira weissflogii*, *J. Phycol.* **19**, 495-507.
Hartley, J., Cairney, J.W.G. and Meharg, A.A. (1999a) Cross-colonization of scots pine (*Pinus sylvestris*) seedlings by ecomycorrhiza fungus *Paxillus involutus* in presence of inhibitory levels of Cd and Zn, *New Phytol.* **142**, 141-149.
Hartley, J., Cairney, J.W.G., Freestone, P., Woody, C. and Meharg, A.A. (1999b) The effects of multiple metal contamination on ecomycorrhizal scots pine (*Pinus sylvestris*) seedlings, *Environ. Pollut.* **106**, 413-424.
Hasegawa, I., Terada, E., Sunairi, M., Wakita, H., Shinmachi, F., Noguchi, A., Nakajima, M. and Yazaki, J. (1997) Genetic improvement of heavy metal tolerance in plants by transfer of the yeast metallothionein gene (CUP1), *Plant and Soil.* **196**, 277-281.
Hill, R.D. (1973) Control and prevention of mine drainage, in *Cycling and Controls of Metals.* Cincinati, Ohio: Natl. Environ. Res. Center, US Environ. Protect. Agency, pp. 91-94.
Hughes, M.N. and Poole, R.K. (1991) Metal speciation and microbial growth-the hard (and soft) facts, *J. Gen. Microbiol.* **137**, 725-734.
Husaini, Y. and Rai, L.C. (1992). pH dependent aluminium toxicity to *Nostoc linckia:* Studies on phosphate uptake, alkaline and acid phosphatase activity, ATP content, photosynthesis and carbon fixation, *J. Plant Physiol.* **39**, 703-707.
Husaini, Y., Rai, L.C. and Mallick, N. (1996a) Impact of aluminium, fluoride and fluoroaluminate complex on ATPase activity of *Nostoc linckia* and *Chlorella vulgaris*, *Biometals* **9**, 277-283.
Husaini, Y., Rai, L.C. and Mallick, N. (1996b) Nutrient uptake and its kinetics in *Nostoc linckia* in presence of aluminium and fluoride at different pH, *J. Gen. Appl. Microbiol.* **42**, 263-270.

Husaini, Y., Singh, A.K., and Rai, L.C. (1991) Cadmium toxicity to photosynthesis and associated electron transport system of *Nostoc linckia*, *Bull. Environ. Contam. Toxicol.* **46**, 146-150.

Hutchinson, T.C. and Czyrska, H. (1975) Heavy metal toxicity and synergism to floating aquatic weeds, V*erh. internal verein. limnol.* **19**, 2102-2111.

Jensen, T.E., Baxter, M., Rachlin, J.W. and Jani, V. (1982) Uptake of heavy metals by *Plectonema boryanum* (Cyanophyceae) into cellular components, especially polyphosphate bodies: an X-ray energy dispersive study, *Environ. Pollut.* Ser. A. **27**, 119-127.

Jentschke, G., Winter, S. and Godbold, D.L. (1999) Ectomycorrhizas and cadmium toxicity in Norway spruce seedlings, *Tree Physiol.* **19**, 23-30.

Jones, A.L. and Harwood, J.L. (1993) Lipid metabolism in the brown marine algae *Fucus vesiculosus* and *Ascophyllum nodosum, J. Exp. Bot.* **44**, 1203-1210.

Jones, R.P. and Gadd, G.M. (1990) Ionic nutrition of yeast-physiological mechanisms involved and implications for biotechnology, *Enzyme Microb. Technol.* **12**, 402-418.

Josef, E., Allewa, L., Josef, S., Jiri, K. and Hiroshi, K. (1999) Diversity and abundance of soil algae in the polar desert, Sverdrup Pass, Central Ellesmere Island, *Polar Record.* **35**, 231-254.

Jung-Ho, S., Jong-Won, Y. and Dong-Seog, K. (1998) Effect of temperature on the accumulation of Pb^{2+} in *Saccharomyces cerevisiae, Environ. Technol.* **8**, 412-415.

Kelly, M. (1988) *Mining and the Freshwater Environment*, BP Elsevier Applied Science, London.

Kessler, A. and Brand, M.D. (1994a) Effects of cadmium on the control and internal regulation of oxidative phosphorylation in potato tuber mitochondria, *Eur. J. Biochem.* **225**, 907-922.

Kessler, A. and Brand, M.D. (1994b) Quantitative determination of the regulation of oxidative phosphorylation by cadmium in potato tuber mitochondria, *Eur. J. Biochem.* **225**, 923-935.

Klein, D.H., Andren, A.W. and Bolton, N.E. (1975) Trace element discharges from coal combustion for power production, *Water Air Soil Pollut.* **5**, 71-77.

Kobayashi, M., Kakizono, T. and Nagai, S. (1993) Enhanced carotenoid biosynthesis by oxidative stress in acetate-induced cyst cells of a green unicellular alga, *Haematococcus pluvialis*, *Appl. Environ. Microbiol.* **59**, 867-873.

Kong, F.-X., Liu, Y. and Cheng, F.-D. (1997) Aluminum toxicity and nutrient utilization in the mycorrhizal fungus *Hebeloma mesophacuss*, *Bull. Environ. Contam. Toxicol.* **59**, 125-131.

Kremer, B. and Markham, J.W. (1982) Primary metabolic effects of cadmium in the brown alga *Laminaria saccharina*. *Zeitschr. Pflanzenphysiol.* **108**, 125-130.

Kriedemann, P.E. Anderson, J.E. (1998) Growth and photosynthetic responses to manganese and copper deficiencies in wheat *Triticum aestivum* and barley *Hordeum glaucum* and *Hordeum leporinum*. *Aust. J. Plant Physiol.* **15**, 429-446.

Kungolos, A. and Aoyama, I. (1992) Using *Saccharomyces cerevisiae* for toxicity assessment including interacting effects and DNA damage, *Wat. Sci. Tech.* **25**, 309-316.

Lawrey. J.D. and Hale, M.E. (1979) Lichen growth responses to stress induced by automobile exhaust pollution, *Science.* **204**, 423-424.

Lee, J., Brooks, R.R., Reeves, R.D. and Jaffre, T. (1977) Chromium-accumulating bryophyte from New Caledonia, *Bryologist.* **80**, 203-205.

Lepp, N. W. (2001) Bryophytes and Pteridophytes, in M.N.V.Prasad (ed), *Metals in the environment analysis by biodiversity*. Marcel Dekker, New York, pp. 159-170

Lepp, N.W. and Roberts, M.J. (1977) Some effects of cadmium on growth of bryophytes, *Bryologist.* **80**, 533-536.

Lepp, N.W. and Salmon, D. (1999) A field study of the ecotoxicology of copper to bryophytes, *Environ. Pollut.* **106**, 153-156.

Lewis, B.G., Johnson, C.M. And Broyer, T.C. (1974) Volatile Selenium In Higher Plants. The Production of dimethyl selenide in cabbage leaves by enzymic cleavage of Se-methyl selenomethionine selenonium salt, *Plant Soil.* **40**, 107-118.

Lewis, B.G., Johnson, C.M. and Delwiche, C.C. (1966) Release of volatile selenium compounds by plants. Collection procedures and preliminary observations, *J. Agric. Food. Chem.* **14**, 638-640.

Leyval, C., Singh, B.R. and Joner, E.J. (1995) Occurrence and infectivity of arbuscular mycorrhizal fungi in some Norwegian soils influenced by heavy metals and soil properties, *Water Air Soil Pollut.* **84**, 203-216.

Lichtenthaler, H.K. (1996) Vegetation stress: an introduction to the stress concept in plants, *J. Plant Physiol.***148**, 4-14.

Lin, J., Uin, H., Jiuchang, R., Chen, F. and Xiaoyan, T. (1999) Effect of Cd on DMS producing marine algae, *Amphidinium hoefleri*, *Huanj. Kex.* **20**, 18-20.

Lu, C.-M. and Zhang, J.-H. (1999) Copper-induced inhibition of PS II photochemistry in cyanobacterium *Spirulina platensis* is stimulated by light, *J. Plant Physiol.* **154**, 173-178.

Mallick, N. and Mohn, F.H. (2001) Heavy metal toxicity on chlorophyll fluorescence characteristics of the green microalga *Scenedesmus obliquus*. (unpublished).

Mallick, N. and Rai, L.C. (1989) Response of *Anabaena doliolum* to bimetallic combinations of Cu, Ni and Fe with special reference to sequential addition, *J. Appl. Phycol.* **1**, 301-306.

Mallick, N. and Rai, L.C. (1990). Effect of heavy metals on the biology of a nitrogen fixing cyanobacterium *Anabaena doliolum, Toxicity Assessment* **5**, 207-219.

Mallick, N. and Rai, L.C. (1992) Metal-induced inhibition of photosynthesis, photosynthetic electron transport chain and ATP content of *Anabaena doliolum* and *Chlorella vulgaris*: Interaction with exogenous ATP, *Biomed. Environ. Sci.* **5**, 241-250.

Mallick, N. and Rai, L.C. (1994) Kinetic studies of mineral uptake and enzyme activities of *Anabaena doliolum* under metal stress, *J. Gen. Appl. Microbiol.* **40**, 123-133.

Mallick, N. and Rai, L.C. (1998) Characterization of Cd-induced low molecular weight protein in a N_2-fixing cyanobacterium *Anabaena doliolum* with special reference to co-/multiple tolerance, *Biometals* **11**, 55-61.

Mallick, N. and Rai, L.C. (1999) Response of the antioxidant systems of the nitrogen fixing cyanobacterium *Anabaena doliolum* to copper, *J. Plant Physiol.* **155**, 146-149.

Meng, L., De-Yong, T., Wang, H-X., Duan, C.-Q., Duan, P.-S. and Gao, S.-Y. (1998) Study of the response of wheat to lead, cadmium and zinc, *J. Environ. Sci.* **10**, 238-244.

Metwally, M. and Abou-Zeid, A. (1996) Effects of toxic heavy metals on grwoth and metabolic activity of some fungi, *Egy. J. Microbiol.* **31**, 115-127.

Missiaen, L., Wuytack, F., De Smedt, H., Vrolix, M. and Casteels, R. (1988) AlF_4^-- reversibly inhibits 'P'-type cation-transport ATPases, possibly interacting with the phosphate-binding site of the ATPase, *Biochem. J.* **253**, 827-833.

Mohanty. P. (1989) Effect of elevated levels of zinc on growth of *Synechococcus* 6301, *Zentralbl. Mikrobiol.* **144**, 531-536.

Monaci, F., Moni, F., Lanciotti, E., Grechi, D. and Bargagli, R. (2000) Biomonitoring of air borne metals in urban environments: new tracers of vehicle emission, in place of lead, *Environ. Pollut.* **107**, 321-327.

Moore, J.W. and Ramamoorthy, S. (1984) *Heavy Metals in Natural Waters: Applied Monitoring and Impact Assessment*, Springer-Verlag, New York.

Morby, A.P., Turner, J.S., Huckle, J.W. and Robinson, N.J. (1993) *SmtB* is a metal regulated repressor of the cyanobacterial metallothionein gene smtA: Identification of a Zn inhibited DNA-protein complex, *Nucleic Acids Res.* **21**, 921-925.

Nagalakshmi, N and Prasad, M.N.V (1998) Copper-induced oxidative stress in *Scenedesmus bijugatus* - protective role of free radical scavengers. *Bull. Envir. Contam. Toxicol.* **61**, 623-628.

Nagalakshmi, N and M.N.V.Prasad (2001) Responses of glutathione cycle enzymes and glutathione metabolism to copper stress in *Scenedesmus bijugatus*. *Plant Science* **160**, 291-299.

Nash, T.H. (1975) Influence of effluents from a zinc factory on lichens, *Ecol. Monog.* **45**, 183-198.

Neto, J.A., Smith, B.J. and Mc Allister, J.J. (2000) Heavy metal concentrations in surface sediments in a near shore environment, Jurujuba Sound, Southeast Brajil, *Environ. Pollut.* **109**, 1-9.

Nriagu, J.O. and Pacyna, J.M. (1988) Quantitative assessment of world wide contamination of air, water and soils by trace metals, *Nature* **333**, 134-139.

Nygard, C. and Ekelund, N.G.A. (1999) Effects of lead ($PbCl_2$) on photosynthesis and respiration of the bladder wrack, *Fucus vesiculosus*, in relation to different salinites, *Water Air Soil Pollut.* **116**, 549-564.

Okamoto, O.K. and Colepicolo, P. (1998) Response of superoxide dismutase to pollutant metal stress in the marine dinoflagellate *Gonyaulax polyedra*. *Comp. Biochem. Physiol. C Pharmacol. Toxicol. Endocrinol.* **119**, 67-73.

Okamoto, O.K., Asano, C.S., Aidar, E. and Colepicolo, P. (1996) Effects of cadmium on growth and superoxide dismutase activity of the marine microalga *Tetraselmis gracilis* (Prasinophyceae), *J. Phycol.* **32**, 74-79.

Olafson, R.W., Loya, S. and Sim, R.G. (1980) Physiological parameters of prokaryotic metallothionein induction, *Biochem. Biophys. Res. Commun.* **95**, 1495-1503.

Olafson, R.W., McCubbin, W.D. and Kay, C.M. (1988) Primary-and secondary-structural analysis of a unique prokaryotic metallothionein from *Synechococcus* sp. cyanobacterium, *Biochem. J.* **251**, 691-699.

Overnell, J. (1975) The effect of heavy metals on photosynthesis and loss of cell potassium in two species of marine algae, *Dunaliella tertiolecta* and *Phaeodactylum tricornutum*, *Marine Biol.* **29**, 99-103.

Park, K.S., Jung, K. and Soon-Yong, C. (1998) Cloning, characterization, and expression of the *CIP2* gene induced under cadmium stress in *Candida* sp, *FEMS Microbiol. Lett.* **162**, 325-330.

Parker, D.L., Rai, L.C., Mallick, N., Rai, P.K. and Kumar, H.D. (1998) Effect of cellular metabolism and viability on metal ion accumulation by cultured biomass from a bloom of the cyanobacterium *Microcystis aeruginosa*. *Appl. Environ. Microbiol.* **64**, 1545-1547.

Passow, H. and Rothstein, A. (1960) The binding of mercury by the yeast cell in relation to changes in permeability, *J. Gen. Physiol.* **43**, 621-633.

Pasternak, K. (1973) The spreading of heavy metals in flowing waters in the region of occurrence of natural deposits and of the zinc and lead industry, *Acta Hydrobiol.* **15**, 145-146.

Patel, K.S., Patel, R.M., Tripathy, A.N., Chandraswashi, C.K., Pandey, P.K., Chinkhalikar, S., Kamavisdar, A. and Aggarwal, S.G. (1999) Concentration of mercury and other heavy metals in central India. in, Mercury Contamination Sites, R. Ebinghons, R.R. Turner, L.C. de Lacerda, O. Vasiliev and W. Salomonos (eds), Springer-Verlag. Berlin, Heidelberg, pp. 487-500.

Pawlik-skowronska, B. and Skowronski, T. (2001) Freshwater algae, in M.N.V Prasad (ed), *Metals in the environment analysis by biodiversity*. Marcel Dekker, New York, pp. 59-94

Pettersson, A., Kunst, L., Bergman, B. and Roomans, G.M. (1985) Accumulation of aluminium by *Anabaena cylindrica* into polyphosphate granules and cell walls: an X-ray energy-dispersive microanalysis study, *J.Gen Microbiol.* **131**, 2545-2548.

Pietilainen, K. (1975) Synergistic and antagonistic effects of lead and cadmium on aquatic primary production, in *International Conference on Heavy Metals in the Environment, Symposium Proceedings*, Vol. II, Toronto, Ontario, Canada, pp. 861-873.

Planas, D. and Healey, F.P. (1978) Effects of arsenate on growth and phosphorus metabolism of phytoplankton, *J. Phycol.* **14**, 337-341.

Prasad, M.N.V. (ed) (2001) *Metals in the Environment - Analysis by biodiversity*. Marcel Dekker Inc. New York. pp. 504

Prasad, M.N.V. (1999) Metallothioneins and metal binding complexes in plants. in, *Heavy Metal Stress in Plants: From Molecules to Ecosystems*, M.N.V. Prasad and J. Hagemeyer (eds), Springer-Verlag. Berlin, Heidelberg, New York. pp. 51-72

Prasad, M.N.V., Drej, K., Skawinska, A. and Strzalka, K. (1998) Toxicity of cadmium and copper in *Chlamydomonas reinhardtii* wild-type (WT 2137) and cell wall deficient mutant strain (CW15). *Bull. Envir. Contam. Toxicol.* **60**, 306-311.

Prasad, M.N.V and Hagemeyer, J. (eds) (1999) *Heavy Metal Stress in Plants: From Molecules to Ecosystems*. Springer-Verlag. Berlin, Heidelberg, New York. pp. xiii+401

Prasad, S.M., Singh, J.B., Rai, L.C. and Kumar, H.D. (1991) Metal-induced inhibition of photosynthetic electron transport chain of the cyanobacterium *Nostoc muscorum*, *FEMS Microbiol. Lett.* **82**, 95-100.

Price, N.M. and Morel, F.M.M. (1990a) Cadmium and cobalt substitution for zinc in a marine diatom, *Nature* **344**, 658-660.

Price, N.M. and Morel, F.M.M. (1990b) Role of extracellular enzymatic reactions in natural waters, in W. Stumm (ed.), *Aquatic Chemical Kinetics*, John Wiley & Sons, New York, pp. 235-257.

Pyatt, F.B. (1979) Lichen ecology of metal spoil tips: Effects of metal ions on ascospore viability, *Bryologist* **79**, 172-179.

Rachlin, J.W., Jensen, T.E. and Warkentine, B. (1984) The toxicological response of the alga *Anabaena flos-aquae* (Cyanophyceae) to cadmium, *Arch. Environ. Contam. Toxicol.* **13**, 143-151.

Rachlin, J.W., Jensen, T.E., Baxter, M. and Jani, V. (1982) Utilization of morphometric analysis in evaluating response of *Plectonema boryanum* (Cyanophyceae) to exposure to eight heavy metals, *Arch. Environ. Contam. Toxicol.* **11**, 323-333.

Rai, L.C. and Raizada, M. (1986) Nickel induced stimulation of growth, heterocyst differentiation, $^{14}CO_2$ uptake and nitrogenase activity in *Nostoc muscorum*, *New Phytol.* **104**, 111-114.

Rai, L.C., Gaur, J.P. and Kumar, H.D. (1981) Phycology and heavy-metal pollution, *Bio. Rev. Cambr. Philos. Soc.* **56**, 99-151.

Rai, L.C., Husaini, Y. and Mallick, N. (1996a) Physiological and biochemical responses of *Nostoc linckia* to combined effects of aluminium, fluoride and pH, *Environ. Exp. Bot.* **36**, 1-12.

Rai, L.C., Jensen, T.E. and Rachlin, J.W. (1990) A morphometric and x-ray energy dispersive análisis approach to monitoring pH altered Cd toxicity in *Anabaena flos-aquae*, *Arch. Environ. Contam. Toxicol.* **19**, 479-487.

Rai, L.C., Mallick, N., Singh, J.B. and Kumar, H.D. (1991) Physiological and biochemical characteristics of a copper tolerant and wild type strain of *Anabaena doliolum* under copper stress, *J. Plant Physiol.* **138**, 68-74.

Rai, L.C., Rai, P.K. and Mallick, N. (1996b) Regulation of heavy metal toxicity in acid-tolerant *Chlorella*: Physiological and biochemical approaches, *Environ. Exp. Bot.* **36**, 99-109.

Rai, L.C., Singh, A.K. and Mallick, N. (1991) Studies on photosynthesis, the associated electron transport system and some physiological variables of *Chlorella vulgaris* under heavy metal stress, *J. Plant Physiol.* **137**, 419-424.

Rai, L.C., Tyagi, B., Mallick, N. and Rai, P.K. (1995) Interactive effects of UV-B and copper on photosynthetic activity of the cyanobacterium *Anabaena doliolum*, *Environ. Exp. Bot.* **35**, 177-185.

Rai, P.K., Mallick, N. and Rai, L.C. (1994) Effect of Ni on certain physiological and biochemical processes of an acid tolerant *Chlorella vulgaris*, *Biometals* **7**, 193-200.

Raizada, M. and Rai, L.C. (1985). Metal-induced inhibition of growth, heterocyst differentiation, carbon fixation and nitrogenase activity of *Nostoc muscorum*: Interaction with EDTA and calcium, *Microb. Lett.* **30**, 153-161.

Ramaswamy, S.N. and Somashekar, R.K. (1982) Ecological studies on algae of electroplating wastes, *Phykos* **21**, 83-90.

Reddy, G.N. and Prasad, M.N.V. (1992a) Chrracterization of cadmium binding protein from *Scenedesmus quadricauda* and Cd toxicity reversal by phyto-chelatin constituting amino acids and citrate. *J. Plant. Physiol.* **140**, 156-162

Reddy, G.N. and Prasad, M.N.V. (1992b) Cadmium-induced potassium efflux from *Scenedesmus quadricauda*. *Bull. Envi. Contam. Toxicol.* **49**, 600-605

Reddy, G.N. and Prasad, M.N.V. (1994) Cadmium toxicity to *Scenedsmus quadricauda* Proteins and protein phosphorylation changes. *Biochem. Arch.* **10**, 185-188

Rejment-Grochowska, I. (1976) Concentration of heavy metals, lead, iron, manganese, zinc and copper in mosses, *J. Hattori Bot. Lab.* **41**, 225-230.

Ren, L., Shi, D., Dai, J. and Ru, B. (1998) Expression of the mouse metallothionein-I gene conferring cadmium resistance in a transgenic cyanobacterium, *FEMS Microbiol. Lett.* **158**, 127-132.

Richardson, D.H.S. (1981) *The Biology of Mosses*. Blackwell, Oxford.

Rijstenbil, J.W., Dehairs, F., Ehrlich, R., Wijnholds, J. A. (1998) Effect of nitrogen status on copper accumulation and pools of metal-binding peptides in the planktonic diatom *Thalassiosira pseudonana*, *Aquatic Toxicol.* **42**, 187-209.

Rijstenbil, J.W., Derksen, J.W.M., Gerringa, L.J.A., Poortvliet, T.C.W., Sandee, A., Vandenberg, M., Vandrie, J. and Wijnholds, J.A. (1994) Oxidative stress induced by copper: Defense and damage in the marine planktonic diatom *Dictylum brightwellii* grown in continuous cultures with high and low zinc levels, *Marine Biol.* **119**, 583-590.

Roberts, T.M. and Goodman, G.T. (1974) The persistence of heavy metals in soils and natural vegetation following closure of a smelter, in D.D. Hemphill (ed.), *Trace Substances in Environmental Health-VIII*, University of Missouri, Columbia, pp. 117-125.

Robinson, N.J. (1989) Algal metallothioneins: Secondary metabolites and proteins, *J. Appl. Phycol.* **1**, 5-18.

Rueter, J.G. (1983) Alkaline phosphatase inhibition by copper: Implications to phosphorus nutrition and use as a biochemical marker of toxicity, *Limnol. Oceanogr.* **28**, 743-748.

Rueter, J.G. and Morel, F.M.M. (1981) The interaction between zinc deficiency and copper toxicity as it affects the silicic acid uptake mechanisms in *Thalassiosira pseudonana*, *Limnol. Oceanogr.* **26**, 67-73.

Salvetat, R., Juneau, P. and Popovic, R. (1998) Measurement of chlorophyll fluorescence by synchronous detection analytical approach for the accurate determination of photosynthesis parameters for whole plants, *Environ. Sci. Technol.* **32**, 2640-2645.

Sanchez, J., Marino, N., Vaquero, M.C., Ansorena, J. and Legorburce, I. (1998) Metal pollution by old lead-zinc mines in Urumea river valley, *Water Air Soil Pollut.* **107**, 303-319.

Santos, C.A.Q. and Gilda, C.R. (1999) Cadmium-induced alterations in the protein profiles of two cyanobacterial isolates, Cd0-10 and Bol-32 (*Chroococcus* spp.), *Asia Life Sci.* **8**, 59-70.

Sarret, G., Alian, M., Damien, C. and Van Haluwyn, C. (1998) Mechanisms of lichen resistance to metallic pollution, *Environ. Sci. Technol.* **32**, 3325-3330.

Schaaper, R.M., Koplitz, R.M., Tkeshelahvili, L.K. and Loeb, L.A. (1987) Metal induced lethality and mutagenesis: Possible role of apurinic intermediates, *Mutation. Res.* **177**, 179-188.

Schaule, B.K. and Patterson, C.C. (1983) Perturbations of the natural lead depth profile in the Sargasso Sea by industrial lead, in C.S. Wong, E. Boyle, K.W. Bruland, J.D. Burton and E.D. Goldberg (eds.), *TraceMetals in Sea Water*, Plenum Press, New York, pp. 487-503.

Schroeder, W.A. and Johnson, E.A. (1995) Singlet oxygen and peroxyl radicals regulate carotenoid biosynthesis in *Phaffia rhodozyma, J. Biol. Chem.* **270**, 18374-18379.

Schutte, J.A. (1977) Chromium in two corticolous lichen from Ohio and West Virginia, *Bryologist* **77**, 279-284.

Shaw, A.J. (1990) Metal tolerance in bryophytes, in A.J. Shaw (ed.), *Heavy Metal Tolerance in Plants:Evolutionary Aspects*, CRC Press, Boca Raton, FL, pp. 133-152.

Shimwell, D.W. and Laurie, A.E. (1972) Lead and zinc contaminations of vegetation in the Southern Pennines, *Environ. Pollut.* **3**, 291-301.

Shoer, I., Nagel, U. and Eggersgluess, D. (1982) Metal contents in sediments from the elbe, weser and enis estuaries and from the german bight (southeastern north sea): grain size effects, *mmitteilungen d. geologisch-padontologischen institut der universitdt hamburg.* **52**, 687-702.

Sicko-goad, L., Ladewski, B.G. and Lazinsky, D. (1986) Synergistic effects of nutrients and lead on the quantitative ultrastructure of *Cyclotella* Bacillariophyceae, *Arch. Environ. Contam. Toxicol.* **15**, 291-300.

Silverberg, B.A., Stokes, P.M. and Ferstenberg, L.B. (1976) Intranuclear complexes in a copper tolerant green alga, *J. Cell Biol.* **69**, 210-214.

Simola, L.K. (1977) The effect of lead, cadmium, arsenate and fluoride ions on the growth and fine structure of *Sphagnum nemoreum* in aseptic culture, *Can. J. Bot.* **55**, 426-435.

Sittig, M. (1975) *Environmental Sources and Emission Handbook. Noyes Data Corporation*, Park Ridge, New Jersey and London, England.

Skaar, H., Ophus, E. and Gullvag, B.M. (1973) Lead accumulation within nuclei of moss leaf cells, *Nature* **241**, 215-216.

Smith, B.J. and Orford, J.D. (1989) Scales of pollution in estuarine sediments around the North Irish Sea, in J. Sweeney (ed.), *The Irish Sea (Special Publication No. 3)*, Irish Geography Society, Dublin, pp. 107-115.

Stanners, D. and Bourdeau, P. (1995) *Europe's Environment*. EEA (European Environment Agency), Copenhagen, pp. 676.

Stauber, J.L. and Florence, T.M. (1990) Mechanisms of toxicity of zinc to the marine diatom *Nitzschia closterium, Mar. Biol.* **105**, 519-524.

Stratton, G.W. and Corke, C.T. (1979) The effect of mercuric, cadmium, and nickel ion combinations on a blue-green alga, *Chemosphere* **8**, 731-740.

Subramanian, V., Itta, P.K. and Griekan, R.V. (1988) Heavy metals in the Ganges Estuary, *Marine Pollut. Bull.* **19**, 290-293.

Sunda, W.G., Barber, R.T. and Huntsman, S.A. (1981) Phytoplankton growth in nutrient rich seawater: Importance of copper-manganese cellular interactions, *J. Mar. Res.* **39**, 567-586.

Tarhanen, S. (1998) Ultrastructural responses of the lichen *Bryoria fuscescens* to simulated acid rain and heavy metal deposition, *Annals of Bot.* **82**, 735-746.

Tease, B.E. and Walker, R. (1987) Comparative composition of the sheath of the cyanobacterium *Gloeothece* ATCC 27152 cultured with and without combined nitrogen, *J. Gen. Microbiol.* **133**, 3331-3339.

Thomine, S., Wang, R., Ward, J.M., Crawfor, N.M. and Schroeder, J. (2000) Cadmium and iron transport by members of a plant metal transporter family in *Arabidopsis* with homology to N ramp genes, *Proc. Nat Acad. Sci. USA* **97**, 4991-4996.

Tiller, K.G., Merry, R.H., Zarcinas, B.A. and Ward, T.J. (1989) Regional geochemistry of metal-contaminated surficial sediments and seagrasses in upper Spencer Gulf, South Australia, *Estuar. Coast. Shelf. Sci.* **28**, 473-493.

Tjahjono, A.E., Hayama, Y., Kakizono, T., Terada, Y., Nishio, N. and Nagai, S. (1994) Hyper-accumulation of astaxanthin in a green alga *Haematococcus pluvialis* at elevated temperature, *Biotechnol. Lett.* **16**, 133-138.

Trebacz, K., Simonis, W. and Schon, K.G. (1994) Cytoplasmic Ca^{2+}, K^+, Cl^- and NO_3^- activities in the liverwort *Conocephalum conicum* L at rest and during action potentials, *Plant Physiol.* **106**, 1073-1084.

Van Assche, F. and Clijsters, H. (1990) Effects of metals on enzyme activity in plants, *Plant Cell Environ.* **13**, 195-206.

Van der Lelie, D and Tibazarwa, C. (2001) Bacteria, in M.N.V. Prasad (ed), *Metals in the environment analysis by biodiversity*, Marcel Dekker, New York, pp. 1-36

Van Tichelen, K.K., Vanstraclen, T. and Colpaert, J.V. (1999) Nutrient uptake by intact mycorrhizal *Pinus sylvestris* seedlings: A diagnostic tool to detect copper toxicity, *Tree Physiol.* **19**, 189-196.

Varrica, D., Aiuppa, A. and Dongarra, G. (2000) Volcanic and anthropogenic contribution to heavy metal content in lichens from Mt. Etna and Vulcano Island (Sicily), *Environ. Pollut.* **108,** 153-162.

Vazquez, M.D., Poschenrieder, C.H., Barcelo, J., Baker, A.J.M., Hatton, P. and Cope, G.H. (1994) Compartmentation of zinc in roots and leaves of the zinc hyperaccumulator *Thlaspi caerulescens, Bot. Acta* **107,** 243-250.

Verma, S.K. and Singh, H.N. (1991) Evidence for energy-dependent copper efflux as a mechanism of Cu^{2+} resistance in the cyanobacterium *Nostoc calcicola, FEMS Microbiol. Lett.* **84,** 291-294.

Weckesser, J., Hofman, K., Jurgens, U.J., Whitton, B.A. and Raffelsberger, B. (1988) Isolation and chemical analysis of the sheaths of the filamentous cyanobacteria *Calothrix parietina* and *C. scopulorum, J. Gen Microbiol.* **134,** 629-634.

Weigel, H.J. (1985) Inhibition of photosynthetic reactions of isolated intact chloroplasts by cadmium, *J. Plant Physiol.* **119,** 179-189.

Wiessenhorn, I., Mench, M. and Leyval, C. (1995) Bioavailability of heavy metals and arbuscular mycorrhiza in a sewage-sludge-amended sandy soil, *Soil Biol. Biochem.* **27,** 287-296.

Xu, S., Jiang, X., Wang, X., Tan, Y., Sun, C., Feng, J., Wang, L., Martens, S. and Gawlik, B.M. (2000) Persistent pollutants in sediments of the Yangtse river, *Bull. Environ. Contam. Toxicol.* **64,** 176-183.

Xuan, T.L., Shinpei, M. and Tamikazu, K. (1999) Responses of *Gonoderma lucidum* to heavy metals, *Mycoscience* **40,** 209-213.

Xylander, M. and Braune, W. (1994) Influence of nickel on the green alga *Haematococcus lacustris* Rostafinski in phases of its life cycle, *J. Plant Physiol.* **144,** 86-93.

Yan, C., Liu, Y., Wang, T.-Z., Tan, Z.-Q., Qu, S.-S. and Ping, S. (1999) Thermochemical studies of the toxic actions of heavy metal ions on *Rhizopus nigricans, Chemosphere* **38,** 891-898.

Yoshimura, E., Kazutoshi, K., Nishizawa, N.K., Satoshi, K.M. and Yamazaki, S. (1998) Accumulation of metals and cellular distribution of aluminum in the liverwort *Scaparia undulata* in acidic and neutral streams in Japan, *J. Environ. Sci. Health* **33,** 671-680.

Yoshimura, E., Nagasaka, S., Sato, Y., Satake, K. and Mori, S. (1999) Extraordinary high aluminum tolerance of the acidophilic thermophilic alga, *Cyanidium caldarium, Sol. Sci. Plant Nutr.* **45,** 721-724.

CHAPTER 6

PHYSIOLOGICAL RESPONSES OF VASCULAR PLANTS TO HEAVY METALS

F. FODOR
Department of Plant Physiology, Eötvös University
Budapest, H-1445 P.O.Box 330, Hungary

1. INTRODUCTION

Toxic heavy metals entering the plant tissues inhibit most physiological processes at all levels of metabolism. The inhibition of growth, photosynthesis, ion and water uptake and nitrate assimilation etc. have been described in many papers. The extent of these effects is greatly dependent on the concentration of the metal ions and the sensitivity or tolerance of the plant. Therefore the concentration is a key factor in interpreting metal actions. Another widely discussed question is the specific symptoms of heavy metal toxicity which is extremely hard to separate from those of other simultaneously affecting or consequently induced stresses.

Literature related to heavy metal effects on enzymes or processes is abundant and widespread but it fails to reliably establish the starting points or sites of action and also the mode of action. Similarly, tolerance mechanisms have been excessively investigated and many reviews have integrated the results but most of them focus on a certain mechanism (e.g. sequestration by phytochelatins) or metal. For these reasons we intend to summarise the recent knowledge in the above mentioned fields and to develop an integrated model for the physiology of heavy metal toxicity in higher plants. In the second part of this paper a comparative review of Pb and Cd toxicity is provided.

2. HEAVY METAL STRESS – A GENERAL APPROACH

2.1. Toxicity – the question of concentration

The group of heavy metals in general consist of essential and nonessential elements for higher plants. The essential ones are Fe, Mn, Zn, Cu, , Mo, Ni. All the rest are neutral or toxic like Cd, Pb, Hg etc. (Marschner, 1995). Al is not a heavy metal but in some respect its effect shows similarities with heavy metal toxicity and for this reason in defined cases examples of Al induced changes will be cited, too. Toxicity is always the question of concentration. The nutrients may be available at an optimal concentration range but they may also be insufficient or oversupplied. For nonessential elements, certainly, there is no optimal concentration range, however they may stimulate plant metabolism at a low concentration. In maize roots growing for 3 hours in the presence of Cd a small growth stimulation was observed (Wójcik and Tukendorf, 1999). In *Fagus sylvatica* low soil Pb concentration resulted in slightly enhanced root elongation, root hair formation and dry matter production (Breckle, 1991).

The optimal concentration of essential metals can be termed 'physiological'. The application of nonessential heavy metals at the same concentration range, however, can hardly be referred to as physiological, although the investigation of their effects can only be comparable to the essential ones if they are applied so. One may raise the question of environmental concentration that is the amount really available for plants in natural habitats. The Pb concentration of the soil solution is usually less than 0.2 µM (Lamersdorf, 1989). According to an estimation by Wagner (1993), non-polluted soil solutions contain Cd concentrations ranging from 0.04 to 0.32 µM. Soil solutions which have a Cd concentration varying from 0.32 to about 1 µM can be regarded as moderately polluted. At Cd concentration in the soil solution that reaches 35 µM only Cd hyperaccumulating species can grow like *Thlaspi caerulescens* (Brown et al., 1994). For comparison, Sanitá di Toppi and Gabbrielli (1999) made a survey of Cd exposure concentrations and exposure times applied in 164 different experiments drawn from about 85 publications. The mean Cd exposure concentration was 223 µM, while the mean exposure time was 5 days. As they conclude there is a lack of direct information regarding the effects of Cd concentrations of environmental relevance (chronic Cd stress). The applied high concentrations may only be useful for producing a certain stress response in the plants (e.g. phytochelatin synthesis) that is further studied and thus provide valuable information but the physiological behaviour of plants in environmental conditions remain obscured. This suggests that there is a need for long term studies in which moderate heavy metal concentrations are applied.

Heavy metals may substantially differ from each other in mobility that may also modify their plant available concentration and consequently their effect. For example root exudates mobilized increasing amounts of the various micronutrients in the order: Cu < Fe < Zn < Mn (Treeby et al., 1989). In an early study Petterson (1976) found that among many heavy metals Mn, Ni and Pb have the largest mobility resulting in the highest shoot/root ratios in cucumber. Thus the question of mobility can be approached from different points of views: e.g. mobility of metals in the soil-plant interface or mobility within the plants. In natural conditions the previous one depends on the chemical form of the metal, the adsorptive properties of soil particles (Gąszczyk, 1989), the soil organic matter horizon (Gawęda, 1991), pH (Gawęda and Capecka, 1995), redox potential, temperature and the concentration of other elements, (McBride, 1994; Haynes, 1990) plant root exudation substances and presence of mycorrhiza (Marschner et al., 1996) while the latter depends on the metal binding capacity of the cell wall, presence and regulation of specific transporters, metal binding substances within the cell and xylem saps and the transpiration rate (Hagemeyer et al., 1986).

Chelating agents either plant-born or artificial may substancially modify plant available metal concentration and mobility in the medium (Norwell, 1991). Organic acids like citrate released by plant roots may form fairly stable complexes with di- and trivalent metal ions (e.g. Fe, Pb, Cd) that may become more mobile in this form. The mugineic acid family substances synthesized by graminaceous species extremely efficiently sweep Fe and other micronutrient metals from the external solution at a broad range of pH values and are reabsorbed by the roots as chelate complexes.The artificial molecules like EDTA (ethylene- diaminetetraacetic acid), DTPA (diethylenetriaminepenta acetic acid) etc., usually applied for Fe mobilisation (which otherwise precipitate at neutral pH values), form chelates the stability values of which are pH dependent (Marschner, 1995). Although chelates alter free ion concentration available for plants in the external medium it has been argued if complexation increases or decreases metal uptake. The accumulation of Cd, Cu, Ni, Pb and Zn from soil in the presence of added synthetic chelates have been reported to dramatically increase. EDTA was the most effective in increasing Pb accumulation in the shoots of *Brassica junicea* (Blaylock et al., 1997; Epstein et al., 1999). In hydroponic solution coordination of Pb by EDTA also enhances the mobility of this otherwise insoluble metal ion (Vassil et al., 1998). In these studies millimolar concentrations of EDTA were applied independently of the nutrients added. When EDTA was applied as Pb complex at 10 µM concentration the accumulation of Pb in the shoot was much lower compared to Pb-citrate or $Pb(NO_3)_2$ and its effect was even stimulatory (Záray et al., 1995).

The old and widely accepted theory of chelate uptake is the split-uptake mechanism where only free metal ions are absorbed by the plant root leaving the chelate in the external solution (Chaney et al., 1972; Marschner et al., 1986). The splitting of the chelate may be due to a simple shift in the equlibrium between chelated and free metal ions caused by the uptake of free ions and may take place by means of reducing agents released from the root (caffeic acid – Olsen et al., 1981, ascorbic acid – Bérczi and Møller, 1998) or plasmamembrane bound reducing enzymes. This is the case during the Fe uptake process of dicotyledonous (Strategy I) plants (Marschner, 1995; Bérczi and Møller, 2000).

The other possibility is that the chelate complex is absorbed by plant roots (Wallace, 1983) and the uptake may occur at breaks in the root endodermis where e.g. new root growth is initiated (Bell et al., 1991). However when double isotope labelling was applied (^{14}C and ^{59}Fe) the results was difficult to explain. Fe deficient plants seemed to accumulate Fe in the shoot leaving the chelating agent behind while Fe sufficient plants take up little amount of Fe producing a $^{14}C/^{59}Fe$ ratio about 1 (Römheld and Marschner, 1981, 1983). In more recent studies the simultaneous accumulation of Pb and EDTA in the shoots of *Brasica junicea* was reported. However we should note that the high concentration of EDTA applied may affect membrane permeability and thus enable the whole chelate complex to enter the roots (Welch, 1995).

Besides these factors the effects of heavy metals are strongly dependent on the age of plant at the time of exposure. *Allium cepa* plants developing from bulbs were found more tolerant to Pb than those developing from seeds which can be connected to the higher Pb content of the seedlings that developed thin roots compared to the thicker adventitious roots of the bulbs. Thus, the seedlings had a higher uptake capacity (Michalak and Wierzbicka, 1998). The young leaves of cucumber plants exposed to Cd were much less affected if the treatment started from the four-leaf-stage of the plants compared to the first-leaf-stage, although they accumulated high amount of Cd (Láng et. al., 1998). The higher resistance of young leaves of more developed plants was supposedly due to the presence of healthy lower leaves which may help the further growth and development serving as an energy, organic matter and stored ion source. In general, the older the plant the greater amount of heavy metals can be tolerated due to accumulation at sites where they are possibly metabolically inactive (e.g. cell walls, vacuole). In contrast, seed germination in certain cases seems to be insensitive to heavy metal toxicity. As high as 10 millimolar concentration of Pb in maize (Obroucheva et al., 1998) and 1 millimolar Pb or 0.1 millimolar Cd in wheat (Titov et al., 1996) failed to inhibit germination which was explained by the exclusion of heavy metals from the young tissues of the radicles. Using germinating bean (*Phaseouls aureus*) seeds and 4 and 6 day-old seedlings it was found that both Cd and Hg inhibit germination but they were much more toxic for 5-day-old seedlings (Cd killing all at 30μM concentration). At germination stage both metals significantly increased the levels of chlorophylls and carotenoids until the 4-day-stage of development (Shaw and Rout, 1998).

2.2. Symptoms of toxicity

In stress physiology there are many different symptoms reported to have been associated with different kind of stresses but only very few of them can be assigned specifically to metal toxicity. This kind of effect can be the plasmodesmatal closure caused by Al (Sivaguru et al., 2000).

In general, heavy metal stress symptoms can be divided to visible and only measurable ones. Concerning phenology, the most usual but one of the least specific symptoms of heavy metal toxicity is the restricted growth compared to the untreated control plants. In a number of studies investigating the effect of Pb growth inhibition is reported to characterize the roots rather then the shoots (Woźny and Jerczyńska, 1991; Kahle, 1993; Godbold and Kettner, 1991). Others report changes in the growth of the whole plant. In sunflower Pb reduces leaf area, dry mass and plant height (Kastori et al. 1998). In bean (*Phaseolus vulgaris*) Pb in 10 μM concentration inhibited the growth of the main root, the number and growth of lateral roots and primary and trifoliate leaves (Woźny and Jerczyńska, 1991).

One of the most usual symptoms of stresses including heavy metal stress is the chlorosis of the leaves due to the decrease of the chlorophyll concentration (Fodor et al., 1996; Láng et al., 1998; Sárvári et al., 1999; Zornoza et al., 1999). Since the chlorophyll concentration may fundamentally influence the functioning of the photosynthetic apparatus and thus affecting the whole plant metabolism it is a really important factor in assessing the impact of heavy metal stress. In more serious cases (more susceptible plants or higher metal concentrations) necrotic spots or whole leaf necrosis appears. In the leaves of three-week-old cucumber treated with 10 µM Ni first yellowish patches appear, then the centre of the patches becomes necrotic. Finally these spots become dry (unpublished). In young sunflower plants 35 µM Ni caused interveinal chlorosis and necrosis (Zornoza et al., 1999).

Heavy metals may cause apparent morphological changes in plants. For example, *Lemna minor* quickly responded to various kinds of heavy metal toxicity by falling apart to separate fronds (Szabados et al., 1983).

Addition of Cd to the nutrient solution significantly decreased shoot and root weight, shoot height, root length and tillers per plants in wheat genotypes but growth inhibition was different between root and shoot and among genotypes (Zhang et al., 2000). It also induced reduction and disappearance of side roots in maize, rye and wheat and browning, fragility and twisting of maize roots (Wójcik and Tukendorf, 1999; Nussbaum et al., 1988).

Metal stress symptoms described only in ultrastructural studies include an increase in synthesis of cell wall polysaccharides and the thickness of the wall in the meristematic cells of the root tip in *Allium cepa* roots in response to Pb (Wierzbicka, 1998).

The physiological symptoms of heavy metal toxicity is not at all specific to the metal but rather specific to the method by which it was observed. In this respect changes in membrane permeability, water and ion uptake, transport and translocation, transpiration, root exudation, enzyme activities, nitrogen metabolism, photosynthetic processes (electron transport, fluorescence induction, phosphorylation, CO_2 fixation) respiration, cell division and expansion, all kinds of synthetic processes and cell homeostasis were mentioned to have been affected by heavy metals. Examples for these taken from the latest literature are given and discussed in the text below.

2.3. Sites and mode of action

There are numerous more or less documented statements of primary effects of heavy metals on different targets like plant organs, organelles or physiological processes in the related literature like the following few examples: One of the specific primary effects of Pb, and possibly Cd too, in maize may be the damage to microtubules in the roots and consequently the inhibition of cell division (Eun et al., 2000). In the leaves of tall fescue (*Festuca arundinacea*) the primary targets of Pb toxicity were the photosynthetic CO_2 fixation and photorespiratory CO_2 evolution (Poskuta and Waclawczyk-Lach, 1995). It is easy to recognize that these kinds of conclusions or statements always focus on a narrow range of the heavy metal action spectrum. Recently Stroiński (1999) has outlined the scheme of Cd stress in plant cells which focuses much on the oxidative stress caused by Cd. His scheme clearly shows the complicated actions of Cd and partly cellular responses, too, but the significance of the processes as compared to their impact and their integration is less obvious. In this paper we attempt to establish a theoretical model for the mode of action of heavy metals in general giving an overview of the different steps and their impact on the plants.

On a whole plant basis the site of action depends on how the heavy metals enter the environment or plant. In contaminated soils first the germinating seeds then the root system of seedlings have to suffer from the presence of nonessential metals. If the contamination is airborn a small portion of metals may directly enter the shoot to a certain extent but the greater portion is incorporated to the soil. Therefore, in most cases, the primary site of heavy metal action is the root system. In this situation, too, there might be a possibility for heavy metals to quickly proceed to the shoot via the apoplastic pathway (Bell et al., 1991) which is facilitated by transpiration. Consequently, the shoot or leaf tissues may as well be primary targets for toxicity stress. For these reasons we put the actions into a specific causal relation order independently of the site of entrance. See the "heavy metal actions column" of the Figure 1).

Figure 1. Stepwise model for the mode of action of heavy metals and plant tolerance mechanisms. The left column represents the building up of heavy metal toxicity, leading to plant death (Stage 5) in cases of acute stress or in the absence of protective responses (e.g. sensitive genotypes). The right column represents plant responses to stress effects at different levels leading to tolerance in the end (Stage 5). Grey arrows represent the direction of changes. Black arrows represent those heavy metal effects which trigger plant responses. Dashed arrows represent plant responses to stress effects at different levels which are triggered along signal transduction pathways and feed back changes of the metabolism. In Stage 1, processes related to heavy metal stress occur in the apoplast and involve C.E.C. saturation, ion interaction, increase in cell wall rigidity, membrane leakage and inhibition of membrane transport processes. Plants give non-specific responses and perhaps signal transduction processes are activated at this level. In Stage 2 metabolic reactions are inhibited in the symplasm and as a response protecting metabolic reactions and prompt detoxification processes are induced in plants. Stage 1 and 2 also occur in leaf tissues after the transport of heavy metals in the xylem. In Stage 3, a wide scale of metabolic and homeostatic processes are inhibited by heavy metals or induced as a stress response of plants. In Stage 4 the toxic effects appear as visible symptoms or if not, effective tolerance mechanisms and strategies can be observed. Stages 3-5 are characterised at the whole plant level.

AOS – Active oxygen species

PHYSIOLOGICAL RESPONSES OF VASCULAR PLANTS

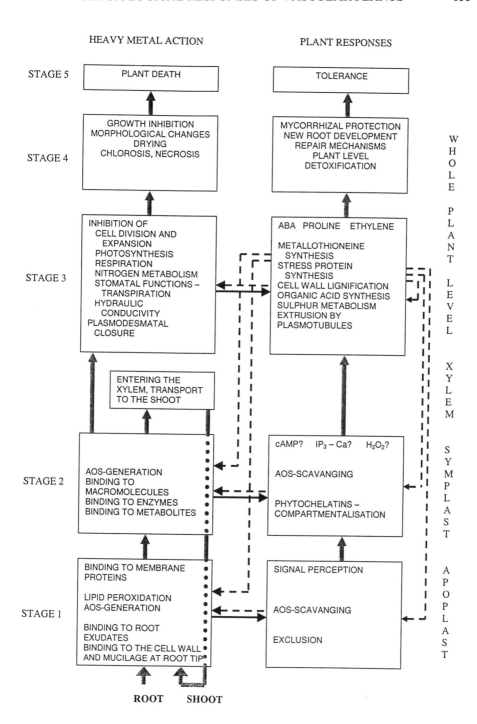

The first action of heavy metals introduced to the plant's rizosphere is the interaction with other ionic components of the soil or culture solutions. They may precipitate with anionic ligands or compete with other cations for binding sites in the cell wall material. Another competition exists between the toxic and essential heavy metals for organic chelating agents (either root exudates or artificial ones) which result in a series of complexes with different stability values. Concerning also the fact that the cell wall itself is an efficient adsorber with considerable cation exchange capacity (C.E.C.) the inhibition or stimulation of cation uptake is inevitably the first effect of heavy metals which may cause significant changes at the higher levels of metabolism (e.g. iron deficiency stress).

In the cell wall, the generation of active oxygen species (AOS) ($O_2^{\bullet-}$ and H_2O_2) may occur by defects in the function of oxidoreductive enzymes (e.g. cell wall oxidases and peroxidases) or the leakage of the electron transport chain caused by heavy metals (Stroiński, 1999). Copper, aluminum, cadmium, zinc and iron damage have all been linked to oxidative stress (Dat et al., 2000). Cell wall peroxidases produce $O_2^{\bullet-}$ at the expense of NADH in a Mn^{2+}-dependent reaction (Elstner and Osswald, 1994) The cell wall constituent, extensin, may be peroxidised e.g. in the presence of H_2O_2 and Fe^{2+} (Ye and Varner, 1991). Iron excess may also cause oxidative damage due to its potential for reacting with $O_2^{\bullet-}$ and H_2O_2 to generate $OH^{\bullet}$ which is considered to be the most toxic of all AOS (Dat et al., 2000).

Heavy metals like Cu, Ni and Zn, just as Al may also cause AOS-mediated lipid peroxidation in the plasmalemma which can lead to serious membrane leakage (Weckx and Clijsters, 1996, 1997; Kochian, 1995; Baccouch et al., 1998). However, the most significant destructive event is supposed to be the reaction with membrane proteins e.g. H^+ATPase and cross-linking thiol groups leading to the inhibition of proper membrane transport processes and/or to the uptake of the toxic metal to the cytoplasm (Meharg, 1993). All these events occur in the apoplastic space and we suggest to consider them as Stage 1 of heavy metal effect.

Stage 2 starts once the toxic metals entered the symplasm and bind to proteins, nonenzymatic macromolecules (e.g. Cd can bind to calmodulin, Rivetta et al., 1997) and other metabolites in which they compete with essential cations. This occurs in parallel inside both the cytoplasm and cell organelles like chloroplasts, mitochondria etc. causing the inhibition of all kinds of metabolic and regulatory reactions and producing oxidative stress again (Stroiński, 1999). It should be noted however that heavy metals have not been reported to directly generate AOS except for a few studies. The short term Cd treatment of potato tuber discs resulted in a rapid accumulation of $O_2^{\bullet-}$ and H_2O_2 in the tissues (Stroiński and Kozłowska, 1997). Zinc treatment increased the H_2O_2 level of bean roots (Weckx and Clijsters, 1997).

It is widely argued (see above) how it happens but heavy metals finally enter the xylem vessels and they are transported to the leaves via the transpiration stream. In the leaf mesophyll heavy metals are again in the apoplasm and Stage 1 and 2 is repeated.

Stage 3 is interpreted at the level of life processes i.e. transport, metabolic and homeostatic events which are negatively affected by toxic heavy metals. Closure of plasmodesmata by the induced synthesis of callose was reported in case of Al toxicity (Sivaguru et al., 2000). Water uptake, and transport decreases due to the decrease of hydraulic conductivity of the roots and root pressure (Cseh et al., 2000), stomatal closure and a resultant decrease in transpiration (Barcelo and Poschenrieder, 1990). This in turn decreases mineral supply for synthetic processes. Nitrogen metabolism decreases due to the decline in nitrate reduction in the leaves (Gouia et al., 2000). Photosynthesis and respiration decrease because of inhibited synthesis and/or destruction of the photosynthetic apparatus, inhibited redox reactions, gas exchange and the damage by oxidative stress as well as inhibited enzymes of glycolysis, pentose phosphate cycle etc. However, it must be noted that the effect on respiration may as well be stimulatory depending on the metal, concentration, treatment time and plant species. More or less specifically cell expansion and cell division are also negatively affected (Woźny and Jerczyńska, 1991; Eun et al., 2000).

The appearance of visible symptoms can be considered as Stage 4 of heavy metal toxicity. Chlorosis of the leaves is one of the first visible signs of metabolic stress referring to decreased chlorophyll and relatively increased carotenoid content. In severe cases almost complete inhibition of Fe transport to the shoot by Cd and the consequent iron deficiency stress and eventually the superimposed photoinhibition may cause necrosis of the leaves. Ni toxicity may cause patchy necrosis and also drying in the necrotic spots (unpublished). Growth inhibition and morphological changes are induced by all the above negative influences of heavy metals which altogether may lead to plant death that is referred to as Stage 5.

All these steps or stages of heavy metal effects in plants are a summary of different individual observations drawn from all kinds of experiments carried out with different metals, different objects and in extremely different conditions. For this reason the generalisation applied here should not be considered as an obligatory stepwise mechanism for heavy metal actions. Instead, it is to be a collection of molecular and structural targets that are potentially attacked by heavy metals and the consequent changes in physiological processes. It must by emphasised that plants have specific and nonspecific responses to heavy metal effects which are activated at different levels that can be correlated to the stages of metal action described above. These responses are integrated to a mechanism which is known as tolerance.

2.4 Tolerance mechanisms

The first protective barrier against heavy metals is of highly non-specific nature (Fig.1: plant responses column). The C.E.C. of cell walls, presence of root exudates (e.g. oxalate) and the root tip mucilage enable the roots to moderate the concentration of metals reaching the cytoplasm by filtering the solution. This filtering capacity is of course very much limited and does not satisfactorily prevent heavy metals from Stage 1 actions. The AOS-scavenging system of the cell wall including the ascorbate cycle and scavenging enzymes may cope with heavy metal action but the scarcity of direct data makes it difficult to draw overwhelming conclusions about specific responses. The third action that occurs in the apoplastic side of the plasmalemma is the binding of heavy metals to hypothetical receptors or the signal perception in the membrane triggering the generation of intracellular signals. Theoretically this could be the key to plant responses to heavy metal stress at higher levels (Fig. 1: Stage 2-3) but unfortunately the components involved in this type of reactions are just suspected at the moment. Similarly to the model for heavy metal effect these responses can be considered altogether as Stage 1 of tolerance mechanisms.

In the cytoplasm heavy metals bind to negatively charged groups on all kinds of molecules but many of them (Hg, Cu, Cd) have especially high affinity to thiolic groups of Cys. There is a group of Cys-rich molecules named phytochelatins with the following structure: $(\gamma-Glu-Cys)_n-Gly$, where n is the number of repetition (usually 2-11) of the unit in brackets. Phytochelatins are able to chelate heavy metals and the resulting stable complexes are carried to the vacuole where they dissociate. The released heavy metals can bind to organic acids there while PCs are transported back to the cytoplasm. PCs are synthesised from glutathione by the enzyme phytochelatin synthase. This mechanism may be the most significant response of plant cells to heavy metal toxicity because the PC synthase is induced promptly and greatly reduces the free heavy metal ion concentration in the cytoplasm (Rauser, 1995).

The AOS scavenging system of the cell is composed of the enzymes superoxide dismutase (SOD), catalase (CAT), ascorbate peroxidase (APX), glutathione reductase and small antioxidant molecules like ascorbic acid, cysteine, glutathione, tocopherol, carotenoids, hydroquinones and polyamines (Stroiński, 1999). Among others, increased activity was reported for SOD in tobacco (Kurepa et al., 1997) and soybean (Chongpraditinun et al., 1992) after elevated Cu treatment and also after Al treatment (Cakmak and Horst, 1991), for catalase and APX in tobacco in iron excess (Kampfenkel et al., 1995). CAT activity in Cd treated bean seedlings (Somashekaraian, et al., 1992) and in oat seedlings treated with excess Cu (Luna et al., 1994) was reported to decrease. Polyamine levels increased in oat, bean (Weinstein, 1986) and potato leaves (Stroiński et al., 1990). Since most of the studies are based on the measurement of activity or concentration of one or more of the above components we can only say that plants may cope with oxidative stress induced directly or indirectly by heavy metals.

The third process co-occurring with the above described ones is perhaps the signal transduction events that trigger metabolic processes by regulating the activity of certain enzymes as a response to heavy metal actions. The control of phytochelatin synthesis and heavy metal accumulation in the vacuole might involve cAMP and IP_3 Ca^{2+} ions signals (Ow, 1996; Stroiński, 1999). H_2O_2 may also take part in signal transduction inducing phytoalexin synthesis and the transduction of Cd signal leading to an intensified glutathione synthesis (Stroiński, 1999). All these would comprise Stage 2 of tolerance mechanisms which correspond to Stage 2 of heavy metal action.

At Stage 3 of tolerance various processes of metabolism and others aimed at the maintenance of homeostasis are induced as affected by different signals. The conversion of glutathione to PCs purges up sulphur metabolism while the accumulation of metals in the vacuole causes a shift in organic acid production for balancing positive charges. Organic acids may play a wider role as they are supposed to be given off to the apoplast for complexing metals there as it was reported for oxalate in case of Pb toxicity (Yang et al., 2000).

In *Allium cepa* two protective mechanisms were described in Pb toxicity: (1) the amount of polysaccharides in the cell wall and the thickness of the wall increased allowing accumulation of Pb in an inactive form and (2) Pb was expelled from the symplast through plasmotubules to the root tip apoplast (Wierzbicka, 1998).

In some cases of heavy metal toxicity the synthesis of new Cys-rich peptides was discovered. These peptides are analogous to the gene-encoded metallothioneins found in animals, fungi and cyanobacteria which have a role in complexing and detoxifying different kinds of heavy metals. In plants the appearance of metallothioneins was reported to be induced by Cu in *Arabidopsis thaliana* (Murphy et al., 1997), Zn in wheat (Lane et al., 1987), Cd in soybean (Blakely et al., 1986).

Far less specific, though of decided importance are the general stress response reactions involving the synthesis of ethylene, proline and ABA. For example, Talanova et al. (2000) found that pretreatment of plants with low concentration of Cd or Pb may have significant effect upon their tolerance to subsequent high concentrations. They conclude that the enhancement of metal resistance in cucumber may be connected with the accumulation of free proline, ABA and soluble proteins in tissues. Similarly, axenic hairy roots of *Daucus carota* treated with 0.1 and 1 mM Cd besides producing PCs showed an increased ethylene production and the synthesis of two new proteins compared to the control which altogether made them apparently insensitive to short term Cd stress.

The relationship between heavy metal effect and ethylene synthesis is not clearly understood so far. It has been suggested that Cd may stimulate and Zn may inhibit the ethylene-forming-enzyme in soybean (Pennazio and Roggero, 1992). However, ethylene itself might be a signal for accelerating lignification (Ievinsh and Romanovskaya, 1991) and cell wall alterations in the vascular system which in turn restricted water and consequently Cd transport into bean leaves (Fuhrer et al., 1981). Ethylene may induce the activity of ascorbate peroxidase and thus stimulates hydrogen peroxide detoxification (Mehlhorn, 1990). Ethylene may also be a factor in regulating the expression of genes encoding metallothioneins (Whitelaw et al., 1997) and stress proteins (Sanitá di Toppi et al., 1999). Finally, it was shown that PC synthesis was inhibited in carrot cell suspension cultures when no ethylene was produced after supplying aminoethoxyvinilglycine (AVG) or using ethylene traps (Sanitá di Toppi et al., 1998).

In *Silene vulgaris* Cd exposure caused proline accumulation. However this finding was shown to be the consequence of Cd-induced water deficit stress and not the direct effect of Cd on the synthesis of proline (Schat et al., 1997). Nevertheless, the role of proline under metal stress conditions was demontrated by several authors. It increses the rigidity of cell walls (Muñoz et al., 1998), has a role in scavenging hydroxyl radicals (Smirnoff and Cumbes, 1989), takes part in enzyme protection under Cd toxicity (Shah and Dubey, 1997) and also in protection of macromolecules preventing denaturation (Schobert and Tschesche, 1978).

The role of ABA in alleviating heavy metal effects is even less discovered. In Cd-treated bush bean plants ABA accumulation was found only after a considerable delay following the exposure to the heavy metal. Then significant increase in the ABA levels in roots, stems and leaves were observed. However, in leaves it was always accompanied by the increase in stomatal resistance (Barcelo et al., 1986; Poschenrieder et al., 1989). ABA may influence the water household and the cell expansion and division although its participation in heavy metal stress responses seems to have secondary importance.

Under high temperature stress eukaryotes including plants produce heat shock proteins (HSPs). This type of proteins turned out to be produced also under cold, water, salt and heavy metal stresses etc. and therefore they are often called as stress proteins. HSPs can be divided into several classes on the basis of their molecular weight in kDa: HSP110, HSP90, HSP70, HSP60 and low molecular weight or small (15-30 kDa) HSPs. Ubiquitin is a small protein also referred to as an HSP (Vierling, 1991; Waters et al., 1996). HSP90, HSP70, HSP60 are believed to function in the stabilisation of proteins in a particular state of folding. Their role is also evidenced in transport of proteins across membranes, assembly of oligomeric proteins and modulation of receptor activities. There is little data available about HSP110 (Vierling, 1991). Moreover, small HSPs facilitate reactivation of chemically denatured enzymes in a nucleotide independent fashion. For this reason HSPs are referred to as a type of molecular chaperones (Waters et al, 1996). Ubiquitin has a high affinity to denatured proteins and is involved in an ATP-dependent, intracellular proteolysis.

Under heavy metal stress specific mRNA transcripts are produced that regulate stress protein synthesis. Cd was shown to regulate the splicing of an intron from the pre-mRNA transcribed by a Gmhsp26-A gene (Czarnecka et al., 1988).

All these processes may be induced by signal transduction initiated by the heavy metals binding to receptors in the plasmamembrane and inside the cell, however none of these pathways have been established, yet. These missing links in the integrated responses of plants to metal stresses leave us in uncertainty concerning the real connections of the individual processes. Nevertheless, it seems to be clear that the interconnection of specific and rather unspecific reactions is built up during the consecutive stages of stress. The metabolic processes triggered at Stage 3 have a feedback enhancement of the mechanisms functioning at lower stages (e.g. purged sulphur metabolism increases glutathione synthesis leading to the compartmentalisation of heavy metals by PCs). On the other hand the integration of these mechanisms can be regarded as the manifestation of whole plant defence strategies at a higher level which can be referred to as Stage 4. At this level the modulated action of detoxification mechanisms may function together (AOS scavenging, compartmentalisation, metallothioneins, organic acid synthesis). Repair mechanisms for decreasing damage and restoring homeostatic balance are switched on (proline and stress proteins synthesis). Resource rearrangement is induced in new root or shoot growth leaving toxic metals in older tissues but mobilizing essential ones. In different rice varieties tolerant or sensitive to Pb the tolerance was associated with the ability to develop adventitious roots that adsorb little amount of Pb (Yang et al., 2000). Oxalate content and exudation in the roots of these tolerant varieties increased as a response to Pb suggesting a mechanism decreasing the bio-availability of Pb.

Plant level tolerance mechanisms may be coupled with the support of mycorrhizal symbiosis. Fungi have been reported to bind heavy metals in their cell walls and produce metallothioneins, PCs (Howe et al., 1997).

As a result of integrated response mechanisms plants are able to acclimate heavy metal toxicity stress in a certain concentration range and exposure time depending on the type of metal and the plant species. We may say, now, corresponding to Stage 5 of heavy metal action that Stage 5 of plant responses is "tolerance".

There is a great intraspecific difference concerning the ability to reach Stage 5. For example, root length and chlorophyll content of *Holcus lanatus* declined rapidly with increasing Pb and Zn concentrations in the nutrient solution. However, root growth and chlorophyll content of genotypes coming from a Pb-Zn mine area was greater than that of the control (Symeonidis and Karataglis, 1992). In contrast to herbaceous metallophytes, tree species generally are not able to adapt to high heavy metal concentrations in the rhizosphere. Consequently, only a few metal-tolerant ecotypes of tree species have evolved (Kahle, 1993). These examples demonstrate that the above described tolerance mechanisms are far from ubiquitous in nature. Ecotypes or genotypes of plants with higher tolerance usually evolve in excessively contaminated environments which create stressful conditions that the plants can still acclimate and during this process the genotypes capable of more efficient detoxification and repair mechanisms may survive and thus be selected. That is what we call adaptation.

The above described model ranking plant responses to heavy metal stresses to Stages 1 to 5 is not meant to be an alternative to other models trying to demonstrate tolerance mechanisms. Recently a well established model was published by Sanitá di Toppi and Gabbrielli (1999) for integrated plant responses to Cd toxicity suggesting that "various mechanisms might operate in response to Cd, both in an additive way and in a mutual–probably more effective–potentiating way". Although the setting of tolerance mechanisms to 5 stages implicates consecutiveness in time, in accordance with the previous model we would like to emphasise the additiveness of stress responses. The stages can be regarded as the level of integration or, to a limited extent, spatial separation (Stages 1 and 2). The main point was to find possibility for generalisation in which specific responses given only to heavy metals and unspecific reactions induced in many types of stresses may be observed together as a whole. The need for this approach is evident if one looks over heavy metal toxicity symptom groups which can be regarded as deficiency stresses, water deficit stress, oxidative stress, light stress etc.

3. THE PHYSIOLOGY OF LEAD AND CADMIUM TOXICITY

The different effects of Pb and Cd on plants have been reviewed excessively (Foy et al., 1978; Koeppe, 1981; Kahle, 1993; Woźny et al., 1990; McLaughlin and Singh, 1999; Sanitá di Toppi and Gabbrielli, 1999). Therefore we only intend to summarize the data accumulated in the fields of uptake and accumulation, effect on growth, photosynthesis, water relations, ion uptake and membrane structures in higher plants during the last decade to give an insight to the latest progress and also to support the statements and theoretical considerations in the previous part (2.) of this chapter.

3.1. Uptake and accumulation

There are two major pathways for the incorporation of heavy metals into plant tissues: the atmospheric deposition from settling dust and soil uptake. The metal load deposited on plant foliage mostly in industrial areas or along heavily used roads can be absorbed through the cuticle and epidermis during precipitation or dew formation (Marschner, 1995). In early environmental studies, the measured leaf heavy metal content was considered predominantly as airborn (e.g. Little and Martin, 1972; Little, 1973) and consequently the uptake through the roots and translocation was underestimated. Others claimed that the main input pathway is the root uptake, the shoot uptake being negligible (Hemphill and Rule, 1975). Recently it has been well established that considerable amount of heavy metals may enter plant tissues via foliar uptake.

In greening etiolated barley leaves excised and incubated in Pb containing solution, lead deposits were detected in the intercellular spaces of mesophyll, in guard cells and in cuticule covering stomata (Woźny et al., 1995). In pea leaves submerged into Cd containing solutions Cd is taken up and translocated from the exposed part of the leaves into petioles and stipules. The foliar uptake depends on the permeability of the cuticular membrane, which is increased by a high intrinsic Cd level which in turn enhances the foliar uptake of Cd in sugar beet. Stomata are not directly involved in the process, instead it is suggested that the uptake occurs through ectodesmata, nonplasmatic channels that are situated particularly in the epidermal cell wall/cuticular membrane system between guard cells and subsidiary cells (Greger et al., 1993).

More recently a stable isotope tracer technique was used to evaluate the translocation of trace metals applied to white spruce foliage at ambient concentrations. Simulated rainfall was applied at segments and it was found that 99% of the excess Pb remained in the application segment, although translocation of Pb away from the application zone mostly towards the shoot tip occurred. The translocation was significantly enhanced in rainfall at pH 4.0 compared with pH 6.0. However, the increases in Pb concentration in any plant section away from the application zone were very small suggesting that foliar uptake is not a major pathway in Pb accumulation in white spruce (Watmough et al., 1999).

In contrast to these considerations, the bulk uptake of heavy metals usually occur from soil solutions and therefore the first accumulating organ is the root system

(Godbold and Kettner, 1991; Neite et al., 1991; Breckle and Kahle, 1992; Hernández et al., 1998; Simon, 1998; Jiang et al., 2000; Zhang et al., 2000).

In the aquatic plant, duckweed (Lemna minor), the highest rate of Pb uptake was found in the basal part of the root. Lead deposits were detected mostly in the cell walls adjacent to the plasma membrane and in the lumen of several endomembrane compartments: the endoplasmic reticulum, dictyosomal vesicles, nuclear envelope and the vacuoles. Lead induced changes of cell ultrasturucture: increase in the number of membranous structures, swelling of ER cisternae and distortion of the dictyosomal cisternae (Kocjan et al., 1996).

In terrestrial plants most of Pb enters the roots at the root hair region and accumulates at the endodermis which serves as a barrier to radial Pb movement. It was found that at the zone of formation of lateral roots and the secondary growth Pb enters the central cylinder (Ksiaźek and Woźny, 1990). In maize seedlings the highest Pb accumulation was found in the root tip meristem, both in the symplast and apoplast (Eun et al., 2000). Cd ions taken up by the plants also accumulate in the roots and are bound to the apparent free space (Cieslinski et al., 1996; Simon, 1998). In bean plants Cd accumulation was observed in the vacuoles (Velazquez et al., 1992).

Most of the accumulated heavy metals remain in the roots. The Cd concentration was 10 times higher in the roots of 16 wheat genotypes (average) than in the shoot and the root retained 78% of the total Cd taken up (Zhang et al., 2000). In sunflower plants both Pb (Kastori et al., 1998) and Cd (Simon, 1998) accumulates more in the roots than in the stems and leaves though their concentration increased in the shoot, too, compared to the control plants. Nevertheless, heavy metals are transported to the shoots but the allocation ratio vary depending on plant species or genotypes and the experimental conditions. According to Vojtechová and Leblová (1991) Pb^{2+} was transported to shoots from roots at a lower rate than Cd^{2+} in maize. Pb was detected in the xylem sap in significant concentrations (Donelly et al., 1990; Záray et al., 1997), implying that transpiration facilatates its accumulation in the shoot. Correspondingly, Cd translocation to the shoot is stimulated by transpiration (Salt et al., 1995).

Substantial differences may exist in the distribution of heavy metals in the shoot. In stinging nettle (*Urtica dioica*) and cucumber (*Cucumis sativus*) Pb concentration decreases towards the shoot tip (Fodor et al., 1996, 1998). The distribution of Pb between the tissues of the shoot was different. In stinging nettle Pb concentration in the shoot decreased in the following order: internodes > petioles > leaves, while in cucumber: petioles > internodes > leaves. However, in cucumber in middle parts of the shoot, that is at the largest, fully expanded leaves the leaf blades had higher Pb concentrations. Similar distribution of Cd was found in cucumber by Moreno-Caselles et al. (2000). Since cucumber has greater leaf surface and significant root pressure compared to stinging nettle, it can be concluded that the accumulation depends on the retention capacity and other transport characteristics of plants.

Leita et al. (1996) investigated the Cd-forms accumulated in the intercellular spaces of the roots and leaves. They found that considerable amount of Cd is bound to the cell walls in the xylem and the water soluble Cd is present mainly in divalent ionic form, They assumed that the uptake and translocation to the leaves also occur in this form. The increasing saturation of binding sites of the cell wall results in increasing Cd concentration in the leaves (Salt et al., 1995). However, concerning the fact that the free ionic heavy metals like Cd and Pb can readily form complexes with ionic ligands like citrate and malate that are also present in the xylem sap it seems reasonable that at least part of the metals are complexed during translocation. If tomato plants *(Lycopersicon esculentum)* were pretreated with citric acid the Cd uptake by the root increased 2-fold, the root-to-shoot transport 5-6-fold and the Cd concentration in the xylem was 6-8-fold compared to the control. Senden et al. (1995) concluded that Cd is complexed by citrate in the xylem sap.

Experiments in which artificial chelating agents were applied support the view that complexation modifies the uptake, transport and accumulation of heavy metals in higher plants. When EDTA was added to Cd-containing nutrient solution, in the presence of Fe(III), the accumulation increased. Following the reduction or uptake of Fe(III) EDTA forms complex with Cd and this decreases accumulation (Srivastava and Appenroth, 1995). In *Brassica juncea* 75-fold concentration of Pb was found in shoot tissue over that in solution when the plants were supplied with both Pb and EDTA (Vassil et al., 1998).

Comparing 11 species/cultivars using both nutrient solutions and Pb-contaminated soils the highest shoot Pb concentration was accumulated by maize. Moreover, addition of EDTA to the Pb-contaminated soil resulted 10.6 mg g^{-1} Pb in the shoot which is the highest value ever reported in the literature (Huang and Cunningham, 1996).

Compared to the significant accumulation of Pb in the shoot the fruits and crops usually contain little contamination (Aarkrog and Lippert, 1971; Baumhardt and Welch, 1972). In contrast Cd was transported to the fruits of cucumber (Moreno-Caselles et al., 2000) and Cd content of the fruits of strawberry plants *(Fragaria anasassa)* was in good correlation with the Cd concentration in the leaves, too (Cieslinski et al., 1996).

3.2. Growth inhibition

The physiological effect of heavy metals is closely connected to their accumulation. In this respect the first organ that is expected to be affected is the root. In Indian mustard Pb has no effect in 10 µM concentration (Titov et al., 1996; Obroucheva et al., 1998) and it slightly inhibits root and shoot growth even at millimolar concentration (Jiang et al., 2000). In this plant there is a significant retention of Pb in the root as the shoot Pb content is only 3 times larger than the trace contamination in the control. The same was found in cucumber. At physiological concentrations (10 µM), Pb does not inhibit shoot growth significantly and does not cause visible morphological changes (Cseh et al., 2000). The reason for it is that Pb is captured in the root apoplast and hence its translocation to the shoot is restricted (Godbold and Kettner, 1991; Breckle and Kahle, 1992).

Cd was found to inhibit root growth and decrease fresh weight and water content in roots and shoots of maize, rye and wheat, the most sensitive of which was maize, although Cd accumulation was lower in this plant (Wójcik and Tukendorf, 1999). Anyhow, root and shoot growth was more severely inhibited by Cd compared to Pb (Titov et al., 1996) and in high-K^+ plants compared to low-K^+ plants showing that the effect of heavy metals depend upon the K^+-status of the plants (Trivedi and Erdei, 1992). Really, Cd seems to be one if not the most toxic heavy metals for plants. Comparing the effects of a series heavy metals (Cd, Pb, Cu, Ni) the highest growth inhibition was found with Cd (Burzyński and Buczek, 1994).

In strawberry seedlings leaf fresh weight decreased more than that of roots as Cd concentration was increased (Cieslinski et al., 1996). In turn in *Sesamum indicum* the decrease in fresh weight was more stressed in the roots (Singh et al., 1997). The reduced elongation of roots under Cd stress may be caused by inhibited mitosis, decreased synthesis of the cell wall components, damaged Golgi apparatus or changes in the metabolism of polysaccharides in the root cap (Punz and Sieghardt, 1993), or inhibited elongation of root cells (Ernst et al., 1992). In maize Pb specifically causes a rapid inhibition of root growth probably due to the inhibition of cell division in the root tip. It perturbs the alignment of microtubules in a concentration-dependent manner, decreases the mitotic index in root apical meristems and the number of dividing cells (Woźny and Jerczyńska, 1991; Eun et al., 2000).

The growth inhibition by Pb and Cd (10 µM) also depends on the presence or absence of Fe-chelates (Fodor et al., 1996). With Fe-EDTA even root growth is stimulated by Pb, with Fe-citrate the inhibition is remarkable by both metals. If Fe was was supplied as $FeCl_3$ (Cseh et al., 2000) inhibition of dry mass accumulation by Pb was only 20 %.

3.3. Photosynthesis

The inhibition of growth is interpreted in many different ways in the related literature. The most frequent conclusion and also one of the most widely investigated questions is the direct or indirect effect of heavy metals on photosynthesis as the primary energy producing process of plants (Krupa and Baszyński, 1995). Decreased photosynthesis can be caused by the decrease in the level of photosynthetic pigments related to breakdown or the inhibition of synthesis. Decreased chlorophyll concentration was found in *Lemna minor* treated with Pb (Garnczarska and Ratajczak, 2000), in sunflower treated with Cd (Gadallah, 1995) and in cucumber treated with either metals (Moreno-Caselles et al., 2000).

In excised leaf segments of pea, Sengar and Pandey (1996) found that Pb lowered photosynthesis specifically by the inhibition of δ-aminolevulinic acid synthesis and the decrease in the 2-oxoglutarate and glutamate pool which may be caused by the competition between the essential ions required for chlorophyll synthesis and Pb^{2+}.

In the presence of Cd under *in vitro* conditions in isolated etioplasts the formation of chlorophyllide from protochlorophyllide was shown to be inhibited (Böddi et al., 1995). Cd decreases the amount of pigment-protein complexes which is prevalent in the PSI and LHCII, the PSII being less sensitive (Láng et al., 1995). The desorganisation of the antenna complexes were observed under both *in vivo* and *in vitro* conditions (Krupa, 1988; Ahmed and Tajmir-Riahi, 1993). The higher sensitivity of PSI may be caused by the inhibition of major antioxidant enzymes by Cd (Gallego et al., 1996).

Applying Cd up to 10 mM concentration to pea (*Pisum sativum*) seedlings caused a sharp decline in chlorophyll content, photosynthetic rates, activity of the photosystems and photosynthetic enzymes (RUBISCO etc.) in 6 days and these effects became even more stressed during the extended treatment (Chugh and Sawhney, 1999).

In Cd-treated plants there can be changes in the hormon balance, too, which in turn may influence the organization of the photosynthetic apparatus (Nyitrai, 1997).

The electron transport system of PSII usually proved to be more sensitive: both acceptor and donor side inhibition was observed (Siedlecka and Baszyński, 1993; Krupa and Baszyński, 1995). In fluorescence induction experiments it was concluded that the electron transport between Q_A and Q_B is slower than in the untreated control (Láng et al., 1998).

In sunflower there was a significant reduction in chlorophyll content and a lesser decrease in photosynthetic O_2 evolution rate and PSII efficiency at low light intensity. In millimolar concentration Pb decreased photochemical quenching and the efficiency of PSII electron transport and significantly affected nonphotochemical fluorescence quenching suggesting the increase in proton gradient across the thylakoid membrane and a decrease of photophosphorylation (Kastori et al., 1998).

In excised leaves of tall fescue (*Festuca arundinacea*) genotypes 100 mM Pb caused inhibition of PSII activity, photosynthetic CO_2 fixation, photorespiration and dark respiration (Poskuta and Waclawczyk-Lach, 1995). In the excised barley leaves Pb reduced chlorophyll content (especially Chl b) and the average number of grana, increased condensed chromatin content in nuclei referring to the increase of transcriptionally non-active DNA content but the number of chloroplasts in mesophyll cells did not change after Pb treatment (Woźny et al., 1995).

The sensitivity of the photosynthetic apparatus depends on the age of plants and the duration of the Cd-treatment. In older plants treated with Cd PSII activity was found to be lower than in the younger ones (Skórzyńska-Polit and Baszyński, 1997).

The differential effects of chelating agents on the photosynthetic apparatus under Pb and Cd stress was investigated by Fodor et al. (1996). Chlorophyll content of Cd-treated plants - independently of the Fe-complex - was very low similarly to the iron deficient plants. Pb is proven to inhibit the chlorophyll accumulation only in plants supplied with Fe-citrate. Cd caused more than 50% inhibition of the *in vivo* $^{14}CO_2$ fixation when applied with Fe-EDTA. With Fe-citrate, the inhibition exceeded 90%. The photosynthetic activities in the Pb-treated plants were not significantly different from the control plants. The amount of chlorophyll containing complexes, especially that of PSI was highly affected by Cd particularly in the lower leaves.

Heavy metal treatment also affect the Calvin cycle (Sheoran et al., 1990; Malik et al.,1992; Krupa et al., 1993; Siedlecka and Krupa, 1996). Under Cd stress the amount of soluble and storage carbohydrates and free amino acids decreased (Gadallah, 1995; Costa and Spitz, 1997).

3.4. Water relations

Photosynthesis is undoubtedly connected to water relations due to gas exchange. The decrease in the rate of photosynthesis was associated with decreased stomatal conductance and transpiration in maize treated with Pb (Stefanov et al., 1993) and in *Bacopa monniera* plants grown *in vitro* and treated with Cd (Ali et al., 2000). Similarly, Cd in barley caused inhibited growth, lower chlorophyll and carotenoid content, decreased photosynthetic rate and accelerated dark respiration rate together with decreased water potential and transpiration rate but the relative water content remained unchanged (Vassilev et al., 1998).

In many experiments the decrease in fresh weight was associated with the decrease in transpiration and the water content of the root and shoot (Hernández et al., 1997; Lozano-Rodriguez et al., 1997; Vassil et al., 1998).

Direct or indirect heavy metal effect on stomatal functions may be responsible for the decline in transpiration. However, the growth inhibition associated with the closure of stomata may result in increased water content in the shoot. For example, water content of the different tissues of cucumber treated with Cd shows varying deviations from the control. Water content of the root decreased but that of the leaves calculated per unit area was elevated (Láng et al., 1998; Sárvári et al., 1999).

Changes in the morphology caused by heavy metals may indirectly but their effects on the cell walls and cell membranes may directly influence the water uptake and transport (Barcelo and Poschenrieder, 1990). In plant membranes the evidence for the existence of water channels put the investigation of heavy metal effect on water permeability of cell membranes into a distinguished position (Steudle and Henzler, 1995; Maggio and Joly, 1995). As a support, in case of mercury it was found that it specifically inhibits water channels (Daniels et al., 1994).

For the plants the water transport through the roots is of fundamental importance because the hydraulic conductivity (Lp) of the root influences not only the water supply but also the ion supply of the plant. In Pb-treated cucumber plants exudation was inhibited to a much larger extent (50 %) than root growth suggesting that Pb inhibited water transport through root cells (Cseh et al., 2000). Despite the growth is relatively unaffected by Pb it can be hypothesized that Pb considerably decrease the water permeability of root cell membranes through its effect on the water and/or ion channels as it is the case in Na-channels in snail neurones (Osipenko et al., 1992) or Ca-activated K-channels in red blood cells (Leinders et al., 1992).

3.5. Ion uptake

Water transport plays a fundamental part in ion transport. The quality and amount of ions and molecules taken up and given off is determined by the interconnected system of transport mechanisms functioning in plant membranes. These transport mechanisms are extremely sensitive to heavy metals (Marschner and Römheld, 1983).

Heavy metals may interact with other toxic and essential macro and microelements. Nature of this interaction depends on the concentration of ions, pH, presence of chelators etc. and thus the results of various investigations are heterogeneos and difficult to compare. In general, toxic metals in certain conditions may cause the deficiency of other elements essential for plants.

In the needles of spruce (*Picea abies*) Pb decreased Ca and Mn concentration (Godbold and Kettner, 1991). In the roots and leaves of beech (*Fagus silvatica*) seedlings, besides Ca and Mn, the concentration of K, Zn, and Fe also decreased (Breckle and Kahle, 1992). However, the Ca and K uptake and translocation of wheat grown on nutrient solutions with different K concentrations was not or just slightly influenced by Pb (Trivedi and Erdei, 1992).

Pb concentration in the exudation sap is approximately the same as Zn concentration but exceeds Fe concentration (Cseh et al., 2000). Xylem sap carries high amount of Mn corresponding to the higher Mn concentration of the shoot and Pb-treatment clearly stimulates Mn-transport (Záray et al., 1997). In contrast in pea *(Pisum sativum)* seedlings Cd most of which accumulated in the roots (90%) almost completely inhibited Mn uptake. Fe concentration and uptake showed less correlation with the Cd treatments (Hernández et al., 1998).

Cd reduced the concentration of S, P, Mg, Mo, Mn, and B and increased Fe in roots and shoots of 16 wheat genotypes. Potassium concentration decreased in roots and increased in shoots while Ca concentration changed the other way round (Zhang et al., 2000). In tomato P, K and Mg levels were affected by Cd in roots and stems and Ca in the leaves. Fe, Mn and Zn contents changed differently depending on the plant part and plant age (Moral et al., 1994). In Cd-treated ryegrass (*Lolium perenne*) and cabbage (*Brassica oleracea*) plants the influx and transport of Cu, Zn, Fe, Mn, Ca and Mg decreased compared to the control (Yang et al., 2000).

Fe deficiency is the most serious and widespread of all that can be caused by the majority of toxic heavy metals just as Pb (Wallace et al., 1992). However, the iron reductase enzyme in Strategy I plants was shown to be unaffected by Pb (Alcantara et al., 1994). zynski and Buczek (1994) *in vitro* observed the inhibition of NADH-ferricyanide-oxidoreductase but *in vivo* there was no effect. The iron deficiency inducible turbo electron transport is inhibited by Cd (Alcantara et al., 1994).

Concerning the effect of chelating agents in iron deficient cucumber seedlings Fe uptake and translocation was stimulated by Pb in the presence of Fe-EDTA but it was inhibited in the presence of Fe-citrate. Cd completely inhibited Fe translocation from the root to the shoot with both chelators and consequently increased Fe content in the root (Fodor et al., 1996).

The idea of heavy metal induced iron deficiency is supported by that many symptoms of Cd toxicity like the changes in pigment-protein patterns or the spectral forms of isolated chloroplasts are very similar to those observed in iron deficient plants (Fodor et al., 1996). Investigating Cd/Fe interactions in young bean seedlings it was concluded that the toxicity targeted the Calvin cycle and the interaction between the effects of Cd and Fe is most probably of indirect nature (Siedlecka and Krupa, 1996). On the other hand, Fe concentrations in the leaves of heavy metal treated plants do not show clear correlations with the decrease in various physiological parameters suggesting that the total Fe content in the tissues is not necessarily the same as the active iron pool (Láng et al., 1998; Sárvári et al., 1999).

Addition of Cd not only decreases the water and nitrate uptake and transport in bean plants but it causes a decrease in nitrate reductase(NR) activation state in short term exposure (24h) while it has no effect in long term treatment (7d) but the level of NR protein decreased by 80% (Gouia et al., 2000). This was also shown by many others for Pb (Burzyński and Buczek, 1994; Singh et al., 1997). Thus it changes the nitrogen metabolism. The direct reason for it can be the inhibition of water transport (Barcelo and Poschenrieder, 1990), since Cd accumulates in the root and the nitrate reduction occurs in the shoot. Cd causes the stomatal closure resulting the decline of transpiration and consequently the nitrate and K transport.

3.6. Membrane structures

Many toxic effects of heavy metals can be attributed to the changes in cell membranes. In tomato the lipid composition of the membranes was changed. The higher decrease in glycolipid content was observed mostly in the leaves, whereas that of the phospholipid and neutral lipid content in the roots (Ouariti et al., 1997). In pea root plasmalemma the amount of phosphatidyletanolamine and phosphatidylcoline decreased but their ratio did not change (Hernández and Cooke, 1997). As a result of cadmium treatment membranes became more rigid and the ATP-ase activity decreased (Fodor et al., 1995). In rice (*Oryza sativa*) plants, Cd applied in 0.1 and 1 mM concentration depolarizes root cell membranes in a few minutes but the membrane potential recovers within 6-8 hours. The membrane permeability increases inducing K^+ efflux from the roots. Cd caused a significant inhibition of root respiration (Llamas et al., 2000).

In young runner bean plants (*Phaselous coccineus*) Cd caused strong reduction of the leaf area, a decreased monogalactosyldiacylglycerol/digalactosyldiacylglycerol ratio and efficiency of the photosythetic apparatus, and in older leaves chlorosis (Skórzyńska-Polit et al., 1998).

Pb applied as Pb-acetate caused some increase of the concentrations of total phospholipids (phosphatidylcholine, phosphatidylethanolamine, phosphatidylinositol, phosphatidylglycerol) content in the leaves of maize (Stefanov et al., 1993) but a decrease in the green leaves bean (*Phaseolus vulgaris*) (Stefanov et al., 1992). The relative concentartions of phosphatidylglycerol and these of phosphatidylcholine increased (Stefanov et al., 1995b).

Phospholipids are important constituents of cell membranes. The change in environmental conditions can lead to modifications of membranes in order to preserve their functionality. The similar changes in maize and bean caused by lead ions are an indication that phospholipids are an important factor for the stabilization of the cell membranes under Pb toxicity.

The amount of saturated fatty acids decreased and that of linolenic acid increased;the sterol composition did not change significantly under Pb stress in either plants (Stefanov et al., 1993, 1995b).

Whereas in *P. vulgaris* the changes in lipid composition appeared almost exclusively in the roots (Stefanov et al., 1992) in *Z. mays* the analogous changes are even more substantial in leaves. In *P. vulgaris* more than 90% of Pb was concentrated in the roots, while in *Z. mays* only about 45% and the rest was in the leaves (Stefanov et al., 1993).

Pb treatment decreased the concentrations of monogalactosyl diacylglycerols and phospholipids in the thylakoid membranes of spinach and increased other glycolipids although the Pb concentrations was lower in thylakoid membrane preparations than in the leaf tissues (Stefanov et al., 1995a).

4. SUMMARY

Toxic heavy metals have different effects on plants depending on their concentration. In polluted areas or naturally occurring metal-rich places high concentrations cause severe toxicity in a number of plants while the constant exposure to low concentrations favour physiological and biochemical adaptations. Each heavy metal have many unspecific and a few specific effects on plant metabolism that are greatly dependent on the sites and mode of their action. The toxicity stress is building up through a stepwise mechanism starting from the accumulation in the apoplast, entering the symplast, then the xylem and finally changing the physiology of the whole plant. Some plants are able to acclimate the negative effects of heavy metals and give specific and unspecific responses which may lead to tolerance. A general model is established in order to find causal relations between these mechanisms and a review of the proceedings in the research of Pb- and Cd-effects on the physiology of higher plants in the last decade is provided.

ACKNOWLEDGEMENTS

This work was supported by the grants OTKA F033076, FKFP P0175 and EEC IC15-CT98-0126.

REFERENCES

Aarkrog, A. and Lippert, J. (1971) Direct contamination of barley with ^{51}Cr, ^{59}Fe, ^{65}Zn, ^{203}Hg and ^{210}Pb, *Radiat. Bot.* **11**, 463–472.
Ahmed, A. and Tajmir-Riahi, H.A. (1993) Interaction of toxic metal ions Cd^{2+}, Hg^{2+}, and Pb^{2+} with light-harvesting proteins of chloroplast thylakoid membranes. An FTIR spectroscopic study, *J. Inorg. Biochem.* **50**, 235–243.
Alcantara, E., Romera, F.J., Canete, M., and Delaguardia, M.D. (1994) Effects of heavy metals on both induction and function of root Fe(III) reductase in Fe-deficient cucumber (*Cucumis sativus* L.) plants, *J. Exp. Bot.* **45**, 1893–1898.
Ali, G., Srivastava, P.E., and Iqbal, M. (2000) Influence of cadmium and zinc on growth and photosynthesis of *Bacopa monniera* cultivated *in vitro*, *Biol. Plant.* **43**, 599–601.
Baccouch, S., Chaoui, A. and El Ferjani, E. (1998) Nickel-induced oxidative damage and antioxidant responses in *Zea mays* shoots, *Plant Physiol. Biochem.* **36**, 689–694.
Barcelo, J., Cabot, C., and Poschenrieder, Ch. (1986) Cadmium-induced decrease of water stress resistance in bush bean plants (*Phaseolus vulgaris* L. cv. *Cotender*). II. Effects of Cd on endogenous abscisic acid levels, *J. Plant Physiol.* **125**, 27–34.
Barcelo, J. and Poschenrieder, Ch. (1990) Plant water relations as affected by heavy metal stress: a review, *J. Plant Nutr.* **13**, 1–37
Baumhardt, G.R. and Welch, L.F. (1972) Lead uptake and corn growth with soil applied lead, *J. Env. Qual.* **1**, 92–94.
Bell, P.F., Chaney, R.L., and Angle, J.S. (1991) Free metal activity and total metal concentrations as indeces of micronutrient availability to barley (*Hordeum vulgare* (L.) '*Klages*'), *Plant Soil*, **130**, 51–62.
Bérczi, A. and Møller, I.M. (1998) NADH-monodehydroascorbate oxidoreductase is one of the redox enzymes in spinach leaf plasma membranes, *Plant Physiol.* **116**, 1029–1036.
Bérczi, A. and Møller, I.M. (2000) Redox enzymes in the plant plasma membrane and their possible roles, *Plant Cell Environ.* **23**, 1287–1302.
Blakeley, S.D., Robaglia, Ch., Brzezinski, R., and Thirion, J.P. (1986) Induction of low molecular weight cadmium-binding compound in soybean roots, *J. Exp. Bot.* **37**, 956–964.
Blaylock, M.J., Salt, D.E., Dushenkov, S., Zakhrova, O., Gussman, C., Kapulnik, Y., Ensley, B.D., and Raskin, I. (1997) Enhanced accumulation of Pb in Indian mustard by soil-applied chelating agents, *Environ. Sci. Technol.* **31**, 860–865.
Böddi, B., Oravecz, A.R. and Lehoczki, E. (1995) Effect of cadmium on organization and photoreduction of protochlorophyllide in dark-grown leaves and etioplast inner membrane preparations of wheat, *Photoshyntetica* **31**, 411-420.
Breckle, S-W. (1991) Growth under stress: heavy metals, in Y. Waisel, A. Eshel and U. Kafkafi (eds.), *Plant roots: the hidden half*, M. Dekker, New York, pp. 351–373.
Breckle, S-W. and Kahle, H. (1992) Effects of toxic heavy metals (cadmium, lead) on growth and mineral nutrition of beech (*Fagus silvatica* L.), *Vegetatio* **101**, 43–53.
Brown, S.L., Chaney, R.L., Angle, J.S., and Baker, A.J.M. (1994) Phytoremediation potential of *Thlaspi coerulescens* and bladder champion for zinc- and cadmium-contaminated soils, *J. Environ. Qual.*, **23**, 1151–1157.
Burzyński, M. and Buczek, J. (1994) The influence of Cd, Pb, Cu and Ni on NO_3^- uptake by cucumber seedlings. II. *In vitro* and *in vivo* effects of Cd, Cu, Pb and Ni on the plasmalemma ATPase and oxidoreductase from cucumber seedlings roots, *Acta Phys. Plant.* **16**, 297–302.
Cakmak, I. and Horst, W.J. (1991) Effect of aluminium on lipid peroxidation, superoxide dismutase, catalase and peroxidase activities in root tips of soybean (*Glycine max*), *Physiol. Plant.* **8**, 463–468.
Chaney, R.L., Brown, J.C., and Tiffin, L.O. (1972) Obligatory reduction of ferric chelates in iron uptake by soybeans, *Plant Physiol.* **50**, 208–213.
Chongpraditinun, P. Mori, S. and Chino, M. (1992) Excess copper induces a cytosolic Cu, Zn-superoxide dismutase in soybean root, *Plant Cell Physiol.* **33**, 239–244.
Chugh, L.K. and Sawhney, S.K. (1999) Photosynthetic activities of *Pisum sativum* seedlings grown in presence of cadmium, *Plant Physiol. Biochem.* **37**, 297–303.

Cieslinski, G., Neilsen, G.H. and Hogue, E.J. (1996) Effect of soil cadmium application and pH on growth and cadmium accumulation in roots, leaves and fruit of strawberry plants (*Fragaria x ananassa* Duch.), *Plant Soil* **180**, 267–276.
Costa, G. and Spitz, E. (1997) Influence of cadmium on soluble carbohydrates, free amino acids, protein content of *in vitro* cultured *Lupinus albus*, *Plant Sci.* **128**, 131–140.
Cseh, E., Fodor, F., Varga, A., and Záray, G. (2000) Effect of lead treatment on the distribution of essential elements in cucumber, *J. Plant Nutr.* **23**, 1095–1105.
Czarnecka, E., Nagao, R.T., Key, J.L., and Gurley, W.B. (1988) Characterization of *gmhsp26-A*, a stress gene encoding a divergent heat shock protein of soybean: heavy-metal-induced inhibition of intron processing, *Mol. Cell. Biol.* **8**, 1113–1122.
Daniels, M.J., Mirkov, T.E., and Chrispeels, M.J. (1994) The plasma membrane of *Arabidopsis thaliana* contains mercury-insensitive aquaporin that is a homolog of the tonoplast water channel protein TIP, *Plant Physiol.* **106**, 1325–1333.
Dat, J., Vandenabeele, S., Vranová, E., Van Montagu, M., Inzé, D., and Van Breusegem, F. (2000) Dual action of the active oxygen species during plant stress responses, *Cell. Mol. Life. Sci.* **57**, 779–795.
Donelly, J.R., Shane, J.B., and Shaberg, P.G. (1990) Lead mobility within the xylem of Red spruce seedlings: Implications for the development of pollution histories, *J. Env. Qual.* **19**, 268–271.
Elstner, E.F. and Osswald, W. (1994) Mechanisms of oxygen activation during plant stress, *Proc. R.. Soc. Edinb. B* **102**, 131–154.
Epstein, A.L., Gussman, C.D., Blaylock, M.J., Yermiyahu, U., Huang, J.W., Kapulnik, Y., and Orser, C.S. (1999) EDTA and Pb-EDTA accumulation in *Brassica junicea* grown in Pb-amended soil, *Plant Soil*, **208**, 87–94.
Ernst, W.H.O., Verkleij, J.A.C., and Schat, H. (1992) Metal tolerance in plants, *Acta Bot. Neerl.* **41**, 229–248.
Eun, S-O., Youn, H.S., and Lee, Y. (2000) Lead disturbs microtubule organization in the root meristem of *Zea mays*, *Physiol. Plant.* **110**, 357–365.
Fodor, F., Cseh, E., Varga, A., and Záray, Gy. (1998) Lead uptake distribution and remobilization in cucumber, *J. Plant Nutr.* **21**, 1363–1373.
Fodor, F. Sárvári, É., Láng, F., Szigeti, Z., and Cseh, E. (1996) Effects of Pb and Cd on cucumber depending on the Fe-complex in the culture solution, *J. Plant Physiol.* **148**, 434–439.
Fodor, E., Szabó-Nagy, A., and Erdei, L. (1995). The effects of cadmium on the fluidity and H$^+$-ATPase activity of plasma mebrane from sunflower and wheat roots, *J. Plant Physiol.* **147**, 87–92.
Foy, C.D., Chaney, R.L., and White, M.C. (1978) The physiology of metal toxicity in plants, *Annu. Rev. Plant Physiol.* **29**, 511–566.
Fuhrer, J., Geballe, G.T., and Fries, C. (1981) Cadmium-induced change in water economy of beans: involvement of ethylene formation, *Plant Physiol.* Lancaster **67 (suppl.)**, 55.
Gadallah, M.A.A. (1995) Effects of cadmium and kinetin on chlorophyll content, saccharides and dry matter accumulation in sunflower plants, *Biol. Plant.* **37**, 233–240.
Gallego, S. M., Benavídes, M-P., and Tomaro, M.L. (1996) Effect of heavy metal ion excess on sunflower leaves: evidence for involvement of oxidative stress, *Plant Sci.* **121**, 151–159.
Garnczarska, M. and Ratajczak, L. (2000) Metabolic responses of *Lemna minor* to lead ions I. Growth, chlorophyll level and activity of fermentative enzymes, *Acta Physiol. Plant.* **22**, 423–427.
Gąszczyk, R. (1989) Desorption of heavy metals in soils under the influence of exchangeable cations. Part II. Podzolic soil formed from weakly loamy sand, *Pol. J. Soil. Sci.* **XXII**, 15–20.
Gawęda, M. (1991) The uptake of lead by spinach *Spinacia oleracea* L. and radish *Raphanus raphanistrum* L. subvar. *radicula* Pers. as affected by organic matter in soil, *Acta Physiol. Plant.* **13**, 167–174.
Gawęda, M. and Capecka E. (1995) Effect of substrate pH on the accumulation of lead in radish (*Raphanus sativus* L. subvar. *radicula*) and spinach (*Spinacia oleracea* L.), *Acta Physiol. Plant.* **17**, 333–340.
Godbold, D.L. and Kettner, C. (1991) Lead influences root growth and mineral nutrition of *Picea abies* seedlings, *J. Plant Physiol.* **139**, 95–99.
Gouia, H., Ghorbal, M.H., and Meyer, C. (2000) Effects of cadmium on activity of nitrate reductase and on other enzymes of the nitrate assimilation pathway in bean, *Plant Physiol. Biochem.* **38**, 629–638.
Greger, M., Johansson, M., Stihl, A., and Hamza, K. (1993) Foliar uptake of Cd by pea (*Pisum sativum*) and sugar beet (*Beta vulgaris*), *Physiol. Plant.* **88**, 563–570.

Hagemeyer, J., Kahle, H., and Breckle, S.-W. (1986) Cadmium in *Fagus silvatica* L. trees and seedlings: leaching, uptake and interconnection with transpiration, *Water Air Soil Poll.* **29**, 347–359.

Haynes, R.J. (1990) Active ion uptake and maintenance of cation–anion balance: a critical example of their role in regulating rizisphere pH, *Plant Soil*, **126**, 247–264.

Hemphill, D.D. and Rule, J. (1975) Foliar uptake and translocation of ^{210}Pb and ^{109}Cd, *Int. Conf. Heavy Metals in the Environment*, Toronto, Ontario 27–31. X. Abstracts: C 239–40.

Hernández, L.E. and Cooke, D.T. (1997). Modification of the root plasma membrane lipid composition of cadmium-treated *Pisum sativum, J. Exp. Bot.* **48**, 1375–1381.

Hernández, L.E., Gárate, A., and Carpenta-Ruiz, R. 1997. Effects of cadmium on uptake, distribution and assimilation of nitrate in *Pisum sativum, Plant Soil* **189**, 97–106.

Hernández, L.E., Lozano-Rodríguez, E., Gárate, A., and Carpenta-Ruiz, R. (1998) Influence of cadmium on the uptake, tissue accumulation and subcellular distribution of manganese in pea seedlings, *Plant Sci.* **132**, 139–151.

Howe, R., Evans, R.L., and Ketteridge, S.W. (1997) Copper-binding proteins in ectomycorrhizal fungi, *New Phytol.* **135**, 123–131.

Huang, J.W. and Cunningham, S.D. (1996) Lead phytoextraction: species variation in lead uptake and translocation, *New Phytol.* **134**, 75–84.

Ievinsh, G. and Romanovskaya, O.I. (1991) Accelerated lignification as a possible mechanism of growth inhibition in winter rye seedlings caused by etephon and 1-aminocycloprapane-1-carboxylic acid, *Plant Physiol. Biochem.* **29**, 327–331.

Jiang, W., Liu, D., and Hou, W. (2000) Hyperaccumulation of lead by roots, hypocotyls, and shoots of *Brassica juniceu, Biol. Plant.* **43**, 603–606.

Kahle, H. (1993) Response of roots of trees to heavy metals, *Environ. Exp. Bot.* **33**, 99–119.

Kastori, R., Plenicar, M., Sakac, Z., Pankovic, D., and Arsenijevic-Maksimovic, I. (1998) Effect of excess lead on sunflower growth and photosynthesis, *J. Plant Nutr.* **21**, 75–85.

Kampfenkel, K., Van Montagu, M., and Inzé, D. (1995) Effects of iron excess on *Nicotiana plumbaginifolia* plants: implications to oxidative stress, *Plant Physiol.* **107**, 725–735.

Kochian, L.V. (1995) Cellular mechanisms of aluminium toxicity and resistance in plants, *Annu. Rev. Plant Physiol. Plant Mol. Biol.* **46**, 237–260.

Kocjan, G., Samardakiewicz, S., and Woźny, A. (1996) Regions of lead uptake in *Lemna minor* plants and localization of this metal within selected parts of the root, *Biol. Plant.* **38**, 107–117.

Koeppe, D.E. (1981) Lead: understanding the minimal toxicity of lead in plants, in N.W. Lepp (ed.), *Effect of Heavy Metal Pollution on Plants Vol 1 Effects of Trace Metals on Plant Function*, Appl. Sci. Publ, London, New Jersey, pp. 55–76.

Krupa, Z. (1988) Cadmium-induced changes in in the composition and structure of the light-harvesting complex II in radish cotyledons, *Physiol. Plant.* **73**, 518–524.

Krupa, Z., and Baszyński, T. (1995) Some aspects of heavy metals toxicity towards photosynthetic apparatus -direct and indirect effects on light and dark reactions, *Acta Physiol. Plant.* **17**, 177–90.

Krupa, Z., Öquist, G. and Huner, N.P.A. (1993) The effects of cadmium on photosynthesis of *Phaseolus vulgaris* L. – a fluorescence analysis, *Physiol. Plant.* **88**, 626–630.

Ksiażek, M. and Woźny, A. (1990) Lead movement in poplar adventitious roots, *Biol. Plant.* **32**, 54–57.

Kurepa, J., Hérouart, D., Van Montagu, M., and Inzé, D. (1997) Differential expression of CuZn- and Fe- superoxide dismutase genes of tobacco during development, oxidative stress and hormonal treatments, *Plant Cell Physiol.* **38**, 463–470.

Lamersdorf, N.P. (1989) The behaviour of lead and cadmium in the intensive rooting zone of acid spruce forest soils, *Toxicol. Environ. Chem.* **18**, 239–247.

Lane, B., Kajoika, R., and Kennedy, R. (1987) The wheat-germ E_c protein is a zinc-containing metallothionein, *Biochem. Cell Biol.* **65**, 1001–1005.

Láng, F., Sárvári, É., Fodor, F., and Cseh, E. (1995) Effects of heavy metals on the photosynthetic apparatus in cucumber, in P. Mathis (ed.), *Photosynthesis: From Light to Biosphere*, Vol. IV., Kluwer Acad. Publ., Dordrecht, pp. 533–536.

Láng, F., Szigeti, Z., Fodor, F., Cseh, E., Zolla, L., and Sárvári, É. (1998) Influence of Cd and Pb on the ion content, growth and photosynthesis in cucumber, in G. Garab (ed.), *Photosynthesis: Mechanisms and Effects*, Vol. IV, Kluwer Acad. Publ., Dordrecht, pp. 2693–2696.

Leinders, T., van Kleef R.G.D.M., and Vijverberg, H.P.M. (1992) Distinct metal ion binding sites on calcium-activated potassium channels in inside out patches of human erythrocytes, *Biochim. Biophys. Acta* **11–12**, 75–82.
Leita, L., de Nobili, M., Cesco, S. and Mondini, C. (1996) Analysis of intercellular cadmium forms in roots and leaves of bush bean, *J. Plant Nutr.* **19**, 527–33.
Little, P. (1973) A study of heavy metal contamination of leaf surfaces, *Environ. Poll.* **5**, 159–172.
Little, P. and Martin, M.H. (1972) A survey of zinc, lead and cadmium in soil and natural vegetation around a smelting complex, *Environ. Poll.* **3**, 241–254.
Llamas, A., Ullrich, C.I., and Sanz, A. (2000) Cd^{2+} effects on transmembrane electrical potential difference, respiration and membrane permeability of rice (*Oryza sativa* L) roots, *Plant Soil* **219**, 21–28.
Lozano-Rodríguez, E., Hernández, L.E., Bonay, P. and Carpenta-Ruiz, R.O. (1997) Distribution of cadmium in shoot and root tissues of maize and pea plants: physiological disturbances, *J. Exp. Bot.* **48**, 123–128.
Luna, C.M., Gonzales, C.A., and Trippi, V.S. (1994) Oxidative damage caused by an excess of copper in oat leaves, *Plant Cell Physiol.* **35**, 11–15.
Maggio, A. and Joly, R.J. (1995) Effects of mercuric chloride on the hydraulic conductivity of tomato root systems, *Plant Physiol.* **109**, 331–335.
Malik, D., Sheoran, I.S., and Singh, R. (1992) Carbon metabolism in leaves of cadmium treated wheat seedlings, *Plant Physiol.* **30**, 223–229.
Marschner, H. (1995) *Mineral Nutrition of Higher Plants*, Acad. Press, London.
Marschner, H. and Römheld, V. (1983). *In vivo* measurement of root-induced pH changes at the soil-root interface: effect of plant spacies and nitrogen source, *Z. Pflanzenphisiol.* **111**, 241–251.
Marschner, H., Römheld, V., and Kissel, M. (1986) Different strategies in higher plants on mobilizing and uptake of iron, *J. Plant Nutr.* **9**, 695–713.
Marschner, P. Godbold, D.L., and Jentschke, G. (1996) Dynamics of lead accumulation in mycorrhizal and non-mycorrhizal Norway spruce (*Picea abies* (L.) Karst.), *Plant Soil* **178**, 239–245.
McBride, M.B. (1994) *Environmental chemistry of soils*, Oxford Univ. Press, New York.
McLaughlin, M.J. and Singh, B.R. (1999) *Cadmium in soils and plants*, Kluwer Acad. Publ., Dordrecht.
Meharg, A.A. (1993) The role of the plasmalemma in metal tolerance in angiosperms, *Physiol. Plant.* **88**, 191–198.
Mehlhorn, H. (1990) Ethylene-promoted ascorbate peroxidase activity protects plants against hydrogen peroxide, ozone and paraquat, *Plant Cell Environ.* **13**, 971–976.
Michalak, E. and Wierzbicka, M. (1998) Differences in lead tolerance between *Allium cepa* plants developing from seeds and bulbs, *Plant Soil* **199**, 251–260.
Moral, R. Navarro Pedreno, I., and Mataix, J. (1994) Effects of cadmium on nutrient distribution, yield and growth of tomato grown in soilless culture, *J. Plant Nutr.* **17**, 953–962.
Moreno-Caselles, J., Moral, R., Pérez-Espinosa, A., and Pérez-Murcia, M.D. (2000) Cadmium accumulation and distribution in cucumber plant, *J. Plant Nutr.* **23**, 243–250.
Muñoz, F.J., Dopico, B., and Labrador, E. (1998) A cDNA encoding a proline-rich protein from *Cicer arietinum*. Changes in expression during development and abiotic stresses, *Physiol. Plant.* **102**, 582–590.
Murphy, A., Zhou, J., Goldsbrough, P.B., and Taiz, L. (1997) Purification and immunological identification of metallothioneins 1 and 2 from *Arabidopsis thaliana*, *Plant Physiol.* **113**, 1293–1301.
Neite, H., Neikes, N., and Wittig, R. (1991) Distribution of heavy meatals in the root area and in organs of herbaceous plnts in beech forests, *Flora (Jena)* **185**, 325–233.
Norwell, W.A. (1991) Reactions of metal chelates in soils and nutrient solutions, in J.J. Mordtvedt (ed.) *Micronutrients in Agriculture* Soil Science of America Inc., Madison, WI., pp. 187–227.
Nussbaum, S., Schmutz, D., and Brunold, C. (1988) Regulation of assimilatory sulphate reduction by cadmium in *Zea mays* L., *Plant Physiol.* **88**, 1407–1410.
Nyitrai, P. (1997) Development of functional thylakoid membranes: regulation by light and hormones, in M. Pessarakli (ed.), *Handbook of Photosynthesis*, Marcel Dekker, New York, pp. 391–406.
Obroucheva, N.V., Bystrova, E.I., Ivanov, V.B., Antipova, O.V. and Seregin I.V. (1998) Root growth responses to lead in young maize seedlings, *Plant and Soil* **200**, 55–61.
Olsen, R.A., Bennett, J.H., Blume, D., and Brown, J.C. (1981) Chemical aspects of the iron stress response mechanism in tomatoes, *J. Plant Nutr.* **3**, 905–921.

Osipenko, O. N., Györi, J., and Kiss, T. (1992) Lead ions close steady-state sodium channels in *Helix* neurones, *Neuroscience* **80**, 483–489.

Ouariti, O., Boussama, N., Zabrouk, M., Cherif, A., and Ghorbal, M.H. (1997) Cadmium- and copper-induced changes in tomato membrane lipids, *Phytochemistry* **45**, 1343–1350.

Ow, D.W. (1996) Heavy metal tolerance genes: prospective tools for bioremediations, *Res. Conserv. Recycl.* **18**, 135–149.

Pennazio, S. and Roggero, P. (1992) Effect of cadmium and nickel on ethylene biosynthesis in soybean, *Biol. Plant.* **34**, 345–349.

Petterson, O. (1976) Heavy-metal ion uptake by plants from nutrient solutions with metal ion, plant species and growth period variations, *Plant Soil* **45**, 445–459.

Poschenrieder, Ch., Gunse, B., and Barcelo, J. (1989) Influence of cadmium on water relations, stomatal resistance and abscisic acid content in expanding bean leaves, *Plant Physiol.* **90**, 1365–1371

Poskuta, J.W. and Wac³awczyk-Lach, E. (1995) *In vivo* responses of primary photochemistry of photosystem II and CO_2 exchange in light and in darkness of tall fescue genotypes to lead toxicity, *Acta Physiol. Plant.* **17**, 233–240.

Punz, W.F. and Sieghardt, H. (1993) The response of roots of herbaceous plant species to heavy metals, *Environ. Exp. Bot.* **33**, 85–98.

Rauser, W.E. (1995) Phytochelatins and related peptides, *Plant Physiol.* **109**, 1141–1149.

Rivetta, A., Negrini, N. and Cocucci, M. (1997) Involvement of Ca^{2+}-calmodulin in Cd^{2+} toxycity during the early phases of radish (*Raphanus sativus* L.) seed germination, *Plant Cell Environ.* **20**, 600–608.

Römheld, V. and Marschner, H. (1981) Effect of Fe stress on utilization of Fe chelates by efficient and inefficient plant species, *J. Plant Nutr.* **3**, 551–560.

Römheld, V. and Marschner, H. (1983) Mechanism of iron uptake by peanut plants. I. Fe^{III} reduction, chelate splitting and release of phenolics, *Plant Physiol.* **71**, 949–954.

Salt, D.E., Prince, R.C., Pickering, I.J., and and Raskin, I. (1995) Mechanisms of cadmium mobility and accumulation in indian mustard, *Plant Physiol.* **109**, 1427–1433.

Sanitá di Toppi, L., and Gabbrielli, R. (1999) Response to cadmium in higher plants, *Environ. Exp. Bot.* **41**, 105–130.

Sanitá di Toppi, L., Lambardi, M., Pazzagli, L., Cappugi, G., Durante, M., and Gabbrielli, R. (1998) Response to cadmium in carrot *in vitro* plants and cell suspension cultures, *Plant Sci.* **137**, 119–129.

Sabitá di Toppi, L., Lambardi, M., Pecchioni, N., Pazzagli, L., Durante, M., and Gabbrielli, R. (1999) Effects of cadmium stress on hairy roots of *Daucus carota*, *J. Plant Physiol.* **154**, 385–391.

Sárvári, É., Fodor, F., Cseh, E., Varga, A., Záray, Gy., and Zolla, L. (1999) Relationship between changes in ion content of leaves and chlorophyll-protein composition in cucumber under Cd and Pb stress, *Z. Naturforsch.* **54c**, 746–753.

Schat, H., Sharma, S.S., and Vooijs, R. (1997) Heavy metal-induced accumulation of free proline in a metal-tolerant and a nontolerant ecotype of *Silene vulgaris*, *Physiol. Plant.* **101**, 477–482.

Schobert., B. and Tschesche, H. (1978) Unusual solution properties of proline and its interaction with proteins, *Biochim. Biophys. Acta* **541**, 270–277.

Senden, M.H.M.N., Verburg, A.J.G.M., van der Meer, T.G., and Wolterbeek, H.Th. (1995) Citric acid in tomato plant roots and its effect on cadmium uptake and distribution, *Plant Soil* **171**, 333–339.

Sengar, R.S. and Pandey, M. (1996) Inhibition of chlorophyll biosynthesis by lead in greening *Pisum sativum* leaf segments, *Biol. Plant.* **38**, 459–462.

Shah, K. and Dubey, R.S. (1997) Effect of cadmium on proline accumulation and ribonuclease activity in rice seedlings: role of proline as a possible enzyme protectant, *Biol. Plant.* **40**, 121–130.

Shaw, B.P. and Rout, N.P. (1998) Age-dependent responses of *Phaseolus aureus* Roxb. to inorganic salts of mercury and cadmium, *Acta Physiol. Plant.* **20**, 85–90.

Sheoran, I. S., Singal, H.R., and Singh, R. (1990) Effect of cadmium and nickel on photosynthesis and the enzymes of the photosynthetic carbon reduction cycle in pigeon pea (*Cajanus cajan* L.), *Photosynth. Res.* **23**, 345–351.

Siedlecka, A. and Baszyński, T. (1993) Inhibition of electron flow around photosystem I in chloroplasts of Cd-treated maize plants is due to Cd-induced iron deficiency, *Physiol. Plant.* **87**, 199–202.

Siedlecka, A. and Krupa, Z. (1996) Interaction between cadmium and iron and its effects on photosynthetic capacity of primary leaves of *Phaseolus vulgaris*, *Plant Physiol. Biochem.* **34**, 833–841.

Simon, L. (1998) Cadmium accumulation and distribution in sunflower plant, *J. Plant Nutr.* **21**, 341–352.

Singh, R.P., Dabas, S., Choudhary, A., and Maheshwari, R. (1997) Effect of lead on nitrate reductase activity and alleviation of lead toxicity by inorganic salts and 6-benzylaminopurine, *Biol. Plant.* **40**, 399–404.

Sivaguru, M., Fujiwara, T., Samaj, J., Baluska, F., Yang, Z.M., Osawa, H., Maeda, T., Mori, T., Volkmann, D., and Matsumoto, H. (2000) Aluminum-induced 1 -> 3-beta-D-glucan inhibits cell-to-cell trafficking of molecules through plasmodesmata. A new mechanism of aluminum toxicity in plants, *Plant Physiol.* **124**, 991–1005.

Skórzyńska-Polit, E. and Baszyński. T. (1997) Differences in sensitivity of the photosynthetic apparatus in Cd-stressed runner bean plants in relation to their age, *Plant Sci.* **128**, 11–21.

Skórzyńska-Polit, E., Tukendorf, A., Selstam, E., and Baszyński, T. (1998) Calcium modifies Cd effect on runner bean plants, *Environ. Exp. Bot.* **40**, 275–286.

Smirnoff, N. and Cumbes, Q.J. (1989) Hydroxyl radical scavenging activity in compatible solutes, *Phytochemistry* **28**, 1057–1060.

Somashekaraian, B.V., Padmaja, K., and Prasad, A.R.K. (1992) Phytotoxicity of cadmium ions on germinating seedlings of mung bean (*Phaseolus vulgaris*): Involvement of lipid peroxides in chlorophyll degradation, *Physiol. Plant.* **85**, 85–89.

Srivastava, A. and Appenroth, K.J. (1995) Interaction of EDTA and iron on the accumulation of Cd^{2+} in duckweeds (*Lemnaceae*), *J. Plant Physiol.* **146**, 173–176.

Stefanov, K., Pandev, S.D., Seizova, K.A., Tyankova, L.A. and Popov, S.S. (1995a) Effect of lead ions on the lipid metabolism in spinach leaves and thylakoid membranes, *Biol. Plant.* **37**, 251–256.

Stefanov, K., Popova, I., Kamburova, E., Pancheva, T., Kimenov, G., Kuleva, L., and Popov, S. (1993) Lipid and sterol changes in *Zea mays* caused by lead ions, *Phytochemistry*, **33**, 47–51.

Stefanov, K., Popova, I., Nikolova-Damyanova, B., Kimenov, G., and Popov, S. (1992) Lipid and sterol changes in *Phaseolus vulgaris* caused by lead ions, *Phytochemistry* **31**, 3745–3748.

Stefanov, K., Seizova, K., Popova, I., Petkov, V., Kimenov, G., and Popov, S. (1995b) Effect of lead ions on the phospholipid in leaves of *Zea mays* and *Phaseolus vulgaris*, *J. Plant Physiol.* **147**, 243–246.

Steudle, E. and Henzler, T. (1995) Water channels in plants: do basic concepts of water transport change?, *J. Exp. Bot.* **290**, 1067–1076.

Stroiński, A. (1999) Some physiological and biochemical aspects of plant resistance to cadmium effect. I. Antioxidative system, *Acta Physiol. Plant.* **21**, 175–188

Stroiński, A., Floryszak-Wieczorek, J., and Woźny, A. (1990) Effects of cadmium on potato leaves and phytophtora infestans relation. I Alterations of potato leaves and phytophtora infestans relations, *Biochem. Physiol. Pflanzen* **186**, 43–54.

Stroiński, A. and Kozłowska, M. (1997) Cadmium-induced oxydative stress in potato tuber, *Acta. Soc. Bot. Pol.* **66**, 189–195.

Symeonidis, L. and Karataglis, S. (1992) The effect of lead and zinc on plant growth and chlorophyll content of *Holcus lanatus* L., *J. Agron. Crop Sci.* **168**, 108–112.

Szabados, M., Horváth, A., and Cseh, E. (1983) Application of *Lemna minor* L. to testing the arsenic content of the soil, *Bot. Kozl.* **70**, 171–177.

Talanova, V.V., Titov, A.F., and Boeva, N.P. (2000) Effect of increasing concentrations of lead and cadmium on cucumber seedlings, *Biol. Plant.* **43**, 441–444.

Titov, A.F., Talanova, V.V., and Boeva, N.P. (1996) Growth responses of barley and wheat seedlings to lead and cadmium, *Biol. Plant.* **38**, 431–436.

Treeby, M., Marschner, H., and Römheld, V. (1989) Mobilization of iron and other micronutrient cations from a calcareous sol by plant-borne, microbial, and synthetic metal chelators, *Plant Soil*, **114**, 217–226.

Trivedi, S. and Erdei, L. (1992) Effects of cadmium and lead on the accumulation of Ca^{2+} and K^+ and on the influx and translocation of K^+ in wheat of low and high K^+ status, *Physiol. Plant.* **84**, 94–100.

Vassil A.D. Kapulnik, Y., Raskin, I., and Salt, D.E. (1998) The role of EDTA in lead transport and accumulation by indian mustard, *Plant Physiol.* **117**, 447–453.

Vassilev, A., Berova, M., and Zlatev, Z. (1998) Influence of Cd^{2+} on growth, chlorophyll content, and water relations in young barley plants, *Biol. Plant.* **41**, 601–606.

Velazquez, M.D., Poschenrieder, Ch., and Barceló, J. (1992) Ultrastructural effects and localization of low cadmium concentrations in bean roots, *New Phytol.* **120**, 215–226.
Vierling, E. (1991) The roles of heat shock proteins in plants, *Annu. Rev. Plant Physiol. Plant Mol. Biol.* **42**, 579–620.
Vojtechová, M. and Leblová S. (1991) Uptake of lead and cadmium by maize seedlings and the effect of heavy metals on the activity of phophoenolpyruvate carboxylase isolated from maize, *Biol. Plant.* **33**, 386–394.
Wagner, G.J. (1993) Accumulation of cadmium in crop plants and its consequences to human health, *Adv. Agron.* **51**, 173–212.
Wallace, A. (1983) One decade update on chelated metals for supplying micronutrients to crops, *J. Plant Nutr.* **6**, 429–438.
Wallace, A., Wallace, G.A., and Cha, J.W. (1992) Some modifications in trace metal toxicities and deficiencies in plants resulting from interactions with other elements and chelating agents – the special case of iron, *J. Plant Nutr.* **15**, 1589–1598.
Waters, E.R., Lee, G.J., and Vierling, E. (1996) Evolution, structure and function of the small heat shock proteins in plants, *J. Exp. Bot.* **47**, 325–338.
Watmough, S.A., Hutchinson, T.C., and Evans, R.D. (1999) The distribution of ^{67}Zn and ^{207}Pb applied to white spruce foliage at ambient concentrations under different pH regimes, *Environ. Exp. Bot.* **41**, 83–92.
Weckx, J.E.J. and Clijsters, H.M.M. (1996) Oxidative damage and defense mechanisms in primary leaves of *Phaseolus vulgaris* as a result of root assimilation of toxic amount of copper, *Physiol. Plant.* **96**, 506–512.
Weckx, J.E.J. and Clijsters, H.M.M. (1997) Zn phytotoxicity induces oxidative stress in primary leaves of *Phaseolus vulgaris*, *Plant Physiol. Biochem.* **35**, 405–410.
Weinstein, L.H., Kaur-Shawney, R., Rajan, M.V., Wettlaufer, S.H., and Galston, A.W. (1986) Cadmium-induced accumulation of putrescine in oat and bean leaves, *Plant Physiol.* **82**, 641–645.
Welch, R.M. (1995) Micronutrient nutrition of plants, *Crit. Rev. Plant. Sci.* **14**, 49–82.
Whitelaw, C.A., Le Huquet, J.A., Thurman, D.A. and Tomsett, A.B. (1997) The isolation and characterisation of Type II metallothionein-like genes from *tomato* (*Lycopersicon esculentum* L.), *Plant Mol. Biol.* **33**, 503–511.
Wierzbicka, M. (1998) Lead in the apoplast of *Allium cepa* L. root tips – ultrastructural studies, *Plant Sci.* **133**, 105–119.
Wójcik, M. and Tukendorf, A. (1999) Cd-tolerance of maize, rye and wheat seedlings, *Acta Phys. Plant.* **21**, 99–107.
Woźny, A. and Jerczyńska E. (1991) The effect of lead on early stages of *Phaseolus vulgaris* L. growth *in vitro* conditions, *Biol. Plant.* **33**, 32–39.
Woźny, A., Schneider, J., and Gwóźdź, E.A. (1995) The effect of lead and kinetin on greening barley leaves, *Biol. Plant.* **37**, 541–552.
Woźny, A., Stroiński, A., and Gwóźdź, E.A. (1990) Plant cell responses to cadmium, Seria Biologia 44, Adam Mickiewicz University, Poznań.
Yang, Y-Y., Jung, J-Y., Song, W-Y., Suh, H-S., and Lee, Y. (2000) Identification of rice varieties with high tolerance or sensitivity to lead and characterization of the mechanism of tolerance, *Plant Physiol.* **124**, 1019–1026.
Ye, Z-H. and Varner, J.E. (1991) Tissue-specific expression of cell wall proteins in developing soybean tissues, *Plant Cell* **3**, 23–37.
Záray, Gy., Dao Thi Phuong, D., Varga, I., Varga, A., Kántor, T., Cseh, E., and Fodor, F. (1995) Influences of lead contamination and complexing agents on the metal uptake of cucumber, *Microchem J.* **51**, 207–213.
Záray, Gy., Varga, A., Fodor, F., and Cseh, E. (1997) Microanalytical investigation of xylem sap of cucumber by total reflection X-ray fluorescence spectrometry, *Microchem. J.* **55**, 64–71.
Zhang, G., Fukami, M., and Sekimoto, H. (2000) Genotypic differences in effects of cadmium on growth and nutrient compositions in wheat, *J. Plant Nutr.* **23**, 1337–1350.
Zornoza, P., Robles, S., and Martin, N., (1999) Alleviation of nickel toxicity by ammonium supply to sunflower plants, *Plant Soil* **208**, 221–226.

CHAPTER 7

PROLINE ACCUMULATION IN HEAVY METAL STRESSED PLANTS: AN ADAPTIVE STRATEGY

P. SHARMILA AND P. PARDHA SARADHI
Department of Environmental Biology, University of Delhi
Delhi 110 007, India

1. INTRODUCTION

In nature plants are exposed to a wide variety of environmental stresses. Heavy metal stress is one of the common stresses that limits plant growth and development. Toxic levels of heavy metals such as Cd^{2+}, Pb^{2+}, Zn^{2+}, Cu^{2+} and Hg^{2+}, occur in some natural as well as agricultural soils due to mining, smelting, some common agricultural practices (such as excessive use of fertilizers) and waste disposal technologies(Steffens,1990; Mejáre and Bülow, 2001).Presently heavy metals have become one of the major environmental hazards as unlike many other pollutants they cannot be degraded/removed either chemically or biologically and hence they are ultimately indestructible (Mejáre and Bülow, 2001). Living systems respond to heavy metal stress by operating various defense strategies such as exclusion, compartmentalization, synthesis of metal binding polypeptides (such as metallotheinins, phytochelatins) formation of complexes and switching on or activating certain metabolic events associated with the/alleviation of toxic effects resulting due to heavy metal stress (Steffens,1990; Alia and Pardha Saradhi, 1991; Ebbs and Kochian, 1998; Clemens et al., 1999; Prasad et al., 1999; Zhu et al., 1999; Cobbett, 2000; Mejáre and Bülow, 2001). Synthesis of compatible solutes like proline seems to be one of the means by which plants combat heavy metal stress (Alia and Pardha Saradhi, 1991; Bassi and Sharma, 1993).

After they are readily absorbed by the root system, heavy metal ions are accumulated in different parts of the plant, resulting in retardation of plant growth as they interfere with the activities of a number of enzymes essential for normal metabolic and developmental processes (Nath, 1986; Van Assche et al., 1988; Clijsters and Van Assche, 1985; Pahlsson, 1989; Van Steveninck et al., 1990; Van Assche and Clijsters, 1990; Rauser, 1990; Zenk, 1996; Zhu et al., 1999).

One of the strategies that plants have evolved to counteract the toxic effects of heavy metal stress is through the accumulation of organic solutes like proline. Proline is an imino acid that accumulates in plants exposed to a wide variety of environmental stresses. Proline accumulation in living organisms under abiotic stress has been reported in a wide variety of organisms ranging from bacteria, algae, crustaceans to higher plants (Delauney and Verma, 1993; Arora and Pardha Saradhi, 1995; Zhang et al., 1995; Puthur et al., 1997). Accumulation of this compatible solute is believed to a) protect plant tissues against osmotic stress by regulating osmotic potential (Aspinall and Paleg, 1981; Wyn Jones and Gorham, 1984, 1990; Ahmad and Hellebust, 1988; Laliberte and Hellebust, 1989a,b; Arora and Pardha Saradhi, 1995); b) protect enzyme denaturation (Paleg et al., 1984; Nikolopoulos and Manetas, 1991); c) act as a reservoir of carbon and nitrogen source (Fukutaku and Yamada, 1984); d) stabilise the machinery of protein synthesis (Kadpal and Rao, 1985); e) scavenge toxic oxygen species such as singlet oxygen and hydroxyl radicals (Smirnoff and Cumbes, 1989; Alia et al., 1991, 1993, 1995, 1997); f) regulate cytosolic pH (Venekamp, 1989; Venekamp et al., 1989); and g) regulate $NAD(P)^+/NAD(P)H$ ratio (Alia and Pardha Saradhi, 1991, 1993; Pardha Saradhi et al., 1993, 1996). Realizing the significance of proline accumulation in increasing the tolerance of plants to stress, over expression of genes capable of increasing its synthesis has been attempted to develop genotypes of plants with enhanced potential to withstand environmental stresses (Delauney and Verma, 1993; Kavi Kishor et al., 1995; Zhang et al., 1995; Hong et al., 2000; Prasad et al., 2000a,b). Inspite of being subject of intensive research during last four decades, the actual reason(s) behind proline accumulation remains controversial.

This review primarily/largely focuses on the role of proline in alleviating toxic effects caused due to heavy metal stress. Major roles assigned to proline are discussed in detail with suitable examples.

2. ROLE IN OSMOREGULATION

Since the first report on the accumulation of proline in *Lolium perenne* under drought conditions in 1954 by Kemble and MacPherson, innumerable reports indicated the existence of positive correlation between proline accumulation and adaptation of various plant species to osmotic stresses such as drought and salinity. Therefore, in general it is presumed that proline accumulation has a key role in osmoregulation in plants under various environmental stresses. Moreover, proline is a small organic molecule with very high solubility in water and infact solution containing proline at concentrations as high as 14 M can be prepared. Proline can significantly contribute towards colligative properties of the cytoplasm by neutralizing differences in osmotic potential.

Proline accumulates in plants that are exposed to toxic levels of heavy metals such as Cd^{2+}, Pb^{2+}, Zn^{2+}, Co^{2+}, Cu^{2+} and Hg^{2+} (Fig. 1). However, amongst these heavy metals Cd^{2+} proved to be the strongest inducer of proline while zinc was the weakest.

Increase in proline accumulation can be correlated with enhanced uptake and accumulation of these heavy metal ions in plants. Heavy metal induced proline accumulation seems to be common to wide variety of plant species viz. *Oryza sativa*, *Triticum aestivum* (Poaceae), *Cajanus cajan*, *Vigna mungo*, *Vigna radiata* (Fabaceae), *Brassica juncea* (Brassicaceae) and *Helianthus annuus* (Asteraceae) belonging to distinct families. However, toxic levels of these heavy metal ions up to a concentration of 1 mM do not bring about any major change in the water/osmotic potential of the growth medium in which plants were grown although these plants accumulated proline. A shift in water potential by only ~-0.05 MPa was observed when the growth medium was supplemented with 5 mM of the heavy metal salt (Table 1). Plants exposed to Pb^{2+} in the growth medium consisting of only distilled water (DW) also accumulated proline several fold higher than that noted in the plants grown in growth medium consisting of essential mineral nutrients (Fig. 1). However, as depicted in Table 1 the water potential of the medium containing Pb^{2+} ranged between -0.001 to -0.06 MPa in comparison to a water potential of -0.125 MPa of mineral growth medium (consisting of macro and micronutrients as per Gamborg's medium (Gamborg et al., 1968). Many fold increase in proline content in parts of the plants exposed to heavy metal stress strongly suggests that stress induced proline accumulation is not related to osmotic adjustment.

Pardha Saradhi and co-workers (1993) noted over three fold higher proline levels in *Oryza sativa* seedlings raised in DW when compared to those grown in Gamborg's medium (Gamborg et al., 1968). These observations also supported the view that stress induced proline accumulation has some other vital role to play rather than in osmoregulation. Excessive production of proline was also recorded in plants exposed to UV radiations and gaseous pollutants such as sulphur dioxide, nitrogen dioxide and ammonia (Anbhazghan et al., 1988; Pardha Saradhi et al., 1993, 1995). In none of these instances proline accumulation can be related with water potential of the medium in which plants were grown. All these independent experimental results negate the opinion that proline accumulation in plants under stress is for the purpose of osmoregulation.

Table 1: The osmotic potential (- MPa) of media supplemented with nitrate salt of heavy metals used for raising seedlings.

Concentration (µM)	$Cd(NO_3)_2$*	$Co(NO_3)_2$*	$Zn(NO_3)_2$*	$Pb(NO_3)_2$#
0	0.125	0.125	0.125	0.001
10	0.125	0.126	0.125	0.001
100	0.128	0.128	0.128	0.002
500	0.132	0.130	0.132	0.008
1000	0.134	0.135	0.135	0.016
2500	0.151	0.153	0.151	0.037
5000	0.174	0.176	0.175	0.066

* Prepared in B5 medium (half concentrated).
# Prepared in distilled water.

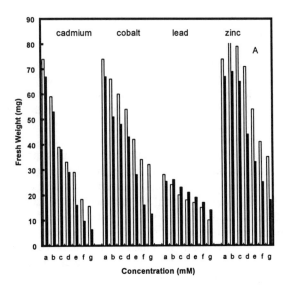

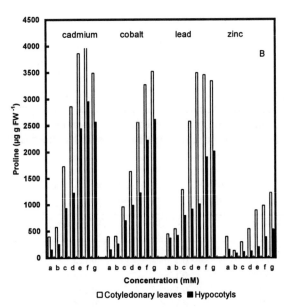

Figure 1. Alteration in fresh weight (A) and proline content (B) of cotyledonary leaves and hypocotyls of Brassica juncea seedlings raised in the presence of varying levels of nitrate salts of cadmium, cobalt, lead and zinc (a, b, c, d. e, f and g represent 0, 10, 100, 500, 1000, 2500 and 5000 µM, respectively)

Since, proline accumulates in wide range of plant species viz. even under stresses (viz. heavy metal stress, mineral deficiency stress, air pollution and UV radiations) which are not hyperosmotic, stress-induced proline accumulation seems to have some other vital role to play rather than in osmoregulation.

3. PROLINE PROTECTS CELLULAR COMPONENTS

3.1 Enzymes/Proteins

Numerous studies in past have demonstrated the potential of proline in ameliorating the deleterious effects of stressful conditions on the structural and functional integrity of enzymes/proteins (Greenway and Osmond, 1972; Treichel et al., 1974; Pollard and Wyn Jones, 1979; Paleg et al., 1981; Nash et al., 1982; Manetas et al., 1986; Paleg et al., 1984; Selinioti et al., 1987; Nikolopoulos and Manetas, 1991). It has been shown that proline exerts protective effects on various enzymes against the effects of salt (Flowers et al., 1978; Pollard and Wyn Jones, 1979; Manetas et al.,1986), heat (Paleg et al., 1981), cold (Krall et al., 1989); dilution (Selinioti et al., 1987; Stamatakis et al., 1988); toxic chemicals like urea (Yancey and Somero, 1979); and PEG induced dehydration/precipitation (Paleg et al., 1984). However, exceptions to this rule exist in the literature reporting the inhibition of activities of certain enzymes by proline (Manetas et al., 1986; Manetas, 1990; Sivakumar et al., 1998, 2000, 2001a,b). Recent work in our laboratory had clearly demonstrated that proline suppresses both carboxylase as well as oxygenase activities of Rubisco from higher plants even when present at concentrations as low as 100 mM. However, proline-induced suppression in oxygenase activity was significantly higher than that of carboxylase activity (Sivakumar et al., 1998; 2000). These studies further demonstrated that proline suppresses Rubisco activity by dissociating small subunits from large subunit octamer core, probably by weakening hydrophobic interaction between them (Sivakumar et al., 2001a). The above findings clearly indicated that stress-induced proline accumulation might have a critical role in lowering the loss in fixed carbon due to salt stress-promoted oxygenase activity of Rubisco.

Although none of the physiochemical theories are capable of explaining the way proline protects proteins/enzymes and other macromolecules or macromolecular structures from structure-randomizing and inactivating stress factors, through certain in vitro experimental evidences it has been realized that proline stabilizes or protects native state of proteins by regulating their water of hydration. Infact, the presence of water is critical to the structural stability of proteins. Thus, proline could be helping in the maintenance of conformational characteristics and integrity of proteins (Paleg et al., 1981). According to Arakawa and Timasheff (1985) compatible solutes can induce preferential hydration of the surface of proteins while simultaneously getting excluded from such surfaces. Moreover, proline like several other compatible solutes can increase surface tension of water and thus can force water-protein interfaces into contact. This promotes proteins to assume/maintain more 'native'/folded configuration. Therefore, compatible solutes like proline help enzymes in often confining them into small fraction of the total cellular volume and strengthening the weak intrinsic forces maintaining their structural integrity against a variety of perturbing microenvironments (see Arakawa and Timasheff, 1985; Papageorgiou and Murata, 1995).

The surface of protein is composed of hydrophilic and hydrophobic domains. The latter domains are vulnerable sites in the protein because the water molecules that are weakly held by these domains are the first to depart when cells encounter a hyperosmotic shock (water deficient environment). By contrast, water is more strongly bound to hydrophilic surface domains and, therefore, such domains are less likely to be affected by a water deficiency. Accordingly a hyperosmotic environment will primarily destabilize the hydrophobic surface domains of proteins. Proline being amphiphilic can bind to hydrophobic surface domains of proteins/enzymes via its hydrophobic moiety and converts them to hydrophilic surfaces. Since much more water is involved in the stabilization of a hydrophobic surface domain than in the stabilization of a hydrophilic surface domain, this conversion means that the cell can now preserve the structural integrity of cytoplasmic proteins under conditions of water deficiency (see Schobert, 1977; Arakawa and Timasheff, 1985).

3.2 Other cellular components

Proline gets accumulated in high quantities in leaves and the chloroplasts have been reported to be the major site for synthesis of proline in plants under stress (Rayapati et al., 1989). L-proline has been demonstrated to enhance photosystem (PS) II and whole chain catalysed electron transport activities of thylakoids. The extent of stimulation in activities was higher in the thylakoids of plants exposed to Cd^{2+} stress in comparison to controls (Table 2). However, proline did not alter PS I mediated photochemical reactions, suggesting that the effect of proline is localised to PS II. The extent of proline mediated stimulation was seen even in the presence of uncoupler NH_4Cl, suggesting that proline induced enhancement in photochemical activities is not related to uncoupling of phosphorylation from electron transport.

The suppression in photochemical activities can be due to i) changes in the PS II-lipid environment specially that of phosphatidyl choline (which is the main acyl lipid associated with the thylakoid domains responsible for PS II activity) (Krupa 1987); ii) decrease in the level of trans-3-hexadecanoic acid (Krupa et al., 1987); iii) diminishing level of unsaturated fatty acid, liolenic acid associated with thylakoids; iv) inhibition of photosynthetic electron transport leading to incomplete reduction of molecular oxygen and increased free-radical production; v) increased formation of alkoxy and peroxy radicals, leading to peroxidative damage to membranes (eg. Cu and Fe ions are effective in conversion of hydroxyperoxides and subsequent lipid peroxidation (De Vos et al., 1989); vi) high affinity of certain heavy metals (Fe and Cu) for thiols (glutathione, thiols of enzymes) which cause oxidative stress (Hendry and Brocklebank, 1985; De Vos et al., 1989); vii) inhibition of enzymes of calvin cycle leading to reduced CO_2 fixation (Rauser, 1978; Stibovora, 1988) and decline in

Table 2 : Effect of proline (1M) without and with 5 mM NH_4Cl on the PS II ($H_2O \rightarrow$ DCPIP) and whole chain $H_2O \rightarrow$ MV) photoreactions in thylakoids isolated from cotyledonary leaves of Brassica juncea seedlings raised in the absence (control) and the presence (treated) of 1 mM $CdCl_2$. Values in parenthesis indicate the percentage stimulation over the respective controls.

Assay	Electron transport rate μ mole O_2 consumed or evolved/mg Chl/h			
	$-NH_4Cl$ $-$Proline	$-NH_4Cl$ $+$Proline	$+NH_4Cl$ $-$Proline	$+NH_4Cl$ $+$Proline
Control				
$H_2O \rightarrow$ MV	69.9±03 (100)	108.0±08 (155)	83.9±09 (100)	135.0±17 (160)
$H_2O \rightarrow$ DCPIP	84.5±05 (100)	119.1±09 (141)	105.7±07 (100)	146.4±18 (139)
Stressed				
$H_2O \rightarrow$ MV	123.0±09 (100)	171.5±10 (139)	184.5±15 (100)	259.2±18 (140)
$H_2O \rightarrow$ DCPIP	140.7±13 (100)	211.0±18 (150)	180.0±12 (100)	274.3±21 (152)

Experiments were carried out at least 3 times.
Data represent mean ± standard deviation.

sugar and starch contents in plant parts under Cd^{2+} stress (Greger and Lindberg, 1986; Malik et al., 1992).

In order to find out the cause of this enhancement, the ability of proline to protect photoinhibitory damages has been investigated. The chloroplast isolated from both control and $CdCl_2$ raised plants were subjected to strong white light in presence and absence of proline. Exposure to high light intensity caused significant reduction in water oxidation capacity of thylakoids in a time dependent manner. The loss in the activity was more pronounced in thylakoids from $CdCl_2$ raised plants than the control plants. Heavy metals are known to substantially enhance the susceptibility of thylakoids to photoinhibition (Neale and Melis, 1989; Atal et al., 1991, 1993). The reduction in photochemical activities in thylakoids exposed to high light intensity could be due to damage of photosystems in particular PS II pigment protein complex (Halliwell and Gutteridge, 1986; Neale and Melis, 1989; Chapman et al., 1990; Sopory et al., 1990). Interestingly, the presence of proline in the incubation medium brought about a significant reduction in the time dependent loss in photochemical activity of thylakoids (from both contol and $CdCl_2$ raised plants) exposed to strong light. Most likely proline enhances photochemical activities of thylakoids possibly by arresting the photoinhibitoy damage to PS II resulting due to excessive production of free radicals which induce thylakoid lipid peroxidation and cross linking of proteins. However, the actual mechanism of proline induced protection of PS II remains to be elucidated.

4. PROLINE PROTECTS CELLULAR COMPONENTS AGAINST FREE RADICALS

Plants exposed to heavy metal stress show alteration in the lipid composition of thylakoid membranes. Lipid peroxidation process involves the peroxidative degradation of poly-unsaturated fatty acids of membrane lipids (Baszynski et al., 1980; Krupa, 1987; Krupa et al., 1987; De Vos and Schat, 1991) and thus, brings about membrane deterioration. Lipid peroxidation capacity gets enhanced markedly in Cd^{2+} treated wheat seedlings compared to control seedlings. This suggests that Cd^{2+} induces the alteration in the structural composition of the membrane. Acording to Skorzynska et al. (1991) enhanced galactolipase activity in Cd^{2+} treated plants leads to an increase in the release of fatty acids due to degradation of galactolipids (mainly MGDG). Somashekaraiah and co-workers (1992) have reported the importance of lipoxygenase in Cd^{2+} induced lipid peroxidation (through the production of free radicals from dioxygenation of membrane lipids and unsaturated fatty acids). In addition, Chl bleaching in concurrent with MDA production was recorded in plants exposed to heavy metals (Heath and Packer, 1968; Kunert, 1984; Somashekaraiah et al., 1992). Hence, heavy metals stress mediated lipid peroxidation leads to the loss in the photosynthetic oxygen evolution (Lee et al., 1976; Sandmann and Boger, 1980a,b; Pick et al., 1987; Siegenthaler et al., 1987; Akabori et al., 1988; Van Assche et al., 1988; Van Assche and Clijsters, 1990; Atal et al., 1993; Alia et al., 1997).

Isolated thylakoids have been reported to readily undergo lipid peroxidation due to the presence of a high proportion of unsaturated fatty acids. The peroxidative process is known to be catalysed by light and thylakoid membrane chlorophylls (Heath and Packer, 1968). Thylakoids from plants under Cd^{2+} stress show higher capacity to produce MDA compared to those from controls. The enhanced lipid peroxidation capacity of thylakoids formed in the seedlings raised in presence of Cd^{2+} suggests a probable change in the lipid composition and hence the functional aspects of the lipid bilayer of thylakoids. The degradation of acyl lipids in thylakoids has been considered to be an indirect reason for Cd^{2+} induced loss in thylakoid mediated photochemical activities (Skorzynska et al., 1991).

Plants exposed to various environmental stresses show acceleration in free radical generation. In general, with increase in the intensity of stress, the plant cells show an increase in the extent of free radical generation as well as proline accumulation (Fig. 2). Such an existence in correlation between free radical generation and proline accumulation provided a clue that proline accumulation is related to non-enzymatic detoxification of free radicals (Alia et al., 1995, 1997). In general, functioning of chloroplast and mitochondria are amongst first to be affected in plant cells exposed to stress. Free radical generation is initiated due to unusual distribution of electrons from electron transport chains of chloroplast and mitochondria to O_2, which results in the production of various toxic oxygen species such as $^1O_2$, $O^{2\cdot-}$, H_2O_2 and OH^-. These toxic oxygen species initiate series

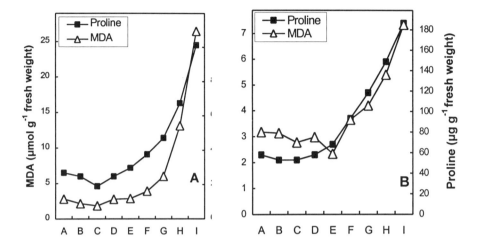

Figure 2. Correlative changes in malondialdehyde (MDA) and proline content in shoots of Brassica juncea (A) and Cajanus cajan seedlings raised in the presence of varying levels of zinc nitrate (a, b, c, d. e, f and g represent 0, 10, 50, 100, 500, 1000, 2500, 5000 and 10000 µM, respectively).

of devastating chain reactions (including peroxidation of lipids, cross-linking/ unusual-splitting of proteins and nucleic acids) which leads leading to total disruption of cellular metabolism.

As chloroplast is the major site for generation of toxic oxygen species and synthesis of proline under stress, it is assumed that proline protects chloroplast functions from stress damages. Isolated chloroplasts generate various forms of toxic oxygen species and other free radicals, when exposed to high light intensities, leading to a quick loss in the activities of photosystems, especially that of photosystem II. Presence of proline in the medium in which chloroplasts are incubated has been shown to significantly lower the loss in light induced photofunctions, suggesting that proline protects photosystems against free radicals/toxic oxygen species that are generated under high light intensities (Alia et al., 1991). The assumption has been confirmed by estimating the malondialdehyde (MDA) (an indicator of free radical production) content in the chloroplasts exposed to high light intensity in presence and absence of proline. The enhancement in MDA levels recorded with chloroplasts exposed to high light intensity in presence of proline was two fold lower than that noted in absence of proline (Alia et al., 1991). However, the amount of MDA produced by thylakoids of plants under heavy metal stress was several times higher than those from controls. The rise in MDA content in the control thylakoids was ~2.3 fold as against ~7 fold in treated (over the respective initial values) after 50 min of light incubation (Fig. 3).

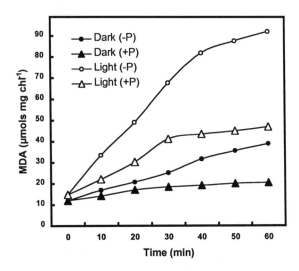

Figure 3. Time dependent change in malondialdehyde (MDA) content of thylakoids (from the cottyledonary leaves of Brassica juncea) exposed to strong white light with a photon flux density of 920 µmol $m^{-2}s^{-1}$ or dark conditions in the presence and absence of 1 M proline.

However, only a little change in the MDA content was seen during dark incubation for the same period. The rise in the MDA level in both the types of thylakoids when exposed to light was reduced when proline was added to the incubation mixture.

Role of proline in scavenging free radicals can also be assessed through the generation of free radicals such as OH˙ under artificial conditions. For example OH˙ can be genearated by mixing ascorbate with hydrogen peroxide and iron or by incubating xanthine oxidase with hypoxanthine and hydrogen peroxide. The extent of OH˙ radical generation can be monitored by simpler methods such as detection of hydroxylation of salicylic acid or loss in the activity of enzymes such as malate dehydrogenase. The rate of hydroxylation of salicylic acid or reduction in the activity of malate dehydrogenase, decreases significantly in presence of proline. Such observations point out that proline has a significant potential to scavenge OH˙ (see Smirnoff and Cumbes, 1989 and references therein).

Thylakoids from stressed seedlings upon exposure to high light intensity generate significantly higher levels of singlet oxygen than those from control thylakoids. Production of $^1O_2$ by illuminated thylakoids has been reported earlier also (Takahana and Nishimura, 1976). Hideg and Vass (1995) reported that singlet oxygen generation from the thylakoids exposed to high intensity illumination results due to photodamage of PS II acceptor site. Besides peroxidation of thylakoids membrane lipids (Takahana and Nishimura, 1976), singlet oxygen is believed to attack important aminoacids such as histidine of the D1 protein or may induce a conformational change in the reaction center complex by damaging its chlorophyll or carotenoid components and thus expose the D1 protein to the proteolytic attack (Hideg

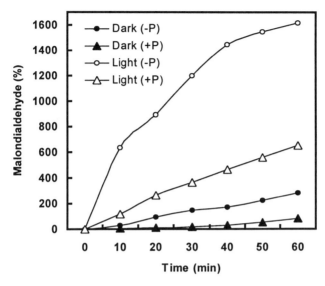

Figure 4. Time dependent change in malondialdehyde content of linolenic acid micelles exposed to UV-radiations or dark conditions in the presence and absence of 1 M proline. Note proline-induced suppression of the UV-mediated peroxidation of linolenic acid micelles.

and Vass, 1995). Using thylakoid as well as linolenic acid micellar systems, we have earlier demonstrated that proline has potential to significantly reduce the level of $^1O_2$ either by lowering its production or by scavenging it (Alia et al., 1997) (Fig. 4). UV-induced lipid peroxidation of linolenic acid micelles is caused due to $^1O_2$ (Bose and Chatterjee, 1993; Alia et al., 1997).

Thus, proline acts as a free radical scavenger mainly by detoxifying and/or decreasing the production of toxic oxygen species like $^1O_2$ and $OH^{\cdot}$. The exact mechanism of scavenging or lowering the production of various toxic oxygen species and other radicals, by proline remains unelucidated. However, it has been suggested that proline could be forming non-toxic hydroxyproline by reacting with $OH^{\cdot}$ (see Smirnoff and Cumbes, 1989).

5. HEAVY METAL INDUCED IRON DEFICIENCY INDUCES PROLINE ACCUMULATION

Heavy metal toxicity has been well documented to result in curtailing the uptake of essential mineral ions resulting in their overall deficiency. Several reports indicated that heavy metal ions replace essential mineral ions such as iron, copper and manganese from the components of electron transport system in chloroplasts and mitochondria and metal ion co-factors such as calcium and zinc from enzymes

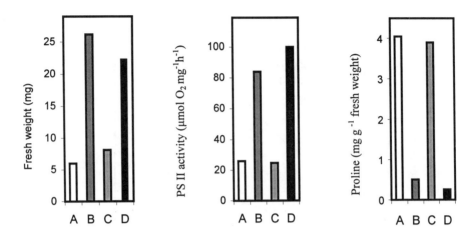

Figure 5. Fresh weight, PS II activity and proline content in shoots of wheat seedlings raised in distilled water (A), Hoagland medium (mineral growth medium) (B), Hoagland medium depleted of iron source (C) and distilled water supplemented with iron source (D).

proteins (Stohs and Bagchi, 1995; Zenk, 1996; Goyer, 1997; Clemens et al., 1999). Mineral elements such as iron, magnesium and calcium are known to play key role in the maintenance of electron transport in chloroplasts and mitochondria (Fork and Sato,1986; Terry and Abadia,1986; Mohanty and Mohanty,1988).

Cadmium induced suppression in length, fresh weight and dry weight of shoots was more prominent in seedlings grown in DW compared to those raised in MGM. While the presence of iron reduced Cd^{2+} induced suppression in the growth of the seedlings raised in DW, its removal from MGM enhanced Cd^{2+} induced suppression in the growth (Fig. 5). Cadmium induced enhancement in the level of proline was significantly higher in plants grown in DW compared to those grown in MGM. The extent of Cd^{2+} stimulated proline accumulation in plants grown in DW was almost similar to those grown in iron depleted MGM. At the same time Cd^{2+} induced increase in proline content in seedlings grown in DW was reduced in the presence of iron (Fig. 5). These results suggested that Cd^{2+} induced proline accumulation is related to iron deficiency. Infact, cadmium has been reported to induce iron deficiency (Agarwala et al., 1977; Thys et al., 1991). Iron deficiency is known to cause chlorosis and suppression in growth (Mattis and Hershey,1992; Sharma and Sanwal,1992).

Earlier, it was shown that Cd^{2+} inhibits the photosynthetic electron transport reactions (Van Assche and Clijsters, 1990; Atal et al., 1991, 1993). Cd^{2+} induced suppression in PSII activity was more prominent in seedlings raised in DW compared to those grown in MGM (Fig. 2). No significant difference in PSII activity was seen between the thylakoids isolated from seedlings raised in iron depleted MGM with Cd^{2+} and in DW with Cd^{2+}. These results suggest that Cd^{2+}

interferes with iron levels in the thylakoids (rendering them deficient in iron) and thus influence various photochemical activities including PSII activity. This could be the main reason behind higher PSII activity of the thylakoids from the seedlings grown in MGM compared to those from MGM depleted of iron during Cd^{2+} stress. This presumption was further strengthened from the reduction in Cd^{2+} induced suppression in PSII activity in thylakoids obtained from the seedlings grown in DW containing iron compared to those grown in absence of iron. Thys and co-workers (1991) had reported that the supply of iron in the growth medium can restore the Cd^{2+} induced loss in chlorophyll content.

6. PROLINE ACCUMULATION DUE TO HEAVY METAL STRESS INDUCED SUPPRESSION IN MITOCHONDRIAL ELECTRON TRANSPORT

Heavy metals such as Cd^{2+} had been reported to suppress mitochondrial electron transport besides photochemical activities in chloroplasts in particular those mediated by PS II (Miller et al. 1973; Bittel et al. 1974; Clijsters and Van Assche 1985; Becerril et al. 1988; Atal et al., 1991; Alia and Pardha Saradhi, 1993; Pardha Saradhi et al., 1993). The major reason behind the suppression in mitochondrial electron transport activity in plants under heavy metal stress is likely be due to heavy metal stress induced iron deficiency as iron is one of the essential components of electron transport system (Marschner, 1986; Greene et al., 1992; Sharma and Sanwal, 1992).

Plants exposed to various environmental stresses including heavy metal stress show a marked decline in the mitochondrial electron transport activity associated with several fold increase in proline levels (Table 3) (Alia and Pardha Saradhi, 1993; Pardha Saradhi et al., 1996).

Amazingly, rice seedlings exposed to various mitochondrial electron transport inhibitors viz. potassium cyanide (KCN), antimycin A and rotenone showed a significantly high accumulation of proline similar to that noted in plants under heavy metal stress (Fig. 6) (Alia and Pardha Saradhi, 1993). These results gave a clue that the accumulation of proline in plants under stress might be related to the suppression in mitochondrial and chloroplastic electron transport systems.

7. PROLINE ACCUMULATION FOR THE MAINTENANCE OF $NAD(P)^+/NAD(P)H$ RATIO

An increase in the level of NADH is expected due to suppression of mitochondrial electron transport activity (Alia and Pardha Saradhi, 1993; Pardha Saradhi et al., 1996). As can be seen in figure 7, the seedlings of rice exposed to heavy metal (Cd^{2+})

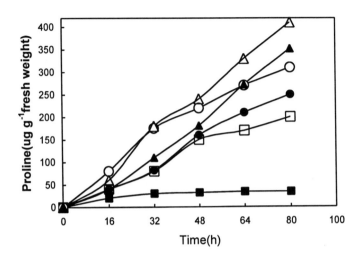

Figure 6. Time dependent change in the level of proline in rice seedlings exposed to mineral growth medium supplemented without (control) (■) and with NaCl (200 mM) (□), CdCl$_2$ (5 mM) (●), rotenone (50 mM) (△), antimycin A (!0 μg/ml) (▲) or KCN (5 mM) (○).

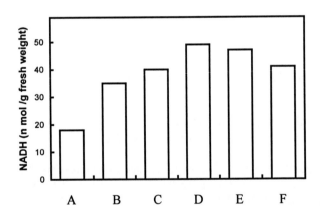

Figure 7. NADH levels in shoots of rice seedlings exposed to mineral growth medium supplemented without (control) (A) and with NaCl (200 mM) (B), CdCl$_2$ (5 mM) (C), rotenone (50 mM) (D), antimycin A (!0 μg/ml) (E) or KCN (5 mM) (F).

Table 3 : Electron transport activity of mitochondria isolated from the shoots of etiolated seedlings of plant species grown raised in mineral growth medium supplemented without (control) and with 1 mM cadmium nitrate. Electron transport activity was measured in terms of O_2 consumed due to the oxidation of NADH and succinate.

Plant species	n mol O_2/ min/ mg/ protein			
	NADH		Succinate	
	control	stressed	control	stressed
Vigna rqadiata	261.4±13.5	109.5±8.5	105.1+5.4	65.5±6.2
Brassica juncea	83.0±7.6	47.1±3.4	37.0±4.1	26.4±3.5
Oryza sativa	175.3±18.5	41.2±4.2	78.5+3.5	45.4±5.1

Experiments were carried out at least 5 times.
Data represent mean ± standard error.

stress, salt stress as well as various inhibitors of mitochondrial transport activity show a significant increase in the level of NADH. Another prominent observation made in plants exposed to abiotic stresses is the significant decline in the carboxylase activity of Rubisco, resulting in curtailing usual progress of Calvin cycle (Sivakumar et al., 1998; 2000, 2001a). This would obviously result in increase in the level of NAD(P)H. Accordingly, an increase in NAD(P)H to NAD(P)$^+$ ratio was recorded invariably in plants subjected to various abiotic stresses such as salt, drought and heavy metal stress (Lawlor and Khanna-Chopra, 1984; Laliberte and Hellebust, 1989; Alia and Pardha Saradhi, 1993; Pardha Saradhi et al., 1996). Such an increase in NAD(P)H might even effect substrate level phosphorylation indirectly besides inhibiting important metabolic reactions that need NAD(P)$^+$. For example conversion of glyceraldehyde-3-phosphate to 1,3-diphosphoglyceric acid mediated by the key enzyme of glycolysis, glyceraldehyde-3-phosphate dehydrogenase, depends on the availability of NAD$^+$ (see Stryer, 1988), the rate limiting step of the pentose phosphate pathway catalysed by glucose-6-phosphate dehydrogenase, is dependent on the availability of NADP$^+$ and is inhibited by NADPH (Laliberte and Hellebust, 1989) and some enzymes of Kreb's cycle depends on the availability of NAD$^+$ (see Stryer, 1989).

In order to let such essential reactions to continue, the living systems have evolved a number of strategies to readily maintain the ratio of NAD(P)H/NAD(P)$^+$. This includes activation of certain reactions which can readily oxidize NAD(P)H. Most pronounced response shown by plants under stress is the significant increase in the activity of various dehydrogenases that can readily maintain NAD(P)$^+$/NAD(P)H ratio as per the following general reaction:

$$X + NAD(P)H + H^+ \xrightleftharpoons[]{\text{Dehydrogenase}} XH_2 + NAD(P)^+$$

Accumulation of organic acids such as malate, lactate, glycollate noted in plant parts exposed to abiotic stress (see Osmond, 1978; Handa et al., 1983; Krampitz and Fock, 1984; Venekamp et al., 1989), most likely results due to the following reactions:

$$\text{Pyruvate} + NAD(P)H + H^+ \xrightleftharpoons[]{\text{ADH}} \text{Ethanol} + CO_2 + NAD(P)^+$$

$$\text{Pyruvate} + NAD(P)H + H^+ \xrightleftharpoons[]{\text{LDH}} \text{Lactate} + NAD(P)^+$$

$$\text{Oxaloacetate} + NAD(P)H + H^+ \xrightleftharpoons[]{\text{MDH}} \text{Malate} + NAD(P)^+$$

$$\text{Glyoxylate} + NAD(P)H + H^+ \xrightleftharpoons[]{\text{GO}} \text{Glycolate} + NAD(P)^+$$

ADH - alcohol dehydrogenase; MDH - malate dehydrogenase; LDH - lactate dehydrogenase; GO - glycolate oxidase

The accumulation of yet another organic acid viz. citrate could result due to limited availability of NAD^+ needed for its metabolism. The accumulation of these organic acids would, however, bring about a disturbance in several metabolic processes in cells by lowering the pH.

The accumulation of proline in plants under stress results due to its active synthesis from glutamate. It is fascinating to know that during synthesis of each molecule of proline from glutamate, two molecules of NAD(P)H are oxidized as per the following reactions:

$$\text{Glutamate} + NAD(P)H + H^+ + ATP \xrightarrow{\text{P5CS}} \gamma\text{-pyrroline-5-carboxylate} + NAD(P)^+ + ADP + Pi$$

$$\gamma\text{-pyrroline-5-carboxylate} + NAD(P)H + H^+ \xrightarrow{\text{P5CR}} \text{Proline} + NAD(P)^+$$

Both P5CS (γ-pyrroline-5-carboxylate synthetase) and P5CR (γ-pyrroline-5-carboxylate reductase) are a kind of dehydrogenases with potential to oxidize NAD(P)H. As proposed earlier, proline accumulation is due to its fresh synthesis from glutamate in order to maintain the ratio of NADH/NAD$^+$ (Alia and Pardha Saradhi, 1991; Pardha Saradhi et al., 1993, 1996). As is clear from above reactions the synthesis of each molecule of proline from glutamte would oxidise two molecules of NAD(P)H (instead of one in the other reactions given above). In addition, in contrary to the reactions (viz. conversion of oxaloacetate to malate, pyruvate to lactate or glyoxylate to glycolate which do not involve any alteration in the number of carboxylic groups associated with the products and their respective substrates, proline is a monocarboxylic acid while glutamate is a dicarboxylic acid. Unlike the organic acids produced in other reactions (associated with the oxidation of NADH), proline is a compatible solute (Venekamp, 1989; Venekamp et al., 1989) whose accumulation does not alter the cellular pH. Infact, it has been proposed that proline accumulation could play a role of redox buffer by storage of excess reductants in a non toxic form (Alia and Pardha Saradhi, 1991, 1993; Pardha Saradhi et al., 1993, 1996). Therefore, we had proposed that excessive synthesis of proline during various environmental stresses is an appropriate adaptive mechanism evolved by the plants to maintain the ratio of NAD(P)$^+$ to NAD(P)H in order to let reactions that are most vital for maintenance of living cells/organisms, to proceed.

REFERENCES

Agarwala, S.C., Bisht, S.S., Sharma, C.P. 1977. Relative effectiveness of certain heavy metals in producing toxicity and symptoms of iron deficiency in barley. *Can J Bot* **55**: 1299-1307.
Ahmad, I., Hellebust, J. A. 1988. The relationship between inorganic nitrogen metabolism and proline accumulation in osmoregulatory responses of two euryhaline microalgae. *Plant Physiol* **88**: 348-354.
Akabori, K., Tsukamoto, H., Tsukihara, J., Nagatsuka, T., Motokawa, O., Toyoshima, Y. 1988. Disintegration and reconstitution of photosystem II reaction center core complex I. Preparation and characterization of three different types of subcomplex. *Biochim Biophys Acta* **932**: 345-357.
Alia, Pardha Saradhi, P. 1991. Proline accumulation under heavy metal stress. *J Plant Physiol* **138**: 554-558.
Alia, Pardha Saradhi, P. 1993. Suppression in mitochondrial electron transport is the prime cause behind stress induced proline accumulation. *Biochem Biophys Res Commun* **193**: 54-58.
Alia, Pardha Saradhi, P., Mohanty, P. 1991. Proline enhances primary photochemical activities in isolated thylakoid membranes of *Brassica juncea* by arresting photoinhibitory damage. *Biochem Biophys Res Commun* **181**: 1238-1244.
Alia, Pardha Saradhi, P., Mohanty, P. 1993. Proline in relation to free radical production in seedlings of *Brassica juncea* raised under sodium chloride stress. *Plant Soil* **156**: 497-500.
Alia, Prasad, K.V.S.K., Pardha Saradhi, P. 1995. Zinc induced changes in the levels of free radicals and proline in *Brassica juncea* and *Cajanus cajan*. *Phytochemistry* **39**: 45-47.
Alia, Pardha Saradhi, P., Mohanty, P. 1997. Involvement of proline in protecting thylakoid membranes against free radical-induced photodamage. *J Photochem Photobiol B: Biology* **38**: 253-257.
Anbazghan, M. Krishnamurthy, R., Bhagwat, K. A. 1988. Proline: an enigmatic indicator of air pollution tolerance in rice cultivars. *J Plant Physiol* **133**: 122-123.
Arakawa, T., Timasheff, S.N. 1985. The stabilization of proteins by osmolytes. *Biophy J* **47**: 411-414.
Arora, S., Pardha Saradhi, P. 1995. Light induced enhancement in proline levels in *Vigna radiata* exposed to environmental stresses. *Aust J Plant Physiol* **22**: 383-386.
Aspinall, D., Paleg, L. D. 1981. Proline accumulation: physiological aspects. In The Physiology and Biochemistry of Drought Resistance in Plants (Paleg, L.G., and Aspinall, D. Eds.), pp.215-228 Academic Press, Sydney.

Atal, N., Pardha Saradhi, P., Mohanty, P. 1991. Inhibition of chloroplast photochemical reactions by treatment of wheat seedlings with low concentrations of cadmium : Analysis of electron transport activity and changes in fluorescence yield. *Plant Cell Physiol* **32**: 943-951.

Atal, N., Pardha Saradhi, P., Mohanty, P. 1993. Effect of iron on photosystem II mediated photochemical activities and proline levels in wheat seedlings during Cd^{2+} stress. Proceedings of DAE Symposium on Photosynthesis and Plant Molecular Biology. Bhabha Atomic Research Centre, Bombay, 1-5.

Bassi, R., Sharma, S.S. 1993. Proline accumulation in wheat seedlings exposed to zinc and copper. *Phytochemistry* **33**: 1339-1342.

Baszynski, T., Wajda, L., Krol, M., Wolinska, D., Krupa, Z., Tukendrof, A. 1980. Photosynthetic activities of cadmium treated tomato plants. *Physiol Plant* **48**: 365-370.

Becerril, J.M., Munoz-Rueda, A., Apricio-Tejo, P., Gonzalez-Murug. 1988.The effects of cadmium and lead on photosynthetic electron transport in Clover and Lucerne. *Plant Physiol Biochem* **26**: 357-363.

Bittell, J.E., Koeppe, D.E., Miller, R.J. 1974. Absorption of heavy metal cations by corn mitochondria and the effects on electron and energy transfer reactions. *Physiol Plant* **30**: 226-230.

Bose, B., Chatterjee, S.N. 1993. Effect of UV-A on the linolenic acid micelles. *Radiat Res* **133**: 340-344.

Chapman, D.J., Wang, W.Q., Barber, J. 1990. Low temperature stress and photoinhibition of photosystem II. In: Sinha, S.K., Sane, P.V., Bhargava, S.C., Agrawal, P.K. (eds.): Proceedings of the International Congress of Plant Physiology. 1, 621-629. Neo Art Press, New Delhi.

Clemens, S., Kim, E.J., Neumann, D., Schroeder, J.I. 1999. Tolerance to toxic metals by a gene family of phytochelatin synthases from plants and yeast. *EMBO J* **18**: 3325-3333.

Clijsters, H., Van Assche, F. 1985. Inhibition of photosynthesis by heavy metals. *Photosynth Res* **7**: 31-40.

Cobbett, C.S. 2000. Phytochelatins and their roles in heavy metal detoxification. *Plant Physiol* **123**: 825-832.

De Vos, C.H.R., Schat, H., Vooijs, R., Ernst, W.H.O. 1989. Copper-induced damage to the permeability barrier in roots of *Silene cucubalus*. *J Plant Physiol* **135**: 164-169.

De Vos, C.H.R., Schat, H. 1991. Free radicals and heavy metal tolerance. In: Ecological responses to environmental stresses. Rozema, J., Verkleij, J.A.C., Eds. pp.22-30, Kluwer Academic Publishers.

Delauney, A.J., Verma, D.P.S. 1993. Proline biosynthesis and osmoregulation in plants. *Plant J* **4**: 215-223.

Ebbs, S.D., Kochian, L.V. 1998. Phytoextraction of zinc by oat (*Avena sativa*), barley (*Hordeum vulgare*), and Indian mustard (*Brassica juncea*). *Environ Sci Technol* **32**: 802-806.

Flowers, T.J., Hall, J.L., Ward, M.E. 1978. Salt tolerance in the halophyte, *Suaeda maritima* (L.) Dum.: properties of malic enzyme and PEP carboxylase. *Ann Bot* **42**: 1065-1074.

Fork, D.C., Satoh, K. 1986. The control of state transitions of the distribution of excitation energy in photosynthesis. Ann Rev Plant Physiol **37**: 335-361.

Fukutaku, Y., Yamada, Y. 1984. Source of proline nitrogen in water stressed soybean (*Glycine max*) II. Fate of ^{15}N-labelled protein. Physiol Plant **61**: 622-628.

Gamborg, D. L., Miller, R. A., Ojima, K. 1968. Nutrient requirements of suspension cultures of soybean root cell. *Exp Cell Res* **50**: 151-158.

Goyer, R.A. 1977. Toxic and essential metal interactions. *Annu Rev Nutr* **17**: 37-50.

Greene, R.M., Geider, R.J., Kolber, Z., Falkowski, P.G. 1992. Iron-induced changes in light harvesting and photochemical energy conversion process in eukaryotic marine algae. *Plant Physiol* **100**: 565-575.

Geenway, H., Osmond, C.B. 1972. Salt responses of enzymes from species differing in salt tolerance. *Plant Physiol* **49**: 256-259.

Greger, M., Lindberg, S. 1986. Effects of Cd^{2+} and EDTA on young sugar beets (*Beta vulgaris*) I Cd^{2+} uptake and sugar accumulation. *Physiol Plant* **66**: 69-74.

Halliwell, B., Gutteridge, J.M.C. 1986. Iron and free radical reactions : Two aspects of antioxidant protection. *Trends Biochem Sci* **11**: 372-375.

Handa, S., Bressan, R.A., Handa, A.K., Carpita, N. C., Hesegawa, P.M. 1983. Solutes contributing to osmotic adjustment in culture cells adapted to water stress. *Plant Physiol* **73**: 834-843.

Heath, R.L., Packer, L. 1968. Photoperoxidation in isolated chloroplasts. I. Kinetics and stoichiometry of fatty acid peroxidation. *Arch Biochem Biophys* **125**: 189-198.

Hendry, G.A.F., Brocklebank, K.J. 1985. Iron-induced oxygen radical metabolism in waterlogged plants. *New Phytol* **101**: 199-206.

Hideg, E., Vass, I. 1995. Singlet oxygen is not produced in photosystem I under photoinhibitory conditions. *Photochem Photobiol* **62**: 949-952.

Hong, H., Lakkineni, K., Zhang, Z., Verma, D.P.S. 2000. Removal of feedback inhibition of Δ^1-pyrroline-5-carboxylate synthetase results in increased proline accumulation and protection of plants from osmotic stress. *Plant Physiol* **122**: 1129-1136.

Kadpal, R. P., Rao, N. A. 1985. Alterations in the biosynthesis of proteins and nucleic acids in finger millet (*Eleucine coracana*) seedlings during water stress and the effect of proline on protein biosynthesis. *Plant Sci* **40**: 73-79.

Kavi Kishor, P.B., Hong, Z., Miao, G., Hu, C.A., Verma, D.P.S. 1995. Overexpression of Δ^1-pyrroline-5-carboxylate synthetase increases proline production and confers osmotolerance in transgenic plants. *Plant Physiol* **108**: 1387-1394.

Kemble, A.R., Mac Pherson, H.T. 1954. Liberation of amino acids in perennial rye grass during wilting. *Biochem J* **58**: 46-49.

Krall, J.P., Edwards, G.E., Andreo, C.S. 1989. Protection of pyruvate, Pi dikinase from maize against cold lability by compatible solutes. *Plant Physiol* **89**: 280-285.

Krampitz, M. J., Fock, H. P. 1984. $^{14}CO_2$ assimilation and carbon flux in the Calvin cycle and the glycollate pathway in water-stressed sunflower and bean leaves. *Photosynthetica* **18**: 329-337.

Krupa, Z., Skorzynska, F., Maksymiec, W., Baszynski, T. 1987. Effect of cadmium treatment on the photosynthetic apparatus and its photochemical activities in greening radish seedlings. *Photosynthetica* **21**: 156-164.

Krupa, Z. 1987. Cadmium-induced changes in the composition and structure of light harvesting chlorophyll binding a/b protein complex in raddish cotyledons. *Physiol Plant* **73**: 518-524.

Kunert, K.J. 1984. The diphenyl-ether herbicide oxyfluorfer: a potent inducer of lipid peroxidation in higher plants. *Z Naturforsch* **39**: 476-481.

Laliberte, G., Hellebust, J. A. 1989a. Regulation of proline content of *Chlorella autotropica* in response to change in salinity. *Can J Bot* **67**: 1959-1965.

Laliberte, G., Hellebust, J.A. 1989b. Pyrroline-5-carboxylate reductase in *Chlorella autotrophica* and *Chlorella saccharophila* in relation to osmoregulation. *Plant Physiol* **91**: 917-923.

Lawlor, D.W., Khanna-Chopra, R. 1984. "Regulation of photosynthesis during water stress". In *Advances in Photosynthesis Research*, IV, pp. 379-382, Sybesma, C., ed. Kluwer Academic Publishers, Dordrescht.

Lee, K.C., Cunningham, B.A., Paulsen, G.M., Liang, G.H., Moore, R.B. 1976. Effects of cadmium on respiration rate and activities of several enzymes in soybean seedlings. *Physiol Plant* **36**: 4-6.

Malik, D., Sheoran, I.S., Singh, R. 1992. Carbon metabolism in leaves of cadmium treated wheat seedlings. *Plant Physiol Biochem* **30**: 223-229.

Manetas, Y., Petropoulou, Y., Karabourniotis, G. 1986. Compatible solutes and their effects on phospho*enol*pyruvate carboxylase of C_4 halophytes. *Plant Cell Environ* **9**: 145-151.

Manetas, Y. 1990. A re-examination of NaCl effects on phosphoenolpyruvate carboxylase at high (physiological) enzyme concentrations. *Physiol Plant* **78**: 225-229.

Marschner H 1986 Mineral nutrition of higher plants. Academic Press Inc. (London) Ltd. 254 p.

Mattis, P.R., Hershey, D.R. 1992. Iron deficiency stress response of *Epipremnum aureum* and *Philodendron scandens* ssp *oxycardium*. *J Plant Physiol* **139**: 498-502.

Mejáre, M., Bülow, L. 2001. Metal binding proteins and peptides in bioremediation and phytoremediation of heavy metals. *T Biotech* **19**: 67-73.

Miller, R.J., Bittell, J.E., Koeppe, D.E. 1973. The effect of cadmium on electron and energy transfer reactions in corn mitochondria. *Physiol Plant* **28**: 166-171.

Mohanty, N., Mohanty, P. 1988. "Cation effects on primary processes of photosynthesis". In *Advances in frontier areas of plant biochemistry*, Singh, R., Sawhney, S.K., eds. pp.1-18. Prentice Hall India, Delhi.

Nash, D., Paleg, L.G., Wiskich, J.T. 1982. The effect of proline, betaine and some other solutes on the heat stability of mitochondrial enzymes. *Aust J Plant Physiol* **9**: 47-57.

Nath, R. 1986. "Biochemical effects". In *Environmental pollution of cadmium, biological physiological and health effects*, Bhatia, B., ed. pp. 60-110, Interprint, New Delhi.

Neale, P.J., Melis, A. 1989. Salinity-stress enhances photoinhibition of photosynthesis in *Chlamydomonas reinhardii*. *J Plant Physiol* **134**: 619-622.

Nikolopoulos, D., Manetas, Y. 1991. Compatible solutes and in vitro stability of *Salsola soda* enzymes: Proline incompatibility. *Phytochemistry* **30**: 411-413.

Osmond, C.B. 1978. Crassulacean acid metabolism: a curiosity in context. *Ann Rev Plant Physiol* **29**: 379-414.

Pahlsson, A.M.B. 1989. Toxicity of heavy metals (Zn, Cu, Cd, Pb) to vascular plants. *Water Air Soil Pollution* **47**: 287-319.
Paleg, L.G., Douglas, T.J., Van Daal, A., Keech, D.B. 1981. Proline and betaine protect enzymes against heat inactivation. *Aust J Plant Physiol* **9**: 47-57.
Paleg, L.G., Stewart, G.R., Bradbeer, J.W. 1984. Proline and glycine betaine influences protein solvation. *Plant Physiol* **75**: 974-978.
Papageorgiou, G.C., Murata, N. 1995. The usually strong stabilizing effects of glycine betaine on the structure and function of the oxygen-evolving photosystem II complex. *Photosyn Res* **44**: 243-252.
Pardha Saradhi, P., Alia, Vani, B. 1993. Inhibition of mitochondrial electron transport is the prime cause behind proline accumulation during mineral deficiency in *Oryza sativa*. *Plant Soil* **156**: 465-468.
Pardha Saradhi, P., Alia, Arora, S., Prasad, K.V.S.K. 1995. Proline accumulate in plants exposed to UV radiation and Protects them against UV induced peroxidation. *Biochem Biophys Res Commun* **209**: 1-5.
Pardha Saradhi, P., Arora, S., Vani B., Puthur, J.T. 1996. Alteration in $NAD^+/NADH$ ratio regulates salt stress induced proline accumulation in *Vigna radiata*. *Proc Nat Acad Sci India* **66(B)**: 89-100.
Pick, V., Weiss, M., Gounaris, K., Barber, J. 1987. The role of different thylakoid glycolipids in the function of reconstituted chloroplast ATP synthetase. *Biochim Biophys Acta* **891**: 28-39.
Pollard, A., Wyn Jones, R. G. 1979. Enzyme activities in concentrated solution of glycine betaine and other solutes. *Planta* **144**: 291-298.
Prasad, K.V.S.K., Pardha Saradhi, P., Sharmila, P. 1999. Correlative changes in the activities of antioxidant enzymes under zinc toxicity in *Brassica juncea*. *Environ Exptl Bot* **42**: 1-10.
Prasad, K.V.S.K., Sharmila, P., Kumar, P.A., Pardha Saradhi, P. 2000a. Transformation of *Brassica juncea* (L.) Czern with bacterial *codA* gene enhances its tolerance to salt stress. *Mol Breed* **6**: 489-499.
Prasad, K.V.S.K., Sharmila, P., Pardha Saradhi, P. 2000b. Enhanced tolerance of transgenic *Brassica juncea* to choline confirms successful expression of the bacterial *codA* gene. *Plant Sci* **159**: 233-242.
Puthur, J.T., Sharmila, P., Prasad, K.V.S.K., Pardha Saradhi, P. 1997. Proline overproduction: a means to improve stress tolerance in crop plants. *Botanica* **47**: 163-169.
Rauser, W.E. 1978. Early effects of phytotoxic burdens of cadmium, cobalt, nickel and zinc in white beans. *Can J Bot* **56**: 1744-1749.
Rauser, W.E. 1990. Phytochelatins. *Annu Rev Biochem* **59**: 61-86.
Rayapati, P. J., Stewart, C. R., Hack, E. 1989. Pyrroline-5-carboxylate reductase is in pea (*Pisum sativum* L.) leaf chloroplasts. *Plant Physiol* **91**: 581-586.
Sandmann, G., Boger, P. 1980a. Copper-mediated lipid peroxidation processes in photosynthetic membranes. *Plant Physiol* **66**: 797-800.
Sandmann, G., Boger, P. 1980b. Copper deficiency and toxicity in Scenedesmus *Z Pflanzenphysiol* **98**: 53-59.
Schobert B. 1977. Is there an osmotic regulatory mechanism in algae and higher plants? *J Theor Biol* **68**: 17-26.
Selinioti, E., Nikolopoulos, D., Manetas, Y. 1987. Organic cosolutes as stabilisers of phosphoenolpyruvate carboxylase in storage: An interpretation of their action. *Aust J Plant Physiol* **14**: 203-210.
Sharma, S., Sanwal, G.G. 1992. Effect of iron deficiency on the photosynthetic system of maize. *J Plant Physiol* **140**: 527-530.
Siegenthaler, P.A., Smutny, J., Rawyler, A. 1987. Involvement of distinct populations of phosphatidyl glycerol and phosphatidyl choline molecules in photosynthetic electron flow activities. *Biochim Biophys Acta* **891**: 85-93.
Sivakumar, P., Sharmila, P., Pardha Saradhi, P. 1998. Proline suppresses Rubisco activity in higher plants. *Biochem Biophys Res Commun* **252**: 428-432.
Sivakumar, P., Sharmila, P., Pardha Saradhi, P. 2000. Proline alleviates salt stress induced enhancement in ribulose 1,5-bisphosphate oxygenase activity. *Biochem Biophys Res Commun* **279**: 512-515.
Sivakumar, P., Sharmila, P., Pardha Saradhi, P. 2001a. Proline suppresses Rubisco activity by dissociating smaller subunits from holoenzyme. *Biochem Biophys Res Commun* **282**: 236-241.
Sivakumar, P., Yadav, S. and P. Pardha Saradhi, P. 2001b. "Rubisco folding - A model system for the functional investigation of molecular chaperones". In *Biophysical Processes in Living Systems*, P. Pardha Saradhi, ed., Scientific Publishers, Inc., USA, p 91-116.
Skorzynska, E., Urbanik-Sypniewska, T., Russa, R., Baszynski, T. 1991. Galactolipase activity of chloroplasts in cadmium-treated runner bean plants. *J Plant Physiol* **138**: 454-459.

Smirnoff, N., Cumbes, Q.J. 1989. Hydroxyl radical scavenging activity of compatible solutes. *Phytochemistry* **28**: 1057-1060.

Somashekaraiah, B.V., Padmaja, K., Prasad, A.R.K. 1992. Phytotoxicity of cadmium ions on germinating seedlings of mung been (*Phaseolus vulgaris*): Involvement of lipid peroxides in chlorophyll degradation. *Physiol Plant* **85**: 85-89.

Sopory, S. K., Greenberg, B. M., Mehta, R. A., Edelman, M., Mattoo, A.K. 1990. Free radical scavengers inhibit light-dependent degradation of the 32 kDa photosystem II reaction center protein. *Z. Naturforsch* **45**: 412-417.

Stamatakis, K., Gavalas, N.A., Manetas, Y. 1988. Organic cosolutes increase the catalytic efficiency of phosphoenol pyruvate carboxylase, from Cynodon dactylon (L.) Pers., apparently through self-association of the enzymic protein. *Aust J Plant Physiol* **15**: 621.

Steffens, J. C. 1990. The heavy metal-binding peptide of plants. *Ann. Rev. Plant Physiol. Plant Mol Biol* **41**: 553-575.

Stibovora, M. 1988. Cd^{2+} ions affect the quaternary structure of ribulose-1,5-biphosphate carboxylase from barley leaves. *Biochem Physiol Pflanzen* **183**: 371-378.

Stohs, S.J., Bagchi, D. 1995. Oxidative mechanisms in the toxicity of metal ions. *Free Radic Biol Med* **18**: 321-336.

Stryer, L. Biochemistry. W. H. Freeman and Company, New York, 1988.

Takahana, U., Nishimura, M. 1976. Effect of electron donors and acceptors, electron transport mediators, and superoxide dismutase on lipid peroxidation in illuminated chloroplast fragments. *Plant Cell Physiol* **17**: 111-118.

Terry, N., Abadia, J. 1986. Function of iron in chloroplasts. *J Plant Nutr* **9**: 609-646.

Thys, C., Vanthomme, P., Schrevens, E., De Proft, M. 1991. Interactions of Cd with Zu, Cu, Mn and Fe for lettuce (*Lactuca sativa* L.) in hydroponic culture. *Plant Cell Environ* **14**: 713-717.

Treichel, S.P., Kirst, G.O., Von Willert, D.J. 1974. NaCl-induced alteration of phosphoenolpyruvate carboxylase activity in halophytes of different habitats. *Z Pflanzenphysiol* **71**: 437-449.

Van Assche, F., Cardinaels, C., Clijsters, H. 1988. Induction of enzyme capacity in plants as a result of heavy metal toxicity: Dose response relations in *Phaseolus vlugaris* L., treated with zinc and cadmium. *Environ Pollut* **52**: 103-115.

Van Assche, F., Clijsters, H. 1990. Effects of metals on enzyme activity in plants. *Plant Cell Environ* **13**: 195-206.

Van Steveninck, R.F.M., Van Steveninck, M.E., Wells, A.J., Fernendo, D.R. 1990. Zinc tolerance and the binding of zinc as zinc phytate in *Lemna minor*; x-ray microanalytical evidence. *J Plant Physiol* **137**: 140-146.

Venekamp, J.H. 1989. Regulation of cytosolic acidity in plants under conditions of drought. *Physiol Plant* **76**: 112-117.

Venekamp, J.H., Lampe, J.E., Koot, J.T.M. 1989. Organic acids as a sources of drought - induced proline synthesis in field bean plants, *Vicia faba* L. *J Plant Physiol* **133**: 654-659.

Wyn Jones, R. G., Gorham, J. "Aspects of salt tolerance in higher plants : An agricultural perspective". In *Genetic Engineering of Plants*, pp. 355-370, Kosuge, T., Meredith, C. P., Hollaender, A., eds., Plenum Press, New York, 1984.

Wyn Jones, R.G., Gorham, J. "Physiological effects of salinity - scope for genetic improvement". In *Proceedings of the International Congress of Plant Physiology*, 2, 943-952, Sinha, S.K., Sane, P.V., Bhargava, S.C., Agrawal, P.K., eds., Society for Plant Physiology and Biochemistry, New Delhi, 1990.

Yancey, P. H., Somero, G. N. 1979. Counteraction of urea destabilization of protein structure by methylamine osmoregulatory compounds of elasmobranch fishes. *Biochem J* **183**: 317-323.

Zenk, M.H. 1996. Heavy metal detoxification in higher plants- a review. *Gene* **179**: 21-30.

Zhang, C., Lu, Q., Verma, D.P.S. 1995. Removal of feedback inhibition of Δ-Pyroline-5-carboxylate synthetase, a bifunctonal enzyme catalyzing the first two steps of proline biosynthesis in plants. *J Biol Chem* **270**: 20491-20496.

Zhu, Y.L., Pilon-Smits, E.A.H., Jouanin, L., Terry, N. 1999. Over expression of glutathione synthetase in Indian Mustard enhances cadmium accumulation and tolerance. *Plant Physiol* **119**: 73-79.

CHAPTER 8

INFLUENCE OF METALS ON BIOSYNTHESIS OF PHOTOSYNTHETIC PIGMENTS

B. MYŚLIWA-KURDZIEL AND K. STRZAŁKA
Department of Plant Physiology and Biochemistry
The Institute of Molecular Biology
Jagiellonian University
7 Gronostajowa Street, PL 30-38, Kraków, Poland

1. INTRODUCTION

Photosynthetic pigments: chlorophylls (Chl), carotenoids and phycobilins play a fundamental role in plants and photosynthetic microorganisms because they enable photosynthesis, the process of solar energy conversion to the energy of chemical compounds. Biosynthesis of photosynthetic pigments is a process prone to several biotic and abiotic stresses including metal stress. Chlorosis and retardation of plant growth that is frequently observed in metal polluted environments indicate that an impairment of photosynthetic pigment biosynthetic pathways is among the earliest targets of heavy metals influence on plant metabolism. This, in turn, has a considerable effect on plastid development, photosynthetic efficiency and general metabolism. Metal-induced decrease in the accumulation of photosynthetic pigments is particularly pronounced during development of seedlings and during the growth of new leaves when active pigment synthesis occurs.

Inhibition of the photosynthetic pigment biosynthesis is one of the primary events in plants during heavy metal stress. As a consequence a delay in the assembly of the photosynthetic apparatus, lower photosynthetic efficiency, slower plant growth and decreased biomass production occurs. Thus, heavy metal pollution may be a serious agricultural problem as it decreases the yield of crop plants and lowers the quality of the plant products due to increased content of toxic metals.

Inhibitory effect on photosynthetic pigment biosynthesis is not restricted to heavy metals only. Other metals (e.g. aluminium) also produce toxic symptoms to plants. This concerns also biometals, necessary for normal plant development and metabolism (e.g. copper), when they appear in an environment in excessive amounts.

From the long list of metals toxic to plants only some were investigated more thorouhly in respect to their influence on biosynthesis of photosynthetic pigments. In these studies much efforts have been undertaken to elucidate the effect of metals biosynthesis and accumulation of chlorophylls, whereas much less attention has been paid to the biosynthesis of carotenoids and phycobilins.This review gives the survey of the present day knowledge on the effect of metal stress on photosynthetic pigment biosynthesis with special emphasis put on the enzymatic steps impaired by the metal treatment.

2. THE INFLUENCE OF METALS ON CHLOROPHYLL BIOSYNTHESIS

2.1 Biosynthesis of chlorophyll

Chlorophyll plays a fundamental role in the process of photosynthesis because of its ability to absorb light. Biosynthetic pathway of this pigment and regulatory mechanisms involved are still intensively studied. Several reviews covering different aspects of the Chl synthesis and its regulation have recently appeared (Senge, 1993; Avissar, 1995; von Wettstein et al., 1995; Reinbothe and Reinbothe, 1996; Rüdiger, 1997 a, b; Beale, 1999). The main steps of this biosynthetic pathway are shown briefly in the Fig. 1. Following the detection of monovinyl- and divinyl-tetrapyrolle intermediates and their biosynthetic interconversion, a multi-branched Chl biosynthesis pathway, leading to the formation of monovinyl- or divinyl-Chl, has also been proposed (reviewed by Rebeiz et al., 1994 and 1999). There is a growing body of evidence that several enzymes of the Chl biosynthesis pathway are the primary targets for metal toxicity, therefore we describe briefly below the most important enzymatic steps involved in this metabolic pathway.

Chlorophyll biosynthesis starts with the synthesis of δ-aminolevulinic acid (ALA), a common precursor of all porphyrins. The C_5 pathway, described for the Chl synthesis in higher plants, algae, cyanobacteria and most of photosynthetic bacteria, involves the activation of a glutamate molecule by a $tRNA^{Glu}$, followed by the enzymatic reduction of glutamyl-$tRNA^{Glu}$ into glutamate-1-semialdehyde. The reaction is catalysed by glutamyl-$tRNA^{Glu}$ reductase, which was found to be light-regulated enzyme in the ALA synthesis. The last step in the ALA synthesis is the transamination of glutamate-1-semialdehyde to ALA.

In the subsequent reaction, which is catalysed by ALA dehydratase (ALAD),two molecules of ALA condense to yield porphobilinogen. Next, four molecules of porphobilinogen are converted into hydroxymethylbilane by porphobilinogen deaminase. This linear tetrapyrolle is finally transformed to protoporphyrin IX in several subsequent reactions including ring closure and isomerisation leading to uroporphyrinogen III, decarboxylation of acetate side chains (coproporphyrinogen III) and two successive oxidations. Protoporphyrin IX is the first coloured species due to the presence of conjugated double bond system.

In the branch of the pathway leading to the protochlorophyllide (Pchlide) synthesis, Mg is incorporated into protoporphyrin IX by Mg chelatase. The Mg-protoporphyrin IX is methylated to the monomethyl ester, which is followed by the formation of the isocyclic ring, characteristic to Chl related pigments. Then, in the reaction catalysed by 8-vinyl reductase, the DV-Pchlide (divinyl-Pchlide) is reduced to MV- Pchlide (monovinyl-Pchlide). However, this step can also follow the Pchlide to chlorophyllide (Chlide) photoreduction, since the 8-vinyl reductase shows a broad range of substrate specificity (von Wettstein et al., 1995; Lebedev and Timko, 1998).

Pchlide to Chlide conversion requires the reduction of the double bond between C_{17} and C_{18} in ring D of Pchlide molecule. This reaction converts the porphyrin molecule,

Pchlide, into a chlorin, Chlide. Two entirely different mechanisms of this reactions are known, one that requires light and is catalysed by a single enzyme, and another that is independent of light but requires at least three polypeptides (Fujita, 1996; Rüdiger, 1997 a, b; Lebedev and Timko, 1998). All Chl-containing organisms can synthesize this pigment in the light dependent way, while angiosperms and some algae cannot use the light independent route.

Light dependent NADPH-protochlorophyllide oxidoreductase (POR) catalyses the Pchlide to Chlide photoreduction in angiosperms (reviewed by Fujita, 1996; Lebedev and Timko, 1998; Schoefs, 1999). Pchlide acts as a photoreceptor chromophore in this reaction, which can be triggered by a single photon. Illumination induces a hydrogen transfer from a NADPH molecule to C_{17} of Pchlide followed by a proton transfer from water or tyrosine of the POR enzyme to C_{18} of Pchlide. Any chemical modification of the rings A and B did not influence the Pchlide photoreducibility, thus, both MV and DV Pchlide can be used as a substrate. Concerning the central metal incorporated into the porphyrin structure, it was found that the Mg of Pchlide could be substituted by Zn without loss of substrate photoreducibility, which is not the case for other metals such as Ni, Cu or Co as well as for pheophorbide (Lebedev and Timko, 1998).

In dark grown angiosperms the Chl synthesis is inhibited at the step of Pchlide formation which is accumulated within the etioplast inner membranes (EPIM). EPIM consist of two different membrane systems: prothylakoids (PT) and one or several prolamellar bodies (PLB) (rev. Selstam and Widell-Wigge, 1993). PLB is a highly regular three-dimensional lattice that is built of tubular membranes. Pchlide pool accumulated in EPIM is spectrally inhomogeneous. Some of Pchlide molecules form ternary complex with POR and NADPH and can be reduced upon illumination (Lebedev and Timko, 1998). This photoactive Pchlide, localized mainly within PLB, has a 77 K fluorescence and absorption maxima at 656 and 650 nm, respectively. The photoinactive Pchlide absorbs maximally at 630 nm and has a fluorescence maximum at 633 nm at 77 K. Based on the gaussian deconvolution of fluorescence emission spectra measured at 77 K, four universal Pchlide forms, having fluorescence maxima at 633, 645, 657 and 670 nm, were identified in dark-grown leaves (Böddi et al., 1992 and 1993). It is now assumed that diversity of Pchlide spectral forms results from different aggregation states of the Pchlide: POR: NADPH complexes (Böddi et al., 1989; Ryberg et al., 1992) as well as from the redox state of NADPH bound as a cofactor (El Hamouri et al., 1981; Franck et al., 1999).

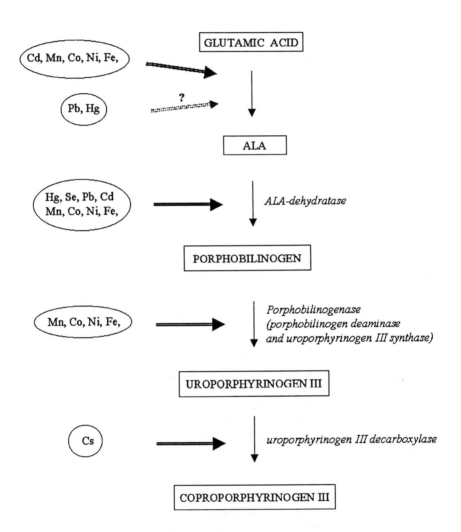

Figure 1. The influence of metal ions on chl biosynthesis. Part 1: Steps from glutamic acid to coproporphyrinogen III

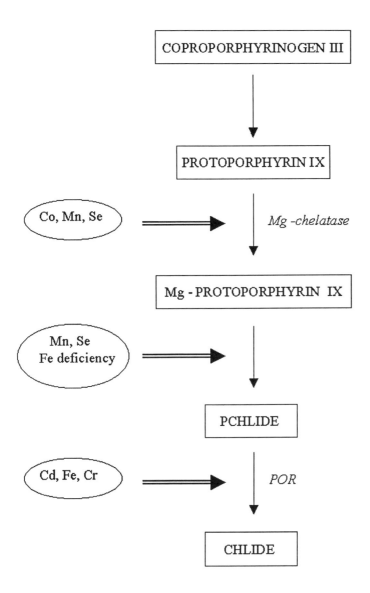

Figure 2. The influence of metal ions on chl biosynthesis. Part 2: Steps from coproporphyrinogen III to Chl

The first stable product of Pchlide photoreduction is Chlide with the absorption maximum at 678 nm and fluorescence maximum at 690 nm (Schoefs, 1999). The spectral parameters of newly formed Chlide undergo some changes; the absorption maximum is shifted towards red and subsequently it exhibits a blue shift, known as a Shibata shift. In greening etiolated leaves, the Shibata shift correlates with the Chlide esterification (Sundqvist and Dahlin, 1997).

The Chlide esterification, is catalysed by the enzyme chlorophyll synthase (reviewed by Rüdiger, 1992 and 1993). This reaction does not change the spectral properties of the pigment but it increases the lipophilic character of the product. Both Chlide a and Chlide b can be converted by the enzyme. Chlide can be esterified with geranylgeranyl diphosphate, which dominates in etiolated plants or with phytyl diphosphate, observed mainly in green plants.

The isoprenoid part of Chl molecule is synthesized in the pathway that starts with isopentenyl diphosphate (C-5 compound) which is condensed with its isomer dimethylallyl diphosphate to geranyl diphosphate (C-10 compound) (Rüdiger, 1997 b). Further condensations lead to farnesyl diphosphate (C-15 compound), and to geranylgeranyl diphosphate (C-20 compound). These reactions are catalysed by geranylgeranyl diphosphate synthase that is a plastid stromal enzyme. It is probably loosely bound to the envelope membrane and catalyses geranylgeranyl diphosphate production from isopentenyl diphosphate without releasing any of the intermediates. Hydrogenation of geranylgeranyl to phytol can occur at two different stages, either before or after Chl esterification and it is catalysed by the hydrogenase which uses NADPH as the hydrogen donor (Rüdiger, 1993).

3. THE INFLUENCE OF METALS ON CHLOROPHYLL ACCUMULATION AND CONTENT

A chlorosis, which is the result of lowered Chl content in leaves, is often observed as the most noticeable symptom of metal treatment of plants. Imai and Siegel (1973) reported eight elements that reduced greening significantly and ordered them as Co>Ni>Cd>others.

An indication that metals can really interfere with pigments and influence the Chl content in plants are heavy metal substituted Chls that were found in some plants during heavy metal stress. The substitution of Mg^{2+} ion in Chl molecules by metal ions such as Cu^{2+}, Zn^{2+}, Cd^{2+}, Hg^{2+}, Pb^{2+} or Ni^{2+} that was observed in water plants was a reason for the breakdown of photosynthesis (Küpper et al., 1996 and 1998). Similar results were described by Kowalewska and Hoffmann (1989) who found Cu substituted Chl in cultures of blue-green alga cultivated in a medium of high Cu concentration and Fe deficiency. An almost proportional relationship between metal toxicity (at a constant concentration) and the rate of Mg substitution in Chl ring were found and the magnitude of this toxic effect could be ordered: $Hg^{2+} > Cu^{2+} > Cd^{2+} > Zn^{2+} > Ni^{2+} > Pb^{2+}$ (Küpper et al., 1996). It was also shown for Lemna trisulca L. that Cu^{2+} produced toxic symptoms at concentrations 1000 fold lower than Cd^{2+} did (Prasad et al., -submitted). Up to 10 mM Cd^{2+} did not induce significant changes in photosynthetic pigment concentration after 48 h treatment, whereas 25 and 50 µM Cu^{2+} induced 50% decrease of Chl content in relation to the untreated plants. For *Chlamydomonas reinhardtii*, 50 and 100 µM concentration of Cu^{2+} was also shown to reduce the Chl content stronger than Cd^{2+} in the same concentration (Prasad et al., 1998).

In older papers, only the effect of metal on total Chl content and accumulation was described, as it was impossible for technical reason to find out which step in Chl biosynthesis was affected. Such a general approach is still present in some ecophysiologically oriented publications. The present state of knowledge about the influence of the selected metals on the total Chl accumulation and the current status of work on the influence of metals on defined steps in Chl biosynthesis is described in various sections in this chapter.

4.1 Cd^{2+} effect

Cadmium is an element that is not necessary as the component of plant mineral nutrition. On the other hand, when present in the soil, it is easily taken up by the roots and transported toward the leaves (Siedlecka and Krupa, 1997). The decrease of Chl content was one of the early reported symptoms of cadmium toxicity (Agarwala et al., 1965). The inhibition of Chl biosynthesis was even suggested to be a primary event in the Cd^{2+} toxicity as compared to the inhibition of photosynthesis (Baszyński et al., 1980). Cd^{2+} was found to reduce the Chl content in cress, radish and lettuce (Czuba and Ormrod, 1973; Naguib et al., 1982), red kidney bean (Imai and Siegel, 1973), rice, wheat and barley (Naguib et al., 1982; Malik et al., 1992; Vassilev et al., 1998), pea (Chugh and Sawhney, 1999) and sunflower (Di Cagno et al., 1999) and many others species that are mentioned below.

The extent of Chl decrease caused by addition of Cd^{2+} to the nutrient solution depends on plant species as well as on the experimental conditions that were applied. It was reported that long-term (2 weeks) treatment of 4-week-old tomato with 20 µM Cd^{2+} resulted in 50% decrease of total Chl content in comparison to the untreated leaves (Baszyński et al., 1980). In another study, Cd^{2+} was found to reduce the total Chl content by about 30% in the second leaves of 3-week-old cucumber growing in the presence of 5 µM Cd^{2+}, whereas 2.5 µM Cd^{2+} did not affect the Chl accumulation (Kłobus and Buczek, 1985). Schlegel et al. (1987) found the decrease of Chl content in both primary leaves and cotyledons of spruce seedlings after a five-week treatment with 1-15 µM Cd^{2+}. On the other hand, 5 µM concentration of Cd^{2+} did not cause any significant differences in Chl content per leaf area comparing to the control in 4-weeks old sugar been seedlings (Greger and Ögren, 1991). The increase of Cd^{2+} concentration to 20 µM caused strong (75%) inhibition of Chl biosynthesis (Greger and Ögren, 1991). The dose dependent decrease of both Chl a and b content was also described for 5-week old sunflower treated with 10-40 µM Cd^{2+} during 2 weeks (Gadallah, 1995). The amount of Chl a was 72-57% of the control at 5-40 µM Cd^{2+}, and that of Chl b was 88-57%, respectively. Significant decrease (35%) of Chl content in Cd-treated maize was observed only for 100 and 250 µM Cd^{2+} concentration (Ferretti et al., 1993). In 40-day old wheat treated with 5 and 10 mM Cd^{2+} for 2-4 days, the decrease of total Chl content in third leaf was approximately two times higher than that observed in the first leaf (Bishnoi et al., 1999). Further decrease of Chl content was found in the fourth leaf. The observed decrease of Chl content was proportional to the accumulation of Cd^{2+} in leaves. Correlation between decrease of the Chl content in barley and the increase of internal Cd^{2+} concentration was shown by Stiborova et al. (1986).

The inhibition of Chl biosynthesis in the presence of Cd^{2+} (0-1 mM) was shown to be independent on the availability of inorganic nitrogen and it was observed both in the presence and absence of nitrogen (Parekh et al., 1990). Inhibition of Chl accumulation also did not depend on the source of iron in the medium (Fodor et al., 1996). The observed inhibitory effect of Cd^{2+} on Chl content in cucumber was the same in the presence of Fe-EDTA and Fe-citrate.

Several factors were found to reduce or partially reverse the inhibitory effect of cadmium on Chl biosynthesis. Addition of manganese to the Cd^{2+} treated tomato partially reversed the decrease of Chl content, however, the amount of accumulated Chl was never as high as in control plants (Baszyński et al., 1980). The effect of cadmium was greatly reduced in the presence of kinetin (Gadallah, 1995) and uniconazole, a triazole derivative (Thomas and Singh, 1996). Uniconazole treatment resulted in enhanced concentrations of Chl and Pchlide at all levels of Cd^{2+} treatment. The effect of uniconazole was even higher in the presence of kinetin (Thomas and Singh, 1996).

Decrease of Chl accumulation was also observed during greening of fragments of etiolated maize and barley leaves floating in the Cd^{2+} solution for several hours (Parekh et al., 1990; Horvath et al., 1996). Nevertheless, the results obtained for these two species were different. For 1 mM Cd^{2+} almost no Chl accumulation was detected in greening etiolated barley leaves (Horvath et al., 1996), whereas this concentration of Cd^{2+} caused about 80% inhibition in the case of maize (Parekh et al., 1990). In lower Cd^{2+} concentration (0.1–0.01 M) the Chl accumulation in etiolated barley leaves became measurable after 10 h of greening (6h for untreated leaves) and the extent of Chl accumulation during greening was strongly reduced. Finally, Chl amounts reached 50 and 13% of the control for 0.01 and 0.1 M of Cd^{2+}, respectively (Horvath et al., 1996). Similar inhibitory effect of Cd^{2+} on Chl synthesis in greening barley was reported by Stobart (1985), who found the inhibition higher than 60% at 0.1 mM and higher than 90% for 10 mM of Cd^{2+}.

Effect of Cd^{2+} on the Chl content accumulated in leaves was found to strongly depend on the growth stage of plants that were treated with cadmium. Bean seedlings treated with Cd^{2+} at the earliest growth stage or during germination showed higher accumulation of Chl expressed on the basis of leaf area than the same plants treated with metal at the later growth stage (Skórzyńska-Polit and Baszyński, 1995; Shaw and Rout, 1998). It was earlier observed by Barcelo (1988) that the first leaves of bush bean plants treated with Cd^{2+} were more green and had a higher Chl content than secondary leaves. Padmaja et al. (1990) also observed an increase in Chl content up to fourth day of germination in Cd^{2+} treated samples, which was similar to the control. Nevertheless, the maximum amount of accumulated Chl depended on the Cd^{2+} concentration within the range 10-50 μM and it was never as high as in untreated samples. Thus, mature leaves having higher organisation level of inner membranes are more sensitive to Cd^{2+} treatment than young developing seedlings.

Mechanisms of chlorosis development caused by the long-term exposure of plants to cadmium are not clear. This may be caused by the indirect effect connected with a cadmium induced Fe and Mg deficiency (Greger and Lindberg, 1987; Siedlecka and Krupa, 1997) or by the direct influence of Cd^{2+} on enzyme(s) of Chl biosynthesis (see section 2.3). It cannot be also excluded that besides the Cd^{2+}-induced Fe deficiency, Chl content in Cd-treated leaves may be lowered due to the accelerated ageing of photosynthetic apparatus (Skórzyńska et al., 1991).

4.2 Cu excess and deficiency

Copper is an essential plant nutrient that is the indispensable component of oxidative enzymes or of particular structural components of cells (reviewed by Maksymiec, 1997). The first visual symptom of Cu^{2+} toxicity is the inhibition of plant growth and chlorosis of leaves (Foy et al., 1978). However, the results concerning the effect of copper on the amount of Chl in leaves differ in the works of different authors.

The decrease of Chl content as the effect of toxic amount of Cu^{2+} was reported for barley, spinach and rice (Agarwala et al., 1977; Foy et al., 1978; Baszyński et al., 1988; Lidon and Henriques, 1991). A 1 mM Cu^{2+} was found to be a minimal concentration that lowered the Chl accumulation in pea leaves growing under photoperiod (Angelov et al., 1993). These results do not fully confirm those of Stiborova et al. (1987) obtained for barley where the decrease of Chl content both for 1 and 0.1 mM Cu^{2+} was observed. Chlorosis of leaves associated with simultaneous destruction of the inner structure of chloroplasts was also observed during a long exposure to Cu^{2+} (throughout the vegetation period) (Eleftheriou and Karataglis, 1989).

On the contrary, an increase in total Chl content was observed in Cu^{2+} tolerant spinach growing in excess of Cu^{2+} (Baszyński et al., 1982). Lolkema and Vooijs (1986) showed that long (30 day) exposure of *Silene cucubalus* to Cu^{2+} resulted in a decrease of Chl content by three times in Cu^{2+} sensitive plants while in Cu^{2+} tolerant plants no changes in Chl level was detected. Beside a specific plant tolerance to Cu^{2+} excess, the effect of Cu^{2+} toxicity strongly depends on the growth stage of plants used in experiments. Excess of Cu^{2+} applied to bean plants at the stationary growth stage of primary leaves resulted in decrease of Chl content, while application of Cu^{2+} at the early growth stage caused the increase of Chl content calculated per leaf area and strong inhibition of leaf expansion (Maksymiec et al., 1994, 1995; Maksymiec and Baszyński, 1996).

Cook et al. (1997) correlated the Chl accumulation in Cu^{2+} treated bean plants (growing under photoperiod) with Cu^{2+} content accumulated in leaves. They observed increasing content of total Chl (expressed per leaf area) with increasing Cu^{2+} concentration in leaves. This increase was more marked for leaf Cu^{2+} concentration greater than 80 mg (Cu)/kg (dry mass). Total Chl concentration in the first and second trifoliate leaves did not show significant variation with increasing Cu^{2+} concentration. In contrast, it was found by Stiborova et al. (1986) that the content of Chl $(a+b)$ in barley was lower for higher internal heavy metal concentration; the inhibition was of 23% for 41 µM internal Cu^{2+} concentration.

Cu deficiency was shown to decrease pigment synthesis and to affect the development of the photosynthetic apparatus (Maksymiec, 1997). The decrease of total Chl content by more than 50% was reported in chloroplasts of Cu-deficient sugar beet (Henriques 1989; Droppa et al. 1984 a, b) as well as in oat and spinach (Baszyński et al., 1978). Mild deficiency caused lower decrease in Chl content (Droppa et al. 1984 a,b)

4.3 Pb^{2+} effect

An impairment of Chl accumulation was observed in cucumber cotyledons treated with lead during greening (Burzyński, 1985). The Chl content measured after 48 h of greening in the presence of 1 mM or 0.1 mM Pb^{2+} reached about 50 and 60% of the control leaves, respectively. However, further kinetics of Chl accumulation was different for the concentrations of Pb^{2+} used in this experiment. In cotyledons treated with 1 mM Pb^{2+}, the Chl content stayed at the same level, whereas for 0.1 mM Pb^{2+} a slight increase in Chl amount was observed up to 72 h of deetiolation.

In mung bean and bajra seedlings treated with 50-250 μM of Pb^{2+} the lower rate of Chl accumulation was observed (Prasad and Prasad, 1987 a, b). For the highest Pb^{2+} concentration that was applied for three days, the amount of Chl reached about 40% of that for untreated seedlings and it remained constant during further treatment.

The effect of Pb^{2+} on the Chl accumulation in cucumber was enhanced in the presence of Fe in the medium. It was also proved that the inhibition of the Chl accumulation caused by Pb^{2+} was higher in plants supplied with Fe-citrate than Fe-EDTA (Fodor et al., 1996). On the contrary, the type of nitrogen source and even a presence of inorganic nitrogen in the medium did not influence the effect of Pb^{2+} on Chl content (Sengar et al., 1996).

Decrease in Chl ($a+b$) content was also observed in 7-day-old barley seedlings grown under natural irradiance and treated with Pb^{2+} since germination. This decrease strongly depended on the internal Pb^{2+} concentration and reached 30% of the control level for the concentration of 108 μM (Stiborova et al., 1986).

4.4 Ni^{2+} effect

Excess of nickel was found to induce some visible symptoms of phytotoxicity, depressing growth and dry matter of plants (Agarwala et al., 1977). The dose-dependent decrease in Chl content was observed in Ni-treated bean plants, however, the ratio of Chl content to leaf area increased for increasing Ni^{2+} concentration (Krupa et al., 1993). The inhibitory effect of Ni^{2+} on the Chl accumulation was observed also in barley and pigeon pea plants (Agarwala et al., 1977; Sheoran et al., 1990 a,b). One week treatment of 4-week old soybean leaves with Ni^{2+} led to the decrease of both Chl a and Chl b for the concentrations of metal higher than 1 μM (Eleiwa and Naguib, 1986)

In 40-day old wheat treated with 5 or 10 mM Ni^{2+} for 2-4 days the decrease of Chl content in the first leaf was 2-3 times higher than that observed for the third leaf

(Bishnoi et al., 1999). Further decrease in relative Chl content was found in the fourth leaf. Results obtained by Bishnoi et al. (1999) showed also that the effect of Ni^{2+} was opposite to Cd^{2+} effect (described in 2.2.1). The Chl content in the top leaf was affected more than in the lower leaf in plant treated with Ni^{2+}, whereas the deleterious effect of Cd^{2+} was more pronounced in the third leaf. However, for both metals, the observed decrease of Chl content was proportional to the accumulation of metals in leaves.

The effect of Ni^{2+} in the concentrations between 1 µM-0.1 M on Chl accumulation during greening of etiolated barley seedlings was investigated by Shalygo et al. (1999). At low Ni^{2+} concentrations, the strong increase of total Chl content in relation to the untreated samples was observed, having maximum for 10 µM concentrations. This stimulatory effect was replaced by progressively drastic inhibition for higher metal concentrations. Finally, leaves treated with 0.1 M Ni^{2+} contained no Chl at all.

4.5. Fe excess and deficiency

It has been shown that iron deficiency in higher plants resulted in chlorosis (De Kock et al., 1960; Agarwala et al., 1977). Moreover, Fe effect was mainly visible in chloroplast, while other Fe-containing organelles were unaffected (Plat-Aloia, 1983). Fe deficiency was related to the decrease of Chls amount, both Chl a and Chl b (Marsh et al., 1963; Terry, 1980). The chlorosis was easily reversible upon Fe resupply (Plat-Aloia, 1983). Fodor et al. (1995) showed that Fe deficiency caused the gradient of Chl content between the first (the oldest) and the third leaf (between 320 and 125 µg/g fresh weight) of cucumber resembling the gradient in Fe content in these leaves. The effect of Fe^{2+} in the broad concentration range, between 1 µM and 0.1 M, on Chl accumulation during greening of etiolated barley seedlings was investigated by Shalygo et al. (1999). The increase of total Chl content for low metal concentrations and a drastic inhibition of its biosynthesis for higher metal concentrations were observed in relation to the untreated plants. The highest Fe^{2+} concentration used suppressed the Chl amount by 94%.

4.6 Hg^{2+} effect

The effect of mercury on Chl accumulation was investigated in bean that was exposed to Hg^{2+} (20-25 µM) during germination or during subsequent developmental stage. The Chl content in leaves increased when Hg^{2+} was applied during germination stage and at the very early stage of seedling development and it remained unchanged when more developed seedlings were treated (Shaw and Rout, 1998).

In spruce seedlings treated with micromolar concentrations of Hg^{2+} for five weeks a decrease in Chl content was observed in both primary leaves and cotyledons (Schlegel et al., 1987).

Lower Chl content was also found in mung bean and bajra seedlings treated with 50-250 µM of Hg^{2+} (Prasad and Prasad, 1987 a, b). The increase of Chl content was observed only for first three days of metal treatment. For 250 μM of Hg^{2+} the maximal Chl amount reached about 40% of those for untreated seedlings.

4.7 Zn^{2+} effect

The effect of zinc on total Chl content was investigated for barley leaves that were grown under natural irradiation in the presence of this metal for 7 days (Stiborova et al., 1986). A slight increase of total Chl content was observed for 1 μM and 1 mM Zn^{2+} concentration, whereas a stronger increase, about 110% of the amount in the untreated leaves, was found for 10 and 100 μM Zn^{2+} concentration in the medium. Agarwala et al. (1977) did not observe any effect on Chl accumulation during one-week treatment of barley with the excess of Zn^{2+}. On the contrary, the decrease of Chl content was found in spruce seedlings after five-week treatment with 30-100 μM Zn^{2+} (Schlegel et al., 1987). The level of Chl decreased with increasing Zn^{2+} concentration in the nutrient solution both for primary leaves and cotyledons.

4.8 Other metals

In this section, we summarise the state of knowledge on chlorophyll accumulation under the influence of other metals than described in the previous chapters.

Shalygo et al. (1999) investigated the effect of cobalt and manganese (located in the same period of the periodic table) in the concentrations between 1 μM and 0.1 M on Chl accumulation during greening of etiolated barley seedlings. Low metal concentrations resulted in the increase of total Chl content in relation to the untreated plants. This increase was higher for Mn^{2+} (111% of Chl in the untreated plants) and almost negligible for Co^{2+}. Progressively drastic inhibition of Chl biosynthesis was observed for higher metal concentrations. Leaves treated with 0.1 M Mn^{2+} or Co^{2+} contained no Chl at all. In another study, Chl synthesis was shown to be stimulated most strongly by K^+, Rb^+, Ni^+ and Zn^{2+}, whereas Cd^{2+} and Mn^{2+} were the main inhibitors (Shalygo and Averina, 1996). The decrease of Chl accumulation in barley caused by Co^{2+} and no effect for Mn^{2+} after a one-week treatment with 1 mM metal concentration was reported earlier by Agarwala et al. (1977). Chatterjee and Chatterjee (2000) compared the toxicity of 0.5 mM cobalt, copper and chromium on Chl content in cauliflower and ordered these metals according to their toxicity as Co>Cu>Cr. What is more, they correlated the toxicity of metals with their translocation efficiency from roots to tops that was the highest for Co and the lowest for Cr. In plants of Nymphaea alba L. growing in the presence of 1-200 μM Cr^{6+}, the Chl content was lowered in relation to the untreated plants (Vajpayee et al., 2000). Mukhopadhyay and Aery (2000) found that Cr^{6+} influenced the Chl content in wheat stronger than Cr^{3+}.

Cesium was found to decrease the Chl accumulation during greening of etiolated barley leaves (Shalygo et al., 1997a). Chl content in 7-day old leaves treated with 10 mM Cs^+ reached 88% and 53% of those in control leaves after 6 and 8 h of greening, respectively.

Vanadium (0.1 μM) was found to hardly affect the Chl accumulation in soybean leaves but it became an inhibitory factor for higher concentrations (Eleiwa and Naguib,

1986). In contrast, strontium (0.1 -100 µM) was reported to stimulate the Chl accumulation in the soybean leaves in the same study (Eleiwa and Naguib, 1986).

Selenium in the concentrations from 12 to 62 µM that was applied to germinating mung bean resulted in lower accumulation of Chl, however, the kinetic of Chl accumulation was similar to that of untreated seedlings (Padmaja et al., 1989 and 1995).

Aluminium (0.5–1 mM) lowered the total Chl content in soybean leaves growing under 12 h photoperiod (300 µE) (Milivojević et al., 2000). In this study the amount of accumulated Chl was lower for higher Al concentration. What is more, the effect of Al was enhanced by phosphorus deficiency.

5. SUMMARY

The effect of metals on Chl content and accumulation has been most intensively studied for cadmium and copper. Investigations were performed for variety of plant species that were treated with metal ions in a broad range of concentration. Metal ions were mainly supplied as a component of the growth medium. Even if the same plant species were used, the growth conditions as well as the duration of metal treatment differed. That is why it is difficult to compare the results obtained by different authors, e.g. it is difficult to determine a universal concentration of a metal producing toxic effect on Chl accumulation. In general, the decrease of Chl content was observed as a result of metal treatment. Young, developing seedlings are less sensitive to the cadmium and copper treatment that well developed plant. In that case an increase of Chl accumulation was observed after metal treatment.

Correlation between the increase of internal metal concentration and the decrease of the Chl content was observed in Cd-treated barley. In contrast, the opposite effect was found in Cu-treated bean plants: the increasing internal Cu concentration correlated with the increase of Chl content. In case of biometals such as copper and iron the deficiency of metal results in decrease of Chl content.

The comparison of metal influence on Chl accumulation can only be undertaken for plants growing and treated with metals in standarized way. It was found that Cu^{2+} and Cd^{2+} in the concentration between 1µM and 1 mM caused higher decrease of Chl content than Pb^{2+} in barley (Stiborova et al., 1986). In another study, Co^{2+} and Ni^{2+} decreased the Chl content in barley stronger than Fe^{3+} (Agarwala et al., 1977). Zn^{2+} and Mn^{2+} did not influence the Chl content. In the study of Shalygo et al. (1999), low concentrations of Mn^{2+}, Co^{2+}, Ni^{2+}, and Fe^{2+} (10^{-6}-10^{-4} M) stimulated Chl accumulation. The highest stimulation for Ni^{2+} and the lowest for Co^{2+} was observed. On the other hand, higher metal concentrations (10^{-3}-10^{-1} M) resulted in the decrease of Chl content.

6. THE EFFECT OF METAL TREATMENT ON DIFFERENT ENZYMATIC STEPS IN CHLOROPHYLL BIOSYNTHESIS

The effects of metals on Chl accumulation described in the previous section can be mediated by various, largely unknown mechanisms. A number of publications in search of these mechanisms focus on the effect of metals on various enzymatic steps in Chl biosynthetic pathway. In this section, we summarise the present-day knowledge about the effects of metals on the activity of selected enzymes engaged in Chl synthesis.

6.1 ALA synthesis

As it was said in the section 2.1 ALA is the first common precursor of all terapyrroles. Control of ALA level was used in some reports to observe the effect of metal ions on Chl biosynthesis.
Cadmium (10-50 μM) inhibited the synthesis of ALA in etiolated bean seedlings (Padmaja et al., 1990). It also inhibited the ALA production in leaves exposed to light, particularly at concentrations higher than 100 μM (Stobart et al., 1985). This effect was suggested as being connected with Cd^{2+} interaction with active thiol groups in ALA synthase.

Burzyński (1985) found the inhibition of ALA synthesis by lead in greening cucumber. In this study, the negative influence of Pb^{2+} was noticed as early as 6 h after deetiolation. During next 42 h of greening, the ability of ALA synthesis increased both in Pb-treated and in control plants, however, for Pb-treated plants the ALA content was reduced by 50 and 80% for 0.1 mM and 1 mM Pb^{2+}, respectively. After 72 h of growth under light, the ability of ALA synthesis decreased both in the control seedlings and those treated with lower Pb^{2+} concentration, but the negative influence of Pb^{2+} persisted.

Influence of metals on ALA synthesis was also proved in experiments with metal treatment of plants in the presence of some Chl precursors. Glutamate, 2-oxoglutarate and magnesium chloride partially reversed the inhibitory effects of Cd^{2+} and Pb^{2+} in greening maize and pea, whereas glycine and succinate did not influence Cd^{2+} and Pb^{2+} treatment (Parekh et al., 1990; Sengar and Padney, 1996). This observation led to the conclusion that Chl biosynthesis was inhibited at the level of availability of 2-oxoglutarate and glutamate for ALA synthesis. These experiments also revealed the complete reversion of the effect of Cd^{2+} or Pb^{2+} by GSH (reduced glutathione) whereas other thiol compounds like cysteine, DTNB (5,5'dithio-bis-2nitrobenzoic acid) and DTT (ditiothreitol) had no effect. GSH is known as the precursor of phytochelatin synthesis. Phytochelatins are synthesized in response to heavy metals stress and they are used for the inactivation of metal ions by their complexing. Next, it was also shown that KH_2PO_4, $CaCl_2$ and KCl partially reversed the inhibitory effect of Pb, while NH_4Cl has no influence (Sengar and Padney, 1996).

ALA content was also monitored in greening seedlings of barley treated with Fe^{2+}, Co^{2+}, Ni^{2+} and Mn^{2+} in the broad range of concentration between 1 μM and 0.1 M (Shalygo et al., 1999). An increase in ALA content was observed for low concentration of these metals. The highest effect was found for 1 μM Ni^{2+} and to the lower extent for 1 μM Co^{2+} as well as for Fe^{2+} and Mn^{2+} (both 1 mM). On the contrary, in treatments with 0.1 M Mn^{2+}, Co^{2+}, and Ni^{2+} the inhibition of ALA synthesis was 18, 33, and 48%, respectively. The effect of 0.1 M Fe was much higher and it accounted for 84% inhibition of ALA synthesis. It was also shown in this study that the investigated metals did not disturb the synthesis of GSA (glutamate-1-semialdehyde aminotransferase), which is one of the enzymes in the ALA synthesis.

Iron deficiency was also found to be a reason for the indirect inhibition of ALA synthesis in higher plant. This deficiency resulted in Mg-protoporphyrin monomethyl ester accumulation, which tended to inhibit the formation of ALA and thus stopped its own synthesis by a feedback inhibition (Spiller et al., 1982; Chereskin and Castelfranco, 1982). In some reports, however, no effects of metals on ALA synthesis were reported. ALA content was unchanged in mung bean and bajra growing in the presence of 50-250 µM of Pb^{2+} or Hg^{2+} or in the presence of 12-62 µM Se^{4+} (Prasad and Prasad, 1987a, b; Padmaja et al., 1989).

6.2 ALA dehydratase activity

ALA dehydratase (ALAD, Porphobilinogen synthase, E.C. 4.2.1.24) catalyses the conversion of ALA to porphobilinogen (see section 2.1). It was shown to be a metal sensitive enzyme requiring Mg^{2+} or Zn^{2+} for its activity (reviewed by Beale, 1999). Thus, in the number of papers, this enzyme was studied as a possible place of metal action in Chl biosynthesis pathway.

The toxic effect of mercury, selenium and lead on the activity of ALAD was demonstrated for mung bean and bajra growing under natural daylight (Prasad and Prasad, 1987 a, b; Padmaja et al., 1989). In these studies, ALAD activity increased with plant growth up to fourth or third day and then it decreased both in metal treated and in control plants. Nevertheless, the maximal activity of ALAD in metal-treated samples was dose dependent and it reached 50% of control samples for 250 µM of Pb^{2+} (or Hg^{2+}) and 62 µM Se^{4+}. The possible mechanism of ALAD inhibition was suggested as being due to the interaction of the heavy metal with SH groups of the enzyme at its active sites. Moreover, the ALAD activity pattern resembled that of Chl accumulation: an increase up to the fourth day and saturation following the fifth day. This supports the view that ALAD is a regulatory enzyme in Chl biosynthesis.

Similar results were obtained for Pb-treated cucumber cotyledons that were grown in darkness, then incubated with lead and finally illuminated in the presence of Pb^{2+} (Burzyński, 1985). In this study, ALAD activity increased during 48 h of greening then started to decrease both in the control and in Pb-treated samples. However, the maximal ALAD activity measured in the presence of 1 mM Pb was reduced by 40% in comparison to the untreated cotyledons.

In cadmium treated (10–50 µM) mung bean growing under natural irradiation ALAD activity also increased with age up to third day both in the control and in Cd-treated samples and then decreased. In this study, the dose dependent inhibition of the activity was found up to 46% of the maximum for 50 µM Cd^{2+} (Padmaja et al., 1990).

Results obtained from the investigation of the inhibitory effect of metals on ALAD activity *in vivo* were confirmed by study of the metal-induced inhibition of the enzyme *in vitro* (Prasad and Prasad, 1987 a, b; Padmaja et al., 1989 and 1990). It was also shown that the activity of ALAD was higher for all Cd^{2+}, Hg^{2+} or Pb^{2+} concentrations in the presence of dithiothreitol (Prasad and Prasad, 1987 b; Padmaja et al., 1990). This confirms that the inhibition of ALAD by these metals both *in vitro* and *in vivo* may be due to the interaction with free SH groups of this enzyme. ALAD was already found to be a metal sensitive enzyme, which requires a thiol group at the substrate binding site (Nandi and Shemin, 1968).

In another study, ALAD activity measured during greening of etiolated barley leaves was found to depend strongly on the concentration of metal. At low concentration of Co^{2+}, Ni^{2+}, Mn^{2+} and Fe^{2+} ALAD activity was stimulated, whereas a higher metal concentration resulted in progressive inhibition (Shalygo et al., 1999). The highest stimulation was observed for 1-100 µM of Fe^{2+} and Mn^{2+} and for 1 mM Co^{2+} and Ni^{2+}. At 0.1 M concentration, ALAD activity was fully suppressed by Co^{2+} and Fe^{2+} and diminished for Mn^{2+} and Ni^{2+} to 6 and 18% of the control, respectively. In enzyme preparation, the inhibition of ALAD activity was observed for all metal concentrations. By use of the antibodies against ALAD, it was found that the metal-induced effects were due to changes in enzyme activity rather than changes in the amount of ALAD for Co^{2+}, Ni^{2+} and Mn^{2+}. On the contrary, the inhibitory effect of Fe^{2+} (0.01-0.1 M) can be caused by the influence of Fe cation on the efficiency of ALAD synthesis.

6.3 Uroporphyrinogen III synthesis

Porphobilinogenase (PBGase) is a complex of two enzymes porphobilinogen deaminase and uroporphyrinogen III synthase that converts porphobilinogen to uroporphyrinogen III. The influence of Mn^{2+}, Fe^{2+}, Co^{2+} and Ni^{2+} on this transformation was investigated by Shalygo et al. (1999). In their study, etiolated barley leaves were incubated in solutions of metal salts. The activity of PBGase was stimulated by 1-100 µM of Mn^{2+}, Fe^{2+}, Co^{2+} and Ni^{2+} although the stimulation was less evident than in the case of ALAD (described in the previous section). On the contrary, the higher doses of metal cations were more inhibitory for PBGase than for ALAD. The most toxic effect was found in the case of Fe^{2+}, especially for 0.01 M concentration for which the total inhibition of PBGase was observed. The weakest effect was found for Ni^{2+}. When the activity of PBGase was checked in enzyme preparation no activation at low metal concentrations was found but, on the contrary, the inhibitory effect was clearly seen. This may be due to the relatively higher concentration and easy access of the metal to the enzyme (no barrier). It may also be that stimulation of this enzyme activity *in vivo* is mediated by the effect of metal on some other processes (including control mechanisms) related to the Chl biosynthesis.

The inhibition of uroporphyrinogen III synthesis by Mn^{2+} was also observed in cyanobacteria (*Anacystis nidulans*) culture (Csatorday et al., 1984). In that case, it was demonstrated as an increase of phycobilin/Chl ratio.

7. CONVERSION OF UROPORPHYRINOGEN III TO COPROPORPHYRINOGEN III

Uroporphyrinogen III decarboxylase (UROD, E.C. 4.1.1.37) was shown to be influenced by cesium (Shalygo et al., 1997a, 1998). Incubation of greening barley with

10 mM Cs^+ lead to the accumulation of excessive amounts of uroporphyrinogen III and to a minor extent of heptacarboxylporphyrinogen III, which is the primary product in the pathway leading from uroporphyrinogen III to coproporphyrinogen III. This was correlated with the drastic decrease of UROD activity.

8. MAGNESIUM PROTOPORPHYRIN SYNTHESIS

Magnesium insertion into protoporphyrin IX is a specific reaction for Chl biosynthesis, which is catalysed by Mg-chelatase. The inhibition of this reaction caused by Co^{2+} (1.7 µM) was observed in cyanobacterium, *Anacystis nidulans*, culture (Csatorday et al., 1984). As a result of Co^{2+} treatment, protoporphyrinogen IX was accumulated. In that case, Chl accumulation was delayed and less efficient. Higher Co^{2+} concentration caused cell growth retardation and prevention of protoporphyrinogen IX transformation (Csatorday et al., 1984).

Different results were obtained for a higher plant treated with cadmium. In etiolated mung beans leaves that were irradiated for 7 h, Cd^{2+} treatment (10-50 µM) did not affect synthesis of protoporphyrin IX as well as its further transformation to Mg-protoporphyrin monomethyl ester and Pchlide (Padmaja et al., 1990).

On the other hand, Mn^{2+} inhibited Chl biosynthesis and enhanced the accumulation of protoporphyrin IX and Mg-protoporphyrin IX monomethyl ester in greening leaves of etiolated barley (Shalygo et al., 1997 b). The influence of Mn^{2+} was fully reversible by Fe^{2+} cations. Similar inhibition of Chl biosynthesis was observed when treated mung bean seedlings with 12–62 µM of Se^{4+} (Padmaja et al., 1995). However, in this study Pchlide was also accumulated besides protoporphyrin IX and Mg-protoporphyrin IX.

9. CONVERSION OF MAGNESIUM-PROTOPORPHYRIN TO PCHLIDE

The accumulation of Mg-protoporphyrin monomethyl ester was observed by Spiller et al. (1982) in iron-deficient corn, bean, barley, spinach and sugarbeet. The amount of accumulated Mg-protoporphyrin monomethyl ester was up to five times higher in Fe-deficient plants in comparison with Fe-supplemented plants. This accumulation was not observed in case of deficiency of other nutrients like magnesium, sulphur, nitrogen. ALA treatment of Fe-deficient corn and bean leaf tissue caused the increase of the amount of accumulated Mg-protoporphyrin monomethyl ester and did not result in Pchlide or Chl production. In contrast, ALA treatment of control plants led to the accumulation of large amount of Pchlide. Similar accumulation was observed for *Potamogeton pectinatus* and *Potamogeton nodosus* growing in anaerobic conditions without metal addition (Spiller et al., 1982).

Besides growing plants on Fe-deficient media, Fe-deficiency can also be induced by chelating the endogenous Fe with reagent such as α,α'-dipyridyl, o-phenantroline or 8-hydroxyquinoline. In that case, an increase in the ratio of Mg-protoporphyrin monomethyl ester to Pchlide or the accumulation of other intermediates of the Mg-branch of tetrapyrrole biosynthesis instead of Chl was observed (Granick,

1961; Duggan and Gassman, 1974; Ryberg and Sundqvist, 1976; Vlcek and Gassman, 1979).

10. PROTOCHLOROPHYLLIDE TO CHLOROPHYLLIDE PHOTOREDUCTION

Until now the influence of metals on this step of Chl biosynthesis was only investigated for plants belonging to angiosperms, in which the Pchlide to Chlide reduction is a light-triggered reaction (see section 1.2). Cadmium is the only metal for which the results of such investigations were published.

The incubation of 8-day old etiolated wheat seedlings with Cd^{2+} (1–100 µM) caused the changes in the 77 K fluorescence emission spectra (Böddi et al., 1995). The spectra of untreated leaves as well as of plant treated with Cd^{2+} concentrations lower than 0.1 mM have two separated maxima: the main one at 655-656 nm and another at 633 nm. The higher Cd^{2+} concentration was applied the bigger amplitude of 633-nm peak was observed. The other peak decreased in intensity and shifted to 653-654 nm. However, the authors could not find any straight correlation between the amplitudes and positions of the emission bands and Cd^{2+} concentrations. Photoreduction of Pchlide was observed as long as the peak at 654 nm was present in the spectra, however, the fluorescence maximum of newly formed Chlide was placed at 690 nm, while it was at 692 nm for the control or for the samples treated with lower Cd^{2+} concentrations. Similar transformation of the photoactive Pchlide form to the non-photoactive as the effect of treatment of etiolated barley seedlings with 1 mM cadmium was observed earlier by Stobart et al. (1985) in the absorption spectra. This observation correlated with the relative fluorescence intensity of newly formed Chlide after illumination, which was lower for Cd-treated leaves than for control leaves indicating lower Chlide amount in the presence of Cd^{2+}. The Cd-induced transformation of the photoactive Pchlide form to the non-photoactive was also confirmed by estimation of the amount of free and photoconvertible Pchlide in acetone extract from leaves before and after illumination. It appeared that the amount of free Pchlide increased whereas those of photoconvertible form decreased (Stobart et al., 1985).

The effect of Cd^{2+} on the photoactive Pchlide: POR: NADPH complexes was more clear in the EPIM suspension. Treatment with 1 mM Cd^{2+} for 10 min. caused similarly strong effect as did the 10 mM Cd^{2+} treatment of intact leaves for 12 h (Böddi et al., 1995). The gaussian resolution of the fluorescence emission spectrum measured for the 440 nm excitation revealed that the increase of the amplitude of 633 nm peak was due to concomitant increase of the 633 form and the appearance of a band at 641.5 nm (having the vibrational band at 707.5 nm). Parameters of other gaussian components were also changed in comparison to the control (Böddi et al., 1992). Two gaussian components in the Chlide spectral region having maximum at 678 and 693 were distinguished in the spectra measured after irradiation. Increase in Cd^{2+} concentration resulted in decrease of intensity and blue shift of these components. Similar spectral changes to those observed in the presence of Cd^{2+}, in particular the appearance of 641.5-nm Pchlide and 678-nm Chlide forms, have already been observed for Pchlide: POR: $NADP^+$ complexes (El Hamouri et al., 1981; Franck et al., 1999) or in case of changes in the PLB structure caused by the lack of NADPH (Ryberg and Sundqvist, 1988).

The POR activity measured in etioplast suspension was almost completely inhibited after a 30 min incubation with 1 mM Cd^{2+}, whereas 97% of the starting activity was still observed in the control (Stobart et al., 1985). It was also found that the enzyme was protected from the inhibition by its substrates Pchlide and NADPH. NADPH added to the suspension of etioplast inner membranes prevented the spectral changes caused by Cd^{2+} when the molar concentration of NADPH was the same or higher than that of Cd^{2+} (Böddi et al., 1995; Berska et al., – submitted). The relative amplitude of 633 nm peak in Cd-treated sample was even smaller than in the control. Other reagents, like glucose-6-phosphate, AMP and NADH did not show any protection of photoactive Pchlide:POR complexes (Berska et al., – submitted).

Another effect of Cd^{2+} treatment on etiolated seedlings is the slower regeneration rate of photoconvertible Pchlide. From *in vivo* spectroscopy of leaves, it appeared that in control leaves 50% of regeneration occurred in 30-40 min, whereas in the presence of 10–100 μM of Cd^{2+} this required 70-100 min. At higher Cd^{2+} concentrations little or no regeneration was observed.

On the contrary, in another study, low temperature fluorescence emission spectra measured for Cd-treated barley showed that neither the synthesis nor the photoconversion of Pchlide was inhibited. However, the blue shift of the main fluorescence emission of Chlide from 685 to 668 nm was found (Horvath et al., 1996). What is more, in this study, Cd^{2+} did not inhibit POR activity if an appropriate amount of substrate was present.

The stability of the photoactive Pchlide: POR: NADPH complexes was also tested in the EPIM suspension treated with Cr^{6+} or Fe^{3+} (Berska et al., - submitted). The fluorescence intensity of the peak at 655 nm decreased with concomitant increase of that at 633 nm for 10 mM concentration of metal ions. These spectral changes were faster for Cr^{6+} than for Fe^{3+}. However, they were slower than in case of Cd^{2+} treatment.

11. SUMMARY

The current state of knowledge about the influence of metal ions on different steps in Chl biosynthesis presented in the previous chapters shows that experiments are still required to make this information completed. Steps of Chl biosynthetic pathway known presently as influenced by ions of metals are shown in Fig 1.

Up to now, the ALA dehydratation that is catalysed by ALAD is the most intensively studied step. This enzyme appeared to be the most susceptible as it is inhibited by a great number of metal ions like Hg^{2+}, Pb^{2+}, Se^{4+}, Cd^{2+}, Fe^{2+}, Mn^{2+}, Co^{2+}, and Ni^{2+}.

Cadmium was shown to interact with Chl biosynthesis at steps of ALA synthesis and dehydratase as well as the photoreduction of Pchlide to Chlide, however, it did not affect the intermediate steps between porphobilinogen and protoporphyrin IX formation. Synthesis and decarboxylation of uroporpyrinogen III was shown to be inhibited by Mn^{2+}, Fe^{2+}, Co^{2+}, Ni^{2+}, Cs^{+}.

12. ACCUMULATION OF CHLOROPHYLL A AND B

There is no evidence in present literature about a direct influence of metal ions on the process of Chl a to Chl b transformation. Nevertheless, there are some results showing changes in the ratio of Chl a to Chl b in metal treated plants.

An increase in Chl a/b ratio was found in cadmium treated maize and soybean seedlings if the Cd^{2+} concentrations were higher than 200 µM (Ferretti et al., 1993). On the other hand, Gadallah (1995) showed that Cd^{2+} concentration higher than 5 µM caused no significant changes in Chl a/b ratio in sunflower whereas a decrease in this ratio was observed for 5 µM Cd^{2+}. A 15% decrease in Chl a/b ratio was also obtained in the study of long-term treatment of tomato with 20 µM Cd^{2+} (Baszyński et al., 1980).

A number of papers report that both excess and deficiency of Cu caused a slight increase in Chl a/b ratio (Angelov et al., 1993; Henriques, 1989; Droppa et al., 1984 a; Baszyński et al., 1978; Stiborova et al., 1986). However, a decrease in Chl a/b ratio was observed in bean plants exposed to Cu at the stationary growth stage of primary leaves (Maksymiec and Baszyński, 1996). During the growth of Cu tolerant spinach in excess of Cu the Chl a/b ratio was unchanged (Baszyński et al., 1982).

Chl *a/b* ratio remained unchanged in bean leaves treated with 100-200 µM nickel and decreased by about 15% for 500 µM Ni^{2+} (Krupa et al., 1993). This suggests that at 500 µM Ni^{2+} Chl *a* was more susceptible than Chl *b* to the Ni^{2+} toxicity. No significant changes in the Chl *a/b* ratio were found in soybean leaves treated with 0.5 – 1 mM aluminium (Milivojevic et al., 2000).

Stiborova et al. (1986) observed an increase in Chl *a/b* ratio in 7-day old barley seedlings treated with Cu^{2+}, Pb^{2+} and Zn^{2+}. The highest effect was observed for Pb^{2+} ions. The content of Chl *b* was considerably lowered by Cu^{2+} and Pb^{2+} ions, however, the amount of Chl *a* did not decline so much. Zn^{2+} did not affect the Chl *b* accumulation and slightly stimulated the biosynthesis of Chl *a*.

Fe deficiency did not cause any changes in Chl *a/b* ratio in sugar beets, although it was correlated with a decrease in total Chl amount (Terry, 1980*)*.

From the data available in the literature, it is difficult to conclude to what extent the changes in the Chl a/b ratio caused by metal stress are the result of the inhibition of the enzymatic activity converting Chl *a* into Chl *b* and to what extent they derive from different rate of degradation of both chlorophyll species.

13. THE INFLUENCE OF METALS ON CAROTENOID CONTENT

In comparison to the high number of papers dealing with the influence of metals on Chl biosynthesis, the effect of metals on carotenoid accumulation is far less studied.

13.1 Changes in the chlorophyll/carotenoids ratio as a result of metal treatment

Metals influence the content and the rate of accumulation of Chls (Chl a and Chl b) and carotenoids (Car) in a different way. Chls to carotenoids (Chl/Car) ratio is often analysed as the indication of the specificity of metal interaction with these two groups of pigments. Besides the type of metal, several factors influence the Chl/Car ratio. One of them is the growth stage of plants used in experiments. A decrease in Chl/Car ratio was observed in bean exposed to Cu at the stationary growth phase of primary leaves (Maksymiec and Baszyński, 1996). On the contrary, after two-week treatment of four-week old spinach and oat plants, a slow rise in the ratio of Chl/Car in Cu–treated plants as compared to the control were observed (Baszyński et al., 1988). This implies a slightly stronger inhibition of the synthesis of carotenoids than that of Chls.

Nickel applied to etiolated bean during greening increased the Chl/Car ratio from 6.75 for the untreated leaves to 8.27 for 0.5 mM of Ni^{2+} (Krupa et al., 1993). This is the evidence of the stronger inhibition of Car accumulation by this metal.

On the contrary, the decrease in Chl/Car was found in soybean leaves treated with aluminium (0.5-1 mM) (Milivojevic et al., 2000). What is more, a decrease in total carotenoid content was found for higher Al concentration. Unlikely the effect of Al on Chl, this effect does not depend on the presence and concentration of phosphorus in the medium. The observed decrease in carotenoid content was probably a reason for lower level of LHC proteins in Al treated plants.

13.2 Carotenoid content in metal treated plants

The influence of cadmium on carotenoid content has been investigated for different plants growing in different conditions. The decrease of total carotenoids content (calculated per dry weight of shoot system) in seedlings of radish, lettuce, wheat and rice growing in the presence of Cd^{2+} (10^{-7}-10^{-4} M) was reported by Naguib et al. (1982).

Two week treatment of 4-week old tomato with 20 µM Cd^{2+} resulted in the decrease of total carotenoid content by 40% (Baszyński, 1980). All carotenoids present in tomato leaves were affected. The strongest effect was observed for β-carotene (45% inhibition), whereas the amount of neoxanthin was lowered by 22% in comparison to the untreated samples. The inhibitory effect of Cd was partially reversed by addition of manganese to the medium, however the carotenoid content never reached the value of control plants (Baszyński, 1980). Uniconazole, a triazole derivative, was also found to reverse partially the inhibitory effect of Cd^{2+} on carotenoid biosynthesis and this effect was even bigger if kinetin was added with uniconazole (Thomas and Singh, 1996).

In soybean seedlings treated with 0.2 mM Cd^{2+}, the amount of β-carotene, calculated per fresh mass, decreased 5 times in comparison to the untreated seedlings (El-Shintinawy, 1999). Smaller changes were obtained for xathophylls. Antheraxanthin, neoxanthin, violaxanthin and lutein+zeaxanthin decreased 3.9, 2.5, 1.5 and 1.25 times, respectively. Degradation of β-carotene raised the ratio of total xanthophylls/β-carotene by 3 fold compared to the control. Addition of 1 mM GSH to 0.2 mM Cd^{2+} recovered the content of pigments close to the control values.

Irradiation of etiolated radish seedlings growing in the presence of Cd^{2+} induced an enhanced accumulation of all carotenoids present in the seedlings (Krupa et al., 1987). The accumulation of β-carotene exceeded that of xanthophylls: lutein, violaxanthin and neoxanthin. Nevertheless, the total amount of carotenoids was always lower in the presence of Cd^{2+} as compared with untreated seedlings, even though the accumulation of these pigments was enhanced during greening.

It was found that 0.1-0.5 mM Ni^{2+} applied to 10 day-old bean leaves during greening caused a decrease in the total carotenoid content (calculated per leaf) by about two times comparing to the untreated leaves (Krupa et al., 1993).

Fe deficiency applied to sugar beet resulted in the decrease of carotenoids, mainly β-carotene and to the lower extent xanthophylls (Terry, 1980). Based on the results published so far, it is difficult to draw any conclusions about the way of metal influence on biosynthesis of carotenoids.

13.3. Phycobilin biosynthesis in the presence of metals

The problem of metal influence on the biosynthesis of photosynthetic pigments in lower plants and cyanobacteria has not been studied as intensively as for higher plants.

It was observed that in culture of cyanobacteria *Anacystis nidulans*, treated with manganese, Chl synthesis was blocked at the level of uroporphyrinogen III synthase (Csatorday et al., 1984). At the same time, the other branch of tetrapyrolle biosynthesis, e.g. to phococyanobilin chromophores, was not blocked. As a result, higher phycobilin/chlorophyll ratio was measured. In the same study, cobalt treatment led to protoporphyrinogen IX accumulation. This indicates that Chl synthesis was blocked at the level of Mg chelatase. Concomitantly, the phycobilin/chlorophyll ratio was unchanged. It means that phycobilin synthesis must be blocked somewhere on its pathway but the exact site of Co^{2+} action is not known.

14. CONCLUSIONS

All the presented results show that the biosynthesis of photosynthetic pigments is one of the sites of metals action on plants. Existing studies have dealt mainly with the influenced of metals on chlorophyll biosynthesis. There is growing body of evidences that several enzymes of the chlorophyll biosynthetic pathway are affected by metal toxicity. Similar knowledge about the effect of metals on particular enzymes that are engaged in Car biosynthesis is not available at present. There is also lack of knowledge concerning the influence of metals on the photosynthetic pigments of lower plants and cyanobacteria.

The problem of metal toxicity has great consequences both for plants as well as for environment. Lower content of photosynthetic pigments results in retardation of the development of the photosynthetic apparatus in young plants and in lower yield of photosynthesis. This can be a reason of lower biomass production that is very important in agriculture. The accumulation of metals in plant tissue has a great significance for production of healthy food. On the other hand, plants that are able to accumulate the high amount of metals without significant growth disturbances can be used to remove them from the environment. Analysis of symptoms of heavy metal stress in plants can also be used as the probe of toxicity level in the environment.

ACKNOWLEDGEMENTS

This work was supported by the grant 6P04A 028 19 from the Committee for Scientific Research (KBN) of Poland.

REFERENCES

Agarwala, S.C., Sharma, C.P., and Farooq, S. (1965) Effect of iron supply on growth, chlorophyll, tissue iron and activity of certain enzymes in maize and radish, *Plant Physiol.* **40**, 493–502.
Agarwala, S.C., Bisht, S.S., and Sharma, C.P. (1977) Relative effectiveness of certain heavy metals in producing toxicity and symptoms of iron deficiency in barley, *Can. J. Bot.* **55**, 1299–1307.
Angelov, M., Tsonev, T., Uzunova, A., and Gaidardjieva, K. (1993) Cu^{2+} effect upon photosynthesis, chloroplast structure, RNA and protein synthesis of pea plants, *Photosynthetica* **28**, 341–350.
Avissar, Y.J. and Moberg, P.A. (1995) The common origins of the pigments of life – early steps of chlorophyll biosynthesis, *Photosynth. Res.* **44**, 221–242.
Barcelo, J., Vazquez, M.D., and Poschenrieder, Ch. (1988) Structural and ultrastructural disorders in cadmium –treated bush plants (*Phaseolus vulgaris* L.), *New Phytol.* **108**, 37–49.
Baszyński, T., Król, M., Krupa, Z., Ruszkowska, M., Wojcieska, U., and Wolińska, D. (1982) Photosynthetic apparatus of spinach exposed to excess copper, *Z. Pflanzenphysiol.* **108**, 385–392.
Baszyński, T., Ruszkowska, M., Król, M., Tukendorf, A., and Wolińska, D. (1978) The effect of copper deficiency on the photosynthetic apparatus of higher plants, *Z. Pflanzenphysiol.* **89**, 207–216.
Baszyński, T., Tukendorf, A., Ruszkowska, M., Skórzyńska, E., and Maksymiec, W. (1988) Characteristics of the photosynthetic apparatus of copper non-tolerant spinach exposed to excess copper, *J. Plant Physiol.* **132**, 708–713.
Baszyński, T., Wajda, L., Król, M., Wolińska, D., Krupa, Z., and Tukendorf, A. (1980) Photosynthetic activities of cadmium-treated tomato plants, *Physiol. Plant.* **48**, 365–370.
Beale, S.I. (1999) Enzymes of chlorophyll biosynthesis, *Photosynth. Res.* **60**, 43–73.
Berska, J., Myśliwa-Kurdziel, B., and Strzałka, K. (in press) Transformation of protochlorophyllide to chlorophyllide in wheat under heavy metal stress.
Bishnoi, N.R., Sheoran, I.S., and Singh, R. (1999) Influence of cadmium and nickel on photosynthesis and water relations in wheat leaves of different insertion level, *Photosynthetica* **28**, 473–479.
Böddi, B., A. Lindsten, M. Ryberg and Sundqvist C. (1989) On the aggregational states of protochlorophyllide and its protein complexes in wheat etioplasts, *Physiol. Plant.* **76**, 135–143.
Böddi, B., Oravecz, A.R., and Lehoczki, E. (1995) Effect of cadmium on organization and photoreduction of protochlorophyllide in dark-grown leaves and etioplast inner membrane preparations of wheat, *Photosynthetica* **31**, 411–420.
Böddi, B., Ryberg, M., and Sundqvist, C. (1992) Identification of four universal protochlorophyllide forms in dark-grown leaves by analyses of 77 K fluorescence emission spectra, *J. Photochem. Photobiol. B: Biol.* **12**, 389–401.
Böddi, B., M. Ryberg and Sundqvist C. (1993) Analysis of the 77 K fluorescence emission and excitation spectra of isolated etioplast inner membranes, *J. Photochem. Photobiol. B: Biol* **21**, 125–133.
Burzyński, M. (1985) Influence of lead on the chlorophyll content and on initial steps of its synthesis in greening cucumber seedlings, *Acta Soc. Bot. Pol.* **54(1)**, 95–105.
Chatterjee, J. and Chatterjee, C. (2000) Phytotoxicity of cobalt, chromium and copper in cauliflower, *Environ. Poll.* **109**, 69–74.
Chereskin, B.M. and Castelfranco, P.A. (1982) Effects of iron and oxygen on chlorophyll biosynthesis. II. Observations on the biosynthetic pathway in isolated etiochloroplasts, *Plant Physiol.* **68**, 112–116.
Chugh, L.K. and Sawhney, S.K. (1999) Photosynthetic activities of *Pisum sativum* seedlings grown in the presence of cadmium, *Plant Physiol. Biochem.* **37**, 297–303.
Cook, C.M., Kostidou, A., Vardaka, E., and Lanaras, T. (1997) Effects of copper on the growth, photosynthesis and nutrient concentrations on phaseolus plants, *Photosynthetica* **34**, 179–193.
Csatorday, K., Gombos, Z., and Szalontai, B. (1984) Mn^{2+} and Co^{2+} toxicity in chlorophyll biosynthesis, *Proc. Natl. Acad. Sci. USA* **81**, 476–478.

Czuba, M. and Ormrod, D. (1973) Effects of cadmium and zinc in ozone-induced phytotoxicity in cress and lettuce, *Can. J. Bot.* **52**, 645–649.

De Kock, P.C., Commisiong, K., Farmer, V.C., and Inkson, R.H.E. (1960) Interrelationships of catalase, peroxidase, hematin, and chlorophyll, *Plant Physiol.* **53**. 206–215.

Di Cagno, R., Guidi, L., Stefani, A., and Soldatini, G.F. (1999) Effect of cadmium on growth of Heliantus annus seedlings: physiological aspects, *New Phytol.* **144**, 65–71.

Droppa, M., Terry, N., and Horvath, G. (1984a) Variation in photosynthetic pigments and plastoquinone contents in sugar beet chloroplasts with changes in leaf copper content, *Plant Physiol.* **74**, 717–720.

Droppa, M., Terry, N., and Horvath, G. (1984b) Effects of Cu deficiency on photosynthetic electron transport, *Proc. Natl. Acad. Sci.* **81**, 2369–2373.

Duggan, J. and Gassman, M. (1974) Induction of porphyrin synthesis in etiolated bean leaves by chelators of iron, *Plant Physiol.* **53**, 206–215.

Eleftheriou, E.P. and Karataglis, S. (1989) Ultrastructural and morphological characteristics of cultivated wheat growing on copper poluted fields, *Bot. Acta.* **102**, 134–140.

Eleiwa, M.E. and Naguib, M.I. (1986) Response of soybean leaves to soil application of nickel, strontium or vanadium, *Egypt. J. Bot.* **29**, 167–180.

El Hamouri, B., M. Brouers and Sironval C. (1981) Pathway from photoinactive P633-628 protochlorophyllide to the P696-682 chlorophyllide in cucumber etioplast suspensions, *Plant Sci. Lett.* **21**, 375–379.

El-Shintinawy, F. (1999) Glutathione counteracts the inhibitory effect induced by cadmium on photosynthetic process in soybean, *Photosynthetica* **36**, 171–179.

Ferretti, M., Ghisi, L., Merlo, F., Dalla Vecchia, F., and Passera, C. (1993) Effect of cadmium on photosynthesis and enzymes of photosynthetic sulphate and nitrate assimilation pathways in maize (*Zea mays* L.), *Photosynthetica* **29**, 49–54.

Fodor, F., Böddi, B., Sarvari, E., Zaray, G., Cseh, E., and Lang, F. (1995) Correlation of iron content,spectral forms of chlorophyll and chlorophyll-proteins in iron deficiend cucumber (*Cucumis sativus*), *Physiol. Plant.* **93**, 750–756.

Fodor, F., Sarvari, E., Lang, F., Szigeti, Z., and Cseh, E. (1996) Effects of Pb and Cd on cucumber depending on the Fe-complex in the culture solution, *J. Plant Physiol.* **148**, 434–439.

Foy, C.D., Chaney, R.L., and White, M.C. (1978) The physiology of metal toxicity in plants, *Annu. Rev. Plant Physiol.* **29**, 511–566.

Franck, F., B. Bereza and Böddi B. (1999) Protochlorophyllide-NADP+ and protochlorophyllide-NADPH complexes and their regeneration after flash illumination in leaves and etioplast membranes of dark-grown wheat, *Photosynth. Res.* **59**, 53–61.

Fujita, Y. (1996) Protochlorophyllide reduction: a key step in the greening of plants, *Plant Cell Physiol.* **37**, 411–421.

Gadallah, M.A.A. (1995) Effects of cadmium and kinetin on chlorophyll content, saccharides and dry matter accumulation in sunflower plants, *Biol. Plant.* **37**, 233–240.

Granick, S. (1961) Magnesium protoporphyrin monoester and protoporphyrin monomethyl ester in chlorophyll biosynthesis, *J. Biol. Chem.* **236**, 1168–1172.

Greger, M. and Lindberg, S. (1987) Effects of Cd^{2+} and EDTA on young sugar beets (*Beta vulgaris*). II. Net uptake and distribution of Mg^{2+}, Ca^{2+} and Fe^{2+}/Fe^{3+}, *Physiol. Plant.* **69**, 81–86.

Greger, M. and Ögren, E. (1991) Direct and indirect effects of Cd^{+2} on photosynthesis in sugar beet (*Beta vulgaris*), *Physiol. Plant.* **83**, 129–135.

Henriques, F.S. (1989) Effects of copper deficiency on the photosynthetic apparatus of sugar beet (*Beta vulgaris* L.), *J. Plant Physiol.* **135**, 453–358.

Horvath, G., Droppa, M., Oravecz, A., Raskin, V.I., and Marder, J.B. (1996) Formation of the photosynthetic apparatus during greening of cadmium-poisoned barley leaves, *Planta* **199**, 238–243.

Imai, I. and Siegel, S.M. (1973) A specific response to toxic cadmium levels in red kidney bean embryo, *Physiol. Plant.* **29**, 118–120.

Kowalewska, G. and Hoffmann, S.K. (1989) Identification of the copper porphyrin complex formed in cultures of blue-green alga *Anabena variabilis*, *Acta Physiol. Plant.* **11**, 39–50.

Kłobus, G. and Buczek, J. (1985) Chlorophyll content, cells and chloroplast number and cadmium distribution in Cd-treated cucumber plants, *Acta Physiol. Plant.* **7**, 139–147.

Krupa, Z., Siedlecka, A., Maksymiec, W., and Baszyński, T. (1993) In vivo response of photosynthetic apparatus of *Phaseolus vulgaris* L. to nickel toxicity, *J. Plant Physiol.* **142**, 664–668.

Krupa, Z., Skórzyńska, E., Maksymiec, W., Baszyński, T. (1987) Effect of cadmium treatment on the photosynthetic apparatus and its photochemical activities in greening radish seedlings, *Photosynthetica* **21**, 156–164.
Küpper, H., Küpper, F., and Spiller, M. (1996) Environmental relevance of heavy metal-substituted chlorophylls using the example of water plants, *J. Exp. Bot.* **47**, 259–266.
Küpper, H., Küpper, F., and Spiller, M. (1998) *In situ* detection of heavy metal substituted chlorophylls in water plants, *Photosynth. Res.* **58**, 123–133.
Lebedev, N. and M. Timko P. (1998) Protochlorophyllide photoreduction, *Photosynth. Res.* **58**, 5–23.
Lidon, F.C. and Henriques, F.S. (1991) Limiting step on photosynthesis of rice plants treated with varying copper levels, *J. Plant Physiol.* **138**, 115–118.
Lolkema, P.C. and Vooijs, R. (1986) Copper tolerance in silene cucubalus, *Planta* **167**, 30–36.
Maksymiec, W. (1997) Effect of copper on cellular processes in higher plants. *Photosynthetica* **34**, 321–342.
Maksymiec, W. and Baszyński, T. (1996) Different susceptibility of runner bean plants to excess copper as a function of the growth stages of primary leaves, *J. Plant Physiol.* **149**, 217–221.
Maksymiec, W., Bednara, J., and Baszyński, T. (1995) Responses of runner bean plants to excess copper as a function of planth growth stages: Effecs on morphology and structure of primary leaves and their chloroplast ultrastructure, *Photosynthetica* **31**, 427–435.
Maksymiec, W., Russa, R., Urbanik-Sypniewska, T., and Baszyński, T. (1994) Effects of excess Cu on the photosynthetic apparatus of runner bean leaves treated at two different growth stages, *Physiol. Plant.* **91**, 715–721.
Malik, D., Sheoran, I.S., and Singh, R. (1992) Carbon metabolism in leaves of cadmium treated wheat seedlings. *Plant Physiol. Biochem.* **30**, 223–229.
Marsh, H.V., Evans, H.J., and Matrone, G. (1963) Investigations of the role of iron in chloroplast metabolism. II. Effects of iron deficiency on chlorophyll synthesis, *Plant Physiol.* **38**, 638–642.
Milivojevic, D.B., Stojanovic, D.D., and Drinic, S.D. (2000) Effects of alluminium on pigments and pigment-protein complexes of soybean, *Biol. Plant.* **43**, 595–597.
Mukhopadhyay, N. and Aery, N.C. (2000) Effect of Cr (III) and Cr (VI) on the growth and physiology of *Triticum aestivum* plants during early seedling growth, *Biologia* **55**, 403–408.
Naguib, M.I., Hamed, A.A., and Al-Wakeel, S.A. (1982) Effect of cadmium on growth criteria of some crop plants, *Egypt. J. Bot.* **25**, 1–12.
Nandi, D.L. and Shemin, D. (1968) δ-aminolevulinic acid dehydratase of *Rhodopseudomonas sphaeroides*, *J. Biol. Chem.* **243**, 1236–1242.
Padmaja, K., Prasad, D.D.K., and Prasad, A.R.K. (1989) Effect of selenium on chlorophyll biosynthesis in mung bean seedlings, *Phytochemistry* **28**, 3321–3324.
Padmaja, K., Prasad, D.D.K., and Prasad, A.R.K. (1990) Inhibition of chlorophyll synthesis in phaseolus vulgaris L. Seedlings by cadmium acetate, *Photosynthetica* **24**, 399–405.
Padmaja, K., Somasekharaiah, B.V., and Prasad, A.R.K. (1995) Inhibition of chlorophyll synthesis by selenium: Involvement of lipoxygenase mediated lipid peroxidation and antioxidant enzymes, *Photosynthetica* **31**, 1–7.
Parekh, D., Puranik, R.M., and Srivastava, H.S. (1990) Inhibition of chlorophyll biosynthesis by cadmium in greening maize leaf segments, *Biochem. Physiol. Pflanzen.* **186**, 239–242.
Plat-Aloia, K.A., Thomson, W.W., and Terry, N. (1983) Changes in plastid ultrastructure during iron nutrition-mediated chloroplast development, *Protoplasma* **114**, 85–92.
Prasad, D.D.K. and Prasad, A.R.K. (1987) Effect of lead and mercury on chlorophyll synthesis in mung bean seedlings. *Phytochemistry* **26**, 881–883.
Prasad, D.D.K. and Prasad, A.R.K. (1987) Altered delta-aminolevulinc acid metabolism by lead and mercury in germinating seedlings of bajra (pennisetum typhoideum). *J. Plant Physiol.* **127**, 241–249.
Prasad, M.N.V., Drej, K., Skawińska, A. and Strzałka, K. (1998) Toxicity of cadmium and copper in *Chlamydomonas reinhardtii* wild type (WT 2137) and cell wall deficient mutant strain (CW15). *Bull. Environ. Contam. Toxicol.* **60** (2), 306–311.
Prasad, M.N.V., Malec, P., Waloszek, A., Bojko, M. and Strzałka, K. (submitted) Physiological responses of *Lemna trisulca* L. (Duckweed) to cadmium and copper bioaccumulation.
Rebeiz, C.A., Ioannides, I.M. Kolossov, V. and Kopetz, K.J. (1999) Chloroplast biogenesis 80. Proposal of a unified multibranched chlorophyll a/b biosynthetic pathway, *Photosynthetica* **36**, 117–128.
Rebeiz, C.A, Parham, R., Fasoula, D.A. and Ioannides, I.I. (1994) Chlorophyll a biosynthetic heterogeneity, in The Proceedings of Ciba Foundation Symposium 180: *"The biosynthesis of the tetrapyrolle pigments"*, Wiley, Chichester, pp. 177–193.

Reinbothe, C. and Reinbothe S. (1996) Regulation of chlorophyll biosynthesis in angiosperms, *Plant Physiol.* **111**, 1–7.
Rüdiger, W. (1992) Last steps in chlorophyll-biosynthesis: Esterification and insertion into the membrane, in J.H. Argyroudi-Akoyunoglou, (ed.), *Regulation of chloroplast biogenesis*, Plenum Press, New New York, pp.183–190.
Rüdiger, W. (1993) Esterification of chlorophyllide and its implication on thylakoid development, in C. Sundqvist and M. Ryberg (eds.), *Pigment-protein complexes in plastids: Synthesis and assembly*, Acad. Press, San Diego, pp. 219–240.
Rüdiger,W. (1997a) Chlorophyll metabolism: from outer space down to the molecular level, *Phytochemistry* **46**, 1151–1167.
Rüdiger,W. (1997b) Chlorophyll biosynthesis and plant development, in H. Greppin, C. Penel and P. Simon (eds), *Travelling shot on Plant development*, University of Geneva, pp. 131–143.
Ryberg, M., Artus, N., Böddi, B., Lindsten, A., Wiktorsson, B. and Sundqvist C. (1992) Pigment-protein complexes of chlorophyll precursors, in J.H. Argyroudi-Akoyunoglou (ed.), *Regulation of Chloroplast Biogenesis*, Plenum Press, New York, pp. 217–224.
Ryberg, M. and Sundqvist, C. (1976) The influence of 8-hydroxyquinoline on the accumulation of porphyrins in dark-grown wheat leaves treated with δ-aminolevulinic acid, *Physiol. Plant.* **36**, 356–361.
Ryberg, M. and Sundqvist C. (1988) The regular ultrastructure of isolated prolamellar bodies depends on the presence of membrane-bound NADPH-protochlorophyllide oxidoreductase, *Physiol. Plant.* **73**, 218–226.
Schlegel, H., Godbold, D.L., and Huttermann, A. (1987) Whole plant aspects of heavy metal induced changes in CO_2 uptake and water relations of spruce (*Picea abies*) seedlings, *Physiol. Plant.* **69**, 265–270.
Schoefs, B. (1999) The light-dependent and light-independent reduction of protochlorophyllide a to chlorophyllide a, *Photosynthetica* **36**, 481–496.
Selstam, E. and Widell-Wigge, A. (1993) Chloroplast lipids and the assembly of membranes, in C. Sundqvist and M. Ryberg (eds.,) *Pigment-protein complexes in plastids: Synthesis and assembly*, Acad. Press, San Diego, pp. 261–267.
Sengar, R.S. and Padney, M. (1996) Inhibition of chlorophyll biosynthesis by lead in greening *Pisum sativum* leaf segments, *Biol. Plant.* **38**, 459–462.
Senge, M.O. (1993) Recent advances in the biosynthesis and chemistry of the chlorophylls, *Photochem. Photobiol.* **57**, 189–206.
Shalygo, N.V. and Averina, N.G. (1996) The effect of metal cations on chlorophyll accumulation in greening barley seedlings, *Dokl. Acad. Nauk Belarus.* **40**, 76–79.
Shalygo, N.V., Averina, N.G., Grimm, B., and Mock, H.P. (1997a) Influence of cesium on tetrapyrolle biosynthesis in etiolated and greening barley leaves, *Physiol. Plant.* **99**, 160–168.
Shalygo, N.V., Kolesnikova, N.V., Voronetskaya, V.V., and Averina, N.G. (1999) Effects of Mn^{2+}, Fe^{2+}, Co^{2+} and Ni^{2+} on chlorophyll accumulation and early stages of chlorophyll formation in greening barley seedlings, *Russ. J. Plant Physiol.* **46(4)** 496–501.
Shalygo, N.V., Mock, H.P., Averina, N.G., and Grimm, B. (1998) Photodynamic action of uroporphyrin and protochlorophyllide in greening barley leaves treated with cesium chloride, *J. Photochem. Photobiol. B: Biol.* **42**, 151–158.
Shalygo, N.V., Voronetskaja, V.V., and Averina, N.G. (1997 b) Effect of Mn^{2+} on porphyrinogenesis in etiolated and greening barley seedlings, *Russ. J. Plant. Physiol.* **44**, 311–316.
Shaw, B.P. and Rout, N.P. (1998) Age-dependent responses of phaseolus aureus Roxb. to inorganic salts of mercury and cadmium, *Acta Physiol. Plant.* **20**, 85–90.
Sheoran, I.S., Aggarwal, N., and Singh, R. (1990a) Effect of cadmium and nickel on *in vivo* carbon dioxide exchange rate of pigeon pea (*Cajanus cajan* L.), *Plant Soil* **129**, 243–249.
Sheoran, I.S., Singal, H.R., and Singh, R. (1990b) Effect of cadmium and nickel on photosynthesis and the enzymes of the photosynthetic carbon reduction cycle in pigeon pea (*Cajanus cajan* L.), *Photosynth. Res.* **23**, 345–351.
Siedlecka, A. and Krupa, Z. (1997) Cd/Fe interaction in higher plants – its consequences for the photosynthetic apparatus, *Photosynthetica* **36**, 321–331.
Skórzyńska, E., Urbanik-Sypniewska, T., Russa, R., and Baszyński, T. (1991) Galactolipase activity in Cd-treated runner bean plants, *J. Plant Physiol.* **138**, 454–459.
Skórzyńska-Polit, E., Baszyński, T. (1995) Photochemical activity of primary leaves in cadmium stressed Phaseolus coccineus depends on their growth stages, *Acta Soc. Bot. Pol.* **64**, 273–279.

Spiller, S.C., Castelfranco, A.M., and Castelfranco, P.A. (1982) Effects of iron and oxygen on chlorophyll biosynthesis. I. *In vivo* observations on iron and oxygen-deficient plants, *Plant Physiol.* **69**, 107–111.

Stiborova, M., Ditrichova, M., and Brezinova, A. (1987) Effect of heavy metal ions on growth and biochemical characteristics of photosynthesis of barley and maize seedlings, *Biol. Plant.* **29**, 453–467.

Stiborova, M., Doubravova, M., Brezinova, A., and Friedrich, A. (1986) Effect of heavy matal ions on growth and biochemical characteristics of photosynthesis of barley, *Photosynthetica* **20**, 418–425.

Stobart, A.K., Griffiths, W.T., Ameen-Bukhari, I., and Sherwood, R.P. (1985) The effect of Cd+2 on the biosynthesis of chlorophyll in leaves of barley, *Physiol. Plant.* **63**, 293–298.

Sundqvist, C. and Dahlin C. (1997) With chlorophyll pigments from prolamellar bodies to light-harvesting complexes, *Physiol. Plant.* **100**, 748–759.

Terry, N. (1980) Limiting factors in photosynthesis. I. Use of iron stress to control photochemical capacity *in vivo*, *Plant Physiol.* **65**, 114–120.

Thomas, R.M. and Singh, V.P. (1996) Reduction of cadmium-induced inhibition of chlorophyll and carotenoid accumulation in *Cucumis sativus* L. by uniconazole (S.3307), *Photosynthetica* **32**, 145–148.

Vajpayee, P., Tripathi, R.D., Rai, U.N., Ali, M.B., and Singh, S.N. (2000) Chromium VI accumulation reduces chlorophyll biosynthesis, nitrate reductase activity and protein content in *Nymphaea alba* L., *Chemosphere* **41**, 1075–1082.

Vassilev, A., Berova, M., and Zlatev, Z. (1998) Influence of Cd^{2+} on growth, chlorophyll content, and water relations in young barley plants, *Biol. Plant.* **41**, 601–606.

Vlcek, L.M. and Gassman, M.L. (1979) Reversal of α,α'-dipyridyl-induced porphyrin synthesis in etiolated and greening red kidney bean leaves, *Plant Physiol.* **64**, 393–397.

Von Wettstein, D., Gough S. and Kannangara C.G. (1995) Chlorophyll biosynthesis, *Plant Cell* **7**, 1039–1057.

CHAPTER 9

HEAVY METAL INFLUENCE ON THE LIGHT PHASE OF PHOTOSYNTHESIS

B. MYŚLIWA-KURDZIEL[a], M.N.V. PRASAD[b] AND K. STRZAŁKA[a]
[a] *Department of Plant Physiology and Biochemistry*
The Institute of Molecular Biology
Jagiellonian University
7 Gronostajowa Street, PL 30-38, Kraków, Poland
[b] *Department of Plant Sciences*
University of Hyderabad, Hyderabad 500046, India

1. INTRODUCTION

Global heavy metal pollution is a serious environmental concern. Among other physiological functions, heavy metals like cadmium, copper, mercury, lead etc. influence the functions of photosynthetic apparatus (Baszyński and Tukendorf, 1984; Clijsters and Van Assche, 1985; Baszyński, 1986; Droppa and Horváth, 1990; Sheoran et al., 1990; Greger and Ögren, 1991; Krupa and Baszyński, 1985 and 1995; Prasad, 1995a and 1997; Krupa, 1999; Prasad and Strzałka, 1999). They may interact with the photosynthetic apparatus at various levels of organization and architecture i.e. accumulation of metals in leaf (main photosynthetic organ), partitioning in leaf tissues like stomata, mesophyll and bundle sheath cells, interaction of metal with cytosolic enzymes and organics. They may also alter the functions of chloroplast membranes and components of photosynthetic electron transport chain, particularly photosystem II (PS II) and photosystem I (PS I). In *Zea mays*, Cd altered the photosynthesis and enzymes of photosynthetic sulphate and nitrate assimilation pathways (Ferretti et al., 1993). Using an FTIR spectroscopy Ahmed and Tajmir-Riahi (1993) showed that cadmium, mercury and lead (0.01-0.1 mM) interacted with the light-harvesting proteins (LHC II) of spinach thylakoid membranes with different efficiency. The strongest effect was found for mercury, which at low concentration induced drastic structural modifications of protein. Cadmium binding that results in major conformational changes of proteins was observed for higher metal concentration. Lead was less effective on protein conformation. Cadmium–induced changes in the structure and composition of LHC II protein were also reported by Krupa (1988).

In the case of biometals like copper (Figure 1), functions of the photosynthetic apparatus are altered both by the excess (Baszyński et al., 1982; Lidon and Henriques, 1993a) and the deficiency of metal ion (Baszyński et al., 1978; Droppa et al., 1984; Henriques, 1989; Droppa and Horváth, 1990; Lidon et al., 1993). The effect of copper excess on the photosynthetic apparatus strongly depends on the growth stage of plants during metal ion application (Maksymiec et al., 1994 and 1995; Maksymiec and Baszyński, 1996a, b and 1999a, b). Copper was found to interfere with antioxidative enzymes in oat (*Avena sativa*) (Luna et al., 1994) and wheat (*Triticum durum*) (Navari-Izzo et al., 1998) leaves.

Another biometal that affects photosynthesis if supplied in deficient amount is iron (Terry and Abadia, 1986; Geider and La Roche, 1994). It was first concluded for sugar beet that iron deficiency reduced the number of photosynthetic units per unit leaf area rather than diminished the efficiency of the photosynthetic energy conversion (Spiller and Terry, 1980). However later, the correlation between the decrease in the efficiency of photosynthetic energy conversion and iron deficiency was described in sugar beet (Morales et al., 1990 and 1991) as well as in cyanobacteria (Guikema, 1985) and eukariotic marine algae (Greene et al., 1992).

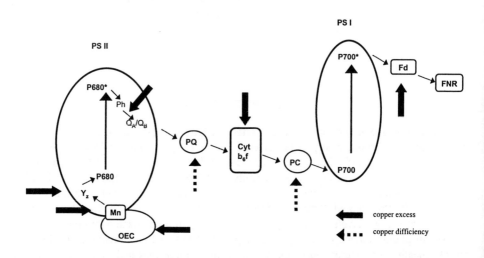

Figure 1: Sites of copper (a biometal) action towards components of the photosynthetic electron transport. OEC, oxygen evolving complex; Mn; manganese cluster; Y_Z ; tyrozine 161 of D1 protein; P680, P700, reaction centre of PS II and PS I, respectively; Q_A/Q_B; plastoquinone A and B; PQ, plastoquinone pool; Cyt b6 f; PC, plastocyanin; Fd, ferredoxin; FNR, ferredoxin-NADP oxidoreductase;

Heavy metals can also influence the photosynthetic activity indirectly by the inhibition of the photosynthetic pigments biosynthesis (see chapter 8) or by the stimulation of the activity of chlorophyllase. Mercury, copper and zinc ions increased the hydrolytic activity of chlorophyllase in rice leaves, with mercury being the most efficient of the three tested metals. Zinc increased the activity of chlorophyllase moderately, and copper affected it the least. Thus, the order of effectiveness of these metal was as follows Hg>Zn>Cu (Drążkiewicz, 1994).

Taking into account that the research carried out in this field, is concentrated on Cd, Cu and Pb toxicity, we described in this chapter mainly the effects of these three metals on the light phase of photosynthesis.

2. METAL UPTAKE AND ACCUMULATION IN CHLOROPLASTS

Metal ions present in soil enter the plant mainly through the root system. Foliar absorption and stem uptake are regarded also as a significant way of metal introduction (see chapter 1). The efficiency of metal uptake depends on plant species. The cuticular permeability of cadmium was demonstrated to be higher for pea (*Pisum sativum*) leaves than for beet (*Beta vulgaris*) (Greger et al., 1993). This process depended on the pH value of the external medium and it was more effective for high pH (5.6) than for low pH (3.6). However, there are experiments, which revealed that stomata are not directly involved in uptake of Cd. On the other hand, leaf cuticles function as weak cation exchangers, due to the negative charge of pectin and non-essential cutin polymers, therefore they act as a barrier for heavy metal ions penetration into leaf.

The accumulation of heavy metals is different for different part of plants. The level of heavy metals detected in leaves frequently did not exceed 10% of the total plant accumulation (Ernst et al., 1992). The level of heavy metals translocated to chloroplasts was estimated to be in the range of 1%. There are, however, a great variations in accumulation of heavy metals depending on plant species, development stage and the type of tissue. For example, spinach plants after 15-days treatment with lead acetate accumulated 45% of Pb in roots and 55% in leaves. Generally, cell walls adsorb more than 90% of heavy metals. The oldest leaves accumulate the highest amounts of heavy metals. Lead level in senescent leaves remains constant throughout the year. However, bioaccumulation of Zn gradually increases in the course of senescence (Ernst et al., 1992).

Leaf epidermis was found to play a significant role in Zn compartmentation in barley. In seedlings growing in the presence of 200 µM Zn^{+2} and in low strength Hoagland solution, the accumulation of this metal was 10-fold in the epidermis, 11-fold in the vacuole, 8-fold in the apoplasm and almost the same in the cytosol in comparison to the plants growing in normal (1 µM) Zn^{2+} concentration (Dietz and Hartung, 1996). It was also found that accumulation of heavy metals in the phytomass of crop plants was different for different plants (Wagner, 1993; Siedlecka, 1995; Prasad, 1996).

Heavy metals show different degree of mobility in plant tissues and their translocation is probably regulated by carrier proteins of the vascular tissues which serve as ligands for transportation (Prasad, 1995a).

Exposure of plants to the excess of a heavy metal may result in the change of uptake or accumulation of other metals. The possible Cu-Ca interactions in higher plants were postulated. It was shown that Cu^{2+} excess can change Ca^{2+} content in plants and, on the other hand, Ca^{2+} can modify the effect of Cu^{2+} excess (Ouzounidou et al., 1995; Maksymiec and Baszyński, 1996b, 1998b and 1999a, b). Similar inhibitory effect for iron uptake was observed for cadmium (Siedlecka and Baszyński, 1993; Siedlecka and Krupa, 1999). Iron uptake was also inhibited in rice plants that were treated with copper (0.002 to 6.25 g m-3 in a nutrient solution) for 30 days (Lidon and Henriques, 1993a, b).

3. INFLUENCE OF HEAVY METALS ON THE PHOTOSYNTHETIC ELECTRON TRANSPORT

In higher plants, heavy metal accumulation in mature leaves reduced net photosynthesis (Barua and Jana, 1986; Angelov et al., 1993; Bishnoi et al., 1993a, b; Rascio et al., 1993; Moya et al., 1993; Abo-Kassem et al., 1995; Prasad, 1995b). The primary effect of heavy metals toxicity on the light reactions of photosynthesis is the inhibition of photosynthetic O_2 evolution, reduction of NADP and photophosphorylation. Secondary effects include the process of chlorophyll (Chl) degradation, changes in Chl a/b ratio and in nitrogen metabolism (Bishnoi et al., 1993a; Gadallah, 1995).

Various sites of inhibition for the reactions of light phase of photosynthesis were described for different heavy metals (Table 1, see also Figures 1 and 2).Cadmium influenced the net photosynthesis in rice (Misra et al., 1989; Moya et al., 1993), maize (Ferretti et al., 1993) and tomato (Baszyński et al., 1980). This metal ions altered also photosynthesis and transpiration of excised silver maple leaves (*Acer saccharum*) (Lamoreaux and Chaney, 1978). Reduction of photosynthesis and transpiration due to Cd treatment was attributed to stomatal closure (Bazzaz et al., 1974b; Greger and Johansson, 1992). Cadmium changed also the light reactions of chloroplasts (Bazzaz and Govindjee, 1974).

The toxic effect of cadmium on photosynthesis as well as on the activity of both PS I and PS II was found to be a dose-dependent and it increased with extended period of exposure (Chugh and Sawhney, 1999). At the beginning of Cd treatment, a more pronounced effect on the PS II activity was observed, however later, PS I was also affected. Interestingly, the impairment of PS II only in chloroplast suspension was observed. High sensitivity to Cd of PS II in isolated chloroplast was also shown by Li and Miles (1975). Skórzyńska-Polit and Baszyński (1995) showed higher susceptibility of plants towards cadmium toxicity at the final growth stage of the primary leaves than in younger plants.

Weigel (1985a, b) demonstrated that the inhibition of photosynthetic reactions in isolated intact chloroplasts treated with Cd was indirectly induced by inhibition of different reaction steps of the Calvin cycle, and not by interaction with photosynthetic reactions in the thylakoid membranes. The membrane-bound photosynthetic reactions in isolated mesophyll protoplasts were not impaired by Cd concentrations which drastically inhibited CO_2 fixation (Weigel, 1985a, b; see also chapter 8 and 10).

Inhibition of photosynthetic CO_2 fixation and Hill reaction activity by cadmium and zinc was reported for isolated spinach chloroplasts (Hampp et al., 1976; Barua and Jana, 1986). Photosynthetic electron transport was inhibited by zinc also in isolated barley chloroplasts (Tripathy and Mohanthy, 1980).

Copper is essential as a micronutrient but at high concentration it is toxic for photosynthetic organisms (Clijsters and Van Assche, 1985; Lidon and Henriques, 1993 a-c; Maksymiec, 1997). At elevated concentrations, Cu inhibits the photosynthetic electron transport (Cedeno-Maldonado et al., 1972; Baron et al., 1995). It is known from *in vitro* studies that PS II is more susceptible to copper toxicity than PS I (Cedeno-Maldonado et al., 1972; Baron et al., 1995; Ouzounidou et al., 1997). Copper is an important constituent of plastocyanin, which enables transport of electrons between cytochrome b_6f complex and PS I (Droppa and Horváth, 1990). Maksymiec and Baszyński (1996a) found the highest sensitivity of the photosynthetic apparatus to Cu excess for plants treated at advanced leaf growth stages.

Deficiency of this metal was found to inhibit the electron transport efficiency between PS II and PS I (Droppa et al., 1984). This was probably due to a decreased level of plastocyanin and to the lower mobility of plastoquinone molecules that resulted from copper–induced changes in the membrane fluidity.

Different sensitivity of photosynthesis as a result of Pb treatment was observed for corn and soybean (Bazzaz et al., 1974b and 1975). Reduced height and biomass were reported for oats (*Avena sativa*) that grown on a contaminated site with excess Cu and Pb in comparison with control plants. It was rather related to the disruption of the photosynthetic function than to the damage of the photosynthetic apparatus (Moustakas et al., 1994).

Mercury was found to inhibit the rate of electron transport through the photosynthetic chain in isolated spinach chloroplasts (Šeršeň et al., 1998). The 50% decrease of DCPIP photoreduction was determined at the concentration of 0.028 mol/m^3 of Hg^{2+}. This value was in a good accordance with the finding of Kimimura and Katoh (1972) and Bernier et al. (1993).

3.1 Photosystem I

PS I is a membrane-bound protein complex that catalyzes oxidation of plastocyanin and reduction of ferredoxin under light conditions (Strzałka and Ketner, 1997). Chlorophyll in the PS I reaction centre (P_{700}) absorbs the photon of light and passes an excited electron to the electron acceptor called A_o and then to ferredoxin by several carriers of increasing redox potential. PS I shows a high degree of homology in all higher plants. Its core complex was found to be highly conserved through evolution from cyanobacteria to higher plants (Almog et al., 1991).

In several investigations the site of heavy metal action was identified close to PS I (Table 1). Cadmium inhibited electron flow on the reducing side of PS I. Siedlecka and Baszyński (1993) compared electron transport activities (DCIP → MV and DCIP → NADP$^+$) in Cd-treated (10, 20 and 30 µM Cd), isolated chloroplasts of 21-day-old maize. The site of Cd inhibition was suggested in PS I between primary electron acceptor X and NADP$^+$. Cadmium treatment caused Fe deficiency, indicating that the light phase of photosynthesis was affected in the treated plants due to Cd-induced Fe deficiency (Krupa and Siedlecka, 1995; Siedlecka and Krupa, 1999). However, in another study on isolated chloroplasts, 3-day treatment with 3, 6, and 9 mM Cd^{2+} did not change significantly the PS I activity compared to PS II (Nedunchezhian and Kulandaivelu, 1995).

Table 1. Sites of heavy metals action on photosynthetic electron transport chain

Site of action	Metals	Reference
Light-harvesting chlorophyll a/b protein complex	Cd, Hg, Pb	Krupa 1988; Krupa and Baszyński 1995; Ahmed and Tajmir-Riahi 1993;
Photosystem II	Cd, Co, Cu, Hg, Ni, Pb, Zn	Miles et al. 1972; Bazzaz and Govindjee 1974; Shioi et al. 1978b; Tripathy and Mohanthy 1980; Clijsters and Van Assche 1985; Baszyński 1986; Hsu and Lee 1988; Samson et al. 1988; Mohanty et al. 1989; Rashid et al. 1991; Yruela et al. 1991 and 1993; Malik et al. 1992; Bernier et al. 1993; Maksymiec et al. 1994; Schröder et al. 1994; Jegerschöld et al. 1995; Chugh and Sawhney 1999;
Oxygen evolving complex	Cd, Cr, Cu, Hg, Zn	Van Duijvendijk-Matteoli and Desmeta 1975; Miller and Cox 1983; Bernier et al. 1993; Ouzounidou et al. 1993; Skórzyńska and Baszyński 1993; Maksymiec et al. 1994; Krupa and Baszyński 1995; Šeršeň et al. 1996 and 1998;
Plastoquinone pool	Cd, Cr, Cu, Zn	Tripathy and Mohanthy 1980; Baszyński et al. 1980 and 1982;
Cytochrome b_6/f	Cu, Ni	Veeranjaneyulu and Das 1982; Lidon and Henriques 1993a; Rao et al. 2000;
Plastocyanin	Cu, Hg, Ni	Kimimura and Katoh 1972; Veeranjaneyulu and Das 1982; Lidon and Henriques 1993;
Ferredoxin / Ferredoxin NADP$^+$ oxidoreductase	Cd, Cu, Hg, Ni, Zn	Shioi et al. 1978b; Honeycutt and Krogmann 1972; Siedlecka and Baszyński 1993; Šeršeň et al. 1996;
Photosystem I	Cu, Pb, Zn, Hg	Miles et al. 1972; Radmer and Kok 1974; Wong and Govindje 1976; Chugh and Sawhney 1999;

Ferredoxin, that is situated on the acceptor side of PS I was found to be one of the places of Cu^{2+} action (Shioi et al., 1978a, b). It was concluded from the EPR study that in *Chlorella vulgaris* copper interacted with ferredoxin and caused the interruption of the electron flow from PS I to the final acceptor $NADP^+$ (Šeršeň et al., 1996). However, the EPR results obtained by Králová et al. (1994) for spinach chloroplasts were different. Copper excess caused the interruption of the electron transport between PS II and PS I and did not damage the electron flow within PS I. Different effect of Cu^{2+} described in the above cited studies can result from some differences between the experimental systems that were used in these investigations: algal and chloroplasts suspensions. Ferredoxin was found to be also the site of mercury action (Honeycutt and Krogmann, 1972). Mercury ions interacted also with the donor side of PS I (Singh et al., 1989), and with plastocyanin (Radmer and Kok, 1974). Based on the EPR study, Šeršeň et al. (1998) found that mercury did not damage neither P700 nor the acceptor side of PS I. However, it caused the P700 oxidation in darkness. There is also a report in the literature on the inhibition of PS I in isolated bundle sheath and mesophyll chloroplasts of *Zea mays* by lead (Wong and Govindjee, 1976).

3.2. Photosystem II

The majority of studies dealing with the effect of heavy metals on the light phase of photosynthesis are devoted to reactions related to PS II. PS II is located in the grana or appressed lamellae. This multi-protein complex consists of at least 25 different subunits (Aro et al., 1993). The photosynthetic electron transport at the level of PS II both at oxidizing (donor) and reducing (acceptor) sides is effectively inhibited by different heavy metals (Table 1). Heavy metals may impair the functions of PS II directly via the plastoquinone pool or indirectly via feedback regulation by inhibition of the photosynthetic carbon reduction cycle enzymes and changes in ATP level.

The excess of copper is toxic for PS II, although its precise site(s) of binding and mechanism(s) of action are still not understood thoroughly (Hsu and Lee, 1988; Figure 1). In an EPR study it was found, that copper inhibited both the donor and the acceptor side of PS II in spinach (*Spinacia oleracea*) (Jegerschöld et al., 1995). In some studies, the Cu^{2+} binding site was located at various parts of the acceptor side of PS II (Mohanty et al., 1989; Yruela et al., 1991 and 1993; Maksymiec et al., 1994). In other investigations, Cu^{2+} was suggested to affect the water-oxidizing site of the donor side of PS II (Shioi et al., 1978a, b; Samson et al., 1988). It was also suggested that copper influenced directly the PS II core (Hsu and Lee, 1988).

Decreased PS II activity was also found to be a result of copper deficiency (Baszyński et al., 1978; Henriques, 1989). Cadmium was found to decrease the PS II activity and net CO_2 exchange in 30-day-old wheat plants (Malik et al., 1992). The inhibition of PS II electron flow was also observed after 10-min dark incubation of chloroplasts from *Zea mays* with 0.2 µM Cd^{2+} (Bazzaz and Govindjee, 1974). Donor side of PS II was found to be a site of zinc interaction (Rashid et al., 1991). Lead inhibited PS II activity in isolated chloroplasts (Miles et al., 1972).

3.3. Chlorophyll fluorescence parameters

Chlorophyll fluorescence measurements are simple and nondestructive. This method can reveal the detailed properties of PS II activity, including the fraction of open PS II centres, energy dissipation via antennae, and photoinhibition of PS II. Fluorescence spectroscopy appeared also a useful tool for studying the toxic functions of heavy metals, specially at low concentrations of environmentally realistic levels (Atal et al., 1991; Maksymiec and Baszyński, 1996a). Whole leaf fluorescence was used as a technique for measuring tolerance of plants to heavy metals (Homer et al., 1980). In *Phaseolus vulgaris* and *Phaseolus coccineus* photosynthetic functions were studied by fluorescence analysis (Krupa et al., 1992 and 1993a, b; Maksymiec and Baszyński, 1999a). This method was also applied as a reliable and rapid screen for Al tolerance in cereals (Moustakas et al., 1993). Chlorophyll fluorescence analysis was used to investigate the effect of manganese toxicity on photosynthesis in leaves of tree species differing in successional traits (Kitao et al., 1997 and 1998).

Copper, lead and mercury supplied to isolated thylakoid membranes resulted in the decrease of fluorescence parameters: F_0 and F_m as well as the lower value of the quantum yield of PS II photochemistry (Boucher and Carpentier, 1999). Ciscato et al. (1999) showed different effect of copper, zinc and cadmium on the fluorescence induction kinetics in primary leaves of bean plants. In this case, metal applications did not cause strong changes in F_v/F_m value, that means only a slight effect of these metals on PS II efficiency. Different fluorescence induction kinetics was rather a consequence of different sites and/or mechanisms of action of the investigated metals.

Variable fluorescence was inhibited by copper (160 µM) in *Silene compacta* (27%, Cu-tolerant) and in *Thlaspi ochroleucum* (81%, Cu-sensitive), which was probably connected with the inhibition of the photooxidizing side of PS II (Ouzounidou, 1993). Modulated fluorescence methods, photoacoustics and leaf absorbance were used to investigate the effects of Cu ions on *in vivo* treated leaves of *Thlaspi ochroleucum* (Ouzounidou et al., 1993). Cu-induced loss of Chl that resulted in the deficiency in light harvesting capacity of PS II and was a reason of the imbalance in energy distribution between the two photosystems. As a result of Cu treatment, the decrease of the maximal fluorescence (F_m) and changes of variable fluorescence (F_v) were also observed. The imbalance between the rate of Q_A reduction by PS II activity and the rate of Q^-_A reoxidation by cyt $b_6 f$ activity resulted in the reduction state of the primary electron acceptor (quinone A, Q^-_A). Treatment with copper decreased also such parameters as R_{fd} (vitality index), photosynthetic energy storage and F_v/F_m values. Measurements of 820 nm absorbance changes in saturating far-red light indicated less pronounced changes in the activity of PS I reaction centre compared with that of PS II in *T. ochroleucum* leaves after Cu exposure (Ouzounidou, 1996).

Chlorophyll fluorescence parameters, like F_m, F_v, and $t_{1/2}$ (the last parameter is a simple indicator of the pool size of electron acceptors on the reducing side of PS II, including the plastoquinone pool) were analyzed for wheat plants (*Triticum aestivum* L. cv. Vergina) that grew in the field on ore bodies containing Cu (3050 µg/g), whereas control plants grew in garden soil with 140 µg/g of Cu. In the ore-grown plants, F_v/F_m was lower and F_0 was higher than in control plants, which was probably connected with a lower pool size of electron acceptors on the reducing side of PS II in the treated plants (Lanaras et al., 1993).

The quantum yield of PS II photochemistry, calculated as the F_v/F_m ratio, measured in dark-adapted leaves in field conditions decreased by 7% in oats (*Avena sativa*) growing on a contaminated site with excess Cu and Pb. The increase of the half-rise time (t1/t2) from the initial (F_0) to maximal (F_m) Chl fluorescence was observed, which was probably related to the decrease of the amount of active pigments associated with the photosynthetic apparatus and with a smaller functional Chl antennae size of the photosynthetic apparatus compared with the control plants (Moustakas et al., 1994).

Yruela et al. (2000) analyzed the Chl fluorescence parameters in thylakoids and PS II membranes of spinach treated with copper. Both the variable fluorescence (F_v) and the F_v/F_m ratio were reduced with increasing Cu^{2+} concentrations (10 – 40 μM), whereas the initial fluorescence (F_0) remained unchanged. For untreated thylakoids, the F_v/F_m ratio was equal 0.79 and it decreased to 0.5 under the influence of 40 μM Cu^{2+}. Similar results were also reported by other authors (Shioi et al., 1978a; Hsu and Lee, 1988; Samson et al., 1988; Mohanty et al., 1989).

Szalontai et al. (1999) investigated the effect of selected heavy metals: cadmium, nickel, copper, lead and zinc on the fluorescence induction kinetics in pea chloroplasts. It was found that Cu^{2+}, Pb^{2+} and Zn^{2+} strongly inhibited the PS II activity, which was observed as a pronounced decrease in both the variable (F_v) and the maximal (F_m) fluorescence. Similar results were earlier obtained by Droppa and Horváth (1990) as well as by Krupa and Baszyński (1995). No effect of Cd^{2+} and Ni^{2+} on Chl fluorescence parameters was observed (Krupa and Baszyński, 1995; Szalontai et al., 1999).

In iron deficient sugar beet (*Beta vulgaris*) the reduction of the PQ pool in darkness was monitored by analysis of Chl fluorescence parameters. The extent of this reduction depended on the degree of Fe-deficiency and the duration of preillumination or dark adaptation (Belkhodja et al., 1998).

Electron transport

Excess of copper was found to inhibit the electron transport through PS II. In some investigations the place of its action was found to be on the donor side of PS II (Figure 1). Treatment of PS II particles with 10 μM Cu^{2+} resulted in the slower kinetics of $P680^+$ reduction, however, charge separation ($P680^+Q_A^-$) remains unchanged (Schröder et al., 1994). It was concluded, that some specific modifications of $TyrZ^+$ and/or its microenvironment was responsible for the inhibition of electron transport to $P680^+$. Arellano et al. (1995) also located the site of Cu^{2+} action at the donor side of PS II, close to the reaction centre. What is more, these authors found that light is essential for Cu damage, and that the effect is irreversible.

A complete disappearance of EPR signal belonging to $TyrZ^+/TyrD^+$ (tyrosine cation radicals from D1 and D2 proteins, respectively), due to the copper treatment was found for *Chlorella vulgaris* (Šeršeň et al., 1996). This effect was concomitant with the decrease in the fluorescence intensity of Chl. In this case the toxic effect of copper can be interpreted as the replacement of Mn^{2+} by Cu^{2+} in OEC, which was confirmed by the presence of the EPR signal belonging to the free Mn^{2+}. The EPR signal from free Mn^{2+} due to the copper treatment was earlier reported by Králová et al. (1994). In some reports, site of Cu^{2+} action was also found on the acceptor side of PS II, between pheophytin and Q_B (Yruela et al., 1991, 1992 and 1993; Mohanty et al., 1989).

Šeršeň et al. (1998) showed in EPR studies, that mercury disrupted the electron transport chain between PS II and PS I in chloroplasts isolated from spinach (*Spinacea oleracea* L.) by interaction with $TyrZ^+/TyrD^+$ intermediates.

3.4 Polypeptide composition

Inhibitory effects of heavy metals on the PS II polypeptides of 33, 23 and 17 kDa has been observed. These extrinsic proteins are needed for long-term stability of the PS II and proper functioning of the oxygen evolving complex (OEC), and regulating its activity.

Yruela et al. (2000) investigated the effect of Cu^{2+} on polypeptide composition in thylakoids and PS II preparation from spinach. For 300 µM Cu^{2+}, that corresponded to the 1400 Cu^{2+} ions per PS II reaction centre unit, the release of 33, 24 and 17 kDa proteins from the membrane was observed. Lower copper concentration (50 µM) did not remove any specific polypeptides of PS II. Copper acts specifically on OEC proteins since no effect on LHC II antenna complex as well as D1 protein was observed.

In *Vigna unguiculata*, degradation of PS II polypeptides of 43, 33, 23 and 17 kDa were observed after 3 days of treatment with 3 - 9 mM Cd^{2+} (Nedunchezhian and Kulandaivelu, 1995). What is more cadmium was found to act at a similar site in chloroplasts as UV-B.

Heavy metals also interfere with proteolitic degradation of the D_1 protein, which takes place during normal functioning of the PS II reaction centre, and the damaged D_1 protein can not be replaced.

3.5. Oxygen evolution activity

The oxygen evolving complex, located on the lumenal side of the thylakoids donates electrons from water to the PS II reaction centre. OEC was reported as the primary target of heavy metals toxicity (Table 1).

OEC was postulated to be the main site of cadmium action (Figure 2). The inhibition was located at the level of manganoprotein of the water splitting system (Van Duijvendijk-Matteoli and Desmeta, 1975). OEC may be destructed either directly or indirectly due to the interaction of cadmium with ions like Mn^{2+}, Ca^{2+} and Cl^- that are necessary for proper functioning of OEC (Maksymiec and Baszyński, 1998a; Skórzyńska and Baszyński, 1993).

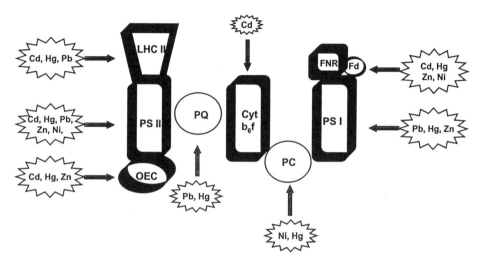

Figure 2: Sites of non-essential heavy metals toxicity in photosynthetic membrane. LHC II, light harvesting complex of PS II; other explanations see Figure 1.

Prasad et al. (2001) compared the toxicity of cadmium and copper for aquatic plant *Lemna trisulca* L. They observed the decrease of oxygen evolution (OE) with increasing Cd concentration (up to 10 mM). Suppression of OE was also observed for 2-50 µM of Cu^{2+}. However, for 1 µM of Cu^{2+} a pronounced stimulation of OE was measured, which correlated with the relative increase of Chl *a* concentration. Stimulation of OE activity was also observed in PS II particles isolated from tobacco for the equimolar Cu^{2+}/ PS II proportion, while higher Cu concentrations resulted in OE inhibition (Burda et al., in press). Yruela et al. (2000) have shown that copper-induced loss of OE activity was faster in PS II membranes than in thylakoid preparation. A 50% inactivation of OE occurred at the concentration of 7 and 27 µM Cu^{2+} for PS II membranes and thylakoids, respectively. The inactivation of OEC seemed to be the primary effect of copper action since no changes in protein composition and integrity were observed for such a low Cu^{2+} concentration. Manganese cluster in the OEC was found to be the place of copper action also in *Chlorella vulgaris* (Šeršeň et al., 1996).

Zinc (1-5 mM) was found to inhibit the oxygen evolution in isolated barley chloroplasts (Tripathy and Mohanty, 1980). Besides the inhibition of oxygen evolution, 10 min treatment of lettuce thylakoids with 0.5-5 mM zinc resulted in the appearance of EPR signal belonging to the free Mn^{2+} (Miller and Cox, 1983). Full inhibition of OE was observed at the stage of the release of 2 manganese atoms per PS II unit.

The release of manganese from OEC was also the result of spinach chloroplasts treatment with mercury (Šeršeň et al., 1998). Bernier et al. (1993) observed the inhibition of OE concomitant with quenching of variable Chl fluorescence as a result of Hg treatment. This inhibitory effect of Hg was reversed by chloride, an inorganic cofactor necessary for the optimal function of PS II. However, Ca, another essential cofactor, showed no reversal capacity (Bernier et al., 1993).

3.6 Photoinhibition

The photosynthetic apparatus is susceptible to damage by light. When the rate of photodamage exceeds the rate of repair, a decrease of photosynthetic yield is observed. This phenomenon is known as photoinhibition (Aro et al., 1993). The main site of photodamage is PS II (Andersson et al., 1992). Two different mechanisms leading to photodamage of PS II known as acceptor side and donor side photoinhibition have been described (Barber and Andersson, 1992). Acceptor side photoinhibition is connected with the formation of the doubly reduced and protonated form of the primary plastoquinone acceptor Q_A (Q_AH_2), which can be irreversibly lost from its binding site on PS II. Donor side photoinhibition is observed when the donor side is unable to deliver electrons to $P680^+$. As a consequence of both types of photoinactivation, the conformational changes of the D1 protein are triggered, which may lead to its degradation. The release into the thylakoid lumen of the extrinsic polypeptides (33, 23 and 17 kDa) and four Mn atoms follows the breakdown of D1 protein. In healthy plants an equilibrium among photoinhibition, D1 degradation and D1 synthesis is maintained and, as a consequence, a high steady-state concentration of active PS II centres is observed. Heavy metal treatment may change this equilibrium and decrease the number of active PS II units, that results in lower photosynthetic yield.

Copper excess was found to slow the PS II repair cycle in the green alga *Chlorella pyrenoidosa* (Vavilin et al., 1995). Similar results were found by Pätsikkä et al. (1998) for bean (*Phaseolus vulgaris* L. cv Dufrix). The authors observed the increase of the quantum yield of photoinhibition resulting in a decrease in the steady-state concentration of the active PS II centres in illuminated leaves. Photoinhibitory damage of PS II caused by copper excess was also reported by Maksymiec et al. (1994) and Yruela et al. (1996).

Synthesis, degradation and assembly of the D1 protein were found to be greatly affected by cadmium (Geiken et al., 1998). In cadmium treated pea (*Pisum sativum*) and broad beans (*Vicia faba*), initially the stimulation of D1 protein synthesis was observed which was followed by its inhibition (Geiken et al., 1998; Franco et al., 1999). However, in these studies the primary effect of cadmium on the PS II complex, which triggered modification of D1 synthesis, has not been revealed. Voigt et al. (1998) showed that sensitivity towards photoinhibition was considerably decreased in the case of the cadmium-tolerant mutant of *Chlamydomonas reinhardtii*.

4. CONCLUSIONS

Heavy metals are effective inhibitors of photosynthesis and they may also induce photoinhibition. Heavy metals can interact directly with different electron carriers and decrease the electron transport efficiency. Another aspect of their toxicity is the inhibition of oxygen evolution (Table 1). Changes in the content or structure of proteins in the photosynthetic apparatus may also be a consequence of heavy metal treatment (Table 1; section *3.2.1*).

Cadmium is an effective inhibitor of plant metabolism, and of the photosynthetic processes in particular. It interacts with most of the components of photosynthetic electron transport chain (Table 1; Figure 2) and influences also net photosynthesis and stomatal closure (Bazzaz et al., 1974a, b, Huang et al., 1974), transpiration (Hagemeyer and Waisel, 1989a, b), mesophyll resistance to CO_2 uptake (Lamoreaux and Chaney, 1978) and dry matter production (Baszyński, 1986).

Heavy metal stress may be also an indirect reason for the impairment of photosynthetic functions because it may induce structural damage to chloroplasts, permanent stomatal closure, increased synthesis of ethylene (promoter of senescence) and imbalance in water relations (Fuhrer, 1988; Pnnazio and Roggero, 1992).

A promising area to study the influence of heavy metals on the photosynthetic apparatus is the investigation of organization and functional properties of the photosynthetic apparatus during the greening of etiolated seedlings under heavy metal stress (Krupa et al., 1987; Woźny et al., 1995; Berska et al., 2001). Horváth et al. (1996) demonstrated that in the greening leaves of barley, cadmium inhibited the incorporation of Chl into pigment-protein complexes that are essential for optimal function of PS II, mainly by interfering with organization of these complexes. Cadmium (0.5-1 mM) also drastically reduced the LHC II accumulation in green bean (*Phaseolus vulgaris*) by reducing the steady state level of *lhcb* transcripts, although the total leaf protein level remained unaffected (Tziveleka et al., 1999).

Copper was found to retard the Chl integration into the photosystems, both PS II and PS I, in greening barley seedlings, which was connected with the persistence of population of unquenched Chl molecules (Caspi et al., 1999). It also accelerated PS II closure and resulted in a larger LHC II antenna (lower Chl a/b ratio). Another effect was a decrease in photosynthetic capacity observed as a suppression of OE (40% suppression for 1 mM Cu^{2+}). However, the toxic effects observed for Cu^{2+} were less severe than those of other heavy metals e.g. Cd^{2+} (Parekh et al., 1990; Horváth et al., 1996). The role of single/multiple metals (essential and non-essential) in etiolated seedlings and their effects on pigment biosynthesis needs critical study. Heavy metal-substituted Chl, particularly in aquatic plants, is of considerable significance for the photosynthetic efficiency (Küpper et al., 1996 and 1998).

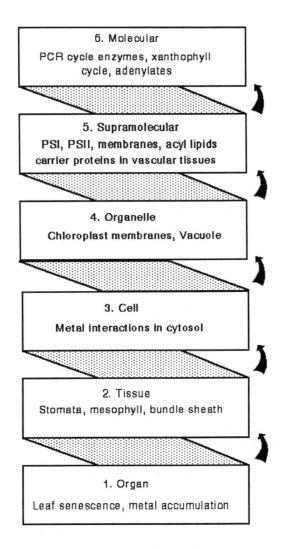

Figure 3. Heavy metals react with photosynthetic apparatus at various levels of organization and architecture i.e. 1. accumulation of metals at organ level i.e leaf; 2. partitioning in leaf tissues like stomata, mesophyll and bundle sheath; 3. metal interactions with cytosolic enzymes and organics; 4. react with chloroplast membranes; 5 supramolecular level action particularly on PS II, PS I, membranes acyl lipids, carrier proteins in vascular tissues; 6. molecular level interactions particularly with photosynthetic carbon reduction cycle enzymes, xanthophyll cycle and adenylates etc.

Investigations on the influence of heavy metals on light reactions of photosynthesis are of great significance. Inhibition of the light photosynthetic reaction that results in a lower yield of photosynthesis is a serious reason for lower biomass production and lower crop yield, which is an important agricultural problem. On the other hand, tolerant plants which are able to carry photosynthesis in the presence of heavy metals in environment can be very useful for phytoremediation purposes.

ACKNOWLEDGEMENTS

This work was supported by the grant 6P04A 028 19 from the Committee for Scientific Research (KBN) of Poland.

Table 2. *Cadmium influence on photosynthesis in angiosperms*

Taxon	Function	Reference
Acer saccharum	PS	Lameroux and Chaney 1978
Beta vulgaris	Cd/EDTA on sugar metabolism Leaf transpiration Foliar uptake	Greger and Landberg 1986 Greger and Johansson 1992 Greger et al. 1993
	Influence of Se on Cu and Cd toxicity and CO_2 assimilation	Landbegr and Greger 1994
Cajanus cajan	PS and enzymes PCR cycle	Sheoran et al. 1990
Cucummis sativus	Chl The influence on NO_3- uptake	Burzyński and Buczek 1989 Burzyński and Buczek 1994
Glycine max	Activities of several enzymes Ultrastructure of developing CP Senescence	Lee et al. 1976 Ghoshrony and Nadakavukaren 1990 Pnnzaio and Roggero 1992
Helianthus annus	Inhibition of PS	Bazzaz et al. 1975
Hordeum vulgare	Biochcmical characteristics of PS RUBISCO structure Chl Formation of PSA during greening	Stiborova et al. 1987 Stiborova 1988 Gadallah 1995a,b Horvath et al. 1996
Medicago sativa	Photosynthetic electron transport Gas exchange functions	Becerril et al. 1989 Becerril et al. 1989
Oryza sativa *Lycopersicon esculentum*	PS PS Release of proteins from TM	Misra et al. 1989 Baszyński et al. 1980 Maksymiec and Baszyński 1998
Oryza sativa	PS and carbohydrate distribution	Moya et al. 1993

Taxon	Function	Reference
Phaseolus coccineus	Chloroplast photochemistry	Tukendorf and Baszyński 1991
		Galactolipase activity
		Skórzyńska *et al.* 1991
	The changes in PSII complex peptide	Skórzyńska and Baszyński 1993
	Interaction on photosynthetic capacity of primary leaves	Krupa and Siedlecka 1995
	Photochemical activity	Skórzyńksa-Polit and Baszyński 1995
	Primary leaf growth	Skórzynksa-Polit *et al.* 1995
Phaseolus vulgaris	Ultrastructural disorders	Barcelo *et al.* 1988
	Induction of enzymes	Van Assche *et al.* 1988al
	Structural and ultrastructural disorders	Poschenreider *et al.* 1989
	Stomatal resistance	Poschenreider *et al.* 1989
Phaseolus vulgaris	Chl degradation in seedlings - involvement of lipid peroxides	Somaseskharaiah *et al.* 1982
	PS - a fluorescence analysis	Krupa *et al.* 1993
Pisum sativum	Foliar uptake	Greger *et al.* 1993
	Influence of Se on uptake and toxicity of Cu and Cd	Landberg and Greger 1994
Raphanus sativa	PS and PR	Krupa *et al.* 1987
	Light harvesting Chl a/b protein complex II	Krupa 1988
Spinacia oleracea	On the inhibitory side of PS II in isolated CP	Van Duijvendijk-Matteolii and Desmata 1975
	Inhibition of photophosphorylation in CP	Lucero *et al.* 1976
	PS and Hill activity of isolated CP	Hampp *et al.* 1976
	PS	Weigel 1985a,b; Baszyński 1986

Taxon	Function	Reference
Trifolium pratense	Photosynthetic electron transport and gas exchange	Becerril *et al.* 1989
Triticum aestivum	Inhibition of CP photochemical reactions in seedlings.	Atal *et al.* 1991
	PS	Bishnoi *et al.* 1993b
	Lipid composition of TM and carbon metabolism	Malik *et al.* 1992a,b
	Influence of Se on uptake	Landberg and Greger 1994 Abo-Kassem *et al.* 1995
Vicia faba	Changes in nitrogen metabolism and enzyme activities	Shalaby and Al-Wakeel 1995
Vigna unguiculata	Changes in polypeptide composition PSI and PSII activities	Nedunchezhian and Kulandaivelu 1995
Zea mays	PEP carboxylase	Stiborova and Leblova 1986
	PS Biochemical characteristics of S Ultrastructure of CP	Stiborova *et al.* 1986 Stiborova *et al.* 1987 Ghoshrony and Nadakavukaren 1990
	PS and enzymes of photosynthetic sulphate and nitrate assimilation Inhibition of electron flow around	Ferretti *et al.* 1993 Siedlecka and Baszyński 1993 PSI in CP due to Cd-induced iron deficiency.
	PS, CP ultrastructure PS, GE functions, Chl and carotenoids Alleviation Cd toxicity by Ca	Rascio *et al.* 1993 Prasad 1995b El-Enany 1995

Table 3. Copper influence on photosynthesis in angiosperms

Taxon	Function	Reference
Alyssum montanum	Chl, oxygen evolution signal, photochemical energy storage (PSII function)	Ouzounidou et al. 1993
Avena sativa	Deficiency on PS	Baszyński et al. 1978
Cucummis coerulea	PS	Maksymiec et al. 1994
Cucummis sativus	The influence on NO_3- uptake	Burzyński and Buczek 1994
Hordeum vulgare	Biochemical characteristics of PS Mechanism of action on RUBISCO	Stiborova et al. 1987 Stiborova 1988
Koeleria splendens	PS Chl	Ouzounidou et al. 1993 Ouzounidou 1995
Minuartia hirsuta	Chl and growth	Ouzounidou et al. 1994
Oryza sativa	Cu PS and related pigments Nitrate to ammonia reduction Photosynthetic electron carriers, RNAse activity and membrane permeability Changes in TM Oxygen toxicity in CP Inhibition of PS	Lidon and Henriques 1991 Lidon and Henriques 1992a-c Lidon and Henriques 1992d Lidon and Henriques 1993a Lidon and Henriques 1993b Lidon and Henriques 1993 Lidon et al. 1993
Phaseolus coccineus	Changes of chl. and its fluorescence	Maksymiec and Baszyński, 1996a,b Gadallah 1995[a]
Pisum sativum	PS, CP structure Influence of Se on uptake and toxicity of and Cd	Angelov et al. 1993 Landberg and Greger 1994

Taxon	Function	Reference
	PSII, electron transport at the secondary quinone acceptor QB	Mohanty et al. 1989b
Silene compacta	Chl, Oxygen evolution signal, photochemical energy storage (PSII functions)	Ouzounidou et al. 1993
	Chl, fluorescence	Ouzounidou 1993
Spinacia oleracea	Deficiency on PS	Baszyński et al. 1978
	Photosynthetic electron transport	Shioi et al. 1978
	Photosynthetic electron transport and	Samuelson and Öquist 1980 Chl-protein complexes PS
	Baszynski et al. 1982	
	Hill reaction and Chl	Barua and Jana 1986
	Characteristics of the PA	Baszyńzki et al. 1988
	Toxicity to PS II	Hsu and Lee 1988
	Changes in acyl lipd and FA composition in TM	Maksymiec et al. 1992
Thlaspi ochroleucum	Chl, fluorescence	Ouzounidou 1993
Triticum aestivum	Influence of Se on uptake and toxicity of Cu and Cd	Landberg and Greger 1994
Zea mays	PEP carboxylase	Stiborova and Leblova 198
	PS	Stiborova et al. 1986
	Biochemical characteristics of PS	Stiborova et al. 1987

REFERENCES

Abo-Kassem, E., Sharaf-el-din, A., Rozema, J.and Foda, E.A. (1995) Synergistic effects of cadmium, NaCl on the growth, photosynthesis and ion content in wheat plants, *Biol. Plant.* 37, 241–249.
Ahmed, A. and Tajmir-Riahi, H.A. (1993) Interaction of toxic metal ions Cd^{2+}, Hg^{2+} and Pb^{2+} with light-harvesting proteins of chloroplst thylakoid membranes. An FTIR spectroscopic study, *J. Inorg. Biochem.* 50, 235—243.
Almog, O., Lotan, O., Shoham, G. and Nechushtai, R. (1991) The composition and organization of photosystem I, *J. Basic Clin. Physiol. Pharmacol.* 2, 123—140.
and cadmium in pea *(Pisum sativum)* and wheat *(Triticum aestivum). Physiol Plant* **90**,
Andersson, B., Salter, A.H., Virgin, I., Vass, I. and Styring, S. (1992) Photodamage to photosystem II- primary and secondary events, *J. Photochem. Photobiol. B:Biol.* 15, 15—31.
Angelov, M., Tsonev, T., Uzunova, A. and Gaidardjieva, K. (1993) Cu^{2+} effect upon photosynthesis, chloroplast structure, RNA and protein synthesis of pea plants, *Photosynthetica* 28, 341—350.
Arellano, J.B., Lazaro, J.J., Lopez-Gorge, J. and Baron, M. (1995) The donor side of photosystem II as the copper-inhibitory binding site. Fluorescence and polarographic studies, *Photosynth. Res.* **45**, 127—134.
Aro, E.M., Virgin, I. and Andersson, B. (1993) Photoinhibition of photosystem II. Inactivation, protein damage and turn-over, *Biochim. Biophys. Acta* **1143**, 113—134.
Atal, N., Saradhi, P.P. and Mohanty P. (1991) Inhibition of the chloroplast photochemical reactions by treatment of wheat seedlings with low concentrations of cadmium. Analysis of electron transport activities and changes in fluorescence yield, *Plant Cell Physiol.* **32**, 943—951.
Barber, J. and Andersson, B. (1992) Too much of a good thing: light can be bad for photosynthesis, *Trends Biochem. Sci.* 17, 61—66.
Barceló, J., Vazquezand, M.D. and Poschenrieder, C. (1988) Structural and ultrastructural disorders in cadmium-treated bush bean plants *(Phaseolus vulgaris* L.). *New Phytologist* 108, 37-49
Baron, M., Arellano, J.B. and Gorge, J.L. (1995) Copper and photosystem II: A controversial relationship, *Physiol. Plant.* **94**, 174—180.
Barua, B. and Jana, N. (1986) Effect of heavy metals on dark induced changes in Hill action activty, chlorophyll and protein contents, dry matter and tissue permeability in detached *Spinacia oleracea* L. leaves, *Photosynthetica* 20, 74—76.
Baszyński, T. (1986) Interference of Cd^{2+} in functioning of the photosynthetic apparatus of higher plants, *Acta Soc. Bot. Pol.* 55, 291—304.
Baszyński, T. and Tukendorf, A. (1984) Copper in the nutrient medium of higher plants and their photosynthetic apparatus activity, *Folia Soc. Sci. Lubliensis* 26, 31—39.
Baszyński, T., Król, M., Krupa, Z., Ruszkowska, M., Wojcieska, U. and Wolińska, D. (1982) Photosynthetic apparatus of spinach exposed to excess copper, *Z. Pflanzenphysiol.* 108, 385—395.
Baszyński, T., Ruszkowska, M., Król, M., Tukendorf, A. and Wolińska D. (1978) The effect of copper deficiency on the photosynthetic apparatus of higher plants, *Z. Pflanzenphysiol.* 89, 207—216.
Baszyński, T., Wajda, L., Król, M., Wolińska, D., Krupa, Z. and Tukendorf, A. (1980) Photosynthetic activities of cadmium treated tomato plants, *Physiol. Plant.* 48, 365—370.
Bazzaz, F.A., Carlson, R.W. and Rolfe, G.L. (1975) Inhibition of corn and sunflower photosynthesis by lead, *Physiol. Plant.* 34, 326—329.
Bazzaz, F.A., Rolfe, G.L. and Carlson, R.W. (1974a) Effect of cadmium on photosynthesis and transpira-tion of excised leaves of corn and sunflower, *Plant Physiol.* 32, 373—376.
Bazzaz, F.A., Rolfe, G.L. and Carlson, R.W. (1974b) Differing sensitivity of corn and soybean photosynthesis and transpiration to lead contamination, *J. Environ. Qual.* 3, 156—158.
Bazzaz, M.B. and Govindjee (1974) Effects of cadmium nitrate on spectral characteristics and light reactions of chloroplasts, *Environ. Sci. Lett.* 6, 1—12.
Becerril, J.M., Gonzalez-Murua, C., Munoz-rueda, R. and De Felipe, M.R. (1989) The changes induced by cadmium and lead in gas exchange and water relations clover and lucerne. Plant *Physiol Biochem* 27, 913-918
Belkhodja, R., Morales, F., Quilez, R., López-Millán, A.F., Abadia, A. and Abadia, J. (1998) Iron deficiency causes changes in chlorophyll fluorescence due to the reduction in the dark of the Photosystem II acceptor side, *Photosynth. Res.* **56**, 265—276.

Bernier, M., Popovic, R. and Carpentier, R. (1993) Mercury inhibition at the donor side of photosystem II is reversed by chloride, *FEBS Lett.* **321**, 19—23.
Berska, J., Myśliwa-Kurdziel, B. and Strzałka, K. (2001) Transformation of protochlorophyllide to chlorophyllide in wheat under heavy metal stress, in Proceedings of the XII International Congress on Photosynthesis, Brisbane, in press.
Bishnoi, N.R., Chug, L.K. and Sawhney, S.K. (1993a) The effect of chromium on photosynthesis, respiration and nitrogen fixation in pea (*Pisum sativum* L.) seedlings, *J. Plant Physiol.* **142**, 25—30.
Bishnoi, N.R., Sheoran, I.S. and Singh, R. (1993b) Influence of cadmium and nickel on photosynthesis and water relations in wheat leaves of differential insertion level, *Photosynthetica* **28**, 473—479.
Boucher, N. and Carpentier, R. (1999) Hg^{2+}, Cu^{2+}, Pb^{2+} - induced changes in photosystem II photochemical yield and energy storage in isolated thylakoid membranes: A study using simultaneous fluorescence and photoacoustic measurements, *Photosynth. Res.* **59**, 167—174.
Burda, K., Kruk, J., Radunz, A., Schmid, G.H. and Strzałka, K. (submitted) Stimulation and inhibition of oxygen evolution in photosystem II by copper (II) ions.
Burzynski, M and Buczek, J. (1994) The influence of Cd, Pb, Cu and Ni on NO_3^- uptake by cucumber seedlings. II. *In vitro* and *in vivo* effects of Cd, Pb, Cu and Ni on the plasmalemma ATPase and oxidoreductase from cucumber seedlings roots. *Acta Physiol Planta* **16**, 297-302
Burzyński, M. and Buczek, J. (1989) Interaction between cadmium and molybdenum affecting the chlorophyll content and accumulation of some heavy metals in the second leaf of *Cucummis sativus* L. *Acta Physiol Planta* **11**, 137-145
Caspi, V., Droppa, M., Horváth, G., Malkin, S., Marder, J.B. and Raskin, V.I. (1999) The effect of copper on chlorophyll organization during greening of barley leaves, *Photosynth. Res.* **62**, 165—174.
Cedeno-Maldonado, A., Swader, J.A. and Heath, R.L. (1972) The cupric ion as an inhibitor of photosynthetic electron transport in isolated chloroplasts, *Plant Physiol.* **50**, 698—701.
Chugh, L.K. and Sawhney, S.K. (1999) Photosynthetic activities of *Pisum sativum* seedlings grown in presence of cadmium, *Plant Physiol. Biochem.* **37**, 297—303.
Ciscato, M., Vangronsveld, J. and Valcke, R. (1999) Effect of heavy metals on the fast chlorophyll fluorescence induction kinetics of photosystem II: a comparative study, *Z. Naturforsch.* **54c**, 735—739.
Clijsters, H. and Van Assche, F. (1985) Inhibition of photosynthesis by heavy metals, *Photosynth. Res.* **7**, 41—40.
Dietz, K.J. and Hartung, W. (1996) Leaf epidermis: its ecophysiological significance, *Prog. Bot.* **57**, 32—53
Drążkiewicz, M. (1994) Chlorophyllase: occurrence, functions, mechanism of action, effects of external and internal factors, *Photosynthetica* **30**, 321—331.
Droppa, M. and Horváth, G. (1990) The role of copper in photosynthesis, *Crit. Rev. Plant Sci.* **9**, 111—123.
Droppa, M., Terry, N. and Horváth, G. (1984) Effects of Cu deficiency on photosynthetic electron transport, *Proc. Natl. Acad. Sci. USA* **81**, 2369—2373.
Ernst, W.H.O., Verkleij, J.A.C. and Schat, H. (1992) Metal tolerance in plants, *Acta Bot. Neerl.* **41**, 229—248.
Ferretti, M., Ghisi, R., Merlo, L., Vecchia, F.D. and Passera, C. (1993) Effect of cadmium on photosynthesis and enzymes of photosynthetic sulphate and nitrate assimilation pathways in maize (*Zea mays* L.), *Photosynthetica* **29**, 49—54.
Franco, E., Alessandrelli, S., Masojidek, J., Margonelli, A. and Giardi, M.T. (1999) Modulation of D1 protein turnover under cadmium and heat stress monitored by [S-35]methionine incorporation, *Plant Sci.* **144**, 53—61.
Fuhrer, J. (1988) Ethylene biosynthesis and cadmium toxicity in leaf tissue of beans *Phaseolus vulgaris* L., *Plant Physiol.* **70**, 162—167.
Gadallah, M.A.A. (1995) Interactive effect of heavy metals and temperature on the growth and chlorophyll, saccharides and soluble nitrogen in Phaseolus plants, *Biol. Plant.* **36**, 373—382.
Gadallah, M.A.A. (1995a) Interactive effect of heavy metals and temperature on the growth and chlorophyll, saccharides and soluble nitrogen in *Phaseolus* plants. *Biol Planta* **36**, 373-382.
Geider, R.J. and La Roche, J. (1994) The role of iron in phytoplancton photosynthesis, and the potential for iron-limitation of primary productivity in the sea, *Photosynth. Res.* **39**, 275—301.
Geiken, B., Masojidek, J., Rizutto, M., Pompili, M.L. and Giardi, M.T. (1998) Incorporation of [S-35]methionine in higher plants reveals that stimulation of the D1 reaction centre II protein turnover accompanies tolerance to heavy metal stress, *Plant Cell Environm.* **21**, 1265—1273.
Ghoshrony, S., and Nadakavukaren, K.J. (1990) Influence of Cadmium on the ultrastructure of developing chloroplast of soybean and corn. *Environ Exp Bot* **30**, 187-192
Greene, R.M., Geider, R.J., Kolber, Z. and Falkowski, P.G. (1992) Iron-induced changes in light harvesting and photochemical energy conversion processes in eukariotic marine algae, *Plant Physiol.* **100**, 565—575.

Greger, M. and Johansson, M. (1992) Cadmium effects on leaf transpiration of sugar beet (*Beta vulgaris*), *Physiol. Plant.* **86**, 465—473.
Greger, M. and Ögren, E. (1991) Direct and indirect effects of Cd^{2+} on photosynthesis in sugar beet (*Beta vulgaris*), *Physiol. Plant.* **83**, 129—135.
Greger, M., and Lindberg, S. (1986) Effects of Cd^{2+} and EDTA on young sugar beets (*Beta vulgaris*). Cd^{2+} uptake and sugar metabolism. *Physiol Planta* **66**, 69-74
Greger, M., and Lindberg, S. (1986) Effects of Cd^{2+} and EDTA on young sugar beets (*Beta vulgaris*). Cd^{2+} uptake and sugar metabolism. *Physiol Planta* **66**, 69-74
Greger, M., Johansson, M., Stihl, A. and Hamza, K. (1993) Foliar uptake of Cd by pea (*Pisum sativum*) and sugar beet (*Beta vulgaris*), *Physiol. Plant.* **88**, 563-570.
Guikema, J.A. (1985) Fluorescence induction characteristics of *Anacystis nidulans* during recovery from iron deficiency, *J. Plant Nutr.* **8**, 891—908.
Hagemeyer, J. and Waisel, Y. (1989a) Excretion of ions (Cd, Li, Na and Cl) by *Tamarix aphylla*, *Physiol. Plant.* **75**, 280—284.
Hagemeyer, J. and Waisel, Y. (1989b) Influence of NaCl, $Cd(NO3)_2$ and air humidity on transpiration of *Tamarix aphylla*, *Physiol. Plant.* **77**, 247—253.
Hampp, R., Beulich, K. and Zeigler, H. (1976) Effects of zinc and cadmium on photosynthetic CO_2 fixation and Hill activity of isolated spinach chloroplasts, *Z. Pflanzenphysiol.* **77**, 336—344.
Henriques, F.S. (1989) Effect of copper deficiency on the photosynthetic apparatus of sugar beet (*Beta vulgaris* L.), *J. Plant Physiol.* **135**, 453—458.
Homer, J.R., Cotton, R. and Evans, E.H. (1980) Whole leaf fluorescence as a technique for measurement of tolerance of plants to heavy metals, *Oecologia* **45**, 88—89.
Honeycutt, R.C. and Krogmann, D.W. (1972) Inhibition of chloroplasts reactions with phenylmercuric acetate, *Plant Physiol.* **49**, 376—381.
Horváth, G., Droppa, M., Oravecz, A., Raskin, V.I. and Marder, J.B. (1996) Formation of the photosynthetic apparatus during greening of cadmium-poisoned barley leaves, *Planta* **199**, 238—243.
Hsu, B. and Lee, J. (1988) Toxic effects of copper on photosystem II of spinach chloroplasts, *Plant Physiol.* **87**, 116—119.
Huang, C.Y., Bazzaz, F.A. and Vanderhoeff, L.N. (1974) The inhibition of soybean metabolism by cadmium and lead, *Plant Physiol.* **54**, 122—124.
Jegerschöld, C., Arellano, J.B., Schroder, W.P., Van-Kan, P.J., Baron, M. and Styring, S. (1995) Copper(II) inhibition of electron transfer through photosystem II studied by EPR spectroscopy, *Biochemistry* **34**, 12747—12754.
Kimimura, M. and Katoh, S. (1972) Studies on electron transport associated with Photosystem I. I. Functional site of plastocyanin: inhibitory effects of $HgCl_2$ on electron transport and plastocyanin in chloroplasts, *Biochim. Biophys. Acta* **283**, 279—292.
Kitao, M., Lei, T.T. and Koike, T. (1997) Effect of manganese toxicity on photosynthesis of white birch (*Betula platyphylla* var. *Japonica*) seedlings, *Physiol. Plant.* **101**, 249—256.
Kitao, M., Lei, T.T. and Koike, T. (1998) Application of chlorophyll fluorescence to evaluate Mn tolerance of deciduous broad-leaved tree seedlings native to northern Japan, *Tree Physiol.* **18**, 135—140.
Králová, K., Šeršeň, F. and Blahová, M. (1994) Effects of Cu (II) complexes on photosynthesis in spinach chloroplasts. Aqua(aryloxyacetato)copper(II) complexes, *Gen. Physiol. Biophys.* **13**, 483—491.
Krupa, Z. (1988) Cadmium-induced changes in the composition and structure of the light harvesting chlorophyll a/b protein complex II in radish cotyledons, *Physiol. Plant.* **73**, 518—524.
Krupa, Z. (1999) Cadmium against higher plant photosynthesis – a variety of effects and where do they possibly come from? *Z. Naturforsch.* **54c**, 723—729.
Krupa, Z. and Baszyński, T, (1995) Some aspects of heavy metal toxicity towards photosynthetic apparatus - direct and indirect effects on light and dark reactions, *Acta Physiol. Plant.* **17**, 177—190.
Krupa, Z. and Baszyński, T. (1985) Effects of cadmium on the acyl lipid content and fatty acid composition in thylakoid membranes isolated from tomato leaves, *Acta Physiol. Plant.* **7**, 55—64.
Krupa, Z. and Siedlecka, A. (1995) Cd/Fe interactions and its effects on photosynthetic capacity of primary bean leaves, in: P. Mathis (ed,) *Photosynthesis: from light to biosphere*, vol 4. pp. 621—624.
Krupa, Z., Öquist, G. and Huner, N.P.A. (1992) The influence of cadmium on primary PS II photochemistry in bean as revealed by chlorophyll fluorescence - a preliminary study, *Acta Physiol. Plant.* **14**, 71—76.
Krupa, Z., Öquist, G. and Huner, N.P.A. (1993) The effects of cadmium on photosynthesis of *Phaseolus vulgaris* L. a fluorescence analysis. *Physiol Planta* **88**, 626-630

Krupa, Z., Öquist, G. and Huner, N.P.A. (1993a) The effects of cadmium on photosynthesis of *Phaseolus vulgaris* L. A fluorescence analysis, *Physiol. Plant.* **88**, 626—630.
Krupa, Z., Siedlecka, A., Maksymiec, W. and Baszyński, T. (1993b) In vivo response of photosynthetic apparatus of *Phaseolus vulgaris* to nickel toxicity, *J. Plant Physiol.* **142**, 664—668.
Krupa, Z., Skórzyńska, E., Maksymiec, W. and Baszyński, T. (1987) Effect of cadmium treatment on the photosynthetic apparatus and its photochemical activities in greening radish seedlings, *Photosynthetica* **21**, 156—164.
Küpper, H., Küpper, F. and Spiller, M. (1996) Environmental relevance of heavy metal substituted chlorophylls using the example of water plants, *J. Exp. Bot.* **47**, 259—266.
Küpper, H., Küpper, F. and Spiller, M. (1998) *In situ* detection of heavy metal substituted chlorophylls in water plants, *Photosynth. Res.* **58**, 123—133.
Lameroux, R.J. and Chaney, W.R. (1978) The effect of cadmium on net photosynthesis, transpiration and dark respiration of excised silver maple leaves. *Physiol Planta* **43**, 231-236
Lamoreaux, R.J. and Chaney, W.R. (1978) The effect of cadmium on net photosynthesis, transpiration and dark respiration of excised silver maple leaves, *Physiol. Plant.* **43**, 231—236.
Lanaras, T., Moustakas, M., Symenoides, L., Diamantoglou, S. and Karataglis, S. (1993) Plant metal content, growth responses and some photosynthetic measurements on field-cultivated growing on ore bodies enriched in Cu, *Physiol. Plant.* **88**, 307—314.
Landberg, T., Greger, M. (1994) Influence of selenium on uptake and toxicity of copper
Li, E.H., Miles, C.D. (1975) Effects of cadmium on photoreaction II of chloroplasts, *Plant Sci. Lett.* **5**, 33—40.
Lidon, F.C. and Henriques, F.S. (1991) Limiting step on photosynthesis of rice plants treated with varying copper levels. *J. Plant Physiol.* **138**, 115-118
Lidon, F.C. and Henriques, F.S. (1992a) Effects of copper on photosynthetic pigments contents in rice plants. *Bot. Bull. of Acad. Sinica* **33**, 141-149
Lidon, F.C. and Henriques, F.S. (1992b) Copper toxicity in rice: a diagnostic criterium and its effect on Mn contents. *Soil Sci* **154**, 130-135
Lidon, F.C. and Henriques, F.S. (1992c) Excess copper reduces photosynthetic pigments. *Rice Biotechnol* **12**, 15-16.
Lidon, F.C. and Henriques, F.S. (1992d) Effects of copper on the nitrate to ammonia reduction in rice plants. *Photosynthetica* **26**, 371-380.
Lidon, F.C. and Henriques, F.S. (1993a) Changes in the contents of the photosynthetic electron carriers, RNAse activity and membrane permeability triggered by excess copper in rice, *Photosynthetica* **28**, 99—108.
Lidon, F.C. and Henriques, F.S. (1993b) Changes in the thylakoid membrane polypeptide patterns triggered by excess Cu in rice, *Photosynthetica* **28**, 109—117.
Lidon, F.C. and Henriques, F.S. (1993c) Copper-mediated oxygen toxicity in rice chloroplasts, *Photosynthetica* **29**, 385—400.
Lidon, F.C., Ramalho, J.C. and Henriques, F.S. (1993) Copper inhibition of rice photosynthesis, *J. Plant Physiol.* **142**, 12—17.
Lucero, H.A., Andreo, C.S. and Vallejos, R.H. (1976) Sulphydryl groups in photosynthetic energy conservation. II. Inhibition of photophosphorylation in spinach chloroplasts by $CdCl_2$. *Plant Sci Lett* **6**, 309-313
Luna, C.M., Gonzalez, C.A. and Trippi, V.S. (1994) Oxidative damage caused by an excess of copper in oat leaves, *Plant Cell Physiol.* **35**, 11—15.
Maksymiec, W. (1997) Effect of copper on cellular processes in higher plants, *Photosynthetica* **34**, 321—342.
Maksymiec, W. and Baszy_ski, T. (1988) The effect of Cd^{2+} on the release of proteins from thylakoid membranes of tomato leaves. *Acta Soc Bot Poloniae* **57**, 465-474.
Maksymiec, W. and Baszyński, T. (1996a) Chlorophyll fluorescence in primary leaves of excess Cu- treated runner bean plants depends on their growth stages and the duration of Cu-action, *J. Plant Physiol.* **149**, 196—200.
Maksymiec, W. and Baszyński, T. (1996b) The role of Ca^{2+} ions in modulating changes induced in bean plants by an excess of Cu^{2+} ions. Chlorophyll fluorescence measurements, *Physiol. Plant.* **105**, 562—568.
Maksymiec, W. and Baszyński, T. (1998a) The effect of Cd^{2+} on the release of proteins from thylakoid membranes of tomato leaves, *Acta Soc. Bot. Pol.* **57**, 465-474.
Maksymiec, W. and Baszyński, T. (1998b) The role of Ca ions in changes induced by excess Cu^{2+} in bean plants. Growth parameters, *Acta Physiol. Plant.* **20**, 411—417.
Maksymiec, W. and Baszyński, T. (1999a) The role of Ca^{2+} ions in modulating changes induced in bean plants by an excess of Cu^{2+} ions. Chlorophyll fluorescence measurements, *Physiol. Plant.* **105**, 562-568.

Maksymiec, W. and Baszyński, T. (1999b) Are calcium ions and calcium channels involved in the mechanisms of Cu^{2+} toxicity in bean plants? The influence of leaf age, *Photosynthetica*, 36, 267—278.
Maksymiec, W., Bednara, J. and Baszyński, T. (1995) Responses of runner bean plants to excess copper as a function of plant growth stages: effects on morphology and structure of primary leaves and their chloroplast ultrastructure, *Photosynthetica* 31, 427—435.
Maksymiec, W., Russa, R., Urbanik-Sypniewska, T. and Baszy_ski, T. (1992) Changes in acyl lipid and fatty acid composition in thylakoids of copper non-tolerant spinach exposed to excess copper. *J .Plant Physiol* 140, 52-55.
Maksymiec, W., Russa, R., Urbanik-Sypniewska, T. and Baszyński, T. (1994) Effect of excess Cu on the photosynthetic apparatus of runner bean leaves treated at two different growth stages, *Physiol. Plant.* 91, 715—721.
Malik, D., Sheoran, I.S. and Singh, R. (1992a) Lipid composition of thylakoid membranes of cadmium treated wheat seedlings. *Indian J Biochem Biophys* 29, 350-354
Malik, D., Sheoran, I.S. and Singh, R. (1992b) Carbon metabolism in leaves of cadmium treated wheat seedlings. *Plant Physiol Biochem* 30, 223-229
Miles, C.D., Brandle, J.R., Daniel, D.J., Chu-Der, O., Schnore, P.D. and Uhlik, D.J. (1972) Inhibition of photosystem II in isolated chloroplasts by lead, *Plant Physiol.* 49, 820—825.
Miller, M. and Cox, R.P. (1983) Effect of Zn^{2+} on photosynthetic oxygen evolution and chloroplast manganese, *FEBS Lett.* 155, 331—333.
Misra, A.K., Sarkunan, V., Miradas, S.K., Nayak, S.K. and Nayar, P.K. (1989) Influence of N, P on cadmium toxicity on photosynthesis in rice, *Curr. Sci.* 58, 1398—1400.
Mohanty, N., Vass, I. and Demeter, S., (1989) Copper toxicity affects photosystem II electron transport at the secondary quinone acceptor Q_B, *Plant Physiol.* 90, 175—179.
Morales, F., Abadia, A. and Abadia, J. (1990) Characterization of the xanthophyll cycle and other photosynthetic pigment changes induced by iron deficiency in sugar beet (*Beta vulgaris* L.), *Plant Physiol.* 94, 607—613.
Morales, F., Abadia, A. and Abadia, J. (1991) Chlorophyll fluorescence and photon yield of oxygen evolution in iron-deficient sugar beet (*Beta vulgaris* L.) leaves, *Plant Physiol.* 97, 886—893.
Moustakas, M., Lanaras, T., Symeonidis, L. and Karataglis, S. (1994) Growth and some photosynthesis characteristics of field grown *Avena sativa* under copper and lead stress, *Photosynthetica* 30, 389—396.
Moustakas, M., Ouzounidou, G. and Lannoye, G. (1993) Rapid screening for aluminum tolerance in cereals by use of chlorophyll fluorescence test, *Plant Breed.* 111, 343—346.
Moya, J.L., Ros, R. and Picazo, I. (1993) Influence of cadmium and nickel on growth, net photosynthesis and carbohydrate distribution in rice plants, *Photosynth. Res.* 6, 75—80.
Navari-Izzo, F., Quartacci, M.F., Pinzino, C., Vecchia, F.D. and Sgherri, C.L.M. (1998) Thylkaoid-bound and stromal antioxidative enzymes in wheat treated with excess copper, *Physiol. Plant.* 104, 630—638.
Nedunchezhian, N. and Kulandaivelu, G. (1995) Effect of Cd and UV-B radiation on polypeptide composition and photosystem activities of *Vigna unguiculata* chloroplasts, *Biol. Plant.* 37, 437—441.
Ouzounidou, G. (1993) Changes in chlorophyll fluorescence as a result of Cu-treatment: Dose-response relations in *Silene* and *Thlaspi*, *Photosynthetica* 29, 455—462.
Ouzounidou, G. (1996) The use of photoacoustic spectroscopy in assessing leaf photosynthesis under copper stress: correlation of energy storage to photosystem II fluorescence parameters and redox change of P-700, *Plant Sci.* 113, 229—237.
Ouzounidou, G. (1997) Sites of copper in the photosynthetic apparatus of maize leaves: kinetic analysis of chlorophyll fluorescence, oxygen evolution, absorption changes and thermal dissipation as monitored by photoacoustic signals, *Aust. J. Plant Physiol.* 24, 81—90.
Ouzounidou, G., Lannoye, R. and Karataglis, S. (1993) Photoacoustic measurements of photosynthetic activities in intact leaves under copper stress, *Plant Sci.* 89, 221—226.
Ouzounidou, G., Moustakas, M. and Lannoye, R. (1995) Chlorophyll fluorescence and photoacoustic characteristics in relationship to changes in chlorophyll and Ca^{2+} content of a Cu-tolerant *Silene compacta* ecotype under Cu treatment, *Physiol. Plant.* 93, 551—557.
Ouzounidou, G., Symeonidis, L., Babalonas, D. and Karataglis, S. (1994) Comparative responses of a copper-tolerant and copper-sensitive population of *Minuatia hirsuta* to copper toxicity. *J .Plant Physiol .*44, 109-115.
Parekh, D., Puranik, R.M., and Srivastava, H.S. (1990) Inhibition of chlorophyll biosynthesis by cadmium in greening maize leaf segments, *Biochem. Physiol. Pflanzen.* 186, 239—242.
Pätsikkä, E., Aro, E.M. and Tyystjärvi, E. (1998) Increase in the quantum yield of photoinhibition contributes to copper toxicity *in vivo*, *Plant Physiol.* 117, 619—627.

Pnnazio, S. and Roggero, P. (1992) Effect of cadmium and nickel on ethylene biosynthesis in soybean, *Biol. Plant.* **34**, 345—349.
Prasad, M.N.V. (1995a) Cadmium toxicity and tolerance in vascular plants, *Environ. Exp. Bot.* **35**, 525—540.
Prasad, M.N.V. (1995b) The inhibition of maize leaf chlorophylls, carotenoids and gas exchange functions by cadmium, *Photosynthetica* **31**, 635—640.
Prasad, M.N.V. (1996) Variable allocation of heavy metals in phytomass of crop plants - human health implications, *Plze. Lék. Sborn. Suppl.* **71**, 19—22.
Prasad, M.N.V. (1997) Trace metals, in: M.N.V. Prasad (ed,) *Plant ecophysiology*, Wiley, New York, pp 207—249.
Prasad, M.N.V. and Strzałka, K. (1999) Impact of heavy metals on photosynthesis, in M.N.V. Prasad and J. Hagemeyer (eds.,) *Heavy metal stress in plants*, Springer-Verlag Berlin, pp. 117—138.
Prasad, M.N.V., Malec, P., Waloszek, A., Bojko, M. and Strzałka, K. (2001) Physiological responses of *Lemna trisulca* L. (Duckweed) to cadmium and copper bioaccumulation, *Plant Science*, in press
Radmer, R. and Kok, B. (1974) Kinetic observation of the photosystem II electron acceptor pool isolated by mercuric ion, *Biochim. Biophys. Acta* **357**, 177—180.
Rao, B.K., Tyryshkin, A.M., Roberts, A.G., Bowman, M.K. and Kramer, D.M. (2000) Inhibitory copper binding site on the spinach cytochrome b_6f complex: implications for Q_o site catalysis, *Biochemistry* **39**, 3285—3296.
Rascio, N., Dallavecchia, F., Ferretti, M., Merlo, I. and Ghisi, R. (1993) Some effects of cadmium on maize plants, *Arch. Environ. Contam. Toxicol.* **25**, 244—249.
Rashid, A., Bernier, M., Pazdernick, L. and Carpentier, R. (1991) Interaction of Zn^{2+} with the donor side of photosystem II, *Photosynth. Res.* **30**, 123—130.
Samson, G., Morisette, J.C. and Popovic, R. (1988) Copper quenching of the variable fluorescence in *Dunaliella tertiolecta*. New evidence for a copper inhibition effect on PS II photochemistry, *Photochem. Photobiol.* **48**, 329—332.
Samuelson, G. and Öquist, G. (1980) Effects of copper chloride on photosynthetic electron transport and chlorophyll-protein complexes of *Spinacea oleracea*. *Plant Cell Physiol* **21**, 455-454.
Schröder, W.P., Arellano, J.B., Bittner, T., Baron, M., Eckert, H.J. and Renger, G. (1994) Flash-induced absorption spectroscopy studies of copper interaction with photosystem II in higher plants, *J. Biol. Chem.* **269**, 32865—32870.
Šerševi, F., Králová, K. and Blahová, M. (1996) Photosynthesis of *Chlorella vulgaris* as affected by diaqua(4-chloro-2-methylphenoxyacetato)copper(II) complex, *Biol. Plant.* **38**, 71—75.
Šerševi, F., Králová, K. and Bumbálová, A. (1998) Action of mercury on the photosynthetic apparatus of spinach chloroplasts, *Photosynthetica* **35**, 551—559.
Shalaby, A.M. and Al-Wakeel, S.A.M. (1995) Changes in nitrogen metabolism enzyme activities of *Vicia faba* in response to aluminum and cadmium. *Biol Planta* **37**, 101-106
Sheoran, I.S., Singal, H.R. and Singh, R. (1990) Effect of cadmium and nickel on photosynthesis and the enzymes of the photosynthetic carbon reduction cycle in pigeon pea (*Cajanus cajan*), *Photosynth. Res.* **23**, 345—351.
Shioi, Y., Tamai, H. and Sasa, T. (1978a) Inhibition of photosystem II in the green alga *Ankistrodesmus falcatus* by copper, *Physiol. Plant.* **44**, 434—438.
Shioi, Y., Tamai, H. and Sasa, T. (1978b) Effects of copper on the photosynthetic electron transport systems in spinach chloroplasts, *Plant Cell Physiol.* **19**, 203—209.
Siedlecka, A and Krupa, Z. (1999) Cd/Fe interaction in higher plants – its consequences for the photosynthetic apparatus, *Photosynthetica* **36**, 321—331.
Siedlecka, A. (1995) Some aspects of interactions between heavy metals and plant minerals, *Acta Soc. Bot. Pol.* **64**, 265—272.
Siedlecka, A.and Baszyński, T. (1993) Inhibition of electron flow around photosystem I in chloroplasts of Cd-treated maize plants is due to Cd-induced iron deficiency, *Physiol. Plant.* **87**, 199—202.
Singh, D.P., Khare, P. and Bisen, P.S. (1989) Effect of Ni^{2+}, Hg^{2+} and Cu^{2+} on growth, oxygen evolution and photosynthetic electron transport in *Cylindospermum* IU 942, *J. Plant Physiol.* **134**, 406—412.
Skórzyńska, E. and Baszyński, T. (1993) The changes in PS II complex peptides under cadmium treatment are they of direct or indirect nature? *Acta Physiol. Plant.* **15**, 263—269.
Skórzynska, E., Urbanik-Sypniewska, T., Russa, R., and Baszyński, T. (1991) Galactolipase activity in Cd-treated runner bean plants. *J Plant Physiol* **138**, 454-459
Skórzyńska-Polit, E. and Baszyński, T. (1995) Photochemical activity of primary leaves in cadmium stressed *Phaseolus coccineus* depends on their growth stage, *Acta Soc. Bot. Pol.* **64**, 273—279.

Somaseskharaiah, B.V., Padmaja, K. and Prasad A.R.K. (1992) Phytotoxicity of cadmium ions on germinating seedlings of mung bean (*Phaseolus vulgaris*): Involvement of lipid peroxides in chlorophyll degradation. *Physiol Planta* **85**, 85-89.
Spiller, S. and Terry, N. (1980) Limiting factors in photosynthesis. II Iron stress diminishes photochemical capacity by reducing the number of photosynthetic units, *Plant Physiol.* **65**, 121—125.
Stiborova, M., Doubravova, M., and Leblova, S. (1986) A comparative study of the effect of heavy metal ions on ribulose-1, 5- bisphosphate carboxylase and phosphoenol pyruvate carboxylase. *Biochemie Physiol der Pflanzen* **181**, 373-379
Strzałka, K. and Ketner, P. (1997) Carbon dioxide, in: M..N.V. Prasad (ed,) *Plant ecophysiology*, Wiley, New York, pp 393—456.
Szalontai, B., Horváth, L., Debreczeny, M., Droppa, M.. and Horváth, G. (1999) Molecular rearrangements of thylakoids after heavy metal poisoning, as seen by Fourier transform infrared (FTIR) and electron spin resonance (ESR) spectroscopy, *Photosynth. Res.* **61**, 241—252.
Terry, N. and Abadia, J. (1986) Function of iron in chloroplasts, *J. Plant Nutr.* **9**, 609—646.
Tripathy, B.C. and Mohanthy, P. (1980) Zinc inhibited electron transport of photosynthesis in isolated barley chloroplasts, *Plant Physiol.* **66**, 1174—1178.
Tukendorf, A., and Baszy_ski, T. (1991) The *in vivo* effect of cadmium on photochemical activities in chloroplasts of runner bean plants. *J Plant Physiol* **138**, 454-459
Tziveleka, L., Kaldis, A., Hegedüs, A., Kissimon, J., Prombona, A., Horváth, G. and Argyroudi-Akoyunoglou, J. (1999) The effect of Cd on chlorophyll and light-harvesting complex II biosynthesis in greening plants, *Z. Naturforsch.* **54c**, 740—745.
Van Assche, F., Cardinales, C. and Clijsters, H. (1988) Induction of enzyme capacity in plants as a result of Heavy metal toxicity, dose response relations in *Phaseolus vulgaris* L.treated with cadmium. *Environmental Pollution* **52**, 103-115.
Van Duijvendijk-Matteoli, M.A. and Desmeta, G.M. (1975) On the inhibitory side of PS II in isolated chloroplasts, *Biochim. Biophys. Acta* **408**, 164—169.
Vavilin, D.V., Polynov, V.A., Matorin, D.N. and Venediktov, P.S. (1995) Sublethal concentrations of copper stimulate photosystem II photoinhibition in *Chlorella pyrenoidosa*, *J. Plant Physiol.* **146**, 609—614.
Veeranjaneyulu, K. and Das, V.S.R. (1982) In vitro chloroplast localization of ^{65}Zn and ^{63}Ni in a Zn-tolerant plant *Ocimum basilicum* Benth, *J. Exp. Bot.* **33**, 1161—1165.
Voigt, J., Nagel, K. and Wrann, D. (1998) A cadmium-tolerant *Chlamydomonas* mutant strain impaired in photosystem II activity, *J. Plant Physiol.* **153**, 566—573.
Wagner, G.J. (1993) Accumulation of cadmium in crop plants and its consequences to human health, *Adv. Agron.* **51**, 172—212.
Weigel, H.J. (1985a) Inhibition of photosynthetic reactions of isolated intact chloroplast by cadmium, *J. Plant Physiol.* **119**, 179—189.
Weigel, H.J. (1985b) The effect of cadmium on photosynthetic reaction of mesophyll protoplast, *Physiol. Plant.* **63**, 192—200.
Wong, D. and Govindjee, (1976) Effects of lead ions on photosystem I in isolated chloroplasts: Studies on the reaction centre P700, *Photosynthetica* **10**, 241—254.
Woźny, A., Schneider, J. and Gwóźdź, E.A. (1995) The effects of lead and kinetin on greening barley leaves, *Biol. Plant.* **35**, 541—552.
Yruela, I., Alfonso, M., Barón, M. and Picorel, R. (2000) Copper effect on the protein composition of photosystem II, *Physiol. Plant.* **110**, 551—557.
Yruela, I., Alfonso, M., Ortiz de Zarate, I., Montoya, G. and Picorel, R. (1993) Precise location of Cu(II) – inhibitory binding site in higher plant and bacterial photosynthetic reaction centres as probed by light-induced absorption changes, *J. Biol. Chem.* **268**, 1684—1689.
Yruela, I., Montoya, G. and Picorel, R. (1992) The inhibitory mechanism of Cu (II) on the photosystem II electron transport from higher plants, *Photosynth. Res.* **33**, 227—233.
Yruela, I., Montoya, G., Alonso, P.J. and Picorel, R. (1991) Identification of the pheophytin-Q_A –Fe domain of the reducing side of the photosystem II at the Cu(II)-inhibitory binding side, *J. Biol. Chem.* **266**, 22847—22850.
Yruela, I., Pueyo, J., Alonso, P.J. and Picorel, R. (1996) Photoinhibition of photosystem II from higher plants, *J. Biol. Chem.* **271**, 27408—27415.

CHAPTER 10

GAS EXCHANGE FUNCTIONS IN HEAVY METAL STRESSED PLANTS

ELŻBIETA ROMANOWSKA
Department of Plant Physiology,
University of Warsaw,
02–096 Warszawa, Miecznikowa 1, Poland

1. INTRODUCTION

The heavy metals such as e.g. Pb^{2+}, Cd^{2+}, Ni^{2+}, Cu^{2+}, Fe^{2+}, Hg^{2+}, Zn^{2+}, Al^{2+}, Co^{2+} are released into the environment by various industrials processes. These contaminants present at an elevated level in the environment, may be taken up by the plants via the root system and/or by foliar absorption and accumulated in different parts of plants, thereby inducing reduced growth (Ernst, 1980; Foy et al., 1978). Plants growing in soils rich in toxic metals are often incapable of taking up water since their roots are short, extending only on the surface, and therefore in no position to take advantage of the underlying water. Plants usually show a higher tolerance to heavy metals than mammals and this fact is responsible for the entry of the metal into the food chain (Baker et al., 1979; Foy et al., 1978).

Heavy metals can be divide into two groups: 1. metals necessary for the normal growth of plants e.g. Zn, Fe, Cu, Mn, however, in higher concentrations hindering plant development; 2. metals not required for plant metabolism, e.g. Cd, Hg, Pb. They are very toxic at much lower concentration than the first group. Heavy metals are widely known inhibitors of plant metabolism. Major mechanism of actions are covalent ion substitution and interaction with SH groups of proteins, resulting in the inactivation or the inhibition of enzymes (Vangronsveld and Clijsters, 1994; Van Assche and Clijsters, 1990).

Photosynthetic CO_2 fixation is a highly sensitive process significantly affected by heavy metals in a number of plant species (Moya et al., 1993; Bishnoi et al., 1993; Romanowska et al., 1998). Inhibition of photosynthesis by heavy metals at several levels of organization is well documented. The degree of heavy metals action on the photosynthesis depends on the growth stage, plant conditions as well as on the duration of the stress. Metals can cause lipid peroxidation in photosynthetic membranes (Sandmann and Boger, 1980), decreased level of pigments, particularly chlorophyll (Malik et al., 1992; Ouzonidou et al., 1995; see also chapters 8 and 9) and induced the destruction of fine structure of chloroplasts (Maksymiec et al., 1995; see also chapters 8 and 9).

The most general symptom of heavy metal toxicity is chlorosis. Chlorosis from excess Zn, Cu, Ni, and Cd appers to be due to a direct or an indirect interaction with foliar Fe, inhibition of Fe uptake (Alva and Chen, 1995; Pich et al., 1995) or by influence of heavy metals at the stage of δ-aminolevulinic acid formation (Stiborova et al., 1986). It has been shown that different heavy metals can interfere with various steps of the photosynthetic electron transport chain (Ouzonidou, 1996; Murthy and Mohanty, 1995; Maksymiec and Baszyński, 1999) but a mechanism of inhibition is still controversial This may partly be due to differences in experimental design. For example, the results of similar experiments could differ, depending on whether metals were applied directly to isolated leaves, cells or organelles, or applied indirectly to intact plants via culture medium.

Many results from *in vitro* experiments on isolated chloroplasts pointed to direct inhibitory effects of heavy metals on PS II reaction center and its redox components (Miles et al., 1972; Samson and Popovic, 1990; Yruela et al., 1992). Inhibition studies were focused on both the donor and acceptor side of PS II (Renganathan and Bose, 1989; Mohanty et al., 1989), at the cytochrome b/f complex and PS I (Bohner et. al., 1980; Baszyński, 1986; Šeršen et al., 1998). Compared to PS II, PS I was found to be less sensitive to metal ion inhibition (Krupa and Baszyński, 1995).

A major mechanism of toxic action of metal is the inhibition of the enzyme Rubisco (Van Assche and Clijsters, 1990; Stiborova et al., 1986). This can occur due to ion substitution, as in the case of Zn (Van Assche and Clijsters, 1986a), -SH interaction, as in the case of Cu and Cd (Stiborova et al., 1986; Stiborova, 1988) or as consequence of enzyme deficiency due to impaired protein biosynthesis (Kremer and Markham, 1982; Angelov et al., 1993a). Disruption of the subunit structure of this enzyme by Cd, Cu, Mn, Ni and Zn should also be considered as the mechanism of the inhibition of its activity (Pierce, 1986).

The principal end products of leaf photosynthesis are starch and sucrose. It has been observed by various groups of researchers that plants under heavy metals stress accumulate sugars and starch, especially sucrose. It has been proposed (Greger et al., 1991) that decrease in utilization of carbohydrates for growth produced by heavy metals is more pronounced than the decrease in CO_2 fixation resulting in an increased accumulation of carbohydrates. Others suggest reduced translocation of sugars out of the leaves (Samarakoon and Rauser, 1979; L'Huillier et al., 1996).

In the opinion of many authors water stress caused by heavy metals may indirectly affect photosynthesis by lowering CO_2 levels available for the Calvin cycle (Bishnoi et al., 1993; Costa et al., 1994). Some heavy metals (Cd or Cu) initiate the premature senescence of plants and visual symptom of that is the degradation of the lamellar structure of the chloroplasts. This influences the function of photosynthesis (Maksymiec et al., 1992; Rascio et al., 1993).

The studies of heavy metals effects on photosynthesis has been largely concentrated on the thylakoid membranes structure and the photosynthetic electron transport chain activity. However, photophosphorylation may be also sensitive to metal ions. Lower levels of photosynthetic ATP production were observed in Cd-treated *Lycopersicom esculentum* (Baszyński et al., 1980) and in Zn-treated *Phaseolus vulgaris* plants in relation to inhibition of the electron flow rate (Van Assche and Clijsters, 1986b). The problem of participation of light and/or dark reactions in inhibition of photosynthesis in heavy metals treated plants still remains open, since it is very difficult to indicate the primary site of metal toxicity.

In contrast to photosynthesis, the stimulation of respiration in response to heavy metal exposure is often observed. This has been observed both in whole plants (Lee et al., 1976), detached leaves (Lamoraux and Chaney, 1978; Poskuta and Wacławczyk, 1995), unicellular green algae (Poskuta et al., 1996), isolated protoplasts (Greger and Ögren, 1991; Parys et al., 1998) and isolated mitochondria (Koeppe and Miller, 1970; Bittell et al., 1974). The mechanism of the enhancement of respiration by heavy metals is not clear. Ernst (1980) suggested that stimulation of respiration could be due to a demand for ATP production through oxidative phosphorylation, compensating for a limited ATP supply from photophosphorylation. However, this seems less likely given that the most sensitive part of photosynthesis seems to be the reactions related ATP consumption and not to ATP production (Siedlecka et al., 1997). The rise in dark respiration in heavy metal treated leaves has also been explained as a result of the oxidation of excess of photosynthetic equivalents which are produced under conditions of limited CO_2 fixation (Poskuta et al., 1996).

The inibition of respiration in response to *in vitro* applications of cadmium to mitochondria isolated from animal (Fluharty and Sanadi, 1962, 1963) and plant cells (Mustafa and Cross, 1971; Reese and Roberts, 1985) was also observed. Experiments with isolated mitochondria showed that effect of heavy metals on respiration was due to used concentration. Inhibitory effect occurred at higher metal concentration and was accompanied by loss of respiratory control due to uncoupling of the phosphorylation (Fluharty and Sanadi, 1962, 1963; Miller et al., 1973; Kleiner, 1974). Uncoupling produced by Cd^{2+} was due to interaction with an essential dithiol grouping in the system. It is well established that the dithiol function is central in the energy trapping reactions in mitochondria and in photosynthetic organelles (Nahar and Tajmir-Riahi, 1995). Ros et al., (1992) have demonstrated ATPase activation by cadmium ions in experiments *in vivo* that suggests a different mechanism (probably indirect), from that proposed for *in vitro*. Inhibition caused by Cd^{+2} *in vitro* could be explained by a direct binding with sulfhydryl groups require for the APTase activity.

These findings suggest that the inhibition of the photosynthetic capacity of plants exposed to heavy metals is due to the inhibition of the both electron transport activities as well as Calvin cycle enzymes. The mechanism(s) of heavy metals stimulation of respiration is still a matter of speculation. The intensity of the CO_2 and O_2 exchange is highly affected by heavy metal ions and is a consequence of a stimulation or inhibition of a number of metabolic processes. Is necessary to know all metabolic disturbance for a better understanding of the effects of heavy metals on gas exchange. The mechanisms of heavy metals injury to the photosynthetic parameters and the way in which the plants respond to this stress are discussed.

2. ESSENTIAL METALS

2.1. Copper (Cu)

Copper is essential element which is present at low concentration in higher plants and algae (Clarkson and Hanson, 1980). It is a component of oxidative enzymes or of particular structural components of cells. Copper deficiency has a direct impact on energy metabolism because it affects the synthesis of the Cu-containing electron carriers: plastocyanin and cytochrome oxidase, whose depletion results in decreased photosynthesis and respiration (Baron and Sandmann, 1988). A trace (greater than 1 µM) element is toxic to algal and higher plant tissues because it directly affects the metabolic processes such as respiration, photosynthesis, CO_2 fixation and gas exchange and other processes (Clijsters and Van Assche, 1985; Vangronsveld and Clijsters, 1994). Copper ions at toxic concentrations act as efficient generators of reactive oxygen species (Kampfenkel et al., 1995; Gallego et al., 1996; Navari-Izzo et al., 1998; Quartacci et al., 2001) exerting inhibitory effect on the activity of both photosystems (Droppa and Horvath, 1990). Copper mediated free radical formation has been demonstrated in isolated chloroplasts (Sandmann and Boger, 1980), in intact roots of *Silene cucubalus* (De Vos et al., 1993), in leaf segments (Luna et al., 1994) and in intact leaves of *Phaseolus vulgaris* (Weckx and Clijsters, 1996). On the other hand, Cu ions increase the activities of antioxidant enzymes (Chongpraditum et al., 1992; Weckx and Clijsters, 1996; Gupta et al., 2000).

The role of cupric ions in photosynthetic organisms depends greatly on their concentration. The inhibition of the rate of the photosynthetic processes by Cu^{2+} has been described in several papers; however, the mechanism and site of inhibitory action are different (Baron et al., 1995). Copper has been shown to inhibit photosynthesis in *Chlorella* (Nielsen et al., 1969), photosynthetic oxygen evolution in *Scenedesmus acutus* (Bohner et al., 1980), photosynthetic electron transport in isolated pea chloroplasts (Renganathan and Bose, 1989) and photosynthetic energy conversion (Uribe and Stark, 1982). Copper absorbed in excess amounts can cause deleterious effects at morphological, physiological and ultrastructural levels.

Excess copper can cause chlorosis (Mocquot et al., 1996), inhibition of root growth and damage of plasma membrane permeability, leading to the ion leakage (Maksymiec, 1997). The significant decrease of quantum yield in the primary photochemistry processes of PS II is responsible for the reduced quantum yield of O_2 evolution (Ouzoniodou, 1994; Boucher and Carpentier, 1999; Szalontai et al., 1999).

The decrease of plant growth under Cu treatment can therefore be related to its effect on photosynthesis. The growth reduction of barley and maize plants was observed as a consequence of Cu toxicity (Stiborova et al., 1987). The root growth was more strongly inhibited by metal ions than the growth of shoots. Low concentration of Cu (1–10 µM) decreased the growth of maize more effectively than that of barley. However, 100 to 1000 µM concentration caused more significant decrease of the barley growth.

The decrease of the maize and barley shoot fresh and dry matter can probably be evolved mainly from the metal-induced reduction of CO_2 fixation *in vivo*, because the ribulose bisphosphate carboxylase (Rubisco) and phosphoenolpyruvate carboxylase (PEPC) activities were significantly reduced. The progressive decrease of Rubisco activity was observed in the rice seedlings with increasing Cu levels (0.01–6.25 mg/l) in the solution medium (Lidon and Henriques, 1991) (Table 1).

Toxic action of Cu as well as Cd and Ni, depending upon the plant growth stage, was recognized (Maksymiec at al., 1994; Luna et al., 1994). In the plants treated at the stationary leaf growth with Cu, excess metal disturbs the relationship between the synthesis of cell constituents and their decomposition. Photosynthetic oxygen evolution by chloroplasts isolated from primary leaves of runner bean plants treated with Cu at different growth stages shows that O_2 evolution decreased significantly with leaf senescence due to a loss in PS II stability, as well as a possible inhibition of the dark phase of photosynthesis (Maksymiec and Baszyński, 1996a, b). It was observed (Cedeno-Maldonado and Swader, 1972) that exposure of the isolated spinach chloroplasts to the Cu ions strongly inhibits O_2 evolution. Inhibition was dependent not only on the concentration of the inhibitor but also on exposure of chloroplasts to light. Dark period reduced the degree of inhibition while short period of light had opposite effects. The authors concluded that light increased inhibition as a result of Cu^{2+} binding to the inhibitory sites on the oxidizing side of photosystem II. Hoganson et al. (1991) show the existence of a metal binding site in oxygen evolving complex (OEC) with higher affinity for Cu^{2+} than for Mn^{2+} suggesting that Cu^{2+} may inhibit photosynthetic oxygen evolution by replacing Mn^{2+}. The decrease of photosynthetic O_2 evolution in Cu-tolerant (*Silene compacta*) and non-tolerant (*Alyssum montanum*) species was observed at high Cu concentrations (30–160 µM) (Ouzonidou et al., 1993). Cu-tolerant plants grown in low Cu concentration revealed an enhancement in O_2 evolution.

Table 1. *Chlorophyll content, Rubisco activities of leaf tissues, electron transport rate in isolated rice chloroplasts 30 days after germination (adapted from Lidon and Henriques, 1991).*

Cu treatment (mg Cu/l)	Chlorophylls mg/g (f.wt.)	Carboxylase	Oxygenase	Photosynthetic electron transport $H_2O \rightarrow DCIP$ µM O_2/mg chl/h
		[activity (nM RuBP/mg protein/min)]		
0.01	3.456	502.07	124.05	159.78
0.05	2.571	375.00	90.51	179.92
0.25	1.707	330.00	75.07	84.16
1.25	0.651	225.07	50.514	75.72
6.25	0.318	210.00	45.027	56.00

The decrease in photosynthetic O_2 evolution may be due to an inhibition of electron transfer and consequently to disturbed photochemistry process. The possible direct Cu action on the metal binding sites in PS II can inhibit whole photosystem activity on the basis of feedback control by accumulation of terminal metabolites of the light phase (Moustakes et al., 1994; Maksymiec and Baszyński, 1996a, b). In the latter case, excessive amounts of ATP and NADPH may accumulate due to inhibition of the Calvin cycle. Then, CO_2 assimilation can be inhibited by poor triose phosphate utilization caused by slow sugar export to the roots tissue as was shown for *Phaseolus vulgaris* plants exposed to excess Ni, Co and Zn (Samarakoon and Rauser, 1979).

Significant decrease in CO_2 assimilation in pea plants treated with Cu^{2+} ions was observed after 24 h treatment with Cu ions in concentration 50 μM and was 50% of the control (Angelov et al., 1993a). After 4 days, photosynthesis at 500 and 1000 μM Cu was only 6% of the activity in the control plants. A possible reason for the decreased photosynthesis was water deficit induced by Cu toxicity; it can be explained by the proline accumulation in copper treated pea plants. It is also reported that copper sharply inhibited CO_2 assimilation in rice seedlings in concentration between 0.01 and 6.25 mg/l (Lidon et al., 1993). Excess copper concentration inhibited net photosynthesis a 13-fold not as a consequence of changes in leaf intracellular CO_2 concentration or stomatal conductance (it was not affected by Cu) but as a result of lower photochemical efficiency of PS II and decreased ATP synthase activity.

Cu ions used in excess can inhibit photophosphorylation due to binding at sites which affect phosphorylation or at essential sulfhydryl or amino groups of some components involved in proton translocation or electron transport (Uribe and Stark, 1982).

There are data on the growth and on the rate of gas exchange for algae growing in an ordinary medium to which Cu was added. Unicellular alga *Chlorella pyrenoidosa* air-grown was more resistant to toxic concentrations of Cu (1–5 mM) than the 1% CO_2 grown. Rates of respiration of the air-grown algae were significantly stimulated by Cu but they were inhibited in the 1% CO_2-grown cells (Poskuta, 1991). Photosynthesis of the 1% CO_2 grown cells was more sensitive to Cu particularly, at the highest concentration tested as compared with the air-grown algae. In contrast to photosynthesis, the cell growth was markedly less inhibited by Cu in used concentration than their gas exchange. According to McBrien and Hassal (1967) Cu inhibited both photosynthesis and respiration in *C. vulgaris* grown in air and even more under anaerobic conditions. Stauber and Florence (1987) noticed however, that Cu depressed not only photosynthesis but cell division, too. The differences in alternative respiration capacity in the air and high CO_2-grown cells of *Chlamydomonas reinhardii* were observed by Goyal and Tolbert (1989). These data indicated that Cu in used concentration might implicate the action on respiratory enzymes of both CN-sensitive and -insensitive respiration in algae. In contrast, green alga *Dunaliella parva* grown in low Cu media reduced thylakoid formation and Cu deficiency caused a decrease in photosynthesis and respiration (Sandmann, 1985; Baron and Sandmann, 1988) as a result of inhibition of synthesis of plastocyanin and cytochrome oxidase.

In the most species excess Cu strongly influence the root development and metabolism, however root Cu can become extremely high before foliar Cu is increased substantially, so it is not clear if Cu in leaves interferes with gas exchange directly. The role of Cu on photosynthesis is very complicated because on the one hand Cu is an essential micronutrient, but on other hand, Cu as a heavy metal strongly affects the photosynthetic process by poisoning the environment. Most of the Cu effects have been observed *in vitro*, since it is still an open question whether these inhibitory mechanisms also operate *in vivo*.

2.2. Nickel (Ni), zinc (Zn), cobalt (Co)

A number of metals, initially studied solely because of their toxic effects, were subsequently found to be biologically essential, their deficiency causing disease. Similarly, toxicity due to metal excess has been observed in many species of wide diversity, though classification of metals as either biologically essential or toxic have required continuous revision. Potentially, every element is toxic when presented to an organism in high enough concentrations. Heavy metals such as Ni, Zn are micronutrients essential for animals, some bacteria, algae and higher plants (Brown et al., 1987a, b) but nickel and zinc in excess are potent inhibitors of development and various metabolic processes in plants (Van Assche and Clijsters, 1990).

Zinc is essential for the function and/or structure of at least three plant enzymes, and is present throughout all phyla. Zn accumulates to a toxic level in H_2O and soil through various emission sources, such as mines and smelters (Buchauer, 1973). Known symptoms of their toxic effects on plants include reduced production and growth.

Most agricultural soils contain an average 25 mg/kg soil dry wt. of Ni (Holmgren et al., 1993). Many plants generally posses mechanisms that permit them to tolerate Ni and develop without phytotoxic problems (Gabbrielli et al., 1990).

There are data that Ni, Zn, Co and Mn cause Fe deficiency in plants (Misra and Ramani, 1991; Terry, 1981) due to Zn-Fe or Ni-Fe interaction and competition (Misra and Ramani, 1991) which results in chlorosis. The inhibitiory effect of Ni on the accumulation of chlorophyll was observed in barley and pigeon pea plants (Agarvala et al., 1977; Sheoran et al., 1990) and for Co in barley plants (Agarwala et al., 1977).

Visible symptoms of Ni and Co phytotoxicity depressed growth and dry matter of plants (Austenfeld, 1979; Punz and Sieghard, 1993; Rauser, 1978) and abnormally high concentration of carbohydrates in the leaves (Samarakoon and Rauser, 1979). Robertson and Meakin (1980) showed that Ni inhibits root growth by inhibiting cellular division in the root apex. The influence of increasing concentration of Cu, Zn, Ni and Cr on wheat seedlings show that metals depressed growth of shoot and root and increased peroxidase activity what indicates an enhanced senescence caused by heavy metals (Karataglis et al., 1991).

Several studies have indicated that Ni inhibits photosynthesis. The photosynthetic reactions, both electron transport as well as CO_2 assimilation, are of primary interest in the context of the toxic effects of this heavy metal on plants (Bishnoi et al., 1993; Krupa et al., 1993). Addition of different concentration of Ni (6–18 µM) to the medium with the intact cell of the cyanobacterium *Spirulina platensis* caused a 26%

inhibition in the whole electron transport activity (Murthy and Mohanty, 1995). Ni was unable to bring appreciable inhibition in the PS II activity even at 18 µM. Zn, even at a relatively low concentration (down to 15 µg/l), inhibits CO_2 fixation in marine

phytoplanktons (Davies and Sleep, 1979). The degree of inhibition was, however, a nonlinear function of the zinc concentration. This observation supports the view that the effect of metals on the growth of phytoplankton is more closely related to the amounts accumulated by the plant cells than to the concentrations in the water surrounding them. *Chlorella pyrenoidosa* grown 24 h in air or 1% CO_2 was treated with Zn (10–40 mM) during 24 h (Poskuta, 1991). The photosynthetic O_2 evolution and dark respiration of both types of cells were significantly inhibited.

In isolated barley chloroplasts, the presence of 2 mM Zn inhibits the electron transport activity of PS II, O_2 evolution and chlorophyll a fluorescence (Tripathy and Mohanty, 1980). Authors suggest that Zn inhibits electron flow at the oxidizing side of PS II, no inhibition of PS I was observed. The photosynthetic O_2 evolution was inhibited about 85 % after 5 min. incubation of chloroplasts with 3 mM Zn and the extent of inhibition increased with the increasing time of incubation. Studies on isolated PS II particles from spinach shown that Zn used in concentration 2–10 mM has strong dissociating action on extrinsic polypeptides of PS II (Rashid et al., 1994). Since dissociation of these polypeptides from OEC not only inhibits the activity of water oxidizing enzyme, but also destabilizes the binding of cofactors, authors conclude that this is to be probable molecular mechanism of action of these metal ions on PS II.

It was shown that thylakoids isolated from bean plants grown 7 days in nutrient solution with added Ni in concentration 0–500 µM exhibit not changes in fluorescence parameters (Krupa et al., 1993). The similar results were obtained when experiments were performed on thylakoids isolated from non-treated plants and incubated with excessive amounts of this metal (1–20 mM). The measurement of photosynthetic electron transport under heavy metal influence both *in vivo* and *in vitro* shows that *in vitro* photosystems are the primary target of the metal toxicity, the lack of an *in vivo* effect of Ni suggests that in this case inhibition has rather indirect character. Authors conclude that the primary target of Ni toxicity in bean plants is the carbon metabolism.

The effects of Ni and Zn ions on the lipid protein association were characterized by means of electron spin resonance (ESR) spectroscopy (Szalontai et al., 1999). The results demonstrated that Ni has no inhibitory effect on the photosynthetic electron transport, but Zn strongly inhibits it, which is related to changes in the protein conformation in the chloroplast thylakoids. A similar effect of Ni was reported by Arndt et al. (1983) for *Chlorella*. Decreasing vitality index proportionally to the metal concentration suggests that CO_2 fixation was diminished.

The toxic effects of nickel on the root and shoot growth were studied in maize seedlings (L'Huillier et al., 1996). Nickel in concentration from 0 to 60 µM reduced significantly root and shoot growth and inhibited, at concentration 60 µM, mitotic activity of the roots.

Ni reduced carbohydrate transport by an inhibition of starch conversion into sucrose and its translocation to the roots. Thus this results in a reduced carbohydrate supply for roots and in consequence depressed mitotic activity in the root meristem. Inhibition of translocation of photoassimilates out of leaves was observed in bean plants exposed to excess Co, Ni and Zn. With time, more carbohydrates accumulated in the leaves of the seedlings exposed to metal, so that after 4 days, the increases were 5.6 fold for Co, 5.7-fold Zn and 8.0-fold for Ni (Samarakoon and Rauser, 1979). These results are consistent with the results obtained for rice plants (Moya et al., 1993). Rice plants were grown 5 or 10 days in a nutrient solution with Ni (0.1 and 0.5 mM). Aplication of Ni produced an inhibition of the transport of carbohydrate from shoot to root and also affected its distribution between organs. Ni decreased the CO_2 fixation (by about 60%) in rice plants. Decrease in utilization of carbohydrate for growth produced by nickel was more pronounced than the decrease in CO_2 fixation.

In the wheat leaves photosynthetic CO_2 fixation was significantly reduced by Ni^{2+} (Bishnoi et al., 1993). The different inhibitory effect of this metal on photosynthesis of various leaves was observed. Two days after the application of 5 and 10 mM Ni the reduction of photosynthesis in the top leaf was 22 and 49%, respectively. Ni slightly affected photosynthesis of the third leaf. Four days after the treatment, 5 and 10 mM Ni diminished CO_2 fixation by 38 and 62% in the first leaf, and by 11 and 21 % in the third leaf, respectively. The different inhibitory effect of this metal on various leaves coincides with its relative concentration in the leaves. The reduction in photosynthetic activity was due to the combined effect of various factors such as decreased chlorophyll content, lower stomatal conductance and reduced transpiration (28%).

Spruce seedlings grown for 5 weeks in 30 and 60 µM Zn had significantly decreased CO_2 assimilation on a dry weight basis (Schlegel et al., 1987). However, on a chlorophyll basis no significant reduction of CO_2 uptake could be demonstrated. Dark respiration was not significantly affected when spruce seedlings were exposed to Zn.

The Co in excess concentration (40–400 µM) caused a reduction of the pea plant fresh and dry masses, water and chlorophyll contents (Angelov et al., 1993b). The rates of CO_2 assimilation and transpiration decreased on the 4 days of Co treatment by 35 and 29%, respectively. Stomatal resistance increased considerably at all tested Co concentration. The above mentioned data suggest that primary effect of Co excess involves a disturbance in plant water regime and stomata closure. The authors also observed the inhibition of Rubisco activity by about 17 % in concentration 40 and 200 µM, and 33 % in 400 µM Co. The inhibition of the activity of Rubisco isolated from barley and PEPC from maize plants was observed after treatment with Zn ions in concentrations of 500 µM (Stiborova et al., 1987). PEPC was more effectively inhibited by Zn than the Rubisco was. Zn^{2+} practically did not affect the activity of both enzymes *in vivo*. Authors suggest that these enzymes can be the target for the toxic action of this metal on photosynthesis.

From presented data we can conclude that essential heavy metals at higher concentration are potential inhibitor of gas exchange in cyanobacteria, green algae and higher plants. Several sites have been proposed to be target of their action such as: degradation of photosystem II, decrease in the level of enzymes associated with the biochemical efficiency of CO_2 fixation, decreased level of chlorophylls, structural changes in the chloroplast membranes, decreased stomatal conductance. According to many investigators, reduced photosynthetic CO_2 assimilation in excess of Ni, Zn or Co is due to stomatal closure and reduction in CO_2 exchange rate.

3. NON-ESSENTIAL METALS

3.1. Lead (Pb)

Lead is widely distributed in the atmosphere, earth, ocean, and ground water. It is absorbed and accumulated and can be identified in most plant and animal tissues to a grater extent than is known for other heavy metals (Hg, Cd) (Vallee and Ulmer, 1972). Studies on the physiology of metal toxicity are difficult and begin with increased metal supply to the roots. However, the transport of Pb from plant roots to shoots is usually very low (Chappelka et al., 1991), which consequently will restrict the influence of root uptake on Pb levels in higher plants. Most works involved non-physiological conditions and study of organelles. The study of isolated organelles has led to better understanding of physiological processes occuring during heavy metal stress but realistic metal toxicity has proven too difficult to assess.

Toxic concentration of lead in the nutrient medium affects several aspects of the physiology of the plant. Major mechanisms of action are covalent ion substitution and interaction with SH groups of proteins, both resulting in the inhibition or inactivation of enzymes (Vangronsveld and Clijsters, 1994; Van Assche and Clijsters, 1990).

Noticeably, the photosynthetic apparatus is affected as a consequence of direct or indirect action of the metal ions. Photosynthesis can be affected at several structural and metabolic levels (Bazzaz et al., 1974, 1975; Clijsters and Van Assche, 1985; Poskuta et al., 1987, 1988). Metal can cause lipid peroxidation in photosynthetic membranes (Sandmann and Boger, 1980) and affect both synthesis and degradation of photosynthetic pigments, particularly chlorophyll (Hampp and Lendzian, 1974; Burzyński, 1985; Stiborova et al., 1986; Woźny et al., 1995). The inhibition of the Rubisco activity has been proposed as a major mechanism of the inhibition of photosynthesis (Stiborova et al., 1986; Clijsters and Van Assche, 1985). Direct effects of lead on the two photosystems have been largely investigated (Becerill et al., 1988; Rashid and Popovic, 1990) but the applicability of these findings to *in vivo* situation is not obvious. Also, lead increases abscisic acid levels in the leaves (Tardieu and Davies, 1992; Parys et al., 1998), that promotes stomatal closure (Bazzaz et al., 1974; Becerril et al., 1988).

Uptake and tolerance of lead in terrestrial and aquatic plants show that the chloroplast should be considered as the primary candidate for lead-associated phytotoxicity (Rebechini and Hanzely, 1974). The lead exposed leaf cells of *Ceratophyllum demersum* show a reduced number of grana stacks and a highly congested fretwork. The amount of stroma in relation to the lamellar system was markedly reduced, leaving little unoccupied matrix within the organelle. Other observable changes involve the starch grains and the conspicuous plastoglobuli. The former was either absent or greatly reduced in size. The plastoglobuli exhibited a total loss of electron density (Reilly and Reilly, 1973; Rebechini and Hanzely, 1974).

Lead accumulated in the leaves affects biosynthesis of porphyrin system in plant tissues, what would be associated with the action of Pb on chlorophyll synthesis (Barua and Jana, 1985; Sharma and Chopra, 1987). It is the general belief that chlorosis due to heavy metal stress is caused by disturbance of iron indispensable in chlorophyll synthesis (Foy, 1978). Synthesis of both types of chlorophyll was reduced in proportion to the increasing amounts of lead ions in the detached leaves of wheat incubated in the $PbCl_2$ solution (Hampp and Lendzian, 1974). At lower concentrations of lead (0–500 µM), synthesis of chlorophyll b was inhibited to a greater extent than that of chlorophyll a. In barley leaves exogenously applied kinetin diminished the inhibitory effect of lead on the chlorophyll content (Woźny et al., 1995), however the mechanism of this process remains unknown. It was shown that when Pb ions were introduced during 24 h into detached pea leaves during greening, photosynthesis was drastically reduced but in the same time respiration rate was increased (Łukaszek and Poskuta, 1998). The authors concluded that respiration was stimulated by Pb^{2+} *per se* not by direct linking to photosynthesis.

Earlier data have indicated that accumulation of lead in leaves of soybean, maize and sunflower causes a reduction in the rates of photosynthesis and transpiration (Bazzaz et al., 1975; Huang et al., 1974). Poskuta et al. (1987, 1988) observed inhibition of apparent photosynthesis, photorespiration and $^{14}CO_2$ uptake in pea and maize leaves treated with lead. The inhibition of CO_2 fixation by shoots was associated with a change in the flow of assimilated carbon into photosynthetic and photorespiratory intermediates and the greatest reduction of the carbon flow occurred at the higher treatment level.

Inhibition of photosynthesis by Pb ions correlated with reduction of stomata opening and limitation of gaseous exchange of the seedlings. They observed a greater inhibition of CO_2 exchange than transpiration.

In contrast to the results of these authors, Bazzaz et al. (1975) noticed higher inhibition of transpiration than photosynthesis in detached leaves of corn and sunflower at any given point in time and any given Pb concentration. These results imply that decreased stomata conductance may be a major factor in reducing gas exchange of Pb-treated leaves. Further evidence to support the effect of Pb on stomata opening has been obtained from experiments in which epidermal peels of sunflower were floated on solution of $PbCl_2$ (Bazzaz et al., 1974). The opening of closed stomata was reduced by as little as 10 µM $PbCl_2$. Lead induced several changes in gas exchange and water relation (Becerril et al., 1988). The direct evidence that restricted CO_2 fixation by lead was mainly due to stomatal closure has been shown for detached pea leaves (Parys et al., 1998) (Fig.1).

The photosynthesis rate after 24 h of Pb treatment (5 mM $Pb(NO_3)_2$) was inhibited 2 times and after 48 h, the rate of net photosynthesis was below the respiration rate. The photosynthetic O_2 evolution by protoplasts isolated from the leaves treated 24 or 48 h with lead was decreased only 10 or 25%, respectively. Thus, the results with protoplasts demonstrate that CO_2 fixation in detached leaves up to 24 h, was restricted mainly by stomatal closure. The stomatal closure was accompanied by an increased abscisic acid levels after 24 and 48 h of Pb treatment, and increased 3 and 7 times, respectively. Nevertheless, it is important to emphasize that not only stomatal closure may affect photosynthesis, but also metal-induced inhibition of photosynthesis may enhance stomatal closure by an increase of the internal CO_2 level. The CO_2 concentration in pea leaf treated 24 h with Pb ions was higher than in the control leaf and the CO_2 compensation point of Pb-treated leaf was increased by about 50%. In contrast to photosynthesis the rates of dark respiration of detached leaves and protoplasts treated for 24 h with this toxicant were stimulated 20 and 40%, respectively.

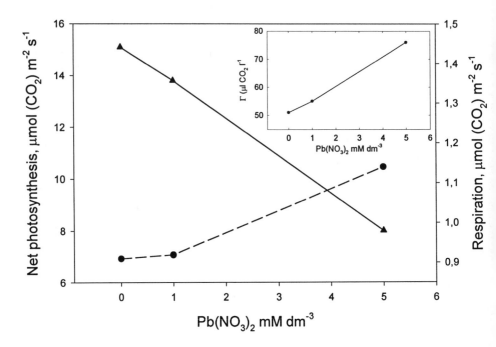

Figure 1. Net photosynthesis (▲) and dark respiration (●) rates of detached pea leaves treated for 24 hours with water or Pb(NO3)2 solution. The inserts shows the effect of this toxicant on the CO_2 compensation point (Γ) (adapted from Parys et al., 1998).

Stimulation of respiration after treatment the whole plants, leaves, protoplasts and mitochondria was observed in numerous studies (Lee et al., 1976; Lamoreaux and Chaney, 1978; Parys et al., 1998; Koeppe and Miller, 1970). It is supposed that under limited CO_2 fixation caused by heavy metals, redox equivalents produced photosynthetically can be transferred to mitochondria and oxidized, and hence the increased respiration (Poskuta and Wacławczyk, 1995; Poskuta et al., 1996). Several studies have indicated that mitochondria could oxidize redox equivalents produced photosynthetically (Krömer et al., 1988; Krömer and Heldt, 1991; Krömer et al., 1993). Earlier it was shown that Pb ions stimulated the oxidation of exogenous NADH in isolated mitochondria (Koeppe and Miller, 1970; Bittel et al., 1974). Experiments with isolated mitochondria showed that effect of heavy metals on respiration was dependent on used concentration. Inhibitory effect occurred at higher metal concentration and was accompanied by loss of respiratory control due to uncoupling of the phosphorylation (Fluharty and Sanahi, 1962, 1963; Miller et al., 1973; Kleiner, 1974).

Data on O_2 consumption or CO_2 release by plants treated with lead are mostly reported only incidentally. There is no stoichiometric relationship between net O_2 or CO_2 exchange and tissue energy consumption.

3.2. Cadmium (Cd)

Cadmium is a major environmental contaminant. It is present in water, air, and soils especially in areas of heavy automobile traffic. Cadmium is absorbed by leaves from surface contamination and may be transported to other plant parts. Most of the Cd is absorbed by roots. Root absorption is found to be the major pathway for cadmium enrichment in plant tissue. Cd has a great mobility in the soil as compared with other heavy metals and the mobility is pH dependent (Baker et al., 1979; Singh and Myhr, 1998). In *Holcus lanatus*, 90% of the total Cd taken up stays in the roots, and the degree of accumulation in the shoots depends on the tolerance of the plant to cadmium (Coughtrey and Martin, 1978). In most plant species the transport of Cd to the shoots usually is directly proportional to the external concentration (Petterson, 1977).

Subcellular fractionation of Cd-containing bean tissues showed that more than 70% of the element was localized in the cytoplasmic fraction, 8–14% was bound to the cell wall fraction or to the organelles. Only trace of Cd could be detected as a free ion (Weigel and Jäger, 1980) and Cd toxicity is related to the free Cd^{2+} contamination (Gutknecht, 1983). However, subcellular distribution of Cd in the unicellular green alga *Chlamydomonas reinhardtii* shows that in purified chloroplasts there was more than 50% cadmium, and cytosol contained only 10% of incorporated Cd^{2+} (Nagel et al., 1996). Vacuolar accumulation of Cd after high level Cd exposure of plants and marine alga *Dunaliella* is suggested from several studies (Hevillet et al., 1986; Krotz et al., 1989; Vögeli-Lange and Wagner, 1990).

It was reported that different effects of cadmium on plant growth was accompanied with various metal concentration found in the leaves of different ages (Chardonnes et al., 1999). Interaction of Cd and gibberellin positively affected growth rate and reduced poisonous effects of Cd on soybean plants (Ghorbanli et al., 1999).

Cd^{2+} was found to reduce chlorophyll content (Stobart et al., 1985; Schlegel et al., 1987) and in elevated levels it inhibited photosynthesis in general and photosynthetic CO_2 fixation in particular (Bazzaz et al., 1974; Hampp et al., 1976).

Net photosynthesis of spruce seedlings, estimated by CO_2 uptake, was markedly inhibited by Cd ions (Schlegel et al., 1987). Carbon dioxide uptake, expressed on a dry weight basis, was reduced to 71 and 33% of the control value after 5 weeks treatment with 1 and 5 µM Cd, respectively. When the CO_2 assimilation was expressed in terms of chlorophyll concentration of the needles, a significant part of the decrease correlated with the decreased chlorophyll level. The rates of dark respiration were not significantly diminished but transpiration rate was reduced by cadmium. At 5 µM Cd the rate of transpiration was decreased to 57% of the control. This suggests that a part of the inhibition of CO_2 uptake may be due to stomatal closure, probably induced by a decreased availability of water in needles.

The reduction of photosynthesis and inhibition of transpiration was observed in excised leaves of corn and sunflower exposed to various concentrations of Cd (Bazzaz et al., 1974). The reduction was dependent on the concentration of $CdCl_2$ solution and became more pronounced with time. In corn leaves CO_2 uptake was inhibited during the first 5 min after the introduction of 18 and 9 mM cadmium, and during 25 min after the introduction of 4.5 mM Cd. Photosynthesis was completely inhibited within 40 min at the high concentration and became 20% of maximum 2 h after the introduction of Cd with the two lower concentrations. In sunflower the photosynthesis was not affected during the first 30 min after the introduction of 9 mM Cd but gradually declined to 30% of maximum within 2 h. The different sensitivity of both plants to Cd was not related to the final Cd concentration in the leaves. Linear correlation found between photosynthesis and transpiration indicate that one of the initial effects of Cd is the induction of stomatal closure. Results of Bishnoi et al. (1993) show that CO_2 assimilation of wheat leaves may be affected to different extent by Cd, depending upon their age and position on the plant. Photosynthesis of the first leaf was reduced by 15% in 10 mM Cd and that of the third leaf by 47%, respectively.

The reduction of photosynthesis under Cd^{2+} treatment is related to the decreased availability of water which, by lowering water potential, causes closure of stomata. Direct or indirect effect of Cd^{2+} on photosynthesis in sugar beet was measured (Greger and Ögren, 1991). The photosynthetic properties were characterized by measurement of CO_2 fixation in intact plants and O_2 evolution by isolated protoplasts. Cd inhibits CO_2 assimilation of leaves (16% at 5 mM Cd) and O_2 evolution of isolated protoplasts when was added directly as well as indirectly (about 40%). Direct Cd^{2+} addition to the protoplasts did not significantly affect the rate of dark respiration. However, when Cd^{2+} was applied during the plants growth, protoplasts isolated from Cd treated plants showed a 20% increase in dark respiration rate.

The increase in dark respiration in response to indirect application of Cd^{2+} is consistent with the suggestion made by Lee et al. (1976) and that respiration increases to compensate for a Cd^{2+}-induced decrease in photophosphorylation. Soybean leaves were analyzed 10 days after Cd^{2+} was added to the culture solution in concentration from 0.045 to 1.35 µM. Respiration rate measured as O_2 uptake increased significantly at cadmium levels of 0.90 and 1.35 µM (about 60%). Increased respiration was positively correlated with the malate dehydrogenase activity. This suggested that the tricarboxylic acid cycle was stimulated by Cd^{2+} for energy production. Excised leaves of silver maple exposed to 0.045–0.18 mM Cd^{2+} exhibited reduced CO_2 assimilation (18%) and transpiration (21%), and increased CO_2 evolution during respiration as much as 193% of the untreated controls after 64 h of cadmium treatment (Lamoreaux and Chaney, 1978) (Tab. 2). The authors suggest that Cd inhibited transpiration by interference with stomatal function and that it inhibited photosynthesis by increasing both stomatal and mesophyll resistance to carbon dioxide uptake. Increased dark respiration may resulted from increased demand for ATP production via substrate-level and/or oxidative phosphorylation. However, this seems less likely given that the most sensitive part of photosynthesis are the reactions related to ATP consumption and not to

Table 2: *Efect of Cd on net photosynthesis, dark respiration and transpiration rates of excised silver maple leaves, 64 h after treatment (adapted from Lamoreaux and Chaney, 1978).*

Cd concent mM	net photosynthesis mg CO_2 dm^{-2} h^{-1}	dark repiration mg CO_2 dm^{-2} h^{-1}	transpiration mg H_2O dm^{-2} h^{-1}
0	3.39	0.45	187.4
0.045	2.54	0.90	128.2
0.090	2.07	1.15	115.5
0.180	0.62	1.32	38.9

ATP production (Siedlecka et al., 1997). The rise in dark respiration in heavy metal treated leaves has also been explained as a result of the oxidation of excess photosynthetic equivalents which are produced under conditions of limited CO_2 fixation (Poskuta et al., 1996). The inhibition of respiration in response to *in vitro* applications of cadmium to mitochondria isolated from animal (Fluharty and Sanahi, 1962, 1963) and plant cells (Mustafa and Cross, 1971; Reese and Roberts, 1985) was also observed. Experiments with isolated mitochondria showed that effect of heavy metals on respiration was dependent on used concentration. Inhibitory effect occurred at higher metal concentration and was accompanied by loss of respiratory control due to uncoupling of the phosphorylation (Fluharty and Sanahi 1962, 1963; Miller et al., 1973; Kleiner, 1974).

It is commonly accepted that Cd inhibited the activity of photosystem II measured by oxygen evolution. At low concentration of cadmium (0.1 and 0.3 mM) the inhibition was 20% however, at 0.5 and 0.7 mM, cadmium inhibited oxygen evolution to 80% in isolated chloroplasts of spinach (Li and Miles, 1975). Low Cd concentration, when methyl viologen was used, brought about an increase of oxygen consumption which could be explained by an uncoupling of the phosphorylation resulting in an increase of the electron transport (Van Duijvendijk-Matteoli and Desmet, 1975). Maximum uncoupling was obtained at 4 mM Cd. Higher Cd concentration caused a decrease of oxygen consumption which should be due to a blockage of the electron transport chain since Cd does not act as an electron acceptor. The observation of Lucerno et al. (1976) that low concentration of cadmium inhibited cyclic and non-cyclic phosphorylation in isolated spinach chloroplasts and depressed coupled electron transport to the basal level were carried out in concentration lower than in experiments Li and Miles (1975).

Inhibitory effect of cadmium on respiratory oxygen consumption as well as on photosynthetic oxygen evolution (De Filippis et al., 1981b) and inhibition of ATP level (De Filippis et al., 1981a) was observed also in *Euglena* cells treated low concentration of cadmium (0.01–50 µM $CdCl_2$). Contrary, Weigel (1985a) observed that Cd ions in concentration 0.1–1.0 mM did not affect electron transport reactions in intact spinach chloroplasts. The light-induced proton gradient across the thylakoid membrane was also not impaired by the metal. CO_2 dependent O_2 evolution was inhibited by Cd about 60%.

Photosynthetic CO_2 fixation by isolated chloroplasts was inhibited at a given Cd concentration and significantly decreased as the Cd concentration increased. Author suggested that in intact chloroplasts Cd affects photosynthesis by inhibition of different steps of the Calvin cycle and not by interaction with photochemical reactions located in the thylakoid membranes (Weigel, 1985b). This finding was confirmed with bean plants. High concentration of Cd added to the nutrient solution (10–50 µM) have very little effect on the primary photochemistry of PS II, as indicated *in vivo* by fast Chl a fluorescence kinetics (Krupa et al., 1992) and modulated fluorescence analysis (Krupa et al., 1993). The authors postulate that, during short term exposure of plants to cadmium in the early stages of growth, the Calvin cycle reactions are more likely than photosystem II to be the primary target of the toxic influence of cadmium. Other authors have concentrated on the inhibitory effect of Cd on CO_2 assimilation, particularly at the level of some Calvin cycle enzymes (Stiborova and Leblova, 1985; Stiborova et al., 1986).

In general, the activities of many photosynthetic carbon reduction cycle (PCR-cycle) enzymes decreased upon the Cd treatment (Sheoran et al., 1990). CO_2 fixation by PCR-cycle acts as a sink for the products of photosynthetic electron transport. Cadmium which diminishes the utilization of photosynthetic energy in carbon metabolism, will also modify the rate of electron transport, by down-regulation of PS II photochemistry. Although the inhibition sites of dark reaction by Cd^{2+} have not been described completely, the discovery that inhibition of photosynthetic CO_2 fixation is determined, above all, by the reactions occurring in the chloroplast stroma is not depreciated. Cadmium inhibits CO_2 fixation in various phases of this process.

Inhibition of net photosynthesis was often accompanied by reduced transpiration and has been attributed to Cd-induced stomata closure. Most information concerning the effect of Cd on photosynthesis are at its molecular level. It is the question if similar mechanism of Cd inhibition occurring in plants in natural environment occurs in plants cultivated on Cd containing medium.

3.3 Aluminum (Al)

Aluminum is a very common metal which although not being a heavy metal could on chemical grounds have similar effects. The solubility of Al is very different from other metals. In soil it is available below pH 4.5 (Foy, 1978), although both, species and cultivars within species, show substantive differences in their response to the presence of Al in solution (Taylor, 1988). Approximately 40% of the world's cultivated lands, and up to 70% of the potentially arable lands, are acidic (Haug, 1984). In many of these acidic soils, aluminum toxicity is a major factor limiting root growth. Some plants accumulate Al in the shoot (Kataraglis, 1987). In mature leaves, Al is localized mainly in the apoplast (Memon et al., 1981) of leaf tissues. The apoplastic location of Al in older leaves may explain why very high aluminum concentrations in such tissues are not phytotoxic. The measurement of shoot Al concentrations to define symptoms are seldom reported, whereas nutrient accumulation varies in different parts of the shoots (Schaedle et al., 1989; Kochian, 1995) and roots (Delhaize and Ryan, 1995).

Aluminum under some conditions, in low concentration, can increase growth. Plants that have shown positive growth response to Al in nutrient cultures include tea, rice, peach, corn and wheat. Possible explanations include: increasing Fe solubility and availability in a calcareous soil, preventing Fe deficiency by blocking negative charges on cell wall sites or promoting P uptake (Foy at al., 1978).

The inhibition of biomass production at high Al concentration in the growth medium has been recognized as a symptom of Al toxicity (Delhaize and Ryan, 1995).

Aluminium has been reported to alter the properties of biological membranes. Membrane fluidity in the microorganism, *Termoplasma acidophilum*, was decreased by Al (Vierstra and Haug, 1978). Also, Zhao et al. (1987) found that Al alters permability of the plasmalemma and tonoplast in the root cortical cells. It was found that the proton transport activity of membrane vesicles isolated from barley roots was inhibited by aluminum.

Al toxicity symptoms include marginal chlorosis and necrosis of younger leaves and spotty chlorosis of older leaves (Pavan and Bingham, 1982).

In vitro studies have shown that Al inhibits ATPase (Suhayda and Haug, 1985) and hexokinase activity (Turner and Copeland, 1981). Each of these effects could be responsible individually or in combination, for Al toxicity to photosynthesis (Hampp and Schnabl, 1975). When Al reaches mitochondria, respiration is inhibited (Roy et al., 1988). It has been shown (Ślaski et al., 1996) that Al can induce activity of some enzymes of the pentose phosphate pathway providing precursors or cofactors for other biosynthetic routes. It is possible that the rapid induction of enzymes of the pentose phosphate pathway play a role in mediating resistance to Al.

Aluminum effects on senescence resulting from production of free radicals in both root and shoot cells has long been studied (Cakmak and Horst, 1991).

As a result of thylakoid degradation, increasing Al toxicity inhibits photosynthesis (Haug, 1983). It was found (Lidon et al., 1999) that in maize leaves Al in concentration of 9 and 81 ppm induced a significant decrease (60%) in photosynthetic electron transport associated with PS I. The CO_2 assimilation rates were inhibited by Al ions by about 20%. Concomitantly plastocyanin concentration showed non-significant changes and cyt f and b563 contents significantly decreased to 71 and 41%, respectively. Earlier investigations have shown that Al ions at concentration of 10 µM caused a significant inhibition of CO_2 fixation, at 500 µM Al the inhibition was 90% (Hamp and Schnabl, 1974). These investigations show a close correlation between inhibition of photosynthetic $^{14}CO_2$-fixation and inhibition of enzyme activities involved in dark reaction of photosynthesis and observed damage of the outer chloroplast membrane.

3.4. Mercury (Hg)

Mercury is widely distributed in the earth's crust, sea, ground, rain water; all phyla and species naturally contain traces varying with location (Vallee and Ulmer, 1972).

In animals, Hg enhanced lipid peroxidation in several organs, and reduced glutathione level (Huang et al., 1996), indicating that the oxidative stress induced lipid peroxidation may be one of the molecular mechanism for cell injury in strong Hg poisoning. In tomato plants mercury altered level of antioxidant enzymes indicating that Hg induced phytotoxicity by oxidative stress (Cho and Park 2000).

Mercury is well known freshwater pollutant. A growth retardation of 50% was found at 150 ppb Hg for macrophytic alga *Chara vulgaris* (Heumann, 1987). Comparatively less sensitive to mercury were the freshwater angiosperms. A 50% plant damage of *Elodea canadensis* was observed by 7.4 ppm mercury (Brown and Rattigan, 1979) and a 50% inhibition of root growth of *Myriophyllum spicatum* occurred at concentration of 3.4 ppm Hg (Stanley, 1974). *Euglena* is very common alga which appers to be very sensitive to solution of heavy metals, and it has been established that in this alga Hg reduces level of pigments (De Filippis et al., 1981a). Therefore the rate of photosynthetic O_2 evolution was reduced (De Filippis et al., 1981b) in sublethal concentration of mercury.

Together with inhibition of photosynthesis, Hg (0.01 µM) induced a severe loss of adenylate pool and rate of photophosphorylation was strongly inhibited (De Filippis et al., 1981c). Reduced ATP/ADP ratio was observed also by other authors. Honeycutt and Krogmann (1972) reported complete inhibition of cyclic photophosphorylation in the presence of 5 µM Hg, while a 50% inhibition of PS II associated reactions resulted from application of 150 µM solution in short term experiments. Under the influence of Hg there is an initial decrease in energy charge, this is due to a retardation of all ATP-dependent process as, e.g., the CO_2 fixation. *Nostoc calciola* cells exposed to Hg (0.05–1 µM) showed strong inhibition of photosynthetic O_2 evolution and CO_2 uptake varied significantly with respect to increasing concentration of Hg (Singh and Singh, 1987).

Photosynthetic O_2 evolution seems to be slightly more sensitive to Hg stress than CO_2 fixation. Hg ions predominantly attack the action sites of Mn^{2+} and ATP generating steps in photosynthesis of the cyanobacterium. Addition of Hg ions to the medium with cyanobacterium *Spirulina platensis* caused the inhibition in electron transport activity (Murthy and Mohanty, 1995). Mercury in concentration 6 µM caused a 48% inhibition of whole electron transport activity; the increase in the concentration to 15 µM brought approximately 71% loss in the activity. 6 µM Hg caused approximately 16% loss in the PS II activity, at 18 µM the loss was 52%.

It was observed using photoacoustic measurements that Hg^{2+}, Cu^{2+} and Pb^{2+} ions change energy storage in spinach thylakoids (Boucher and Carpentier, 1999). A maximal inhibition of photosynthetic energy storage of 50 and 80% was obtained with Cu^{2+} and Hg^{2+}, respectively. Energy storage was insensitive to Pb ions. Authors suggest that the sensitive portion of energy storage is attributed to the possible recurrence of cyclic electron transport around PS II that would depend on the extent of inhibition produced on the acceptor side by the heavy metal ion used.

It was shown that in chloroplasts of spinach Hg ions interact with some sites in the photosynthetic electron transport chain situated in D1 and D2 proteins and with the manganese cluster in the oxygen evolving complex. The binding of Hg ions to the chloroplast proteins and the formation of organometallic complexes were documented by the quenching of fluorescence emission of aromatic amino acids (Šerševi et al., 1998). Spruce seedlings exposed to Hg at concentration 0–0.1 µM showed a reduced chlorophyll content of the primary needles to 70–74% of the control (Schlegel et al. 1987). In the needles of Hg-treated seedlings the decrease in chlorophyll content was independent of the Hg concentration or chemical form supplied. The CO_2 assimilation rate on a dry weight basis was significantly reduced to 91% of the control rate. The lowered CO_2 assimilation cannot be fully explained by a decrease in chlorophyll level. However, examination of the rates of transpiration suggests that a part of the inhibition of CO_2 uptake may be due to a stomatal closure. Exposure to 0.01 µM Hg depressed the transpiration rate to 51% of the control, whereas 0.1 µM Hg had no effect. In summary, influence of Hg on gas exchange may be explained mostly by a decreased availability of water in the leaves.

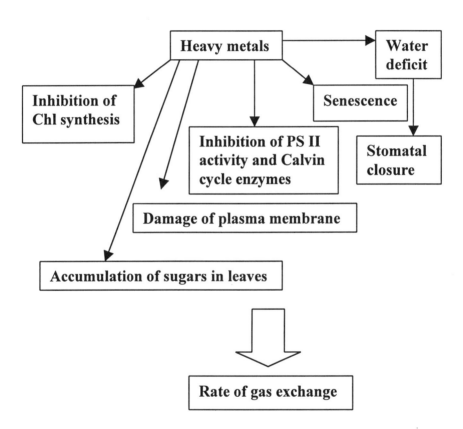

Figure 2. Influence of heavy metals on processes affecting gas exchange.

Table 3. The efect of heavy metals on gas exchange rates.

Heavy metal	Plant species	Metal conc.	Photosynthesis (control 100%)	Method	References
Cd	Zea mays	18 mM	38%	O_2 exchange	Bazzaz et al., 1974
Cd	Helianthus annuus (excised leaves)	18 mM	90%	CO_2 exchange	Bazzaz et al., 1974
Cd	Glycine max	500 µM	60%	CO_2 exchange	Huang et al., 1974
Cd	Acer saccharinum (after 64 h, detached leaves)	5 mM	18%	CO_2 exchange	Lamoreaux and Chaney, 1978
Cd	Spinacia oleracea (chloroplasts)	500 µM	44%	O_2 electrode	Weigel 1985
Cd	Beta vulgaris	20 µM	55%	CO_2 exchange	Greger, Ögren 1991
		10 µM	40%	O_2 electrode	Greger, Ögren 1991
Pb	Glycine max	500 µM	60%	CO_2 exchange	Huang et al., 1974
Pb	Helianthus annuus	5 mM	70%	CO_2 exchange	Bazzaz et al., 1975
Pb	Zea mays	5 mM	60%	CO_2 exchange	Bazzaz et al., 1975
Pb	Pisum sativum	400mg/l	50%	$^{14}CO_2$ uptake	Poskuta et al., 1987
Pb	Zea may	400mg/l	40%	$^{14}CO_2$ uptake	Poskuta et al., 1988
Pb	Festuca arundinacea (detached leaves after 30 min.)	100 mM	80%	CO_2 exchange	Poskuta, Wacławczyk, 1995
Pb	Pisum sativum	5 mM	53%	CO_2 exchange	Parys et al., 1998
Zn	Picea abies (after 5 weeks)	60 µM	67%	CO_2 exchange	Schlegel et al., 1987
Zn	Chlorella pyrenoidosa (after 48 h)	10 mM	54%	O_2 electrode	Poskuta, 1991
Cu	Dunaliella parva	10^{-8} M	43%	O_2 electrode	Sandmann, 1985
Cu	Chlorella pyrenoidosa (after 48 h)	2.5 mM	52%	O_2 electrode	Poskuta, 1991
Hg	Picea abies (after 5 weeks)	0.01µM	27%	CO_2 exchange	Schlegel et al., 1987
Hg	Nostoc calciola	0.5 µM	25%	$^{14}CO_2$ uptake	Singh, Singh, 1987
		0.20 µM	50%	O_2 electrode	Singh, Singh, 1987
Ni	Oryza sativa	0.5 mM	41%	CO_2 exchange	Moya et al., 1993
Ni	Triticum aestivum (after 2 days, top leaf)	10 mM	49%	CO_2 exchange	Bishnoi et al., 1993
Cd	Vicia faba	0.90µM	163%	O_2 manometry	Lee et al. 1976
Cd	Vallisneria spiralis	10 µM	93%	O_2 electrode	Jana and Choudhuri 1982
Cd	Nicotiana tabacum	44.5µM	93%	O_2 manometry	Reese, Roberts 1985
Pb	Pisum sativum	5mM	120%	CO_2 exchange	Parys et al. 1998
Cu	Euglena gracilis (24 treatment)	200 µM	180%	O_2 electrode	Edjlali, Calvayrac 1991
Cu	Chlorella pyrenoidosa (after 24h)	2.5mM	171%	O_2 electrode	Poskuta 1991
Hg	Potamogeton pectinatus	0.1 mM	97%	O_2 exchange	Jana and Choudhuri 1982
Hg	Picea abies	0.1µM	105%	CO_2 exchange	Schlegel et al. 1987

4. CONCLUSIONS

The effects of heavy metals on gas exchange are quite complex and depend on metal species and concentration, exposure time and plant species. Even differences between population or cultivars of the same species have been observed. Heavy metals have serious adverse effects on growth and metabolic processes in plants, including reduction in chlorophyll content, degradation of chloroplasts, reduction of photosynthesis and inhibition of enzyme activities.

In general, most heavy metals preferentialy inhibit PS II activity (Yruela et al., 1992; Renganathan and Bose, 1989). In isolated chloroplasts and mitochondria it was shown that Cd, Cu, Pb, Zn inhibit ATP synthase activity (Kleiner, 1974; Baszyński et al., 1980; Uribe and Stark, 1982; Van Assche and Clijsters, 1986b). Cd, Pb, Ni, Cu, Zn lead to decline in CO_2 fixation rates and have pronounced effects on Calvin cycle enzymes.

Heavy metals slow down Calvin cycle reactions and limit ATP and NADPH consumption, which leads to feedback inhibition of the photosynthetic electron transport. These findings suggest that the inhibition of the photosynthetic capacity of plants exposed to heavy metals is due to the inhibition of both electron transport activities as well as Calvin cycle enzymes.

Decreased transpiration or increased stomatal resistance have been observed in many studies on plants exposed to toxic metals (Angelov et al., 1993a, b; Bazzaz et al., 1974; Becerril et al., 1988; Barcelo and Poschenrieder, 1990; Bishnoi et al., 1993).

Stomatal closure may be caused by a direct interaction of the metals with guard cells, but in whole plants the increased stomatal resistance may be a consequence of early toxicity effects in roots and stems, leading to decreased water availability in leaves.

The importance of heavy metals induced stomatal closure in relation to photosynthesis is not clear. Some authors have suggested that stomatal closure substantially contributes to the decrease of photosynthesis observed in metal stressed plants, while others indicate that metals act primarily on the light and dark reaction of photosynthesis, so an increase of stomatal resistance is of a secondary importance.

Metal induced reduction of photosynthesis may enhance stomatal closure by an increase of the internal CO_2 level or by changes in the starch and energy metabolism. The reduced photosynthesis observed in heavy metal stressed plants seems to conflict with the increase in the carbohydrates content determined in some experiment (Samarakoon and Rauser, 1979). It has been proposed that the decrease in utilization of carbohydrates for growth produced by heavy metals is more pronounced than the decrease in CO_2 fixation, resulting in an increased accumulation of carbohydrates (Greger et al., 1991). Some data on O_2/CO_2 consumption/release by plants treated with heavy metal, are compiled in Table 3. Values are expressed as percent of control. The influence of heavy metals on processes related to gas exchange are presented in Fig.2. Despite extensive studies conducted on the effects of various heavy metals on different photosynthetic parameters in growing plants, our knowledge is still incomplete. Various changes in physiological and biochemical events occurring at different sites during photosynthesis are tightly linked so that even slight inhibition of one of them may result in the series of processes limiting gas exchange in the plants.

ACKNOWLEDGEMENTS

This work was supported by Polish Committee for Scientific Research Grant 6 P04C 06019

REFERENCES

Agarvala, S.C., Bisht, S.S., and Sharma, C.P. (1977) Relative effectiveness of certain heavy metals in producing toxicity and symptoms of iron deficiency in barley, *Can. J. Bot.* **55**, 1299–1307.

Alva, A.K. and Chen, E.Q. (1995) Effects of external cooper concentration on uptake of trace elements by citrus seedlings, *Soil Sci.* **159**, 59–64.

Angelov, M., Tsonev, T., Dobrinova, K., and Velikova, V. (1993b) Changes in some photosynthetic parameters in pea plants after treatment with cobalt, *Photosynthetica* **28**, 289–295.

Angelov, M., Tsonev, T., Uzunova, A., and Gaidardjiewa, K. (1993a) Cu^{2+} effect upon photosynthesis, chloroplast structure, RNA and protein synthesis of pea plants, *Photosynthetica* **28**, 341–350.

Arndt, U., Wystricil, G., and Wyzgolik, K. (1983) Heavy metals and their effects to primary processes of photosynthesis. Reactions of the Kautsky-effect, in *Proc. 6th Word Congress on Air Quality*, Paris, vol. 2, pp. 515–521.

Austenfeld, F.A. (1979) Nettophotosynthese der Primär- und Folgeblätter von *Phaseolus vulgaris* L. unter dem Einfluß von Nickel, Cobalt und Chrom, *Photosynthetica* **13**, 434–438.

Baker, D.E., Amacher, M.C., and Leach, R.M. (1979) Sewage sludge as a source of cadmium in soil-plant-animal systems, *Environ. Health Perspect.* **28**, 45–49.

Barcelo, J. and Poschenrieder, Ch. (1990) Plant water relations as affected by heavy metal stress: a review, *J. Plant Nutr.* **13**, 1–37

Baron, M., Arellano, B., and Lopez Gorge, J. (1995) Copper and photosystem II: A controversial relationship, *Physiol. Plant.* **94**, 174–180.

Baron, M. and Sandmann, G. (1988) The role of Cu in respiration of pea plants and heterotrophically growing *Scenedesmus* cells, *Z. Naturforsch.* **43c**, 438–442.

Barua, B. and Jana, S. (1985) Effects of heavy metals on dark induced changes in Hill reaction activity, chlorophyll and protein contents, dry matter and tissue permeability in detached *Spinacia oleracea* L. leaves, *Photosynthetica* **20**, 74–76.

Baszyński, T. (1986) Interference of Cd^{2+} in functioning of the photosynthetic apparatus of higher plants, *Acta Soc. Bot. Polon.* **55**, 291–304.

Baszyński, T., Wajda, L., Król, M., Wolińska, D., Krupa, Z., and Tukendorf, A. (1980) Photosynthetic activities of cadmium-treated tomato plants, *Physiol. Plant.* **48**, 365–370.

Bazzaz, F.A., Carlson, R.W., and Rolfe, G.L. (1975) Inhibition of corn and sunflower photosynthesis by lead, *Physiol. Plant.* **34**, 326–329.

Bazzaz, F.A., Rolfe, G.L., and Carlson, R.W. (1974) Effect of Cd on photosynthesis and transpiration of excised leaves of corn and sunflower, *Physiol. Plant.* **32**, 373–376.

Becerril, J.M., Munoz-Rueda, A., Aparcio-Tejo, P., and Gonzalez-Murua, C. (1988) The effects of cadmium and lead on photosynthetic electron transport in clover and lucerne, *Plant Physiol. Biochem.* **26**, 357–363.

Bishnoi, N.R., Sheoran, I.S., and Singh, R. (1993) Influence of cadmium and nickel on photosynthesis and water relation in wheat leaves of different insertion level, *Photosynthetica* **28**, 473–479.

Bittel, J.F., Koeppe, E., and Miller, R.J. (1974) Sorption of heavy metal cations by corn mitochondria and the effects on electron and energy transfer reactions, *Physiol. Plant.* **30**, 226–230.

Bohner, H., Böhme, H., and Böger, P. (1980) Reciprocal formation of plastocyanin and cytochrome c-553 and the influence of cupric ions on photosynthetic electron transport, *Biochim. Biophys. Acta* **592**, 103–112.

Boucher, N. and Carpentier, R. (1999) Hg^{2+}, Cu^{2+}, and Pb^{2+}-induced changes in photosystem II photochemical yield and energy storage in isolated thylakoid membranes: a study using simultaneous fluorescence and photoacoustic measurements, *Photosynth. Res.* **59**, 167–174.

Brown, B.T. and Rattigan, B.M. (1979) Toxicity of soluble copper and other metal ions to *Elodea canadensis*, *Environ. Pollut.* **20**, 303–314.

Brown, P.H., Welch, R.M., and Cary, E.E. (1987a) A micronutrient essential for higher plants, *Plant Physiol.* **85**, 8801–803.

Brown, P.H., Welch, R.M., and Cary, E.E. (1987b) Beneficial effects of nickel on plant growth, *J. Plant Nutr.* **10**, 2125–2135.

Buchauer, M.J. (1973) Contamination of soil and vegetation near a zinc smelter by zinc, cadmium, copper, and lead, *Environ. Sci. Technol.* **7**, 131–135.

Burzyński, M. (1985) Influence of lead on the chlorophyll content and on initial step of its synthesis in greening cucumber seedlings, *Acta Soc. Bot. Polon.* **54**, 95–105.

Cakmak, I. and Horst, W.J. (1991) Effect of aluminum on lipid peroxidation, superoxide dismutase, catalase, and peroxidase activities in root tips of soybean (*Glycine max*), *Physiol. Plant.* **83**, 463–468.

Cedeno-Maldonado, A. and Swader, J.A. (1972) The cupric ion as an inhibitor of photosynthetic electron transport in isolated chloroplasts, *Plant Physiol.* **50**, 698–701.

Chappelka, A.H., Kush, J.S., Runion, G.B., Meier, S., and Kelly, W.D. (1991) Effects of soil-applied lead on seedling growth and ectomycorrhizal loblolly pine, *Environ. Pollut.* **72**, 307–316.

Chardonnes, A.N., Tenbookum, W.M., Vellinga, S., Schat, H., Verkleij, J.A.C., and Ernst, W.H.O. (1999) Allocation patterns of zinc and cadmium in heavy metal tolerant and sensitive *Silene vulgaris*, *J. Plant Physiol.* **155**, 778–787.

Cho, U-H. and Park, J-O. (2000) Mercury-induced oxidative stress in tomato seedlings, *Plant Sci.* **156**, 1–9.

Chongpraditnun, P., Mori, S., and Chino, M. (1992) Excess cooper induces a cytosolic Cu, Zn-superoxide dismutase in soybean root, *Plant Cell Physiol.* **33**, 239–244.

Clarkson, D.T. and Hanson, J.B.(1980) The mineral nutrition of higher plants, *Annu. Rev. Plant Physiol.* **31**, 329–398.

Clijsters, H. and Van Assche, F. (1985) Inhibition of photosynthesis by heavy metals, *Photosynth. Res.* **7**, 31–40.

Costa, G., Michaut, J.C., and Morel, J.L. (1994) Influence of cadmium on water relations and gas exchanges in phosphorus-deficient *Lupinus albus*, *Plant Physiol. Biochem.* **32**, 105–114.

Coughtrey, P.J. and Martin, M.H. (1978) Cadmium uptake and distribution in tolerant and non-tolerant populations of *Holcus lanatus* L. grown in solution culture, *Oikos* **30**, 555–560.

Davies, A.G. and Sleep, J.A. (1979) Photosynthesis in some British coastal water may be inhibited by zinc pollution, *Nature* **277**, 292–293.

De Filippis, L.F., Hampp, R., and Zeigler, H. (1981a) The effect of zinc, cadmium and mercury on *Euglena*. I. Growth and pigments, *Z. Pflanzenphysiol* **101**, 37–47.

De Filippis, L.F., Hampp, R., and Zeigler, H. (1981b) The effect of zinc, cadmium and mercury on *Euglena*. II. Respiration, photosynthesis and photochemical activity, *Arch. Microbiol.* **128**, 407–411.

De Filippis, L.F., Hampp, R., and Ziegler, H. (1981c) The effect of sublethal concentrations of zinc, cadmium and mercury on *Euglena*. Adenylates and energy charge, *Z. Pflanzenphysiol.* **103**, 1–7.

De Vos, C.H.R., Schat, H., Vooijs, R., and Ernst, W.H.O. (1993) Effect of cooper on fatty acid composition and peroxidation of lipids in the roots of cooper tolerant and sensitive *Silene cucubalus*, *Plant Physiol. Biochem.* **31**, 151–158.

Delhaize, E. and Ryan, P.R. (1995) Aluminum toxicity and tolerance in plants, *Plant Physiol.* **107**, 315–321.

Droppa, M. and Horvath, G. (1990) The role of copper in photosynthesis, *Crit. Rev. Plant. Sci.* **9**, 111–123.

Edjali, M. and Calvayrae, R. (1991) Effects des ions métalliques sur l'intensité respiratoire et sur les capacités catalitiques chez *Euglena gracilis*, *Z. CR Acad.Sci.*, Paris 312 III, pp.177–182.

Ernst, W.H.O. (1980) Biochemical aspects of cadmium in plants, in J.O. Nriagu (ed.), *Cadmium in The Environment*, J. Wiley & sons, New York, pp. 639–653.

Fluharty, A.L. and Sanadi, D.R. (1962) On the mechanism of the oxidative phosphorylation. IV. Mitochondrial swelling caused by arsenite in combination with 2,3-dimercaptopropanol and by cadmium ion, *Biochemistry*. **1**, 276–281.

Fluharty, A.L. and Sanadi, D.R. (1963) On the mechanism of the oxidative phosphorylation. VI. Localization of the dithiol in oxidative phosphorylation with respect to the oligomycin inhibition site, *Biochemistry*. **2**, 519–521.

Foy, C.D., Chaney, R.L., and White, M.C. (1978) The physiology of metal toxicity in plants, *Annu. Rev. Plant Physiol.* **29**, 511–566.

Gabbrielli, R., Pandolfini, T., Vergnano, O., and Palandri, M.R. (1990) Comparison of two species with different nickel tolerance strategies, *Plant Soil* **122**, 271–277.

Gallego, S.M., Benavides, M.P., and Tomaro, M. L. (1996) Effect of heavy metal ion excess on sunflower leaves: evidence for involvement of oxidative stress, *Plant Sci.* **121**, 151-159.

Ghorbanli, M., Kaveh, S.H., and Sepehr, M.F. (1999) Effects of cadmium and gibberellin on growth and photosynthesis of *Glycine max*, *Photosynthetica* **37**, 627-631.

Goyal, A. and Tolbert, N.E. (1989) Variations in the alternative oxidase in *Chlamydomonas* grown in air or high CO_2, *Plant Physiol.* **89**, 958-962.

Greger, M., Brammer, E., Lindberg, S., Larsson, G., and Idestam-Almquist, J. (1991) Uptake and physiological effects of cadmium in sugar beet (*Beta vulgaris* L.) related to mineral provision, *J. Exp. Bot.* **42**, 729-737.

Greger, M. and Ögren, E. (1991) Direct and indirect effects of Cd^{2+} on photosynthesis in sugar beet (*Beta vulgaris*), *Physiol. Plant.* **83**, 129-135.

Gupta, M., Cuypers, A., Vangronsveld, J., and Clijsters, H. (2000) Copper affects the enzymes of the ascorbate-glutathione cycle and its related metabolites in the roots of *Phaseolus vulgaris*, *Physiol. Plant.* **106**, 262-267.

Gutknecht, J. (1983) Cadmium and thallous ion permeabilities through lipid bilayer membranes, *Biochim. Biophys. Acta* **735**, 185-188.

Hampp, R., Beulich, K. and Ziegler, H. (1976) Effects of zinc and cadmium on photosynthetic CO_2-fixation and Hill activity of isolated spinach chloroplasts, *Z. Pflanzenphysiol.* **77**, 336-344.

Hampp, R. and Lendzian, K. (1974) Effect of lead ions on chlorophyll synthesis, *Naturwissenschaften*, **61**, 218-219.

Hampp, R. and Schnabl, H. (1975) Effect of aluminum ions on $^{14}CO_2$-fixation and membrane system of isolated spinach chloroplasts, *Z. Pflanzenphysiol.* **76**, 300-306.

Haug, A. (1983) Molecular aspects of aluminum toxicity, *CRC Crit. Rev. Plant. Sci.* **1**, 345-373.

Heumann, H.-G. (1987) Effects of heavy metals on growth and ultrastructure of *Chara vulgaris*, *Protoplasma* **136**, 37-48.

Hevillet, E., Moreau, A., Halpern, S., Jeanne, N., and Piseux-Dao, S. (1986) Cadmium binding to a thiol-molecule in vacuoles of *Dunalliela bioculata* contaminated with $CdCl_2$: electron probe analysis, *Biol. Cell.* **58**, 79-86.

Hoganson, C.W., Casey, P.A., and Hansson, O. (1991) Flash photolysis studies of manganese-depleted Photosystem II: evidence for binding of Mn^{2+} and other transition metal ions, *Biochim. Biophys. Acta* **1057**, 399-406.

Holmgren, G.G.S., Meyer, M.W., Chaney, R.L., and Idestam-Almquist, J. (1993) Cadmium, lead, zinc, copper and nickel in agricultural soil of the United States of America, *J. Environ. Qual.* **22**, 335-348.

Honeycutt, R.C. and Krogmann, D.W. (1972) Inhibition of chloroplast reactions with phenyl mercuric acetate, *Plant Physiol.* **49**, 376-380.

Huang, C., Bazzaz, F.A., and Vanderhoef, L.N. (1974) The inhibition of soybean metabolism by cadmium and lead, *Plant Physiol.* **54**, 122-124.

Huang, Y.L., Cheng, S.L., and Lin, T.H. (1996) Lipid peroxidation in rats administered with mercuric chloride, *Biol. Trace Elem. Res.* **52**, 193-206.

Jana, S. and Choudhuri, M.A. (1982) Senescence in submerged aquatic angiosperms: effects of heavy metals, *New Phytol.* **90**, 477-484.

Kampfenkel, K., van Montagu, M., and Inze, D. (1995) Effects of iron excess on *Nicotiana plumbaginifolia* plants, *Plant Physiol.* **107**, 725-735.

Karataglis, S. (1987) aluminum toxicity in *Avena sativa* cv. Kassandra and a comparison with the toxicity caused by some other metals, *Phyton* **27**, 1-14.

Karataglis, S., Moustakas, M., and Symeonidis, L. (1991) Effect of heavy metals on isoperoxidases of wheat, *Biol. Plant.* **33**, 3-9.

Kleiner, D. (1974) The effect of Zn^{2+} ions on mitochondrial electron transport, *Arch. Biochem. Biophys.* **165**, 121-125.

Kochian, L.V. (1995) Cellular mechanisms of aluminium toxicity and resistance in plants, *Annu. Rev. Plant Physiol. Plant Mol. Biol.* **46**, 237-260.

Koeppe, D.E. and Miller, R.J. (1970) Lead effects on corn mitochondrial respiration, *Science* **167**, 1376-1377.

Kremer, B. and Markham, J.W. (1982) Primary metabolic effects of cadmium in the brown alga, *Laminaria saccharina*, *Z. Pflanzenphysiol.* **108**, 125-130.

Krömer, S. and Heldt, H. (1991) Respiration of pea leaf mitochondria and redox transfer between the mitochondrial and extramitochondrial compartment, *Biochem. Biophys. Acta* **1057**, 42-50.

Krömer, S., Malmberg, G., and Gardeström, P. (1993) Mitochondrial contribution to photosynthetic metabolism. A study with barley (*Hordeum vulgare*) leaf protoplasts at different light intensities and CO_2 concentrations, *Plant Physiol.* **102**, 947–955.

Krömer, S., Stitt, M., and Heldt, H. (1988) Mitochondria oxidative phosphorylation participating in photosynthetic metabolism of a leaf cell, *FEBS Lett.* **226**, 352–356.

Krotz, R.M., Evangelou, B.P., and Wagner, J.G. (1989) Relationships between cadmium, zinc, Cd-peptide, and organic acid in tobacco suspension cells, *Plant Physiol.* **91**, 780–787.

Krupa, Z. and Baszyński, T. (1995) Some aspects of heavy metals toxicity towards photosynthetic apparatus – direct and indirect effects on light and dark reactions, *Acta Physiol Plant.* **17**, 177–190.

Krupa, Z., Öquist, G., and Hunner, N.P.A. (1992) The influence of cadmium on primary photosystem II photochemistry in bean as revealed by chlorophyll a fluorescence – a preliminary study, *Acta Phsysiol. Plant.* **14**, 71–76.

Krupa, Z., Siedlecka, A., Maksymiec, W., and Baszyński, T. (1993) In vivo response of photosynthetic apparatus of *Phaseolus vulgaris* L. to nickel toxicity, *J. Plant Physiol.* **142**, 664–668.

L'Huillier, L., d'Auzac, J., Durand, O., and Michaud-Ferriere, N. (1996) Nickel effects on two maize (*Zea mays*) cultivars: growth, structure, Ni concentration, and localization, *Can. J. Bot.* **74**, 1547–1554.

Lamoreaux, R.J. and Chaney, W.R. (1978) The effect of cadmium on net photosynthesis, transpiration, and dark respiration of excised silver maple leaves, *Physiol. Plant.* **43**, 231–236.

Lee, K.C., Cunningham, B.A., Paulsen, G.H., Liang, G.M., and Moore, R.B. (1976) Effects of cadmium on respiration rate and activities of several enzymes in soybean seedlings, *Physiol. Plant.* **36**, 4–6.

Li, E.H. and Miles, C.D. (1975) Effects of cadmium on photoreaction II of chloroplasts, *Plant Sci. Lett.* **5**, 33–40.

Lidon, F.C., Barreiro, M.G., Ramalho, J.C., and Lauriano, J.A. (1999) Effects of aluminum toxicity on nutrient accumulation in maize shoots: implications on photosynthesis, *J. Plant Nutr.* **22**, 397–416.

Lidon, F.C. and Henriques, F.S. (1991) Limiting step on photosynthesis of rice plants treated with varying copper levels, *J. Plant Physiol.* **138**, 115–118.

Lidon, F.C., Ramalho, J.C., and Henriques, F.S. (1993) Copper inhibition of rice photosynthesis, *J. Plant Physiol.* **142**, 12–17.

Lucerno, H.A., Carlos, S.A. and Vallejos, R.H. (1976) Sulphydryl groups in photosynthetic energy conservation III. Inhibition of photophosphorylation in spinach chloroplasts by $CdCl_2$, *Plant Sci. Lett.* **6**, 309–313.

Luna, C.M., Gonzalez, C.A., and Trippi, V.S. (1994) Oxidative damage caused by an excess of copper in oat leaves, *Plant Cell Physiol.* **35**, 11–15.

Łukaszek, M. and Poskuta, J.W. (1998) Development of photosynthetic apparatus and respiration in pea seedlings during as influenced by toxic concentration of lead, *Acta Physiol. Plant.* **20**, 35–40.

Maksymiec, W. (1997) Effect of copper on cellular processes in higher plants, *Photosynthetica* **34**, 321–342.

Maksymiec, W. and Baszyński T. (1996a) Chlorophyll fluorescence in primary leaves of excess Cu-treated runner bean plants depends on their growth stages and the duration of Cu-action, *J. Plant Physiol.* **149**, 196–200.

Maksymiec, W. and Baszyński T. (1996b) Different susceptibility of runner bean plants to excess copper as a function of the growth stages of primary leaves, *J. Plant Physiol.* **149**, 217–221.

Maksymiec, W. and Baszyński, T. (1999) The role of Ca^{2+} ions in modulating changes induced in bean plants by an excess of Cu^{2+} ions. Chlorophyll fluorescence measurements, *Physiol. Plant.* **105**, 562–568.

Maksymiec, W., Bednara, J., and Baszyński, T. (1995) Responses of runner bean plants to excess copper as a function of plant growth stages: effects on morphology and structure of primary leaves and their chloroplast ultrastructure, *Photosynthetica* **31**, 427–435.

Maksymiec, W., Russa, R., Urbanik-Sypniewska, T., and Baszyński, T. (1992) Changes in acyl lipid and fatty acid composition in thylakoids of copper non-tolerant spinach exposed to excess copper, *J. Plant Physiol.* **140**, 52–55.

Maksymiec, W., Russa, R., Urbanik-Sypniewska, T., and Baszyński, T. (1994) Effect of excess Cu on the photosynthetic apparatus of runner bean leaves treated at two different growth stages, *Physiol. Plant.* **91**, 715–721.

Malik, D., Sheoran, I.S., and Singh, R. (1992) Carbon metabolism in leaves of cadmium treated wheat seedlings, *Plant Physiol. Biochem.* **30**, 223–229.

McBrien, D.C.H., and Hassal, K.A. (1967) The effect of toxic doses of copper upon respiration, photosynthesis and growth of *Chlorella vulgaris*, *Physiol. Plant.* **20**, 113–117.

Memon, A.R., Chino, M., and Yatazawa, M. (1981) Microdistribution of aluminum and manganese in the tea leaf tissue as revealed by X-ray microanalysis, *Commun. Soil. Sci. Plant Anal.* **12**, 441–452.

Miles, C.D., Brandle, J.R., Daniel, D.J., Chu-Der, O., Schnare, P.D., and Uhlik, D.J. (1972) Inhibition of photosystem II in isolated chloroplasts by lead, *Plant. Physiol*, **49**, 820–825.

Miller, R.J., Bittell, J.E., and Koeppe, D.E. (1973) The effect of cadmium on electron and energy transfer reactions in corn mitochondria, *Physiol. Plant.* **28**, 166–171.

Misra A. and Ramani S. (1991) Inhibition of iron absorption by zinc induced Fe-deficiency in Japanese mint, *Acta Physiol. Plant.* **13**, 37–42.

Mocquot, B., Vangronsveld, J., Clijsters, H., and Mench, M. (1996) Copper toxicity in young maize (*Zea mays* L.) plants: effects on growth, mineral and chlorophyll contents, and enzyme activities, *Plant Soil* **182**, 287–300.

Mohanty, N., Vass I., and Demeter, S. (1989) Copper toxicity affects photosystem II electron transport at the secondary quinone acceptor, Q_B, *Plant Physiol.* **90**, 175–179.

Moustakes, M., Lanaras, T., Symeonidis, L., and Karataglis, S. (1994) Growth and some photosynthetic characteristics of field grown *Avena sativa* under copper and lead stress, *Photosynthetica* **30**, 389–396.

Moya, J.L., Ros, R., and Picazo, I. (1993) Influence of cadmium and nickel on growth, net photosynthesis and carbohydrate distribution in rice plants, *Photosynth. Res.* **36**, 75–80.

Murthy, S.D.S. and Mohanty, P. (1995) Action of selected heavy metal ions on the photosystem 2 activity of the cyanobacterium *Spirulina platensis*, *Biol. Plant.* **37**, 79–84.

Mustafa, M.G. and Cross, C.E. (1971) Pulmonary alveoral macrophage. Metabolism of isolated cells and effect of cadmium ions on electron and energy transfer reactions, *Biochemistry* **10**, 4176–4185.

Nagel, K., Adelmeier, U., and Vpigt, J. (1996) Subcellular distribution of cadmium in the unicellular green alga *Chlamydomonas reinhardtii*, *J. Plant Physiol.* **149**, 86–90.

Nahar, S. and Tajmir-Riahi, H.A. (1995) Do metal ions alter the protein secondary structure of light-harvesting complex of thylakoid membranes, *J. Inorg. Biochem.* **58**, 223–234.

Navari-Izzo, F., Quartacci, M.F., Pinzino, C., Vecchia, F.D., and Sgherri, L.M. (1998) Thylakoid-bound and stromal antioxidative enzymes in wheat treated with excess copper, *Physiol. Plantarum* **104**, 630–638.

Nielsen, E.S., Kamp-Nielsen, L., and Wium-Andersen, S. (1969) The effect of deleterious concentrations of copper on the photosynthesis of *Chlorella pyrenoidosa*, *Physiol. Plant.* **22**, 1121–1133.

Ouzonidou, G. (1994) Copper-induced changes on growth, metal content and photosynthetic functions of *Alyssum montanum* L. plants, *Environ. Exp. Bot.* **34**, 165–172.

Ouzonidou, G. (1996) The use of photoacoustic spectroscopy in assessing leaf photosynthesis under copper stress: correlation of energy storage to photosystem II fluorescence parameters and redox change of P700, *Plant Sci.* **113**, 229–237.

Ouzounidou, G., Lannoye, R., and Karataglis, S. (1993) Photoacoustic measurements of photosynthetic activities in intact leaves under copper stress, *Plant Sci.* **89**, 221–226.

Ouzonidou, G., Moustakes, M., and Lannoye, R. (1995) Chlorophyll fluorescence and photoacoustic characteristics in relationship to changes in chlorophyll and a Ca^{2+} content in a Cu-tolerant *Silene compacta* ecotype under Cu treatment, *Physiol. Plant.* **93**, 551–557.

Parys, E., Romanowska, E., Siedlecka, M., and Poskuta, J.W. (1998) The effect of lead on photosynthesis and respiration in detached leaves and in mesophyll protoplasts of *Pisum sativum*, *Acta Physiol. Plant.* **20**, 313–322.

Pavan, M.A. and Bingham, F.T. (1982) Toxicity of aluminum to coffee seedlings grown in nutrient solution, *Soil. Sci. Soc. Amer. J.* **45**, 993–997.

Petterson, O. (1977) Differences in cadmium uptake between plant species and cultivars, *Swed. J. Agric. Res.* **7**, 21–24.

Pich, A., Scholz. G., and Stephan, U. W. (1995) Iron dependent changes of heavy metals, nicotinamine, and citrate in different plant organs and in the xylem exudate of the two tomato genotypes. Nicotinamine as possible copper translocator, in J. Abadia, (ed.), *Iron Nutrion in Soil and Plants*, Kluver Acad. Publ., Dordrecht, pp. 51–58.

Pierce, J. (1986) Determinants of substrate specificity and the role of metal in the reactions of Ribulosebisphosphate Carboxylase/Oxygenase, *Plant Physiol.* **81**, 943–945.

Poskuta, J.W. (1991) Toxicity of heavy metals to growth, photosynthesis and respiration of *Chlorella pyredoinosa* cells grown either in air or 1% CO_2, *Photosynthetica* **25**, 181–188.

Poskuta, J.W., Parys, E., and Romanowska, E. (1996) Toxicity of lead to photosynthesis, accumulation of chlorophyll, respiration and growth of *Chlorella pyrenoidosa*. Protective role of dark respiration, *Acta Physiol. Plant.* **18**, 165–171.

Poskuta, J., Parys, E., and Romanowska, E. (1987) The effects of lead on the gaseous exchange and photosynthetic carbon metabolism of pea seedlings, *Acta Soc. Bot. Polon.* **56**, 127–137.
Poskuta, J.W., Parys, E., Romanowska, E., Gajdzis-Gujdan, H., and Wróblewska, B. (1988) The effects of lead on photosynthesis, ^{14}C distribution among photoassimilates and transpiration of maize seedlings, *Acta Soc. Bot. Polon.* **57**, 149–155.
Poskuta, J.W. and Wacławczyk, E. (1995) *In vivo* responses of primary photochemistry of photosystem II and CO_2 exchange in light and in darkness of tall fescue genotypes to lead toxicity, *Acta Physiol. Plant.* **17**, 233–240.
Punz, W.F. and Sieghardt, H. (1993) The response of roots of herbaceous plant species to heavy metals, *Environ. Exp. Bot.* **33**, 85–98.
Quartacci, M.F., Cosi, E., and Navari-Izzo, F. (2001) Lipids and NADPH-dependent superoxide production in plasma membrane vesicles from roots of wheat grown under copper deficiency or excess, *J. Exp. Bot.* **52**, 77–84.
Rascio, N., Dalla Vecchia F., Ferretti, M., Merlo, L. and Ghisi, R. (1993) Some effects of cadmium on maize plants, *Arch. Environ. Contam. Toxicol.* **25**, 244–249.
Rashid, A., Camm, E.L., and Ekramoddoullah, K.M. (1994) Molecular mechanism of action of Pb^{2+} and Zn^{2+} on water oxidizing complex of photosystem II, *FEBS Lett.* **350**, 296–298.
Rashid, A. and Popovic, R. (1990) Protective role of $CaCl_2$ against Pb^{2+} inhibition in photosystem II, *FEBS Lett.* **271**, 181–184.
Rauser, W.E. (1978) Early effects of phototoxic burdens of cadmium, cobalt, nickel and zinc in white beans, *Can. J. Bot.* **56**, 1744–1749.
Rebechini, H.M. and Hanzely, L. (1974) Lead-induced ultrastructural changes in chloroplasts of the hydrophyte, *Ceratophyllum demersum*, *Z. Pflanzenphysiol.* **73**, 377–386.
Reese, R.N. and Roberts, L.W. (1985) Effects of cadmium on whole cell and mitochondrial respiration in tobacco cell suspension cultures (*Nicotiana tabacum* L. var. *xanthi*), *J. Plant. Physiol.* **120**, 123–130.
Reilly, A. and Reilly, C. (1973) Copper induced chlorosis in *Beciu homblei* (De Wild) Duvign and Plancke, *Plant Soil*, **38**, 671–674.
Renganathan, M. and Bose, S. (1989) Inhibition of primary photochemistry of photosystem II by copper in isolated pea chloroplasts, *Biochim. Biophys. Acta* **974**, 247–253.
Robertson, A.I. and Meakin, M.E.R. (1980) The effect of nickel on cell division and growth of *Brachystegia spiciformis* seedlings, *Kirkia*, **12**, 115–125.
Romanowska, E., Parys, E., Słowik, D., Siedlecka, M., Piotrowski, T., and Poskuta, J. (1998) The toxicity of lead to photosynthesis and respiration in detached pea and maize leaves, in G. Garab (ed.), *Photosynthesis: Mechanisms and Effects*, vol. IV, Kluwer Acad. Publ, pp. 2665–2668.
Ros, R., Morales, A., Segura, J., and Picazo, I. (1992) *In vivo* and *in vitro* effects of nickel and cadmium on the plasmalemma ATPase from rice (*Oryza sativa* L.) shoots and roots, *Plant Sci.* **83**, 1–6.
Roy, A.K., Sharma, A., and Talukder, G. (1988) Some aspects of aluminum toxicity in plants, *Bot. Rev.*, **54**, 145–178.
Samarakoon, A.B. and Rauser, W.E. (1979) Carbohydrate levels and photoassimilate export from leaves of *Phaseolus vulgaris* exposed to excess cobalt, nickel, and zinc, *Plant Physiol.* **63**, 1165–1169.
Samson, G. and Popovic, R. (1990) Inhibitory effects of mercury on photosystem II photochemistry in *Dunaliella tetriolecta* under *in vivo* conditions, *J. Photochem. Photobiol.* **35**, 303–310.
Sandmann, G. and Boger, P. (1980) Copper mediated lipid peroxidation processes in photosynthetic membranes, *Plant Physiol.* **66**, 797–800.
Sandmann, G. (1985) Photosynthetic and respiratory electron transport in Cu^{2+}-deficient *Dunaliella*, *Physiol. Plant* **65**, 481–486.
Schaedle, M., Thornton, F.C., Raynal, D.J., and Tepper, H.B. (1989) Response of tree seedlings to aluminum, *Tree Physiology* **5**, 337–356.
Schlegel, H., Douglas, L., and Hüttermann, A. (1987) Whole plant aspects of heavy metal induced changes in CO_2 uptake and water relations of spruce (*Picea abies*) seedlings, *Physiol. Plant.* **69**, 265–270.
Šeršeň, F., Kralova, K., and Bumbalova, A. (1998) Action of mercury on the photosynthetic apparatus of spinach chloroplasts, *Photosynthetica* **35**, 551–559.
Sharma, S.D. and Chopra, R.N. (1987) Effect of lead nitrate and lead acetate on growth of the moss *Semibarbula orientalis* (Web.) Wijk. et Marg. grown *in vitro*, *J. Plant Physiol.* **28**, 243–249.
Sheoran, I.S., Aggarwal, N., and Singh, R. (1990) Effects of cadmium and nickel on *in vivo* carbon dioxide exchange rate of pigeon pea (*Cajanus cajan* L.), *Plant Soil*, **129**, 243–249.

Siedlecka, A., Krupa, Z., Samuelsson, G., Öquist, G., and Gardeström, P. (1997) Primary carbon metabolism in *Phaseolus vulgaris* plants under Cd/Fe interaction, *Plant Physiol. Biochem.* **35**, 951–957.
Singh, B.R. and Myhr, K. (1998) Cadmium uptake by barley as affected by Cd sources and pH levels, *Geoderma* **84**, 185–194.
Singh, C.B. and Singh, S.P. (1987) Effect of mercury on photosynthesis in *Nostoc calcicola:* role of ATP and interacting heavy metal ions, *J. Plant. Physiol.* **129**, 49–58.
Stauber, J.L. and Florence, T.M. (1987) Mechanism of copper toxicity to algae, *Mar. Biol.* **94**, 511–519.
Stiborová, M. (1988) Cd^{2+} ions affect the quaternary structure of ribulose-1.5-bisphosphate carboxylase from barley leaves, *Biochem. Physiol. Pflanzen.* **183**, 371–378.
Stiborová, M., Ditrichová, M., and Březinová, A. (1987) Effect of heavy metal ions on growth and biochemical characteristics of photosynthesis of barley and maize seedlings, *Biol. Plant.* **29**, 453–467.
Stiborová, M., Doubravová, S., Březinová, A., and Friedrich, A. (1986) Effect of heavy metal ions on growth and biochemical characteristics of photosynthesis of barley, *Photosynthetica* **20**, 418–425.
Stiborová, M., Doubravová, S., and Leblová, A. (1986) A comparative study of the effect of heavy metal ions on ribulose-1,5-bisphosphate carboxylase and phosphoenolpyruvate carboxylase, *Biochem. Physiol. Pflanzen* **181**, 373–379.
Stiborová, M. and Leblová, S. (1985) Heavy metal inactivation of maize (*Zea mays* L.) phosphoenolpyruvate carboxylase isoenzymes, *Photosynthetica* **19**, 500–503.
Stobart, A.K., Griffiths, W.T., Ameen-Bukhari, I., and Sherwood, R.P. (1985) The effect of Cd^{2+} on the biosynthesis of chlorophyll in leaves of barley, *Physiol. Plant.* **63**, 293–298.
Suhayda, C.C. and Haug, A. (1985) Citrate chelation as a potential mechanism against aluminum toxicity in cells: the role of calmodulin, *Can. J. Biochem. Cell Biol.* **63**, 1167–1175.
Szalontai, B., Horvath, I.L., Debreczeny, M., Droppa, M., and Horvath, G. (1999) Molecular rearrangements of thylakoids after heavy metal poisoning, as seen by Fourier transform infrared (FTIR) and electron spin resonance (ESR) spectroscopy, *Photosynth. Res.* **61**, 241–252.
Ślaski, J.J., Zhang, G., Basu, U., Stephens, J.L., and Taylor, G.J. (1996) aluminum resistance in wheat (*Triticum aestivum*) is assiciated with rapid, Al-induced changes in activities of glucose-6-phosphate dehydrogenase and 6-phosphogluconate dehydrogenase in root apices, *Physiol. Plant.* **98**, 477–484.
Stanley, R.A. (1974) Toxicity of heavy metals and salts to Eurasian watermilfoil (*Myriophyllum spicatum* L.), *Arch. Environ. Contam. Toxicol.* **2**, 331–341.
Taylor, G.J. (1988) The physiology of aluminium phytotoxicity, in H. Siegl (ed.), *Metal Ions in Biological Systems: aluminium and Its Role in Biology*, Marcel Dekker, New York, pp. 123–163.
Tardieu, F. and Davies, W.J. (1992) Stomatal response to abscisic acid is a function of current plant water status, *Plant Physiol.* **98**, 540–545.
Terry, N. (1981) Physiology of trace element toxicity and its relation to iron stress, *J. Plant Nutr.* **3**, 561–578.
Tripathy, B.C. and Mohanty, P. (1980) Zinc-inhibited electron transport of photosynthesis in isolated barley chloroplasts, *Plant Physiol.* **66**, 1174–1178.
Turner, J.F. and Copeland, L. (1981) Hexokinase II of pea seed, *Plant Physiol.* **68**, 1123–1127.
Uribe, G.E. and Stark, B. (1982) Inhibition of photosynthetic energy conversion by cupric ion, *Plant Physiol.* **69**, 1040–1045.
Vallee, B.L. and Ulmer, D.D. (1972) Biochemical effects of mercury, cadmium and lead, *Annu. Rev. Biochem.* **41**, 91–128.
Van Assche, F. and Clijsters, H. (1986a) Inhibition of photosynthesis in *Phaseolus vulgaris* by treatment with toxic concentration of zinc: effect on ribulose-1,5-bisphosphate carboxylase/oxygenase, *J. Plant Physiol.* **125**, 355–360.
Van Assche, F. and Clijsters, H. (1986b) Inhibition of photosynthesis in *Phaseolus vulgaris* by treatment with toxic concentration of zinc: effects on electron transport and photophosphorylation, *Physiol. Plant.* **66**, 717–721.
Van Assche, F. and Clijsters, H. (1990) Effects of metals on enzyme activity in plants, *Plant Cell Environ.* **13**, 195–206.
Van Duijvendijk-Matteoli, M.A. and Desmet, G.M. (1975) On the inhibitory action of cadmium on the donor side of photosystem II in isolated chloroplasts, *Biochim. Biophys. Acta* **408**, 164–169.
Vierstra, R. and Haug, A. (1978) The effect of Al^{3+} on the physical properties of membrane lipids in *Thermoplasma acidophilium*, *Biochem. Biophys. Res. Commun.* **84**, 138–143.
Vögeli-Lange, R. and Wagner, G.J. (1990) Subcellular localization of cadmium and cadmium–binding peptides in tobacco leaves, *Plant Physiol.* **92**, 1086–1093.

Vangronsveld, J. and Clijsters, H. (1994) Toxic effects of metals, in M.E. Farago (ed.), *Plants and Chemical Elements. Biochemistry, Uptake, Tolerance and Toxicity*, VCH Publishers, New York, pp.150–177.

Weckx, J.E.J. and Clijsters, H.M.M. (1996) Oxidative damage and defense mechanisms in primary leaves of *Phaseolus vulgaris* as a result of root assimilation of toxic amounts of cooper, *Physiol. Plant.* **96**, 506–512.

Weigel, H.J. (1985a) Inhibition of photosynthetic reactions of isolated intact chloroplasts by cadmium, *J. Plant Physiol.* **119**, 179–189.

Weigel, H.J. (1985b) The effect of Cd^{2+} on photosynthetic reactions of mesophyll protoplasts, *Physiol. Plant.* **63**, 192–200.

Weigel, H.J. and Jäger, H.J. (1980) Subcellular distribution and chemical form of cadmium in bean plants, *Plant Physiol.* **65**, 480–482.

Woźny, A., Schneider, J., and Gwóźdź, E.A. (1995) The effects of lead and kinetin on greening barley leaves, *Biol. Plant.* **37**, 541–552.

Yruela, I., Montoya G., and Picorel, R. (1992) The inhibitory mechanism of Cu (II) on the Photosystem II electron transport from higher plants, *Photosynth. Res.* **33**, 227–233.

Zhao, X-J., Sucoff, E., and Stadelmann, E.J. (1987) Al^{3+} and Ca^{2+} alteration of membrane permability of *Quercus rubra* cortex cells, *Plant Physiol.* **83**, 159–162.

CHAPTER 11

HEAVY METAL INTERACTIONS WITH PLANT NUTRIENTS

Z. KRUPA, A. SIEDLECKA, E. SKÓRZYNSKA-POLIT AND W.MAKSYMIEC

Department of Plant Physiology, Maria Curie-Skłodowska University
Akademicka 19, PL 20-033 Lublin, Poland

1. INTRODUCTION

According to the classical definition "*Interactions between nutrients occur when the supply of one nutrient affects the absorption, distribution or function of another nutrient. Thus, depending on nutrient supply, interactions between nutrients can either induce deficiencies or toxicities and can modify growth response*" (Robson and Pitman, 1983). There are many nonspecific as well as specific interactions between mineral nutrients of plants (Robson and Pitman, 1983; Marschner 1988). When contents of any mineral nutrients are near the deficiency range the importance of interactions between two mineral nutrients increases. Specific interactions, *e.g.* competition between nutrients at the cellular level or replacement of one nutrient by another, are also important in evaluating critical toxicity contents (Foy et al., 1978; Marschner, 1988).

Competitive interactions can also occur between essential nutrients and other elements in the environment, particularly with heavy metals. Heavy metals, aggressive and ufortunately still very common environmental pollutants, are easily taken up by plants and easily compete with many nutrient elements. These phenomena suggest that more attention should be paid to the relationships between heavy metals and other essential plant nutrients.

2. HEAVY METALS UPTAKE AND DISTRIBUTION

From the physiological point of view we divide heavy metals into two groups (Siedlecka, 1995):
- metals necessary for plant metabolism as enzyme activators or regulators, *e.g.* Fe, Cu, Mn, Mo, which may become toxic if supplied in excess;

Accumulation of heavy metals is greatly related to the form in which they are present in the nutrient medium. De Kock and Mitchell (1957) have shown differences in divalent and trivalent cations uptake by mustard and tomato plants, depending on their availability in inorganic or chelated forms:
- divalent cations (for instance Co, Ni, Zn, Cu) were much easier taken up by plants when present in ionic form than chelates, particularly with EDTA. Their uptake was diminished even when EDTA was present in the nutrient medium;
- trivalent cations, like Cr, Ga or In behaved differently, *i.e.* were easily taken up and translocated to shoots if present in chelated form. It was postulated that probably the electric charge of the chelates was the major factor controlling absorptions of these cations by the roots. Trivalent cations present in the ionic form were taken up only in trace amounts.

Heavy metals also differ substantially according to their accumulation in plant roots and shoots (Hara et al., 1976 a,b; Jastrow and Koeppe 1980).

As seen in Figure 1, the first group consists of elements to more extent accumulated in roots than in shoots (Cd, Fe, Cu, Co, Mo). The second group accumulates in roots with very low amounts transported to shoots - Pb, Sn, Ti, Ag, Cr, Zr, V and Ga. Elements of the third group are uniformly distributed between roots and shoots (Zn, Mn, Ni).

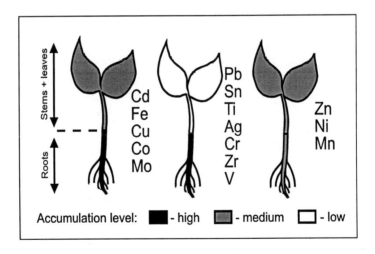

Figure 1. Heavy metals distribution between plant organs (Siedlecka, 1995)

Mechanisms of heavy metals uptake and distribution are very complex and not yet fully elucidated. Cadmium, very strongly accumulated and one of the most toxic heavy metals, appears to be the most frequently studied. When in low supply Cd is taken up metabolically and non-metabolically when supply is high (Greger, 1989). According to Cutler and Rains (1974) three steps of non-metabolic uptake can be distinguished:

1. exchange absorption, *i.e.* reversible Cd binding to the root surface. Heavy metals like Zn, Cu and Hg are non-selectively bound and could be easily removed by desorption solutions (diluted acids or diluted chelators);

2. non-metabolic binding as the primary accumulation mechanism. Heavy metal binds to the sites located on cell wall constituents or macromolecules within the cell thus maintaining a concentration gradient sufficient to accumulate Cd by diffusion;

3. diffusion responsible for Cd entering into the symplast.

It is, however, difficult to agree with strictly diffusional way of Cd transport into the plant. More likely, Cd forms two fractions as suggested by Smeyers-Verbeke et al. (1978). One fraction would be non-metabolically bound, free diffusible and exchangeable, whilst the other is taken up by metabolic processes and its absorption is inhibited, probably by competition, by Zn and Cu.

The pattern of Cu uptake seems to follow the one described above for Cd. Mechanisms of active and passive Cu uptake probably operate simultaneously: the importance of the passive uptake increases with increased level of the metal in the growth medium. At low Cu concentrations in the environment the active absorption of this element predominates (Fernandes and Henriques, 1991). Moreover, the specific ionophore, probably a polypeptide carrier, was also postulated by these authors. The mechanisms of other divalent cations uptake (Co, Mn, Ni, Pb) follow, in general, the same pattern as postulated for Cd and Cu (Mishra and Kar, 1974; Mukopadhyay and Sharma, 1991; Woźny and Krzesłowska, 1993; Palit et al., 1994).

In the last few years more and more attention has been paid to membrane transporters involved in the transport of heavy metals in different classes of organisms, including plants. There are, in general, three classes of such transporters - the heavy metal ATPases (Cpx-type), the natural resistance associated macrophage proteins (Nramp) and members of the cation diffusion facilitators family (CDF). Since this subject was recently very comprehensively reviewed by Foulkes (2000) and Williams et al. (2000), we only point to several interesting aspects of these transporters.

P-type heavy metal ATPases, belonging to large family of P-type ATPases using ATP to pump charged molecules across membranes, can transport potentially toxic elements like Cd, Cu, Zn or Pb. They have been defined as Cpx-ATPases because they share the common feature of conserved intramembranous motif of Cys-Pro-Cys, Cys-Pro-Ser or Cys-Pro-His (Cpx motif). *Arabidopsis thaliana* is the first higher plant in which this type of ATPase was reported. However, as yet there is no data supporting its role in metal transport so the possible physiological role of P-type ATPase in higher plants can only be speculated (Williams et al., 2000).

Highly conserved Nramp family of proteins, involved in the transport of divalent metal ions, have been found in a wide range of organisms, including higher plants. Investigations of this family of transporter proteins have been largely restricted to rice. However, genes showing similarity to Nramps have been recently identified in *Arabidopsis thaliana* (Williams et al., 2000). These proteins could probably transport distinct but related metal ions required for particular physiological processes in different plant tissues or organs.

The family of CDF proteins is involved in the transport of Zn, Co and Cd and its members have been identified both in procaryotic and eucaryotic organisms, including higher plants. Some of them may function in heavy metals uptake, whilst others catalyse efflux. Zinc transporter belonging to CDF proteins has been recently found in *Arabidopsis thaliana* (Williams et al., 2000).

As it was mentioned before, the mechanisms that contribute to heavy metal uptake and distribution still remain to be elucidated. Presumably, at the cellular level specific transporters could play a crucial role in the uptake and secretion of heavy metals, as well as in their homeostasis. The *Arabidopsis* genome sequencing studies may allow us to identify proteins and mechanisms responsible for heavy metals uptake, transport and distribution in higher plants.

3. HEAVY METALS *VERSUS* PLANT NUTRIENTS

It is known that the unfavorable effects of heavy metals on plants are manifested, among others, by inhibiting the normal uptake and utilization of mineral nutrients (Foy, 1983; Biddappa et al., 1987; Burzyński, 1987; Goodbold and Kettner, 1991; Trivedi and Erdei, 1992). Although many heavy metals such as Fe, Cu, Zn, Mn, Ni or Co are essential for plants as micronutrients, both these elements present in excess as well as non-essential heavy metals like Cd, Hg, Pb are extremely toxic for the organism (Foy et al., 1978; Prasad, 1997). One of the crucial factors of heavy metals influence on plant metabolism and physiological processes are their relationships with other mineral nutrients (Wallace and Romney, 1977). Symeonidis and Karataglis (1992) divided plant responses to combinations of metals in the growth medium into three groups:
1. additive, when plant growth under conditions of multiple metal stress was equal to that observed in the presence of metals supplied separately;
2. antagonistic, when plant growth parameters under multiple metal stress exceeded those observed when metals were supplied separately;
3. synergistic, when plant growth under multiple metal stress was diminished comparing with separate supply.

This was also observed earlier by Fernandes and Henriques (1991) who found Cu to be synergistic with Zn and Ni, antagonistic with Cd and additive with Co.

The examples presented in Table 1 show that excess of heavy metals in the growth environment influences, mostly negatively, levels of indispensable plant mineral components. The order of susceptibility of essential elements to heavy metals appears to be: Fe>>Zn=Cu (Table 1, Siedlecka, 1995). Disturbances in K and Ca levels were also

Table 1. Some examples of the influence of heavy metals toxic levels on plant mineral composition (based on Siedlecka, 1995)

Heavy metal	Plant species	Plant organ	Level of mineral nutrients decreased	Level of mineral nutrients increased
Cu	Halimione portulacoides	roots	Zn	Fe
	Avena sativa	stems	N,P,K	Ca,Mg
	Koeleria splendens	roots	Fe,Mg,Ca,K	
	Koeleria splendens	shoots	Fe,Mg,Ca,K	
	Halimione portulacoides	leaves	Fe,Zn	
Cd	Raphanus sativa	roots	Fe,Zn	
	Cucumis sativus	roots	Fe,Mg,K	
	Cucumis sativus	cotyledons	Fe,K,Ca	
	Raphanus sativa	shoots	Fe,Zn,Cu	
	Zea mays	leaves	Fe	
	Cucumis sativus	leaves	Fe,Zn,Cu	Mn
Ni	Avena sativa	stems	N	P,K,Ca,Mg
Co	Avena sativa	stems	K	N,P,Ca,Mg
Mn	Avena sativa	stems	Ca,Mg	N,P
Zn	Halimione portulacoides	roots	Cu	Fe
	Avena sativa	stems	N	P,Ca,Mg
	Halimione portulacoides	leaves	Fe,Cu	
Pb	Raphanus sativa	roots	Fe,Zn,Cu	
	Cucumis sativus	roots	Fe,K,Ca	Mg
	Cucumis sativus	cotyledons	Mg,K,Ca	Fe
	Raphanus sativa	shoots	Fe,Zn,Cu	

Nutrient decrease - maximum 95% of control, nutrient increase - minimum 105% of control. Data taken from Hunter and Vergnano, 1953; Burzyński and Buczek, 1989; Khan and Frankland, 1983; Burzyński 1987; Siedlecka and Baszyński, 1993; Reboredo 1994; Ouzounidou, 1995.

observed. It has to be noted that K/Ca ratio is crucial for water balance in plants. However, there are also observations that the contents of essential elements (P, K, Mg, Fe, Zn, Mn, Cu, Ca) were not significantly modified by Cd treatment, even at relatively high concentrations of this heavy metal in the growth environment (Agriffoul et al., 1998).

As it was mentioned above, heavy metals interfere mostly with Fe, leading to strong deficiency of this crucial element. Fe deficiency may seriously affect plant metabolism because of very complex role of this metal (for review see Abadia, 1986; Siedlecka and Krupa, 1999). Numerous data confirm Fe deficiency in plants accumulating increased amounts of Cd, Cu, Co, Cr, P, Ni, Mn and Zn (Hunter and Vergnano, 1953; Tanaka and Navasero, 1966; Terry, 1981; Khan and Khan, 1983; Agarwal et al., 1987; Greger and Lindberg, 1987; Vazquez et al., 1987; Walker et al., 1987; Misra and Ramani, 1991, Siedlecka and Baszyński, 1993; Ouzounidou, 1995; Schmidt et al., 1997; Lagriffoul, 1998).

Dependent on the concentration, period of application and Fe status of the plants, heavy metals can decrease the availability of Fe in the medium, inhibit induction of Fe-stress response or block either external or internal Fe-binding sites. Although there are many data showing that heavy metals induce Fe deficiency, leading to the most visible symptoms, *i.e.* iron chlorosis, it has to be noted that excess heavy metals might or might not lead to decreased Fe levels in the leaves (Agarwala et al., 1977; Foy et al., 1978; Lolkema and Vooijs, 1986; Reboredo and Henriques, 1991; Pich and Scholz, 1996, Hernández et al., 1998).

It is well known that there are two evolutionary developed strategies of Fe uptake by higher plants. Strategy I, evolutionary older and most abundant (all dicots and monocots except grasses), is a three-phase process, complicated and very susceptible to unfavourable environmental conditions (for review see Guerinot and Yi, 1994; Moog and Brüggemann, 1994; Siedlecka and Krupa, 1999). The main disadvantage of this strategy is that all steps of Fe^{3+} uptake and reduction take place in the rhizosphere. This makes the whole process very sensitive to, for instance, heavy metals toxicity. Interference of Cd, Cu, Ni, Cr with strategy I Fe^{3+} uptake and reduction was reported frequently (Misra and Ramani, 1991; Ros et al., 1992; Alcántara et al., 1994; Obata et al., 1996; Schmidt, 1996; Schmidt et al., 1997; Siedlecka and Krupa, 1999).
Strategy II plants (Gramineae), evolutionary much younger, are also better adjusted to deal with polluted environment (Römheld, 1987; Maas et al., 1988; Brown and Jolley, 1989; Scholz and Stephan, 1990; Ma and Nomoto, 1996). Strategy II is less sensitive to unfavourable environmental conditions, probably mainly due to the fact that most of crucial steps occurs in root cells. However, this strategy is also disturbed by heavy metals, *e.g.* Cu, Co, Zn and Pb (Ma and Nomoto, 1996; Varga et al., 1997).
Cu/Fe interaction is one of the most interesting and complicated. These two essential metals depend on each other for proper cellular metabolism (Harris, 1994). A subtle cooperation between them is certainly more synergistic than antagonistic, although it is very hard to determine precisely where in the organism the two metals interact, or if there is a third component somehow holding control over the two.

The antagonism between Zn and Fe might explain why both low and high Fe levels depressed plant growth (Adriano et al., 1971; House, 1999). Retarded plant growth might as well be the effect of low Fe level in the nutrient medium or inhibited Zn uptake at high Fe supply. Zn, however, affected Fe translocation more than its absorption. Misra and Ramani (1991) confirmed that Zn toxicity induces Fe deficiency due to Fe/Zn interaction and competition that might be explained by similar ionic radii of hydrated cations and cell regulatory mechanisms. However, Rosen et al. (1977) reported that in Zn-poisoned plants the symptoms of chlorosis were not correlated with Fe deficiency. Additional Fe supply diminished these symptoms, probably by limiting Zn transport to leaves. Similar phenomenon was observed for Fe/Mn relations where Fe level regulates Mn uptake and utilization in plants (Tanaka et al., 1966; Gupta and Chipman, 1976).

The main function of so-called iron plaque (coatings of iron oxides or hydroxides) formed on the roots of waterlogging plants is to provide a proper Fe supply and to protect plants against some heavy metals toxicity. It has been reported that the formation of plaque can improve growth and ameliorate Cu and Ni toxicity under mildly toxic conditions of Cu and Ni (Greipsson and Crowder, 1992; Greipsson, 1994). However, the opposite results have been reported by Ye et al. (1998) for *Typha latifolia* seedlings, with and without plaque, exposed to Cu and Ni in the nutrient medium. These differences may also relate to plant species and the metal investigated (Alloway, 1995; Ye et al., 1997a,b).

Cd, Cu, Pb and Zn show a wide range of interactions. Cadmium interacts with Cu and Zn at the root surface, whilst Pb interacts with Zn and Cd within the plant. Some of these interactions are antagonisms (Zn with Cu and Cd), others are synergistic (Cd with Zn at the root surface, Pb with Cd within the plant) (Kabata-Pendias and Pendias, 2001). Synergism and additivity of Cd-Zn interaction might be related to inadequate compartmentation of Cd and Zn, and is generally unfavorable to plant growth (Chaoui et al., 1997).

Increased supply of some essential metallic elements may appear dangerous for heavy metal-treated plants. Nickel toxicity was increased in the presence of Co, Zn, Mn and Mo, leading to intensified Ni-type chlorosis (Hunter and Vergnano, 1953). For plants cultivated at phytotoxic Mn concentrations tissue ratio of Mo/Mn appeared to be the most important factor in maintaining good physiological conditions (Le Bot et al., 1990; Goss and Carvalho, 1992). The effect of added Cu was to increase the toxicity of Zn (Luoy and Rimmer, 1995). Recently, Nakazawa et al. (2000) showed that Cd uptake was promoted by the coexistence of Ni in the medium, and Ni uptake was inhibited by the coexistence of Cd in the nutrient solution.

Heavy metal uptake might be also inhibited or limited by some essential nutrients. Usually Ca, P, Zn, Cu, Mn and Fe are supposed to act antagonistically against Cd. However, it seems to depend on plant species and variety. This antagonism could be due to competition for available absorption sites at the root surface or to the formation of non-absorbable complexes in the soil solution. Jarvis et al. (1976) reported that Cd uptake by ryegrass is depressed by Ca, Mn and Zn. Zn is as well able to lower the Cd content of plants as to increase it (Girling and Peterson, 1981; Sadana and Bijah, 1987). Manganese is widely recognised as an antidotum to elevated uptake and distribution of some heavy metals. Baszyński et al. (1980) reported a protective role of elevated Mn contents in plants against Cd toxicity towards the photosynthetic apparatus. Harrisson et al. (1983) showed that Cd uptake was reduced in the presence of Mn, assuming a competition between Mn and Cu for a common uptake site.

Wallace (1982) noticed a protective effect of Cu on beans when high Cd doses were applied. Cataldo et al. (1983) found that Cu, Fe and Zn competitively inhibited Cd absorption by soybean seedlings, whilst Thys et al. (1991) showed that additional Fe supply significantly diminished Cd content in lettuce leaves with parallel increase in K and Zn content. The limiting effect of excess Fe supply on Cd uptake and distribution was also confirmed for bean plants by Siedlecka and Krupa (1996a). Also Zn was reported to decrease Pb uptake (Symeonidis and Karataglis, 1992).

Recently, it is more and more widely recognized that examining the effects of heavy metals in various combinations is more representative than single metal studies, because of the real environmental problems faced by plants. The environmental loading of heavy metals very often involves several elements. Thus, organisms face a multiple metal stress. Because of high degree of complexity higher plants are not yet very often used in such experiments. The algal models are widely used to study the interactions of multiple heavy metals in terms of their action, being either additive, antagonistic or synergistic (Rosko and Rachlin, 1975, 1977; Rachlin et al., 1982, 1983; Taylor, 1989; Taylor and Stadt, 1990; Visviki and Rachlin, 1991). Rachlin and Grosso (1993), using the growth response of *Chlorella vulgaris* as a model system, showed that combination of Cd, Co and Cu yields a toxicity series Cd>Cu>Co. The combinations of Cd+Cu and Cu+Co acted synergistically, while Cd+Co and the trimetallic combination Cd+Cu+Co resulted in antagonistic interactions.

4. SOME PHYSIOLOGICAL AND METABOLIC CONSEQUENCES OF HEAVY METAL INTERACTIONS WITH PLANT NUTRIENTS

Heavy metal interactions with plant nutrients affect on the large scale all physiological and metabolic processes occurring in plants. Here we present some examples of such phenomena.

The metabolism of nitrogen and sulfate is strongly dependent on heavy metal presence in the environment. It has been recently shown that excess Cd affects sulfate assimilation enzymes and nitrogen metabolism in higher plants (Lee and Leustek, 1999; Chien and Kao, 2000).

One of the most pronounced effects has the interaction between heavy metals and Fe. Iron, being one of the crucial elements in plants, is also very susceptible to the competition by other metallic elements, including toxic metals. It is well known that the photosynthetic apparatus, in its all aspects, is especially exposed to damages related to heavy metal-induced Fe deficiency. Cadmium, one of the most toxic heavy metals, is actually the most aquainted as the opponent of Fe. Its effect results, first of all, in Fe deficiency which was documented, for instance, by Wallace et al. (1992), Siedlecka (1995) and Siedlecka and Krupa (1996a). Besides growth and chlorophyll synthesis, Cd-induced Fe deficiency evidently affects the photosynthetic electron transport, mostly because of its many Fe-containing constituents (Siedlecka and Baszyński, 1993).

Therefore, it has the negative effect on all further steps of photosynthesis, including the Calvin cycle (for details see Siedlecka and Krupa, 1996b, 1999; Siedlecka et al., 1997).

Calcium occupies a pivotal position in plant cell signal transduction. The plant signals thought to be transferred by Ca include touch, wind, temperature shock, fungal elicitors, wounding, oxidative stress. red light, blue light, anaerobiosis, ABA, applied electrical fields, osmotic stress and last but not least, mineral stress. One cannot forget the crucial role of Ca ions in the regulation of the oxygen evolving complex in photosynthesis. Increasing content of Cu or Cd in the environment leads to decreased Ca contents in plant organs, especially in leaves (Jensen and Adalsteinsson, 1989; Maksymiec and Baszyński, 1998). Since Ca is indispensable for normal functioning of the photosynthetic apparatus, not only for the Photosystem II but also for the Calvin cycle (Kreimer et al., 1988, Ghanotakis and Yocum, 1990), the whole photosynthesis becomes less efficient under such conditions (Maksymiec and Baszyński, 1998, 1999a,b; Skórzyńska-Polit et al., 1998). Moreover, the level of damages related to heavy metal/Ca interactions strongly depends on plant growth stage, with younger plants being less susceptible (Skórzyńska-Polit et al., 1998; Maksymiec and Baszyński, 1999a).

It is well known that photosynthetic rates depend on external inorganic phosphate (P_i) supply. Under insufficient P_i provision photosynthesis is inhibited, and the rate of this process can be increased after feeding P_i to leaf tissue (Sivak and Walker, 1986; Kleczkowski 1994). Heavy metals, widely known inhibitors of plant metabolism, may also decrease intracellular levels of many essential nutrient elements, including phosphorus (Leeper, 1978; Wallace, 1989; Verkleij and Prast, 1989; Verma et al., 1993). Recently, Krupa et al. (1999) reported that Cd affects levels of P_i in rye leaves leading not only to disturbed photosynthetic electron transport and diminished yield of the Calvin cycle, but also to very deep effects considering the level of both large and small subunits of the crucial enzyme - ribulose-bisphosphate-carboxylase/oxygenase (Rubisco). However, the moderate extra P_i supply to leaves could, to some extent, reverse Cd toxicity.

One of the most spectacular effects of heavy metal interactions with essential nutrient elements is the substitution of the Mg ion in the chlorophyll molecule by certain toxic heavy metals. It was first reported for green algae and lichens (Gross et al., 1970; Puckett, 1976; De Filippis, 1979). Complex formation rate (= the tendency of an ion to be bound in the centre of the chlorophyll molecule) was found to be as follows: Hg>Cu>Cd>Zn>Ni>Pb (Küpper et al., 1996, 1998). Most such chlorophylls are unsuitable for photosynthesis because of their much lower fluorescence quantum yields compared to Mg-chlorophylls. Hence, the formation of these compounds leads to a complete breakdown of photosynthesis.

5. CONCLUSION

The phenomena of heavy metal interactions with plant nutrients are very complex and yet not fully elucidated. First of all, we must realize the complexity of the higher plant organism and all the aspects of its physiology and metabolism. None of these processes are isolated, they all are interrelated and control one another. Therefore, any disturbance, any invasion of a toxic element into an organism leads to a series of very complex and complicated responses on both cellular and tissue and organ level.. Here, we presented some of them, having in mind that there is no simple answer to the basic questions - why and how?

ACKNOWLEDGEMENTS

This work is supported by the Polish Committe for Scientific Reseasrch, grant 6P04C.064.15.

REFERENCES

Adriano, D.C., Paulsen, G.M. and Murphy, L.S. (1971) Phosphorus-iron and phosphorus-zinc relationships in corn (*Zea mays*). *Agron. J.* **63**, 36-39.

Agarwal K, Sharma, A. and Talukder, G. (1987) Copper toxicity in plant cellular systems. *Nucleus* **30**, 131-158.

Agarwala, S.C., Bisht, S.S. and Sharma, C.P. (1977) Relative effectiveness of certain heavy metals in producing toxicity and symptoms of iron deficiency in barley. *Can. J. Bot.* **55**, 1299-1307.

Agriffoul, A., Mocquot, B., Mench, M., and Vangronsveld, J. (1998) Cadmium toxicity on growth, mineral and chlorophyll contents, and activities of stress related enzymes in young maize plants (*Zea mays* L.). *Plant Soil* **200**, 241-250.

Alcántara, E., Romera, F.J., Canete, M., and De la Guardia, M.D. (1994) Effects of heavy metals on both induction and function of root Fe(III) reductase in Fe-deficient cucumber (*Cucumis sativus* L.) plants. *J. Exp. Bot.* **45**, 1893-1898.

Alloway, B.J. (1995) *Heavy Metals in Soils*, Blackie, Glasgow.

Baszyński, T., Wajda, L., Król, M., Wolińska, D., Krupa, Z., and Tukendorf, A. (1980) Photosynthetic activities of cadmium-treated tomato plants. *Physiol. Plant.* **48**, 365-370.

Biddappa, C.C., Khan, H.H., Joshi, O.P., and Manikandan, P. (1987) Effect of root feeding of heavy metals on the leaf concentration of P, K, Ca and Mg in coconut (*Cocos nucifera* L.). *Plant Soil* **101**, 295-297.

Brown, J.C., and Jolley, V.D. (1989) Plant metabolic responses to iron-deficiency stress. A variety of mechanisms grouped into two major strategies, make iron available from the soil. *BioScience* **39**, 546-551.

Burzyński, M. (1987) The influence of lead and cadmium on the absorption and distribution of potassium, calcium, magnesium and iron in cucumber seedlings. *Acta Physiol. Plant.* **9**, 229-238.

Burzyński, M., and Buczek, J. (1989) Interaction between cadmium and molybdenum affecting the chlorophyll content and accumulation of some heavy metals in the second leaf of *Cucumis sativus* L. *Acta Physiol. Plant.* **11**, 137-145.

Chaoui, A., Ghorbal, M.H., and El Ferjani, E. (1997) Effects of cadmium-zinc interactions on hydroponically grown bean (*Phaseolus vulgaris* L.). *Plant Sci.* **126**, 21-28.

Chien, H.F., and Kao, C.H. (2000) Accumulation of ammonium in rice leaves in response to excess cadmium. *Plant Sci.* **156**, 111-115.
Cutler, J.M., and Rains, D.W., (1974) Characterization of cadmium uptake by plant tissue. *Plant Physiol.* **54**, 67-71.
De Filippis, L.F., (1979) The effect of heavy metals on the absorption spectra of *Chlorella* cells and chlorophyll solutions. *Z. Pflanzenphysiol.* **93**, 129-137.
De Kock, P.C., and Mitchell, R.L. (1957) Uptake of chelated metals by plants. *Soil Sci.* **84**, 55-62.
Fernandes, J.C., and Henriques, F.S. (1991) Biochemical, physiological, and structural effects of excess copper in plants. *Bot. Rev.* **57**, 246-273.
Foulkes, E.C. (2000) Transport of toxic metals across cell membranes. *Proc. Soc. Exp. Biol. Med.* **223**, 234-240.
Foy, C.D. (1983) The physiology of plant adaptation to mineral stress. *Iowa State J. Res.* **57**, 355-391.
Foy, C.D., Chaney, R.L., and White, M.C. (1978) The physiology of metal toxicity in plants. *Annu. Rev. Plant Physiol.* **29**, 511-566.
Ghanotakis, D.F,, and Yocum, C.F. (1990) Photosystem II and the oxygen-evolving complex. *Annu. Rev. Plant Physiol. Plant Mol. Biol.* **41**, 255-278.
Girling, C.A., and Peterson, P.J, (1981) The significance of the cadmium species in the uptake and metabolism of cadmium in crop plants. *J. Plant Nutr.* **3**, 707-720.
Goodbold, D.L., and Kettner, C. (1991) Leaf influences root growth and mineral nutrition of *Picea abies* seedlings. *J. Plant Physiol.* **139**, 95-99.
Goss, M.J., and Carvalho, J.M.G.P.R. (1992) Manganese toxicity: the significance of magnesium for the toxicity of wheat plants. *Plant Soil* **139**, 90-98.
Greger, M. (1989) *Cadmium effects on carbohydrate metabolism in sugar beet (Beta vulgaris)*. Doctoral dissertation. Academitryck AB, Edsbruck, 9-11.
Greger, M., and Lindberg, S. (1987) Effects of Cd^{2+} and EDTA on young sugar beets (*Beta vulgaris*). II. Net uptake and distribution of Mg^{2+}, Ca^{2+} and Fe^{2+}/Fe^{3+}. *Physiol. Plant.* **68**, 81-86.
Greipsson, S. (1994) Effects of iron plaque on roots of rice on growth and metal concentration of seeds and plant tissues when cultivated in excess copper. Commun. *Soil Sci. Plant Anal.* **25**, 2761-2769.
Greipsson, S., and Crowder, A.A. (1992) Amelioration of copper and nickel toxicity by iron plaque on roots of rice (*Oryza sativa*). *Can. J. Bot.* **70**, 824-830.
Gross, R.E,, Pugno, P. and Dugger, W.M. (1970) Observations on the mechanism of copper damage in *Chlorella*. *Plant Physiol.* **46**, 183-185.
Guerinot, M.L., and Yi, Y. (1994) Iron: nutritious, noxious, and not readily available. *Plant Physiol.* **104**, 815-820.
Gupta, U.C., and Chipman, E.W. (1976) Influence of iron and pH on the yield of iron, manganese, zinc, and sulfur concentrations of carrots grown on sphagnum peat soil. *Plant Soil* **44**, 559-566.
Hara, T., Sonoda, Y., and Iwai, I. (1976a) Growth response of cabbage plants to transition elements under water culture conditions. I. Titanium, vanadium, chromium, manganese and iron. *Soil Sci. Plant Nutr.* **22**, 307-315.
Hara, T., Sonoda, Y., and Iwai, I .(1976b) Growth response of cabbage plants to transition elements under water culture conditions. II. Cobalt, nickel, copper, zinc and molybdenum. *Soil Sci. Plant Nutr.* **22**, 317-325.
Harris, E.D. (1994) Iron-copper interactions: some new revelations. *Nutr. Rev.* **52**, 311-319.
Harrison, S.J., Lepp, N.W., and Phipps, D.A. (1983) Copper uptake by excised roots. III. Effect of manganese on copper uptake. *Z. Pflanzephysiol.* **109**, 285-289.
Hernández, L.E., Lozano-Rodriguez, E.., Gárate, A., and Carpena-Ruiz, R. (1998) Influence of cadmium on the uptake, tissue accumulation and subcellular distribution of manganese in pea seedlings. *Plant Sci.* **132**, 139-151.
House, W.A. (1999) Trace elements bioavailability as axemplified by iron and zinc. *Fields Crop Res.* **60**, 115-141.
Hunter, J.G., and Vergnano, O. (1953) Trace elements toxicities in oat plants. *Ann. Appl. Biol.* **40**, 761-777.
Jarvis, S.C., Jones, L.H.P., and Hopper, M.J. 1976 Cadmium uptake from solution by plants and its transport from roots to shoots. *Plant Soil* **44**, 179-191.

Jastrow, J.D., and Koeppe, D.E. (1980) Uptake and effects of cadmium in higher plants *In Cadmium in the Environment. Part I*: Ecological Cycling, Ed. JO Nriagu, John Wiley & Sons, 608-638.
Jensen, P., and Adalsteinsson, S. (1989) Effects of copper on active and passive Rb^+ influx in roots of winter wheat. *Physiol. Plant.* **75**, 195-200.
Kabata-Pendias, A., and Pendias, H. (2001) *Trace Elements in Soils and Plants*, CRC Press, Boca Raton.
Khan, D.H., and Frankland, B. (1983) Effects of cadmium and lead on radish plants with particular reference to movement of metals through soil profile and plants. *Plant Soil* **70**, 335-345.
Khan, S., and Khan, N. (1983) Influence of lead and cadmium on the growth and nutrient concentration of tomato (*Lycopersicum esculentum*) and egg plant (*Solanum melongea*). *Plant Soil* **74**, 387-394.
Kleczkowski, L.A. (1994) Inhibitors of photosynthetic enzymes/carriers and metabolism. *Annu. Rev. Plant Physiol. Plant Mol. Biol.* **45**, 339-367.
Kreimer, G., Melkonian, M., Holtum, J.A.M., and Latzko, E. (1988) Stromal free calcium concentration and light-regulated activation of chloroplast fructose-1,6-bisphosphatase. *Plant Physiol.* **86**, 423-428.
Krupa, Z., Siedlecka, A., and Kleczkowski, L.A. (1999) Cadmium-affected level of inorganic phosphate in rye leaves influences Rubisco subunits. *Acta Physiol. Plant.* **21**, 257-261.
Küpper, H., Küpper, F., and Spiller, M. (1996) Environmental relevance of heavy metal-substituted chlorophylls using the example of water plants. *J. Exp. Bot.* **47**, 259-266.
Küpper, H., Küpper, F., and Spiller, M. (1998) *In situ* detection of heavy metal substituted chlorophylls in water plants. *Photosynth. Res.* **58**, 123-133.
Lagriffoul, A., Mocquot, B., Mench, M., and Vangronsveld, J. (1998) Cadmium toxicity effects on growth, mineral and chlorophyll contents, and activities of stress related enzymes in young maize plants (*Zea mays* L.). *Plant Soil* **200**, 241-250.
Le Bot, J., Goss, M.J., Carvalho, M.J.G.P.R., Van Beusihem, M.L., and Kirkby, E.A. (1990) The significance of the magnesium to manganese ration in plant tissues for growth and alleviation of manganese toxicity in tomato (*Lycopersicon esculentum*) and wheat (*Triticum sativum*) plants. *Plant Soil* **124**, 205-210.
Lee, S., and Leustek, T. (1999) The effect of cadmium on sulfate assimilation enzymes in *Brassica juncea*. *Plant Sci.* **141**, 201-207.
Leeper, G.W. (1978) Forms of heavy metals *In Managing the Heavy Metals on the Land, Pollution Engineering and Technology 6*, Ed. GW Leeper, Marcel Dekker, New York, 5-39.
Lolkema, P.C., and Vooijs, R. (1986) Copper tolerance in *Silene cucubalus*. Subcellular distribution of copper and its effects on chloroplasts and plastocyanin synthesis. *Planta* **167**, 30-36.
Luo, Y., and Rimmer, D.L. (1995) Zinc-copper interaction affecting plant growth on a metal-contaminated soil. *Environ. Pollut.* **88**, 79-83.
Ma, J.F., and Nomoto, K. (1996) Effective regulation of iron acquisition in graminaceous plants. The role of mugineic acids as phytosiderophores. *Physiol. Plant.* **97**, 609-617.
Maksymiec, W., and Baszyński, T. (1998) The role of Ca ions in changes induced by excess Cu^{2+} in bean plants. Growth parameters. *Acta Physiol. Plant.* **20**, 411-417.
Maksymiec, W., and Baszyński, T.(1999a) Are calcium ions and calcium channels involved in the mechanisms of Cu^{2+} toxicity in bean plants? The influence of leaf age. *Photosynthetica* **36**, 267-278.
Maksymiec, W., and Baszyński, T. (1999b) The role of Ca^{2+} ions in modulating changes induced in bean leaves by an excess of Cu^{2+} ions. Chlorophyll fluorescence measurements. *Physiol. Plant.* **105**, 562-568.
Maas, F.M., Van de Weterin, D.A.M., Van Beusichem, M.L., and Bienfait, H.F. (1988). Characterization of phloem iron and its possible role in regulation of Fe-efficiency reactions. *Plant Physiol.* **87**, 167-171.
Marschner, H. (1988) *Mineral Nutrition of Higher Plants*, Academic Press, London-San Diego-New York.
Mishra, D., and Kar, M. (1974) Nickel in plant growth and metabolism. *Bot. Rev.* **40**, 395-452.
Misra, A., and Ramani, S. (1991) Inhibition of iron absorption by zinc induced Fe-deficiency in Japanese mint. *Acta Physiol. Plant.* **13**, 37-42.
Moog, P.R., and Brüggemann, W. (1994) Iron reductase systems on the plant plasma-membrane - a review. *Plant Soil* **165**, 241-260.
Mukhopadhyay, M.J., and Sharma, A. (1991) Manganese in cell metabolism of higher plants. *Bot. Rev.* **57**, 117-149.
Nakazawa, R., Ozawa, T., Naito, T., Yano, Y., and Takenaga, H. (2000) Interactions between cadmium and nickel in the growth and heavy metal uptake in cultivated tobacco cells. *Man Environ.* **26**, 124-129.
Obata, H., and Inoue, N. (1996) Effects of cadmium on plasma membrane ATPase from plant roots differing in tolerance to cadmium. *Soil Sci. Plant Nutr.* **42**, 361-366.

Ouzounidou, G. (1995) Cu-ions mediated changes in growth, chlorophyll and other ion contents in a Cu-tolerant *Koeleria splendens. Biol. Plant.* **37**, 71-78.
Palit, S., Sharma, A., and Talukder, G. (1994) Effect of cobalt on plants. *Bot. Rev.* **60**, 149-181.
Pich, A., and Scholz, G. (1996) Translocation of copper and other micronutrients in tomato plants (*Lycopersicon esculentum* Mill.): nicotianamine-stimulated copper transport in the xylem. *J. Exp. Bot.* **47**, 41-47.
Prasad, M.N.V. (1997) Trace metals. *In Plant Ecophysiology.* Ed. MNV Prasad, John Wiley and Sons Inc., New York, 207-249.
Puckett, K.J. (1976) The effects of heavy metals on some aspects of lichen physiology. *Can. J. Bot.* **54**, 2695-2703.
Rachlin, J.W., Jensen, T.E., Baxter. M., and Jani, V. (1982) Utilization of morphometric analysis in evaluating response of *Plectonema boryanum* (Cyanophyceae) to exposure to eight heavy metals. *Arch. Environ. Contam. Toxicol.* **11**, 323-333.
Rachlin, J.W., Jensen, T.E., and Warkentine, B. (1983) The growth response of the diatom *Navicula inverta* to selected concentrations of the netals: cadmium, copper, lead and zinc. *Bull. Torr. Bot. Club* **110**, 217-223.
Rachlin, J.W., and Grosso, A. (1993) The growth response of the green alga *Chlorella vulgaris* to combined divalent cation exposure. *Arch. Environ. Contam. Toxicol.* **24**, 16-20.
Reboredo, F. (1994) Interaction between copper and zinc and their uptake by *Halimione portulacoides* (L.) *Aellen. Bull. Environ. Contam. Toxicol.* **52**, 598-605.
Reboredo, F., and Henriques, F. (1991) Some observations on the leaf ultrastructure of *Halimione portulacoides* (L.) Aellen grown in a medium containing copper. *J. Plant Physiol.* **137**, 717-722.
Robson, A.D., and Pitman, M.G. (1983) Interactions between nutrients in *higher plants In Encyclopedia of Plant Physiology*, Eds. A Läuchli and RL Bieleski, Springer Verlag, Berlin-Heidelberg-New York-Tokyo, Vol. 15A, 147-180.
Ros, R., Morales, A., Segura, J., and Picazo, I. (1992) *In vivo* and *in vitro* effects of nickel and cadmium on the plasmalemma ATPase from rice (*Oryza sativa* L.) shoots and roots. *Plant Sci.* **83**, 199-202.
Rosen, J.A., Pike, C.S., and Golden, M.L. (1977) Zinc, iron, and chlorophyll metabolism in zinc-toxic corn. *Plant Physiol.* **59**, 1085-1087.
Rosko, J.J., and Rachlin, J.W. (1975) The effect of copper, zinc, cobalt and manganese on the growth of the marine diatom *Nitzschia closterium. Bull. Torr. Bot. Club* **102**, 100-106.
Rosko, J.J., and Rachlin, J.W. (1977) The effect of cadmium, copper, mercury, zinc and lead on cell division, growth, and chlorophyll a content of the chlorophyte *Chlorella vulgaris. Bull. Torr. Bot. Club* **104**, 226-233.
Römheld, V. (1987) Different strategies of iron acquisition in higher plants. *Physiol. Plant.* **70**, 231-234.
Sadana, U.S., and Bijah, S. (1987)Effect of zinc application on yield and cadmium content of spinach (*Spinacia oleracea* L.) grown in cadmium polluted soil. *Ann. Bot.* **3**, 59-60.
Siedlecka, A. (1995) Some aspects of interactions between heavy metals and plant mineral nutrients. *Acta Soc. Bot. Pol.* **64**, 265-272.
Siedlecka, A., and Baszyński, T. (1993) Inhibition of electron flow around photosystem I in chloroplasts of Cd-treated maize plants is due to Cd-induced iron deficiency. *Physiol. Plant.* **83**, 199-202.
Siedlecka, A., and Krupa, Z. (1996a) Interaction between cadmium and iron. Accumulation and distribution of metals and changes in growth parameters of *Phaseolus vulgaris* L. seedlings. *Acta Soc. Bot. Pol.* **66**, 1-6.
Siedlecka, A., and Krupa, Z. (1996b) Interaction between cadmium and iron and its effects on photosynthetic capacity of primary leaves of *Phaseolus vulgaris. Plant Physiol. Biochem.* **34**, 833-842.
Siedlecka, A., Krupa, Z., Samuelsson, G., Öquist, G., and Gardeström, P. (1997) Primary carbon metabolism in *Phaseolus vlgaris* plants under Cd/Fe interaction. *Plant Physiol. Biochem.* **35**, 951-957.
Siedlecka, A., and Krupa, Z. (1999) Cd/Fe interaction in higher plants - its consequences for the photosynthetic apparatus. *Photosynthetica* **36**, 321-331.
Sivak, M.N., and Walker, D.A. (1986) Photosynthesis *in vivo* can be limited by phosphate supply. *New Phytol.* **102**, 499-512.
Schmidt, W. (1996) Influence of chromium (III) on root-associated Fe (III) reductase in *Plantago lanceolata* L. *J. Exp. Bot.* **47**, 805-810.

Schmidt, W., Bartels, M., Tittel, J., and Fühner, C. (1997) Physiological effects of copper on iron aquisition processes in plantago. *New Phytol.* **135**, 659-666.

Scholz, G., and Stephan, U.W. (1990) Nicotianamine concentration in iron sufficient and iron deficient sunflower and barley roots. *J Plant Physiol.* **136**, 631-634.

Skórzyńska-Polit, E., Tukendorf, A., Selstam, E., and Baszyński, T. (1998) Calcium modifies Cd effect on runner bean plants. *Environ. Exp. Bot.* **40**, 275-286.

Smeyers-Verbeke, J., De Graev, M., Franois, M., De Jaegere, R., and Massart, D.L. (1978) Cd uptake by intact wheat plants. *Plant Cell Environ.* **1**, 291-296.

Symeonidis, L., and Karataglis, S. (1992) Interactive effects of cadmium, lead and zinc on root growth of two metal tolerant genotypes of *Holcus lanatus* L. *Biometals* **5**, 173-178.

Tanaka, A., and Navasero, S.A. (1966) Interaction between iron and manganese in rice plants. *Soil Sci. Plant Nutr.* **12**, 29-33.

Tanaka, A., Loe, R., and Navasero, S.A. (1966) Some mechanisms involved in the development of iron toxicity symptoms in the rice plants. *Soil Sci. Plant Nutr.* **12**, 32-38.

Taylor, G.J. (1989) Multiple metal stress in *Triticum aestivum*. Differentiation between additive, multiplicative, antagonistic, and synergistic effects. *Can. J. Bot.* **67**, 2272-2276.

Taylor, G.J., and Stadt, K.J. (1990) Interactive effects of cadmium, copper, manganese, nickel and zinc on root growth of wheat (*Triticum aestivum*) in solution *In Plant Nutrition - Physiology and Applications*, Ed. ML Van Beusichem, Kluwer Academic Publishers, Dordrecht, The Netherlands, 317-322.

Terry, N. (1981) Physiology of trace elements toxicity and its relation to iron stress. *J. Plant Nutr.* **3**, 561-578.

Terry, N., and Abadia, J. (1986) Function of iron in chloroplasts. *J. Plant Nutr.* **9**, 609-646.

Thys, C., Vanthome, P., Schrevens, E., and De Proft, M. (1991) Interaction of Cd with Zn, Cu, Mn and Fe for lettuce (*Lactuca sativa* L.) in hydroponic culture. *Plant Cell Environ.* **14**, 713-717.

Trivedi, S., and Erdei, L. (1992) Effects of cadmium and lead on the accumulation of Ca^{2+} and K^+ and on the influx and translocation of K^+ in wheat of low and high K^+ status. *Physiol. Plant.* **84**, 94-100.

Varga, A., Zaray, G., Fodor, F., and Cšeh, E. (1997) Study on interaction of iron and lead during their uptake process in wheat roots by total-reflective X-ray fluorescence spectrometry. *Spectrochim. Acta Bot.* **52**, 1027-1032.

Vazquez, M.D., Poschenrieder, C.H., and Barceló, J. (1987) Chromium VI induced structural and ultrastructural changes in bush bean plants (*Phaseolus vulgaris* L.). *Ann. Bot.* **59**, 427-438.

Verkleij, J.A.C., and Prast, J.E. (1989) Cadmium tolerance and co-tolerance in *Silene vulgaris* (Moench.) Garcke [=*Silene cucubalus* (L.) Wib.]. *New Phytol.* **111**, 637-645.

Verma, S.K., Singh, R.K., and Singh, S.P. (1993) Copper toxicity and phosphate utilization in the cyanobacterium *Nostoc calcicola*. *Bull. Environ. Contam. Toxicol.* **50**, 192-198.

Visviki, I., and Rachlin, J.W. (1991) The toxic action and interactions of copper and cadmium to the marine alga *Dunaliella minuta*, in both acute and chronic exposure. *Arch. Environ. Contam. Toxicol.* **24**, 16-20.

Wallace, A. (1982) Additive, protective and synergistic effects on plants with excess trace elements. *Soil Sci* **133**, 319-323.

Wallace, A. (1989) Effects of zinc when manganese was also varied for bush beans grown in solution culture. *Soil Sci.* **147**, 444-448.

Wallace, A., and Romney, E.M. (1977) Synergistic trace metal effects in plants. *Commun. Soil Sci. Plant Anal.* **8**, 699-707.

Wallace, A., Wallace, G.A., and Cha, J.W. (1992) Some modifications in trace metal toxicities and deficiencies in plants resulting from interactions with other elements and chelating agents. *J. Plant Nutr.* **15**, 1589-1598.

Walker, W.M., Miller, J.E., and Hassett, J.J. (1987) Effect of lead and cadmium upon the calcium, magnesium, potassium and phosphorus concentration in young corn plants. *Soil Sci.* **124**, 145-151.

Williams, L.E., Pittman, J.K., and Hall, J.L. (2000) Emerging mechanisms for heavy metal transport in plants. Biochim. Biophys. Acta **1465**, 104-126.
Woźny A and Krzesłowska M (1993) Plant cell responses to lead. Acta Soc. Bot. Pol. **62**, 101-105.
Ye, Z.H., Baker, A.J.M., Wong, M.H., and Willis, A.J. (1997a) Zinc, lead and cadmium tolerance, uptake and accumulation by *Typha latifolia*. New Phytol. **136**, 469-480.
Ye, Z.H., Baker, A.J.M., Wong, M.H., and Willis, A.J. (1997b) Copper and nickel uptake, accumulation and tolerance in *Typha latifolia* with and without iron plaque on the root surface. New Phytol. **136**, 481-488.
Ye, Z.H., Baker, A.J.M., Wong, M.H. and Willis, A.J. (1998) Zinc, led and cadmium accumulation and tolerance in *Typha latifolia* as affected by iron plaque on the root surface. *Aquat. Bot.* **61**, 55-67.

CHAPTER 12

FUNCTIONS OF ENZYMES IN HEAVY METAL TREATED PLANTS

A. SIEDLECKA AND Z. KRUPA
*Department of Plant Physiology, Maria Curie-Skłodowska University
Akademicka 19, 20-033 Lublin, Poland*

1. INTRODUCTION

Heavy metals influence interfere with a variety of processes in higher plant cells. Usually in research work, as well as in reviews, this stress is considered in the context of the main processes affected such as growth, photosynthesis, respiration, adaptation to stress conditions and so on. We almost never focus our attention on the fact that in most cases the proteins that are both the targets for heavy metal damage and the mechanism by which plants combat heavy metal stress are proteins with different names but a common -*ase* endings – *i.e.* they are enzymes. Their functioning is regulated by many factors during homeostatic metabolism. Under conditions of heavy metal stress this homeostasis has been disrupted. Numerous changes in plant metabolism, reflecting both stress disturbances as well as plant efforts leading to a new homeostasis have their reflection in altered enzyme function. The general approach in understanding heavy metal toxicity is: toxic = inhibitory. The truth is different. There are numerous genes and enzymes stimulated by heavy metal stress. Usually they are involved in widely understood tolerance and adaptation mechanisms. This chapter will review enzyme function in heavy metal-affected plant metabolism as a whole. Details of the impact of heavy metals on particular processes, such as photosynthesis, or on particular groups of proteins, such as phytochelatins, are described in other chapters of this book.

2. ENZYMES AS PROTEINS PERSISTING IN HEAVY METAL AFFECTED CELLS

Enzymes are proteins and as such can be affected by external factors at all stages – from coding, through synthesis, to regulation of activity and, finally, degradation. Two main effects can be described in this aspect of the impact of heavy metals on plant metabolism. The first is down-regulation, which is the most widely known. However, there is also up-regulation, which is even more interesting although not so well recognised. Both effects of heavy metals on enzyme synthesis and stability in plant cells are described below.

2.1. Heavy metals-dependent down-regulation of enzymes presence in plant cells

Protein synthesis is the first step in the production of enzymes and it can be a primary site of heavy metal activity (Fig.1). In both animals and plants, elements such as Cu, Cd, Cr, Ni, Pb, and Zn, have been reported to induce disturbances in cytokinesis, chromatin structure, and mutagenesis as well as increased activity of DNase and lowered DNA synthesis and stability (Fig.1, Gabara et al. 1992, Nicolas et al. 1997, Wierzbicka 1999, Zoroddu et al. 2000, reviewed in: Johnson 1998, Müller et al. 2000). Heavy metals can induce mutagenesis by chemical modification of the nucleobases, by oxidative damage of the nucleobases by the metal itself or by the generation of reactive oxygen species, through mistakes during the DNA repair process and through DNA cross-linking and deformation (reviewed in: Müller et al. 2000). All these processes lead to base mispairing in DNA and thus to modifications in gene functioning. In plants, reduced efficiency of DNA synthesis, weaker DNA protection from damaged chromatin proteins (histones) and increased deoxyribonuclease (DNase) activity have been reported for Cd, Cu, Cr, Ni, Pb, Hg, Pt, and Zn (Wojtyła-Kuchta and Gabara 1991, Gabara et al. 1992, Stecka et al. 1995, Espen et al. 1997, Zoroddu et al. 2000, reviewed in: Maksymiec 1997, Müller et al. 2000). Usually several different metal effects work in parallel during the damage process. DNA has been shown to be damaged by Cu due to Cu^{2+} binding to N-terminal histydyl groups of chromatin proteins, and due to Cu^+ formation with parallel formation of superoxide radicals in the presence of additional redox factors, such as nucleus-located Mn (Prütz et al. 1990, Prütz 1990, 1994, Ueda et al. 1994, reviewed in: Maksymiec 1997).

RNA synthesis and stability, the next step leading to protein formation, is also affected by heavy metals. Heavy metals can directly or indirectly repress different genes, resulting in a decrease in the transcription of important enzymes, as has been reported for *rbcS* (small Rubisco subunit) and *Lhcb* (light harvesting complex II chlorophyll binding protein) in presence of Cd, and root membrane located Fe^{3+} reductase following exposure to Cd, Cu, Cr and Ni stress (Fig.1, Alcántara et al. 1994, Schmidt 1996, Schmidt et al. 1997, Stefanov et al. 1997, Tziveleka et al. 1999). Elements such as Cu, Ni, Cd, and Pb have been reported to decrease RNA synthesis and to activate ribonuclease (RNase) activity, leading to further decreases in RNA content (Fig.1., Lidon and Henriques 1993a, Costa and Morel 1994, Espen et al. 1997, reviewed in: Siedlecka and Krupa 1999). Decreases in the amount and activity of nucleoli has been reported for Cd-, Cr-, and Pb-treated *Pisum sativum* roots (Gabara et al. 1995).

The heavy metal-dependent down-regulation in RNA synthesis and stability appears to be more complicated and it is not simply direct and indirect inhibition. In pea roots, RNA content decreased in the nucleolus but increased in the nucleus, and was much higher in the cytoplasm during exposure to Cd, Cr, and Pb toxicity (Łbik-Nowak and Gabara 1997). However, increased RNA content did not lead to increased protein synthesis but to decreased protein content, as well as imbalances in protein distribution between the cytoplasm and the nucleus (Wojtyła-Kuchta and Gabara 1991, Stecka et al. 1995). This suggests that the assembly of the translation machinery from RNA and ribosomes, as well as its functioning – the next step in protein synthesis – may also be a target of heavy metals toxicity, although this possibility is yet to be properly investigated (Fig.1.)

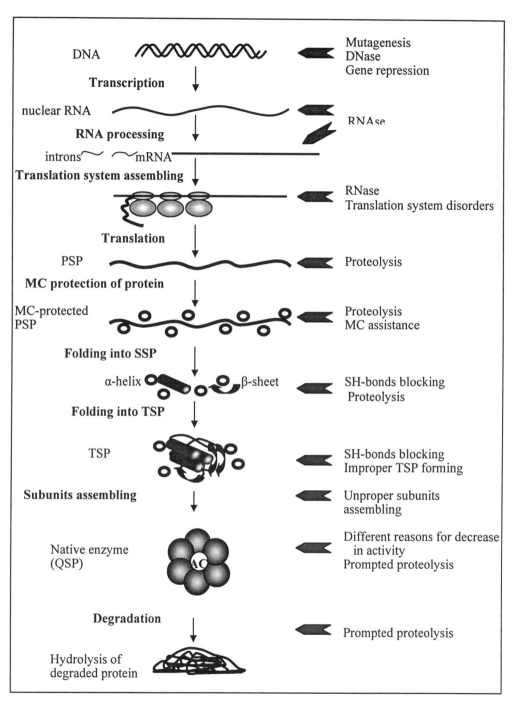

Figure 1. Heavy metal – dependent down-regulation of enzyme synthesis and stability in plant cells.

Abbreviations: AC, active center; DNase, deoxyribonuclease; MC, molecular chaperone; PSP, primary structure of protein (amino acids sequence); QSP, quaternary structure of protein (subunits composition); RNase, ribonuclease; SSP, secondary structure of protein; TSP, tertiary structure of protein; α-helix and β-sheet, two basic types of SSP. Black rings represent presence and assistance of MCs during different stages of protein maturation. Thin arrows show stages of protein persistence within cells (data taken from Lewin 1997, Lacroix et al. 1999). Thick arrows show heavy metal interference with the process (see text for details and reference data).

In plant cells, proteins that have been classified as ribosome inactivation proteins or ribotoxins can be activated, *i.e.* by viral or fungal infections. These proteins are toxic N-glycosidases that deactivate ribosomes by depurination of the sarcin loop of large rRNA resulting in the transcript system failing to assemble (reviewed in: Nielsen and Boston 2001). One of these proteins, gleonin, was reported to be stimulated by Zn and to have additional deoxyribonuclease activity (Nicolas et al. 1997). Inhibition of *de novo* protein synthesis has frequently been reported for numerous metals, including Cd, Ni, and Zn (Fig.1, Angelov et al. 1993, Chakravarty and Srivastava 1997, Espen et al. 1997, Boussama et al. 1999, Nagoor and Vyas 1999, Chen et al. 2001, reviewed in: Siedlecka and Krupa 1999).

Protein folding into secondary (*i.e.* α-helix, β-sheet), tertiary (globular structure) and quaternary (forming native enzymes from subunits) structures can also be affected by heavy metals (Fig.1). Unfortunately this aspect of heavy metal stress has so far been poorly investigated and research has mainly been performed *in vitro* – on isolated proteins, organelles or even on artificial systems. Davies et al. (1999) reported substantial changes in artificial metalloenzyme activity caused by even small disturbances in the precise attachment of the metal ligand within the protein cavity. The influence of heavy metals on native protein complexes appears to be a result of both direct and indirect affects. Cu, Pb, and Zn toxicity cause changes in the surface properties of thylakoid membrane proteins, leading to increased lipid affinity and an increased tendency to form protein-lipid complexes instead of protein-protein complexes. However, Cd and Ni did not have this effect (Szalontai et al. 1999). Nevertheless, Cd has been shown to disturb the structure of the thylakoid protein complexes (reviewed in: Sanitá Di Toppi and Gabbrielli 1999, Krupa 1999). Another widely reported important aspect of heavy metal toxicity to protein structure is their affinity for disulfide bonds and thiol groups, which may result in damage to the secondary structure of proteins (Fig.1, reviewed in: Van Assche and Clijsters 1990, Krupa and Baszyński 1995, Marschner 1995, Krupa 1999). However, disulfide bonds cannot form spontaneously during protein folding. This requires the activity of redox enzymes and the presence of an oxidant capable of accepting the electrons released during the process. System coupling oxidative disulfide bond formation with respiratory electron transport chain has recently been described in detail for bacteria (Bader et al. 1999, reviewed in: Glockhuber 1999) and its function in heavy metal-treated plants still remains unclear. However, the general susceptibility of redox systems to heavy metals makes it likely that this is the next important target of the stress.

Protein degradation is also an important consequence of heavy metal stress and it is connected both with the metals toxicity and plant stress adaptation (Fig.1). This aspect of the protein life cycle is discussed in detail in a latter part of this chapter (4.2.1. Ubiquitin – benefits of degradation).

2.2. Heavy metals-dependent enhancement of enzymes longevity in plant cells

Heavy metals can cause gene activation (Fig.2) in two main ways. One is the uptake and distribution within the plant of metallic elements which, like Cu and Fe, are necessary plant

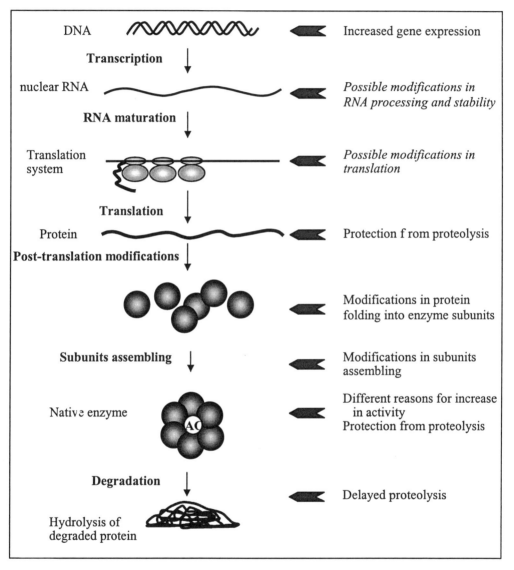

Figure 2. Heavy metal - induced up-regulation of enzyme synthesis and stability in plant cells. Abbreviations: AC, active center. Thin arrows show stages of protein persistence within cells (data taken from Lewin 1997, Lacroix et al. 1999). Thick arrows show heavy metal interference on the process. Text in italics refers to hypothetical mechanisms (see chapter text for details and reference data).

nutrients but that have to be distributed within plant in a way that protects proteins from the oxidizing properties of the metals. This requires special proteins that are needed to take up and transport the metals within plant cells and between organelles (Alcántara et al. 1994, Herbik et al.1996, Himelblau and Amasino 2000, reviewed in: Siedlecka 1995, Siedlecka and Krupa 1999, Siedlecka et al. 2001. See also Chapter 11). A family of transporter genes has been found for Zn (Grotz et al. 1998, reviewed in: Guerinot 2000). These genes are believed to be activated by metal deficiency but they may be also up-regulated when this deficiency results from other heavy metal toxicity, such as was reported for root Fe (III) reductase upon moderate Cu and Cr stress (Schmidt 1996, Schmidt et al. 1997). The second reason for genes to be up-regulated during heavy metal stress is the widely understood process of stress adaptation and stress tolerance. This includes such well-known phenomena as the activation of genes for γ-glutamylcysteine synthase (γECS), γ-Glu-Cys dipeptydyltranspeptidase (known as phytochelatine synthase), peroxidases, and catalases, as well as newly described genes such as the Zn transporter in Zn-hyperaccumulator, *Thlaspi caerulescens* (Vansuyt et al. 1997, Lasat et al. 2000, reviewed in: Rabinovitch and Fridovich 1983, Van Assche and Clijsters 1990, Vangronsveld and Clijsters 1994, Zenk 1996, Siedlecka et al. 2001). Heavy metals are known to activate some genes involved in other stress response, such as heat-shock or pathogenesis-related proteins (reviewed in: Marschner 1995, Maksymiec 1997, Sanitá Di Toppi and Gabbrielli 1999). The genes encoding structural proteins, such as proline-rich cell wall proteins, can also be activated by heavy metals. In *Phaseolus vulgaris*, green-tissue specific activation of expression of the gene family for a proline-rich cell wall proteins was observed during exposure to As, Cd, Hg and Zn but not Cu (Chai et al. 1998). The same protein gene family remained unaffected in roots under heavy metal stress (Chai et al. 1998). Nonetheless, due to translational regulation, gene activation is not necessarily connected to increases in protein content (Fig.2). Mechanisms of transcriptional and post-transcriptional regulation of gene expression may even be metal-specific, *e.g.* in tobacco Ag, Cd, Co, Cu, and Zn have stimulated transcription of the CBP20 gene (~20 kDa pathogen and wound-inducible antifungal protein) but the mature protein appeared only in Cd- and Zn-treated plants (Hensel et al. 1999). Strong activation of the ATP sulfurylase gene (AS) was observed in *Brassica juncea* treated with Cd, Cu, Hg, Pb and Zn. AS catalyses the first reaction responsible for bringing sulfur into plant metabolism. It activates sulfate by forming 5'-adenylylsulfate, which is very important for heavy metal adaptation because it has a direct impact on the availability of cysteine – a substrate for phytochelatine synthesis. Under heavy metal treatment a 3-fold increase in AS mRNA was observed and a synchronous increase in enzyme activity was also observed. The only exception was Cd treatment, which resulted in a severe lag in enzyme expression despite increased mRNA content (Lee and Leustek 1999).

Post-transcriptional activation of enzymes is a new discovery in plant heavy metals research (Fig.2). Rapid Cd activation of *Arabidopsis* γECS, an enzyme catalysing the synthesis of glutathione (GSH – a key component of plant oxidant and heavy metals defences), was reported by May et al. (1998). Before this report, γECS was thought to be activated by Cd and Zn at the transcriptional stage.

3. DIFFERENT MECHANISMS OF CHANGES IN ENZYMES ACTIVITY - HEAVY METALS IMPACT ON NATIVE PROTEINS

Protein persistence in heavy metal-treated plants is a very important aspect of enzyme function in stressed cells. Nonetheless, the period after synthesis and before degradation, *i.e.* the period it functions, is the most important period of a proteins existence. There are many complicated mechanisms involved in changes to enzymatic activity, including both direct and indirect effects of the stressing metal as well as different (and sometimes even more complex) ways of plant adaptation and recovery (Fig.3).

3.1 Mechanisms of inhibition

It is well known that enzymes carrying out primary metabolic reactions are sensitive to heavy metals. This is particularly true of photosynthetic proteins but it is also the case for proteins responsible for mitochondrial respiration (reviewed in: Van Assche and Clijsters 1990, Krupa and Baszyński 1995, Marschner 1995, Siedlecka 1995, Siedlecka and Krupa 1995, Prasad 1996, Das et al. 1997, Sanitá Di Toppi and Gabbrielli 1999, Krupa 1999, Siedlecka et al. 2001). In general, the mechanisms of this inhibition can be divided into three main groups (Fig.3):
- *i)* damage to the active centre
- *ii)* damage to the protein (or proteins) assembled into active complexes
- *iii)* dissociation of active complexes (or enzyme).

Plant age seems to be a factor in switching between the last two mechanisms of enzyme damage. In young plants, the predominant stress response is the susceptibility of individual proteins within active complexes, while in mature plants it is the induction of senescence and the dissociation of complex subunits that is the major cause of the deterioration of active complexes.

3.1.1 Damage of active centre

One of the most common ways that heavy metals affect enzyme activity is through damage to the active centre (Fig.3). In enzymes utilizing divalent cations the natural catalytic component is frequently replaced by the toxic heavy metal. The scale of this type of damage can be demonstrated by keeping in mind that Zn alone is an essential catalytic component of over 300 plant enzymes and plays an important structural role for many others, including the transcriptional machinery ("zinc-fingers") (Grotz et al. 1998). Rubisco is particularly susceptible to this type of inhibition. Rubisco is activated by divalent cations, preferably Mg^{2+}, but Ca^{2+} may also be an activator. This low specificity, together with the high reactivity of other heavy metal ions, results in a strong inhibition of Rubisco activity by numerous heavy metals, particularly Cd, Cu, Fe, Mn, Pb, and Zn, (Van Assche and Clijsters 1986, Stiborova et al. 1986, Houtz et al. 1988, Stiborova 1988, Malik et al. 1992, Siedlecka et al. 1997, 1999, reviewed in: Van Assche and Clijsters 1990, Siedlecka et al. 1998, 2001, Krupa 1999, Siedlecka and Krupa 1999). Replacement of Mn in the oxygen evolving complex by Cd has been reported to be one of reason for photosystem II inhibition (Baszyński et al. 1980).

3.1.2. Damage of protein(s) grouped in native enzyme or active complex

Individual protein components of active complexes, such as subunits of the ATPase, photosystem proteins and so on, show differential sensitivity to heavy metal stress. Each such

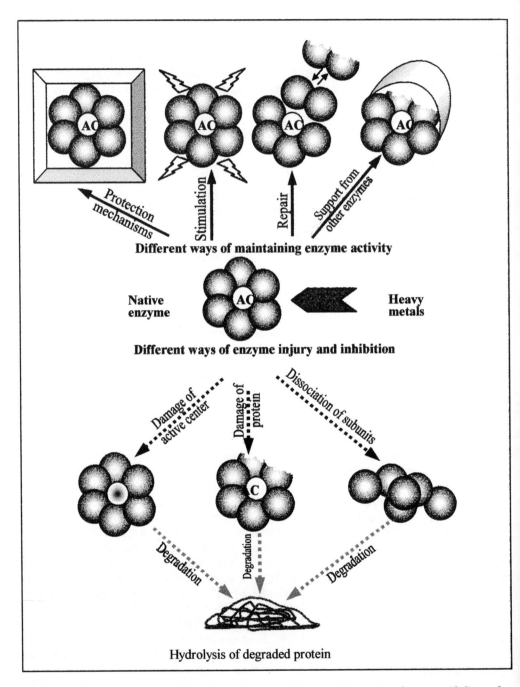

Figure 3. Heavy metal influence on enzyme activity. Solid arrows show possibilities for increase or maintaining enzyme activity. Dashed arrows show possible ways of enzyme injury leading to inhibition of activity and degradation of enzyme protein. All arrows represent both direct as well as indirect effects. See text for details.

complex usually has at least one target protein more sensitive than the others but its damage results in a decrease in the activity of the whole complex (Fig.3). Among different plant species exposed to various heavy metals (Cd, Cu, Ni, Pb, and Zn), preferential damage to individual proteins was observed for: the D1 protein subunit of the photosystem II reaction centre, the 17- and 23-kDa polypeptides of the oxygen evolving complex, ferredoxin and plastocyanin associated with photosystem I in the photosynthetic electron transport chain, and the coupling factor in chloroplast ATPase complex (Veeranjaneyulu and Das 1982, Maksymiec and Baszyński 1988, Chiang and Dilley 1989, Skórzyńska-Polit and Baszyński, 1993, 1997, Lidon and Henriques 1993a,b, Maksymiec et al. 1994, Rashid et al. 1994, Nedunchezhian and Kulandaivelu 1995, reviewed in: Krupa and Baszyński 1995, Prasad 1996, Krupa 1999, Siedlecka et al. 2001).

3.1.3. Dissociation of active complex (native enzyme) subunits

Stiborova (1988) reported a rapid dissociation of the large and small subunits of ribulose-1,5-bisphospahte carboxylase/oxygenease (Rubisco), followed by their immediate proteolysis in the presence of Cd. Disturbances in the thylakoid lipid matrix that resulted in disturbances in the light harvesting complex II oligomerization and dissociation of other thylakoid polypeptide components, such as constituents of the photosystem II core and the oxygen evolving complex have been reported to result from Cd, Cu, Ni, Pb, V, and Zn stress (Khan and Malhorta 1987, Krupa 1988a,b, Maksymiec and Baszyński 1988, Krupa and Baszyński 1989, Skórzyńska et al. 1991, Maksymiec et al. 1992, 1994, Stefanov et al. 1992, 1993, 1995, Lidon and Henriques 1993 Skórzyńska-Polit and Baszyński 1993, 1997, Rashid et al. 1994, Nedunchezhian et al. 1995, Quartacci et al. 2000). In all these experiments the protein complexes were particularly labile in mature parts of the leaves, in which heavy metal stress resulted in the acceleration of senescence.

3.2. *Mechanisms of increase or maintaining of enzymes activity*

All of the factors described above result in limitations in enzyme activities. However, plants have developed numerous mechanisms enabling them to maintain or restore the activity of many enzymes, thus enabling them to create a new homeostasis and to adapt to the heavy metal stress.

3.2.1. Protection to stress

The protection of enzymes against heavy metal stress may come from two mechanisms:
- i) general mechanisms of stress avoidance when heavy metals are kept outside the symplast
- ii) enzyme protection from damage by molecular chaperones, heat shock proteins, and so on.

Proline accumulates in plants under unfavourable environmental conditions, including heavy metal treatment, and it has been described as biomarker of stress (Alia and Saradhi 1991, Ernst and Peterson 1994). The main stimulation for proline accumulation comes from a water imbalance in metal-treated plants (Schat et al. 1997). However, a recent report by Chen et al. (2001) may provide a more detail explanation for this phenomenon. They suggest that rather than a water imbalance, it is a Cu-dependent increase in abscisic acid (ABA) content that was the main cause for proline accumulation. Since ABA is well known to increase in water-deficient plants, the mechanism proposed by Chen and co-authors points to a signalling pathway leading to increased proline synthesis. An additional role of proline accumulation may be nitrogen and carbon storage in situations where proteolysis, and thus nitrogen turnover, is inhibited by heavy metal ions, and where *de novo* protein synthesis is inhibited, as reported for Pb-, and Cu- treated plants (Angelov et al. 1993, Mohan and Hosetti 1997). It has been demonstrated by *in vitro* experimetns that proline can protect important enzymes such as nitrate reductase and glucose-6-phosphate dehydrogenase against Cd and Zn toxicity (Sharma et al. 1998). Proline also protected Rubisco activity during salinity stress but oxygenase activity was suppressed as a side effect (Sivakumar et al. 2000). Similar effects may be expected for heavy metals. Under Cu stress Angelov et al. (1993) observed high resistance of Rubisco to stress, while a very high sensitivity to Cu and a Cu- induced shift to oxygenation has been widely reported (reviewed in: Droppa and Horváth 1990, Siedlecka et al. 2001).

Another group of proteins involved in heavy metal protection are small cysteine-rich proteins called metallothioneins and phytochelatins. These proteins remove from the most toxic form of the heavy metal, their free ions, from the cytoplasm by bonding them to cysteine thiol groups. This would be enough to provide the enzymes with protection but the phytochelatine-heavy metal complexes are then transported to the vacuole for stronger heavy metal immobilization by complexing with inorganic S or organic acids. The phytochelatine is then recycled. The contribution of phytochelatins and metallothioneins to plant heavy metal stress resistance has been described in another part of this chapter (4.2.4. Metallothioneins and phytochelatins).

3.2.2. Stimulation of activity

We used to think that heavy metals always inhibited enzyme activities. However, some enzymes are also stimulated during plant responses to heavy metal stress. First, the activity increases among enzymes directly involved in stress protection, such as phytochelatin synthase. An increase in activity of "assistant enzymes", such as carbonic anhydrase, that supports Rubisco activity, has been also described (3.2.4. Support from other enzymes). Since photosynthesis seems to be the most susceptible process in plant metabolism to heavy metals, plant cells have to deal with problems of carbon and energy imbalance. Limitations in cellular ATP and NADPH content under moderate Cd, Ni and Pb stress can be compensated for by increased activity of respiratory and photorespiratory enzymes (Reese et al. 1986, Mattioni et al. 1997, Siedlecka et al. 1999b, reviewed in: Van Assche and Clijsters 1990, Siedlecka et al. 2001). All enzymes involved in mechanisms of stress adaptation are also stimulated by heavy metals (see in this chapter part 4.2. Specific enzymes, particular functions).

3.2.3. Repair

Some proteins of critical importance for carrying any process are known to have a very high turnover rate – for example the D1 protein, one of reaction centre proteins in photosystem II, carrying quinone-binding sites. These proteins need to have a very efficient system of release from the native complex and their replacement by newly synthesized proteins. If this type of protein becomes a target of heavy metal toxicity it can be replaced quickly and the activity of the whole complex is restored after a brief inhibition (Fig.3). In fact, antagonism between efficiency and inhibition in D1 protein turnover upon Cd and Cu stress appears to be a very important aspect of PSII function under stress conditions, and plants with an increased D1 turnover rate were much less susceptible to heavy metal stress (Giradi et al. 1997, Geiken 1998, Pätsikkä et al. 1998, Franco et al. 1999, reviewed in: Barón et al. 1995, Krupa and Baszyński 1995, Krupa 1999).

3.2.4. Support from other enzymes

When an important metabolical step is inhibited by stress, the plant response may be to replace this activity through the activation of alternative pathways or to restore the activity through support from associated enzymes (Fig.3). The most common characteristic from this type of stress responses is the activation of the alternative respiratory pathway in cyanide-resistant cells. For heavy metal stress the mechanism of support for key enzymes coming from associated enzymes has not been thoroughly investigated but it has been shown for the carboxylation process. Enzymes catalysing the first step of carbon metabolism in both C_3 and C_4 photosynthesis, Rubisco and phosphoenolopyruvate carboxylase (PEPC), are the most sensitive targets of heavy metal stress (Van Assche and Clijsters 1986, Stiborova et al. 1986, Houtz et al. 1988, Stiborova 1988, Malik et al. 1992, Siedlecka et al. 1997, 1999a, Stein and Ohlenbush 1997, reviewed in: Van Assche and Clijsters 1990, Krupa and Baszyński 1995, Siedlecka et al. 1998, 2001, Krupa 1999). All plant metabolism depends on carboxylation and this is the mechanism under particular control in both types of photosynthesis (Rodermel 1999, Spreitzer 1999, reviewed in: Chollet et al. 1996, Nimmo 2000). However, in C_4 plants, despite high susceptibility of PEPC to heavy metals, photosynthesis as a whole is more resistant to stress than in C_3 plants. In this case, support comes from the two-step carboxylation. PEPC is responsible only for preliminary CO_2 assimilation, while Rubisco and the Calvin cycle are the final patchway for carbon acquisition by metabolism. As long as there is some PEPC activity, the CO_2 supply to the Calvin cycle can be efficient (reviewed in: Siedlecka et al. 2001). The more susceptible situation for C_3 plants, which do not have this "internal mechanism of CO_2 concentration", means that a different mechanism is needed to support Rubisco. Rubisco activity in these plants has been regulated by two cooperating enzymes: carbonic anhydrase (CA) and Rubisco activase (RA) (reviewed in: Portis 1990, 1992, Siedlecka et al. 1998, 2001) It has been shown that under moderate Cd stress an increase in CA activity results in a more efficient CO_2 supply to Rubisco, leading to increased activation and catalysis. This supports the maintenance of Rubisco activity despite a decrease in Rubisco content (Siedlecka et al. 1997, 1999a, reviewed in Siedlecka et al. 1998, 2001). A similar function of CA in support of Rubisco has been reported for other stresses such as osmotic stress, senescence promoted by abscisic acid and methyl jasmonate, and Zn imbalance (Kicheva et al. 1997, Lazova et al. 1999, Salama et al. - personal communication). It is noteworthy that this mechanism is limited to moderate Cd-, and probably Zn-toxicity, at least at Fe and Cu excess both CA and Rubisco activities were severely inhibited (Siedlecka et al. 1997, Siedlecka and Krupa - unpublished results).

4. FUNCTIONS OF ENZYMES IN HEAVY METALS TREATED PLANTS

The effect of heavy metal stress on the function of plant enzymes has two main aspects. The first is inhibition and recovery (or maintenance) of activity by enzymes involved in key processes, such as photosynthesis. The second is the induction of specific enzymes involved in either plant protection against heavy metal stress, such as phytochelatine synthase, or with protection against secondary damage, such as free radicals scavenging enzymes.

4.1. Primary metabolism – creating new homeostasis by unusual functioning of ordinary systems

Until recently the main interest in heavy metals research was focused on the function of individual enzymes and on selected groups of metabolites. However, it is now time to look at enzyme function under stress conditions in a more comprehensive or holistic way. This type of research has been reported mainly for Cd, probably because Cd is one of the most toxic and therefore most interesting heavy metals.

Nitrate assimilation by *Zea mays* plants has been shown to be drastically decreased by Cd (Boussama et al. 1999). The decrease in nitrate content was coupled to inhibition of enzymes responsible for the transformation of inorganic nitrate into organic amino groups: *e.g.* nitrate reductase, nitrite reductase, glutamine synthetase, ferredoxin-glutamate synthetase and NADH-glutamate synthase. Additionally, for glutamate dehydrogenase induction was observed, correlated with an increase in NADH-dependent aminating activity and decrease in NAD(+)-related deaminating activity. Cadmium treatment also resulted in an increase in protease activity and the accumulation of ammonia, as well as a decrease in total protein and amino acids in roots and shoots. Changes in the amino acids content were followed by changes in their quantity – *e.g.* glutamate, glycine, lysine, methionine and proline accumulate in Cd-treated plants. All these changes in enzyme activities as well as in amino acid patterns showed clearly a Cd-dependent shift in the pathways of nitrogen assimilation from normal metabolism to increased stress protection. An increase in glutamate concentration, as result of a stimulation in the activity of NADH-glutamate dehydrogenase, leads to an increase in the amount of substrate for the synthesis of heavy metal stress-defence compounds such as proline, glutathione, γ-glutamylcysteine and, therefore, phytochelatins, while proline itself protects enzymes against heavy metal toxicity.

Similarly, an increase in the capacity for protection by a shift in sulfur metabolism has been reported by Lee and Leustek (1999). The expression of genes encoding key enzymes involved in sulfur assimilation, ATP sulfurylase and APS reductase (APS: 5´-adenylsulfate), increased under Cd-stress conditions, while this effect was blocked by feeding plants cysteine or glutathione.

4.2. Specific enzymes, particular functions

4.2.1. Ubiquitin – benefits of degradation

Ubiquitin is a small protein, consisting of 76 amino acids, and it is one of the most abundant and most highly conserved proteins in eucaryotic cells. Together with three additional enzymes it prepares proteins for degradation:

- i) E1 – ATP-dependent ubiquitin activating enzyme, by formation of high-energy thioester bond between ubiquitin and E1,
- ii) E2 – ubiquitin-conjugated enzyme, a carrier protein transferring activated ubiquitin to the target protein and,
- iii) E3 – ubiquitin-protein ligase, forms a bond between internal lysine residues of the target protein and active ubiquitin.

Additional ubiquitin molecules are then bound to the ubiquitinated lysine, forming "chain-like" structures called polyubiquitines. Ubiquitinated or polyubiquitinated proteins are degraded into short peptides, with the subsequent release of the ubiquitin monomers (which can be reused) by a multisubunit protease complex called a proteasome (Genschik et al. 1994,

Feussner et al. 1997, Wang et al. 2000, reviewed in: Belknap and Garbarino 1996, Craig and Tyers 1999, Callis and Vierstra 2000). In the nucleus, additional proteins such as constitutive photomorphogenic 1 protein (COP1), are involved in targeting nuclear proteins into nuclear bodies for degradation through the ubiquitin-proteasome system (Von Arnim and Deng 1994, Holm and Deng 1999, reviewed in: Reyes 2001). An increase in ubiquitin or polyubiquitin gene expression in Cd-, and Hg-stressed higher plants (*Nicotiana sylvestris, Phaseolus vulgaris, Zea mays*) was associated with stress adaptation, while ubiquitin-less mutants were hypersensitive to Cd (Genschink et al. 1992, 1994, Didierjean et al. 1996, Feussner et al. 1997, Chai and Znahg 1999, reviewed in: Sanitá Di Toppi and Gabbrielli 1999). Why is more efficient protein degradation an adaptation to heavy metal stress? The most likely reason is that a faster recycling of amino acids from old or stress-damaged proteins is beneficial to cells in which heavy metal toxicity has caused a nitrogen imbalance. Under Pb toxicity protein degradation is slower due to the formation of complexes of Pb^{2+} with protein -COOH groups, preventing proteolysis. Under such conditions, the normal mechanism for protein degradation blocked and proteins are degraded by any alternative pathway resulting in proline accumulation (Sen et al. 1994, Mohan and Hosetti 1997).

Gitan and Eide (2000) reported another benefit from protein ubiquitination and degradation. In yeast cells they found that immediate ubiquitination and degradation of the zinc-transporter protein protects cells against Zn stress. A similar mechanism has not been reported in plant cells, but it seems probable (reviewed in: Guerinot 2000).

4.2.2. Heat shock proteins and molecular chaperones

The plant heat shock protein (HSP) family is a large group of proteins that are not only induced by in heat-shock (reviewed in: Vierling 1991, Waters 1995, Waters et al. 1996, Sanitá Di Toppi and Gabbrielli 1999, Eriksson 2000). Some of HSPs are constitutive and play a very important role as molecular chaperones (MCs). MCs are members of the so-called high molecular weight HSP family. There is also an abundant family of small HSPs with a molecular mass ranging from 17,000 to 30,000 daltons. Ubiquitin is considered to be an additional class of HSPs.

Molecular chaperones (MCs) are proteins that control the folding of other proteins and after the assembly of the mature protein they dissociate from the complex. Chaperones never bind, or remain bound, to properly folded proteins (Fig.1., reviewed in: Lewin 1997, Eriksson 2000). Under stress conditions chaperones are involved in two processes: one is proper folding of newly synthesized proteins, the second is protection from proteolysis and renaturation of denatured proteins. When the refolding process is unsuccessful, MCs assist ubiquitination, and thus degradation, of the damaged protein (reviewed in: McClellan and Frydman 2001). The classification of MCs is based on the different structures of HSPs. There are 3 classes of MCs (reviewed in: Lewin 1997): type HSP 70 (the most ubiquitous family of ~ 70kDa proteins), type HSP 60 (a large complex of proteins called chaperonins – functioning as chaperone, and co-chaperonins – not active, but necessary for chaperonin activity), and type HSP 90 (cooperating with certain transcription factors). HSPs are known to protect thylakoid membranes against heavy metal stress, particularly when they have already accumulated in the plant cells due to heat pre-treatment (Neumann et al. 1994).

4.2.3. Enzymes of oxidative stress – dealing with a secondary problem

The reduction of O_2 from O_2 to H_2O by mitochondria, as well chloroplast O_2 formation from H_2O during photosynthesis, has been known to generate intermediates – reactive oxygen species, such as singlet oxygen ($^1O_2$), superoxide radicals ($O_2^{\cdot-}$), hydrogen peroxide (H_2O_2), and the most toxic hydroxyl radicals ($OH^\bullet$). In young and healthy plant cells this process is kept under control but unfavourable environmental conditions may led to the generation of oxidative stress. Heavy metals, especially those with redox properties, such as Cu, Mn, and Fe, or those that are highly reactive, such as Cd, are known to increase the amount of reactive oxygen species in plant tissues, leading to development of secondary oxidative stress (Luna et al. 1994, Yruela et al. 1996, Mazhoudi et al. 1997, Moran et al. 1997, Lidon and Teixeira 2000, reviewed in: Siedlecka et al. 2001). Reactive oxygen species are capable of oxidizing thiol groups, making them dangerous to protein structure and the activity of thiol-regulated proteins, such as most of the Calvin cycle enzymes (reviewed in: Noctor and Foyer 1998). Defence against this stress involves antioxidants (substances capable of scavenge reactive oxygen species without itself conversion into radicals) with small molecular weight, such as ascorbate, glutathione, phenolic compounds, and α-tocopherol. Moreover, a set of antioxidative enzymes is involved. Superoxide dismutases (SODs) and catalases (CATs) are common to all living organisms. Also peroxidases (POXes), including plant-specific ascorbate peroxidase (AP), are present in various subcellular compartments, including the cytosol, chloroplasts, peroxisomes and glyoxysomes (Moran et al. 1997, Patra et al. 1997, Vansuyt et al. 1997, reviewed in: Noctor and Foyer 1998, Siedlecka et al. 2001). SODs convert superoxide and hydroxyl radicals into H_2O_2, which is degraded into water and molecular oxygen by CATs or POXes. The difference between the last two groups of enzymes is in their functioning: CATs operate alone, have low substrate affinity and need two molecules of H_2O_2 per cycle, while POXes have a much higher affinity to H_2O_2 and require a reductant. In plant cells ascorbate is the predominant reductant and AP is the main POX (reviewed in: Noctor and Foyer 1998). All these enzymes are affected by heavy metal stress. In general, as long as the stress is not too strong for the plant defence capacity, the main response is an increase in SODs and POXes activities along with a decrease in CATs – which is less efficient than POXes in H_2O_2 scavenging. These results were obtained for a number of heavy metals, including Cd, Cu, Mn, Ni, Pb, and Zn and for numerous plant species, although the response varied depending on the plants sensitivity, specific organ, selected metal and its concentration, as well as the duration of the stress (Chongpraditnun et al. 1992, Samashekaraiah et al. 1992, Przymusiński et al. 1995, Chaoui et al. 1997, Mohan and Hosetti 1997, Weckx and Clijsters 1997, Gupta et al. 1999, Rucińska et al. 1999, Rao and Sresty 2000, reviewed in: Rabinovitch and Fridovich 1983, Van Assche and Clijsters 1990, Vangronsveld and Clijsters 1994, Siedlecka et al. 2001). An increase in CAT activity under Cd and Hg stress has been reported in some *Phaseolus* species (Shaw 1995, Mazhoudi et al. 1997, Drążkiewicz et al. 1999, Skorzyńska-Polit et al. 1999, 2000).

Recent reports have shown some very interesting data on the interactions between heavy metals and oxidative stress. Lidon and Teixeira observed that in Mn-treated rice plants, both under Mn-deficiency as well as at Mn excess, there is a significant increase in chloroplast SOD and AP activities, while at physiological Mn supply CAT was the predominant oxy-radical scavenger (Lidon and Teixeira 2000). Vansuyt et al. (1997) reported that Fe excess resulted in the rapid accumulation of AP transcript in *Brassica napus* seedlings. Moreover, this gene regulation was Fe-dependent and other changes in plant redox status had no influence on this activation. In *Lemna minor* it was not the AP located in chloroplasts and cytoplasm but CAT together with vacuolar guaiacol peroxidase (GP) that were the most inducible antioxidant enzymes under Cu stress (Mazhoudi et al. 1997, Teiseire and Guy 2000). Very efficient cooperation between AP and GP in hydrogen peroxide scavenging, as well as their parallel induction by heavy metal stress and acid rain has been also described (Koricheva et al. 1997, Yamasaki and Grace 1998).

Interactions between heavy metals and reactive oxygen species are not simply the induction of oxidative stress with a resultant increase in metal toxicity. The adaptative response to oxidative stress can also protect plants against the effects of heavy metal stress, especially those connected with heavy metal mutagenicity. A decrease in Hg-induced aberrations and micronuclei formation after pre-treatment of plants with H_2O_2 and paraquat (a redox-cycling agent causing intracellular oxidative stress) was reported for *Allium cepa* and *Hordeum vulgare*, despite the extremely genotoxic form of methyl mercuric chloride used in the experiment (Patra et al. 1997). Among eucaryotes, the mechanisms of this pre-adaptation remians unclear. However, prokaryotes have been shown to induce about 40 proteins in response to oxidative stress, including DNA-repair enzyme endonuclease IV (Chan and Weiss 1987, reviewed in: Demple 1991).

4.2.4.Metallothioneins and phytochelatins

Metallothioneins (MTs) are small cysteine-rich, and thus effective heavy-metal binding, proteins present in animals. They participate in Zn transport and distribution and play a key-role in animal tolerance to heavy metals (reviewed in: Andrews 2000). They have also been detected in some plant species (Rauser and Curvetto 1980, reviewed in: Zenk 1996, Siedlecka et al. 2001). Phytochelatins (PCs) are plant-specific heavy metal chelating peptides, rich in cysteine and glutamine. They are synthesized from glutathione and are involved both in Zn and Cu distribution and delivery to metal-demanding enzymes, as well as in the detoxication of numerous heavy metals, including Au, Ag, Bi, Cd, Cu, Hg, Pb, and Zn, (Grill et al. 1985 1989, reviewed in: Zenk 1996, Siedlecka et al. 2001). The main difference between MTs and PCs is in their synthesis. MTs are gene coded, while PCs are enzymatically polymerised from glutathione or related peptides by γ-glutamylcysteine synthetase (γ-ECS). Moreover, glutathione is always present in plant cells due to its participation in sulfur metabolism, in scavenging reactive oxygen species, and in protection against oxidative stress. What consequences may these differences between MTs and PCs have for plant responses to heavy metal stress? Until recently the answer appeared very difficult because MTs are not common in plants. However, in a unique experiment with *Brassica juncea,* a plant which has both forms of metal-binding peptides, Cu stress resulted in significant activation of the γ-ECS gene with a parallel down-regulation of MTs transcript level (Schäffer et al. 1997). This means a shift to an easier and more effective system of Cu chelating in *Brassica* cells under Cu stress.

Glutathione and thiol metabolism, metallothioneins and plant-specific metal chelating peptides and proteins have been described in details in Chapters 2, 6, and 7 of this book.

5. COPPER AND CADMIUM LIKE PATTERNS – ARE THEY TWO DIFFERENT KINDS OF TOXIC INFLUENCE AND PLANT STRESS RESPONSE?

The toxic impact of Cd and Cu on higher plants has numerous common features: growth inhibition, accelerated senescence, chlorosis, different plant response with respect to growth stage in which the metals are applied (*via* senescence induction in mature plants and by Calvin cycle inhibition in young ones), damage to organelle structure, and increased susceptibility of photosynthesis in comparison to respiration (reviewed in: Droppa and Horváth 1990, Krupa and Baszyński 1995, Krupa 1999, Prasad 1996, Siedlecka et al. 2001).

Also common are the most general mechanisms of heavy metal stress defence, such as phytochelatin synthesis, proline accumulation, changes in activity of antioxidative enzymes (reviewed in: Krupa and Baszyński 1995, Prasad 1996, Sanitá Di Toppi and Gabbrielli 1999, Siedlecka et al. 2001). However, there is one very important difference between these two metals: Cu can change its redox state. Even the fact that Cu is an essential micronutrient, whilst Cd is not necessary for plant metabolism, seems to be of only minor importance. Excessive Cu is poisonous and when we compare the toxic effects of Cd and Cu a strong relationship between the redox properties of the metal and the metabolic affects of Cu are apparent in different aspects of lipid and protein damage (Tab.1., Ouariti et al. 1997).

Table 1. Comparison of two different types of plant response to heavy metal stress: Cd-, and Cu-like. Abbreviations: CBP20 - ~20 kDa pathogen and wound-inducible antifungal protein

Enzyme/protein/process	Cd-type response	Cu-type response	References
mechanism of DNA damage	histone binding	oxidative damage	Maksymiec 1997
protein affinity for lipids	unchanged	increased	Szalontai et al. 1999
protein thiol groups and disulfide bonds	bound and blocked	oxidized	Viarengo 1985, Marschner 1995
root Fe(III) reductase expression	inhibited	activated	Schmidt 1996, Schmidt et al. 1997, Siedlecka et al. 2001
ATP sulfurylase expression	inhibited	activated	Lee and Leustek 1999
proline-rich cell wall protein synthesis	increased	unaffected	Chai et al. 1998
CBP20 synthesis after heavy metal-stimulated transcription	active	blocked	Hensel et al. 1999
Rubisco activity regulation to:	carboxylation	shift to oxygenation	Siedlecka et al. 2001
Carbonic anhydrase at moderate stress	activated	inhibited	Siedlecka et al. 1997, 1999a, 2001

Unfortunately, most research has been focused on one heavy metal and comparisons of just Cd and Cu represent only a fraction of the papers reporting more complex experiments. According to our present understanding, the function of the two ways of stress affecting and plant response: Cd-like, and Cu-like, are only note-worthy observation and need further investigations.

ACKNOWLEDGEMENTS

We are grateful to Dr. Vaughan Hurry (Umeå Plant Science Centre, Department of Plant Physiology, Umeå University, Sweden) for helpful comments on the manuscript. The financial assistance of the Polish Committee for Scientific Research, grant 6P04C.064.15, is gratefully acknowledged.

REFERENCES

Alcántara, E., Romera, F.J., Cañete, M., and De la Guardia, M.D. (1994) Effects of heavy metals on both induction and function of root Fe (III) reductase in Fe-deficient cucumber (*Cucumis sativus* L.) plants, *J. Exp. Bot.* **45**, 1893-1898.
Alia, and Saradhi, P.P. (1991) Proline accumulation under heavy-metal stress, *J. Plant Physiol.* **138**, 554-558.
Andrews, G.K. (2000) Regulation of metallothionein gene expression by oxidative stress and metal ions, *Biochemical Pharmacol.* **59**, 95-104.
Angelov, M., Tsonev, T., Uzunova, A., and Gaidardjieva, K. (1993) Cu(2+) effect upon photosynthesis, chloroplast structure, RNA and protein-synthesis of pea-plants, *Photosynthetica* **28**, 341-350.
Bader, M., Muse, W., Ballou, D.P., Gassner, C., and Bardwell, J.C.A. (1999) Oxidative protein folding is driven by the electron transport system, *Cell* **98**, 217-227.
Barón, M., Arellano, J., and Lòpez Gorgé, J. (1995) Copper and photosystem II: a controversial relationship, *Physiol. Plant.* **94**, 174-180.
Baszyński, T., Wajda, L., Król, M., Wolińska, D., Krupa, Z., and Tukendorf, A. (1980) Photosynthetic activities of cadmium-treated tomato plants, *Physiol. Plant.* **48**, 365-370.
Belknap, W.R. and Garbarino, J.E. (1996) The role of ubiquitin in plant senescence and stress responses, *Trends Plant Sci.* **1**, 331-335.
Boussama, N., Ouariti, O., Suzuki, A., and Ghorbal, M.H. (1999) Cd-stress on nitrogen assimilation, *J. Plant Physiol.* **155**, 310-317.
Callis, J. and Vierstra, R.D. (2000) Protein degradation in signalling, *Curr. Op. Plant Biol.* **3**, 381-386.
Chai, T.Y., Didierjean, L., Burkard, G., and Genot, G. (1998) Expression of a green tissue specific 11 kDa proline-rich protein gene in bean in response to heavy metals, *Plant Sci.* **133**, 47-56.
Chai, T.Y. and Zhang, Y.X. (1999) Expression analysis of polyubiquitin genes from bean in response to heavy metals, *Acta Bot. Sin.* **41**, 1052-1057.
Chakravarty, B. and Srivastava, S. (1997) Effects of genotype and explant during *in vitro* response to cadmium stress and variation in protein and proline contents in linseed, *Annals Bot.* **79**, 487-491.
Chan, E. and Wiess, B. (1987) Endonuclease IV of *Escherichia coli* is induced by paraquat. *Proc. Natl. Acad. Sci. USA* **84**, 3189-3193.
Chaoui, A., Mazhoudi, S., Ghorbal, M.H. and Ferjani, E.E. (1997) Cadmium and zinc induction of lipid peroxidation and effects on antioxidative enzyme activities in bean (*Phaseolus vulgaris* L.), *Plant Sci.* **127**, 139-147.
Chen, C.T., Chen, L.M., Lin, C.C., and Kao, C.H. (2001) Regulation of proline accumulation in detached leaves exposed to excess copper, *Plant Sci.* **160**, 283-290.
Chiang, G.G. and Dilley, R.A. (1989) Intact chloroplasts show Ca^{2+}-gated switching between localized and delocalized proton gradient energy coupling (ATP formation), *Plant Physiol.* **90**, 1513-1523.
Chollet, R., Vidal, J., and O'Leary, M.H. (1996) Phosphoenolopyruvate carboxylase: a ubiquitous highly regulated enzyme in plants, *Annu. Rev. Plant Physiol. Plant Mol. Biol.* **47**, 273-298.
Chongpraditnun, P., Mori, S., and Chino, M. (1992) Excess copper induces cytosolic Cu, Zn-superoxide dismutase in soybean roots, *Plant Cell Physiol.* **33**, 239-244.
Costa, G. and Morel, J.L. (1994) Water relations, gas exchange and amino acid content in Cd-treated lettuce, *Plant Physiol. Biochem.* **32**, 561-570.
Craig, K.L. and Tyers, M. (1999) The F-box: a new motif for ubiquitin dependent proteolysis in cell cycle regulation and signal transduction, *Progr. Biophys. Mol. Biol.* **72**, 299-328.
Das, P., Samantaray, S., and Rout, G.R. (1997) Studies on cadmium toxicity in plants: a review. *Environ. Poll.* **98**, 29-36.
Davies, R.R., Kuang, H., Qi, D., Mazhary, A., Mayaan, E., and Distefano, M.D. (1999) Artificial metalloenzymes based on protein cavities: exploring the effects of altering the metal ligand attachment position by site directed mutagenesis, *Bioorg. Med. Chem. Lett.* **9**, 79-84.
Demple, B. (1991) Regulation of bacterial oxidative stress genes, *Annu. Rev. Genet.* **25**, 315-337.
Didierjean, L., Frendo, P., Nasser, W., Genot, G., Marivet, J., and Burkard, G. (1996) Heavy-metal-responsive genes in maize: identification and comparison of their expression upon various forms of abiotic stress, *Planta* **199**, 1-8.

Drążkiewicz, M., Skórzyńska-Polit, E., Krupa, Z. (1999) Reactive forms of oxygen and activity of antioxidant enzymes in *Arabidopsis thaliana* treated with excess copper, *Acta Physiol. Plant.* **21 (suppl.)**, 40.
Droppa, M. and Horváth, G. (1990) The role of copper in photosynthesis, *Crit. Rev. Plant Sci.* **9**, 111–123.
Eriksson, M.-J. (2000) *ClpB/HSP 100 proteins in cyanobaterium Synechococcus. Insigts into regulation and function or some like it hot!* Doctoral dissertation. Solfjärden Offset AB, Umeå.
Ernst, W.H.O. and Peterson, P.J. (1994) The role of biomarkers in environmental assesment.4.Terrestial plants, *Ecotoxicology* **3**, 180–192.
Espen, L., Pirovano, L., Cocucci, S.M. (1997) Effects of Ni^{2+} during the early phases of radish (*Raphanus sativus*) seed germination, *Environ. Exp. Bot.* **38**, 187–197.
Feussner, K., Feussner, I., Leopold, I., and Wasternack, C. (1997) Isolation of cDNA coding for an ubiquitin-conjugating enzyme UBC1 of tomato - the first stress-induced UBC of higher plants, *FEBS Lett.* **409**, 211–215.
Franco, E., Alessandrelli, S., Masojídek, J., Margonelli, A., Giradi, M.T. (1999) Modulation of D1 protein turnover under cadmium and heat stress monitored by [^{35}S] methionine incorporation, *Plant Sci.* **144**, 53–61.
Gabara, B., Krajewska, M., and Stecka, E. (1995) Calcium effects on number, dimension and activity of nucleoli in cortex cells of pea (*Pisum sativum* L.) roots after treatment with heavy metals, *Plant Sci.* **111**, 153–161.
Gabara, B., Wojtyła-Kuchta, B., and Tapczyńska, M. (1992) The effects of calcium on DNA synthesis in pea (*Pisum sativum* L.) roots after treatment with heavy metals, *Folia Histochem. Cytobiol.* **30**, 69–74.
Geiken, B., Masojídek, J., Rizzuto, M., Pompili, M.L., and Giradi, M.T. (1998) Incorporation of [^{35}S] methionone in higher plants reveals that stimulation of the D1 reaction center II protein turnover accompanies tolerance to heavy metal stress, *Plant Cell Environ.* **21**, 1265–1273.
Genschik, P., Marbach, J., Uze, M., Feurman, M., Plesse, B., and Fleck, J. (1994) Structure and promoter activity of a stress and developmentally regulated polyubiquitin-encoding gene of *Nicotiana tabacum*, *Gene* **148**, 195–202.
Genschik, P., Parmentier, Y., Durr, A., Marbach, J., Criqui, M.C., Jamet, E., and Fleck, J. (1992) Ubiquitin genes are differentially regulated in protoplast-derived cultures of *Nicotiana sylvestris* and in response to various stresses, *Plant Mol. Biol.* **20**, 897–910.
Giradi, M.T., Masojídek, J., and Godde, D. (1997) Effects of abiotic stresses on the turnover of the D_1 reaction centre protein, *Physiol. Plant.* **101**, 635–642.
Gitan, R.S., Eide, D.J. (2000) Zinc-regulated ubiquitin conjugation signals endocytosis of the yeast ZRT1 zinc transporter. *Biochem. J.* **346**, 329–336.
Glockshuber, R. (1999) Where do the electrons go?, *Nature* **410**, 30–31.
Grill, E., Löffler, S., Winnacker, E.-L., and Zenk, M.H. (1989) Phytochelatins, the heavy-metal-binding peptides of plants, are synthesized from glutathione by a specific γ-glutamylcysteine dipeptydyl transpeptidase (phytochelatin synthase), *Proc. Natl. Acad. Sci. USA* **86**, 6838–6842.
Grill, E., Winnacker, E.-L., and Zenk, M.H. (1985) Phytochelatins: the principal heavy-metal complexing peptides of higher plants, *Science* **230**, 674–676.
Grotz, N., Fox, T., Connolly, E., Park, W., Guerinot, M.L., and Eides, A. (1998) Identification of a family of zinc transporter genes from *Arabidopsis* that respond to zinc deficiency, *Proc. Natl. Acad. Sci. USA* **95**, 7220–7224.
Guerinot, M.L. (2000) The ZIP family of metal transporters. *Biochim. Biophys. Acta* **1465**, 190–198.
Gupta, M., Cuypers, A., Vangronsveld, J., and Clijsters, H. (1999) Copper affects the enzymes of the ascorbate-glutathione cycle and its related metabolites in the roots of *Phaseolus vulgaris*, *Physiol. Plant.* **106**, 262–267.
Hensel, G., Kunze, G., and Kunze, I. (1999) Expression of the tobacco gene CBP20 in response to developmental stage, wounding, salicylic acid and heavy metals, *Plant Sci.* **148**, 165–174.
Herbik, A. Giritch, A., Hortsmann, C., Becker, R., Balzer, H.-J., Bäumlein, H., Stephan, U.W. (1996) Iron and copper nutrition-dependent changes in protein expression in a tomato wild type and the nicotianamine-free mutant *chloronerva*, *Plant Physiol.* **111**, 533–540.
Himelblau, E. and Amasino, R.M. (2000) Delivering copper within plant cells, *Curr. Op. Plant Biol.* **3**, 205–210.
Holm, M. and Deng, X.W. (1999) Structural organization and interactions of COP1, a light-regulated developmental switch. *Plant Mol. Biol.* **41**, 151–158.
Houtz, R.L., Nable, R.O. and Cheniae, G.M. (1988) Evidence for effects on the *in vivo* activity of ribulose-bisphosphate carboxylase/oxygenase during development of Mn toxicity in tobacco, *Plant Physiol.* **86**, 1143–1149.
Johnson, F.M. (1998) The genetic effects of environmental lead, *Mut. Res.* **410**, 123–140.
Khan, A.A., and Malhotra, S.S. (1987) Effects of vanadium, nickel and sulphur dioxide on polar lipid biosynthesis in jack pine, *Phytochemistry* **26**, 1627–1630.
Kicheva, M.I. and Lazova, G.N. (1997) Response of carbonic anhydrase to polyetylen glycol-mediated water stress in wheat, *Photosynthetica* **34**, 133–135.
Koricheva, J., Roy, S., Vranjic, J.A., Haukioja, E., Hugens, P.R., and Hänninen, O. (1997) Antioxidant responses to stimulated acid rain and heavy metal deposition in birch seedlings, *Environ. Poll.* **95**, 249–258.
Krupa, Z. (1988a) Cadmium-induced changes in the composition and structure of light-harvesting chlorophyll *a/b* protein complex II in radish cotyledons, *Physiol. Plant.* **73**, 518–524.

Krupa, Z. (1988b) Acyl lipids in the supramolecular chlorophyll-protein complexes of photosystems – isolation artefacts or integral components regulating their structure and function?, *Acta Soc. Bot. Pol.* **57**, 401–418.

Krupa, Z. (1999) Cadmium against higher plant photosynthesis – a variety of effects and where do they possible come from?, *Z. Naturforsch. – J. Biosci.* **54c**, 723–729.

Krupa, Z. and Baszyński, T. (1989) Environmental stresses as factors modifying the structure of the light-harvesting chlorophyll *a/b* complex II, *Photosynthetica* **23**, 695–698.

Krupa, Z. and Baszyński, T. (1995) Some aspects of heavy metal toxicity towards photosynthetic apparatus – direct and indirect effects on light and dark reactions, *Acta Physiol. Plant.* **17**, 177–190.

Lacroix, E., Kortemme ,T., Lopez de la Paz, M., and Serrano, L. (1999) The design of linear peptides that fold as monomeric β-sheet structures, *Curr. Op. Struct. Biol.* **9**, 487–493.

Lasat, M.M., Pence, N.S., Garvin, D.F., Ebbs, S.D., and Kochian, L.V. (2000) Molecular physiology of zinc transport in the Zn hyperaccumulator *Thlaspi caerulescens*, *J. Exp. Bot.* **51**, 71–79.

Lazova, G.N., Kicheva, M.I., and Popova, L.P. (1999) Effect of abscisic acid and methyl jasmonate on carbonic anhydrase activity in pea. *Photosynthetica* **36**, 631-634.

Lee, S. and Leustek, T. (1999) The affect of cadmium on sulphate assimilation enzymes in *Brassica juncea*, *Plant Sci.* **141**, 201–207.

Lewin, B. (1997) *Genes VI*. Oxford Univ. Press, Oxford.

Lidon, F.C. and Henriques, F.S. (1993a) Changes in the contents of the photosynthetic electron carriers, RNase activity and membrane permeability triggered by excess copper in rice. *Photosynthetica* **28**, 99–108.

Lidon, F.C. and Henriques, F.S. (1993b) Changes in the thylakoid membrane polypeptide patterns triggered by excess Cu in rice, *Photosynthetica* **28**, 109–117.

Lidon, F.C. and Teixeira, M.G. (2000) Oxy radicals production and control in the chloroplast of Mn-treated rice, *Plant Sci.* **152**, 7–15.

Luna, C.M., Gonzalez, C.A., and Trippi, V.S. (1994) Oxidative damage caused by an excess of copper in oat leaves. *Plant Cell Physiol.* **35**, 11–15.

Łbik-Nowak, A., Gabara, B. (1997) Effects of calcium on RNA content in meristemic cells of pea (*Pisum sativum* L.) roots treated with toxic metals, *Folia Histochem. Cytobiol.* **35**, 231–235.

Maksymiec, W. (1997) Effect of copper on cellular processes in higher plants, *Photosynthetica* **34**, 321–342.

Maksymiec, W. and Baszyński, T. (1988) The effects of Cd^{2+} on the release of proteins from membranes of tomato leaves, *Acta Soc. Bot. Pol.* **57**, 465–474.

Maksymiec, W., Russa, R., Urbanik-Sypniewska, T., and Baszyński, T. (1992) Changes in acyl lipid and fatty acid composition in thylakoids of copper non-tolerant spinach exposed to excess copper, *J. Plant Physiol.* **140**, 52–55.

Maksymiec, W., Russa, R., Urbanik-Sypniewska, T., and Baszyński, T. (1994) Effects of excess Cu on the photosynthetic apparatus of runner bean leaves treated at two different growth stages, *Physiol. Plant.* **91**, 715–721.

Malik, D., Sheoran, I.S., Singh, R. (1992) Carbon metabolism in leaves of cadmium treated wheat seedlings, *Plant Physiol. Biochem.* **30**, 223–229.

Marschner, H. (1995) *Mineral Nutrition of Higher Plants,*. Acad. Press, London.

Mattioni, C., Gabbrielli, R., Vangronsveld, J., and Clijsters, H. (1997) Nickel and cadmium toxicity and enzymatic activity in Ni-tolerant and non-tolerant populations of *Silene italica* Pers., *J. Plant Physiol.* **150**, 173–177.

May, M.J., Vernoux, T., Sánchez-Fernández, R., Van Montagu, M., and Inzé, D. (1998) Evidence for posttranscriptional activation of γ-glutamylcysteine synthase during plant stress responses, *Proc. Natl. Acad. Sci. USA* **95**, 12049–12054.

Mazhoudi, S., Chaoui, A., Ghrobal, M.H., Ferjani, E.E. (1997) Response of antioxidant enzymes to excess copper in tomato (*Lycopersicon esculentum* Mill.). *Plant Sci.* **127**, 129–137.

McClellan, A.J. and Frydman, J. (2001) Molecular chaperones and the art of recognizing a lost cause, *Nature Cell Biol.* **3**, E1–E3.

Mohan, B.S. and Hosetti, B.B. (1997) Potential phytotoxicity of lead and cadmium to *Lemna minor* grown in sewage stabilization ponds, *Environ. Poll.* **98**, 233–238.

Moran, J.F., Klucas, R.V., Grayer, R.J. Abian, J., and Becana, M. (1997) Complexes of iron with phenolic compounds from soybean nodules and other legume tissues: prooxidant and antioxidant properties, *Free Radical Biol. Med.* **22**, 861–870.

Müller, J., Sigel, R.K.O. and Lippert, B. (2000) Heavy metal mutagenicity: insights from binorganic model chemistry, *J. Inorg. Biochem.* **79**, 261–265.

Nagoor, S. and Vyas, A.V. (1999) Physiological and biochemical responses of cereal seedlings to graded levels of heavy metals. III – Effects of copper on protein metabolism in wheat seedlings, *J. Env. Biol.* **20**, 125–129.

Nedunchezhian, N. and Kulandaivelu, G. (1995) Effects of Cd and UV-B radiation on polypeptide composition and photosystem activities of *Vigna unguiculata* chloroplasts, *Biol. Plant.* **37**, 437–441.

Neumann, D., Lichtenberger, O., Gunther, D., Tschiersch, K., and Nover, L. (1994) Heat-shock proteins induce heavy-metal tolerance in higher plants, *Planta* **194**, 360–367.

Nicolas, E., Beggs, J.M., Haltiwanger, B.M., and Taraschi, T.F. (1997) Direct evidence for the deoxyribonuclease activity of the plant ribosome inactivating protein gleonin, *FEBS Lett.* **406**, 162–164.

Nielsen, K. and Boston, R. (2001) Ribosome-inactivating proteins: a plant perspective, *Annu. Rev. Plant Physiol. Plant Mol. Biol.* **52**, 785–(in press)

Nimmo, H.G. (2000) The regulation of phosphoenolopyruvate carboxylase in CAM plants, *Trends Plant Sci.* **5**, 75–80.

Noctor, G. and Foyer, C.H. (1998) Ascorbate and glutathione: keeping active oxygen under control, *Annu. Rev. Plant Physiol. Plant Mol. Biol.* **49**, 249–279.

Ouariti, O., Boussama, N., Zarrouk, M., Cherif, A., and Ghorbal, M.H. (1997) Cadmium- and copper-induced changes in tomato membrane lipids, *Phytochemistry* **45**, 1343–1350.

Quartacci, M.F., Pinzino, C., Sgherri, C.L.M., Vecchia, F.D., and Navari-Izzo, F. (2000) Growth in excess copper induced changes in the lipid composition and fluidity of PSII-enriched membranes in wheat, *Physiol. Plant.* **108**, 87–93.

Patra, J., Panda, K.K., and Panda, B.B. (1997) Differential induction of adaptative responses by paraquat and hydrogen peroxide against the genotoxicity of methyl mercuric chloride, maleic hydrazide and ethyl methane sulfonate in plant cells *in vivo*, *Mutation Res.* **393**, 215–222.

Pätsikkä, E., Aro, E.M., Tyystjärvi, E. (1998) Increase in the quantum yield of photoinhibition contributes to copper toxicity *in vivo*. *Plant Physiol.* **117**, 619–627.

Portis, A.R. Jr. (1990) Rubisco activase, *Biochim. Biophys. Acta* **1015**, 15–28.

Portis, A.R. Jr. (1992) Regulation of ribulose-1,5-bisphosphate carboxylase/oxygenase activity, *Annu. Rev. Plant Physiol. Plant Mol. Biol.* **43**, 415–437.

Prasad, M.N.V. (1996) Trace metals, in M.N.V. Prasad (ed.), *Plant Ecophysiology*, John Wiley and Sons, New York, 207–249.

Przymusiński, R., Rucińska, R., Gwóźdź, E.A. (1995) The stress-stimulated 16 kDa polypeptide from lupin roots has activity of cytosolic Cu:Zn-superoxide dismutase, *Environ. Exp. Bot.* **35**, 485–495.

Prütz, W.A. (1990) The interaction between hydrogen peroxide and the DNA-Cu(I) complex: effects of pH and buffers, *Z. Naturforsch. – J. Biosci.* **45c**, 1197–1206.

Prütz, W.A. (1994) Interaction between glutathione and Cu(II) in the vicinity of nucleic acids. *Biochem. J.* **320**, 373–382.

Prütz, W.A., Butler, J., Land, E.J. (1990) Interaction of copper (I) with nucleic acids. *Int. J. Radiat. Biol.* **58**, 215–234.

Rabinovitch, H.D., Fridovich, I. (1983) Superoxide radicals, superoxide dismutases and oxygen toxicity in plants. *Photochem. Photobiol.* **37**, 679–690.

Rashid, A., Camm, E.L., and Ekramoddoyullah, A.K.M. (1994) Molecular mechanism of action of Pb^{2+} and Zn^{2+} on water oxidizing complex of photosystem II, *FEBS Lett.* **350**, 296–298.

Rao, K.V.M. and Sresty, T.V.S. (2000) Antioxidative parameters in the seedlings of pigeonpea (*Cajanus cajan* (L.) Millspaugh.) in response to Zn and Ni stress, *Plant Sci.* **157**, 113–128.

Rauser W.E. and Curvetto N.R. (1980) Metallothionein occurs in roots of *Agrostis* tolerant to excess copper, *Nature* **287**, 563–564.

Reese, R.N., Mc Call, R.D., Roberts, L.W. (1986) Cadmium-induced ultrastructural changes in suspension-cultured tobacco cells (*Nicotiana tabacum* L. var. Xanthi), *Environ. Exp. Bot.* **26**, 169–173.

Reyes, J.C. (2001) PML and COP1 – two proteins with much in common, *Trends Biochem. Sci.* **26**, 18–20.

Rodermel, S. (1999) Subunit control of Rubisco biosynthesis – a relic of an endosymbiotic past?, *Photosynth. Res.* **59**, 105–123.

Rucińska, R., Waplak, S., and Gwóźdź, E.A. (1999) Free radical formation and activity of antioxidant enzymes in lupin roots exposed to lead, *Plant Physiol. Biochem.* **37**, 187–194.

Samashekaraiah, B.V., Padmaja, K., and Prasad, A.R.K. (1992) Phytotoxicity of cadmium ions on germinating seedlings of mung bean (*Phaseolus vulgaris*): involvement of lipid peroxides in chlorophyll degradation, *Physiol. Plant.* **85**, 85–89.

Sanitá Di Toppi, L., Gabbrielli, R. (1999) Response to cadmium in higher plants, *Environ. Exp. Bot.* **41**, 105–130.

Schat, H., Sharma, S.S., Vooijs, R. (1997) Heavy metal-induced accumulation of free proline in a metal-tolerant and a nontolerant ecotype of *Silene vulgaris*, *Physiol. Plant.* **101**, 477–482.

Schäffer, H.J., Greiner, S., Rausch, T., and Haag-Kerwer, A. (1997) In seedlings of the heavy metal accumulator *Brassica juncea* Cu^{2+}differentially affects transcript amounts for γ-glutamylcysteine synthetase (γ-ECS) and metallothionein (MT2), *FEBS Lett.* **404**, 216–220.

Schmidt, W. (1996) Influence of chromium(III) on root-associated Fe(III) reductase in *Plantago lanceolata* L., *J. Exp. Bot.* **47**, 805–810.

Schmidt, W., Bartels, M., Tittej, J., and Fuhner, C. (1997) Physiological effects of copper on iron acquisition process in *Plantago*, *New Phytol.* **135**, 659–666.

Sen, A.K., Mandal, N., and Mandal, S. (1994) Studies on the biochemical changes in *Salvinia natans* to chromium toxicity, *Environ. Ecol.* **12**, 279–283.

Sharma, S.S., Schat, H. and Vooijs,, R. (1998) *In vitro* alleviation of heavy metal-induced enzyme inhibition by proline, *Phytochemistry* **49**, 1531–1535.
Shaw, B.P. (1995) Effect of mercury and cadmium on the activities of antioxidative enzymes in the seedlings of *Phaseolus aureus*. *Biol. Plant.* **37**, 5578–596.
Siedlecka, A. (1995) Some aspects of interactions between heavy metals and plant mineral nutrients, *Acta Societ. Bot. Pol.* **64**, 265–272.
Siedlecka, A., Gardeström, P., Samuelsson, G., Kleczkowski, L.A., and Krupa, Z. (1999a) A relationship between carbonic anhydrase and Rubisco in response to moderate cadmium stress during light activation of photosynthesis, *Z. Naturforsch. – J. Biosci.* **54c**, 759–763.
Siedlecka, A., Kleczkowski, L.A., and Krupa, Z. (1999b) Activity of the peroxisomal hydroxypyruvate reductase (HPR1) in the higher plants under Cd stress conditions, *Folia Histochem. Cytobiol.* **37 (suppl.)**, 66.
Siedlecka, A. and Krupa, Z. (1999) Cd/Fe interaction in higher plants – its consequences for the photosynthetic apparatus, *Photosynthetica* **36**, 321–331.
Siedlecka, A., Krupa, Z., Samuelsson, G., Öquist, G. and Gardeström, P. (1997) Primary carbon metabolism in *Phaseolus vulgaris* plants under Cd/Fe interaction, *Plant Physiol. Biochem.* **35**, 951–957.
Siedlecka, A., Samuelsson, G., Gardeström, P., Kleczkowski, L.A., and Krupa, Z. (1998) The "activatory model" of plant response to moderate cadmium stress – relationship between carbonic anhydrase and Rubisco, in G. Garab (ed.) *Photosynthesis: Mechanisms and Effects*, Kluwer Acad. Publ., Netherlands, Vol. 4, 2677–2680.
Siedlecka, A., Tukendorf, A., Skórzyńska-Polit, E., Maksymiec, W., Wójcik, M., Baszyński, T., and Krupa, Z. (2001) Selected Angiosperms Families response to heavy metals in the Environment, in M.N.V. Prasad (ed.), *Metals in the Environment: Analysis by Biodiversity*, Marcel Dekker Inc., New York, 171-217.
Sivakumar, P., Sharmila, P., and Saradhi, P.P. (2000) Proline alleviates salt-stress-induced enhancement in ribulose-1,5-bisphosphate oxygenase activity, *Biochem. Biophys. Res. Comm.* **279**, 512–515.
Skórzyńska, E., Urbanik-Sypniewska T., Russa, R., and Baszyński, T. (1991) Galactolipase activity of chloroplasts in cadmium-treated runner bean plants, *J. Plant Physiol.* **138**, 454–459
Skórzyńska-Polit, E. and Baszyński, T. (1993) The changes in PSII complex polypeptides under cadmium treatment – are they of direct or indirect nature?, *Acta Physiol. Plant.* **15**, 263–269.
Skórzyńska-Polit, E. and Baszyński, T. (1997) Differences in sensitivity of the photosynthetic apparatus in Cd-stressed runner bean plants in relation to their age, *Plant Sci.* **128**, 11–21.
Skórzyńska-Polit, E., Drążkiewicz, M., and Krupa, Z. (1999) Activity of some enzymatic antioxidants in *Arabidopsis thaliana* upon Cd stress. *Acta Physiol. Plant.* **21 (suppl.)**, 72.
Skórzyńska-Polit, E., Drążkiewicz, M., and Krupa, Z. (2000) Activity of glutathione-ascorbate cycle enzymes in Cd-treated Arabidopsis thaliana, Abstracts of 12th Congress of the Federation of European Societies of Plant Physiology. 2000 August 21–25, Budapest, S19-132, 216.
Spreitzer, R.J. (1999) Questions about the complexity of chloroplast ribulose-1,5-bisphosphate carboxylase/oxygenase, *Photosynth. Res.* **60**, 29–42.
Stecka, E., Krajewska, M., and Gabara, B. (1995) Calcium effects on the content of DNA and NYS-stained nuclear, nucleolar and cytoplasmic proteins in cortex cells of pea (*Pisum sativum* L.) roots treated with heavy metals, *Acta Societ. Bot. Pol.* **64**, 239–244.
Stefanov, I., Frank, J., Gedamu, L., and Misra, S. (1997) Effect of cadmium treatment on the expression of chimeric genes in transgenic tobacco seedlings and calli, *Plant Cell Rep.* **5**, 291–294.
Stefanov, K., Popova, I., Kamburova, E., Pancheva, T., Kimenov, G., Kuleva, L., and Popov, S. (1993) Lipid and sterol changes in *Zea mays* caused by lead ions, *Phytochemistry* **33**, 47–51.
Stefanov, K., Popova, I., Nikolova-Damyanova, B., Kimenov, G., and Popov, S. (1992) Lipid and sterol changes in *Phaseolus vulgaris* caused by lead ions, *Phytochemistry* **31**, 3745–3748.
Stefanov, K., Seizova, K., Popova, I., Petkov, V., Kimenov, G., and Popov, S. (1995) Effect of lead ions on the phospholipid composition in leaves of *Zea mays* and *Phaseolus vulgaris*, *J. Plant Physiol.* **147**, 243–246.
Stein, K. and Ohlenbush, G. (1997) Inhibition of the enzyme phosphoenolopyruvate-carboxylase (PEPC) by different pollutants, *Talanta* **44**, 475–481.
Stiborova, M. (1988) Cd^{2+} ions affects the quaternary structure of ribulose-1,5-bisphosphate carboxylase from barley leaves. *Biochem. Physiol. Pflanz.* **183**, 371–378.
Stiborova, M., Hromadkova, R., and Leblova, S. (1986) Effects of ions of heavy metals on the photosynthetic characteristics of maize (*Zea mays* L.), *Biologia* (Bratislava) **41**, 1221–1228.
Szalontai, B., Horváth, L., Debreczeny, M., Droppa, M., and Horváth, G. (1999) Molecular rearrangements of thylakoids after heavy metal poisoning, as seen by Fourier transform infrared (FTIR) and electron spin resonance (ESR) spectroscopy, *Photosynth. Res.* **61**, 241–252.
Teisserie, H. and Guy, V. (2000) Copper-induced changes in antioxidant enzymes activities in fronds of duckweed (*Lemna minor*), *Plant Sci.* **153**, 65–72.
Tziveleka, I., Kaldis, A., Hegedüs, A., Kissimon, J., Prombona, A., and Horváth, G., Argyroudi-Akoyunoglou A. (1999) The effects of Cd on chlorophyll and light-harvesting complex II biosynthesis in greening plants, *Z. Naturforsch. C – J. Biosci.* **54c**, 740–754.

Ueda, J., Shimazu, Y., and Ozawa, T. (1994) Oxidative damage induced by Cu(II)-oligopeptide complexes and hydrogen peroxide, *Biochem. Mol. Biol. Int.* **34**, 801–808.

Van Assche, F., Clijsters, H. (1986) Inhibition of photosynthesis in *Phaseolus vulgaris* by treatment with toxic concentration of zinc: effect on ribulose-1,5-bisphosphate carboxylase/oxygenase, *J. Plant Physiol.* **125**, 355–360.

Van Assche, F., Clijsters, H. (1990) Effects of metals on enzyme activity in plants, *Plant Cell and Environ.* **13**, 195–206.

Vangronsveld, J. and Clijsters, H. (1994) Toxic Effects of Metals, in M.E. Farago (ed.) *Plants and the Chemical Elements. Biochemistry, Uptake, Tolerance and Toxicity.* VCH, Weinheim, 150–177.

Vansuyt, G., Lopez, F., Inzé, D., Briat, J.-F., and Fourcroy, P. (1997) Iron triggers a rapid induction of ascorbate gene expression in *Brassica napus*, *FEBS Lett.* **410**, 195–200.

Veeranjaneyulu, K., Das, V.S.R. (1982) Intrachloroplast localization of ^{65}Zn and ^{63}Ni in a Zn-tolerant plant, *Ocimum basilicum* Beneth, *J. Exp. Bot.* **33**, 1161–1165.

Viarengo, A. (1985) Biochemicals effects of trace metals, *Mar. Poll. Bull.* **16**, 153–158.

Vierling, E. (1991) The roles of heat shock proteins in plants, *Annu. Rev. Plant Physiol. Plant Mol. Biol.* **42**, 579–620.

Von Arnim, A.G., Deng, X.W. (1994) Light inactivation of *Arabidopsis* photomorphogenic repressor COP1 involves a cell-specific regulation of its nucleocytoplasmic partitioning, *Cell* **79**, 1035–1045.

Wang, J., Jiang, J., and Oard, J.H. (2000) Structure, expression and promoter activity of two polyubiquitin genes from rice (*Oryza sativa* L.), *Plant Sci.* **156**, 201–211.

Waters, E.R. (1995) The molecular evolution of the small heat-shock proteins in plants, *Genetics* **141**, 785–795.

Waters, E.R., Lee G.J. and Vierling, E. (1996) Evolution, structure and function of the small heat-shock proteins in plants, *J. Exp. Bot.* **47**, 325–338.

Weckx, J.E.J. and Clijsters, H.M.M. (1997) Zn phytotoxicity induces oxidative stress in primary leaves of *Phaseolus vulgaris*, *Plant Physiology and Biochemistry* **35**, 405–410.

Wierzbicka, M. (1999) Comparison of lead tolerance in *Allium cepa* with other plant species, *Environ. Poll.* **104**, 41–52.

Wojtyła-Kuchta, B. and Gabara, B. (1991) Changes in the content of DNA and NYS-stained nuclear, nucleolar and cytoplasmic proteins in cortex cells of pea (*Pisum sativum* cv De Grace) roots treated with cadmium, *Biochem. Physiol. Pflanz.* **187**, 67–76.

Yamasaki, H. and Grace, S.C. (1998) EPR detection of phytophenoxyl radicals stabilized by zinc ions: evidence for the redox coupling of plant phenolics with ascorbate in the H_2O_2-peroxidase system, *FEBS Lett.* **422**, 377–380.

Yruela, I., Pueyp, J.J., Alonso, P.J., and Picorel, R. (1996) Photoinhibition of photosystem II from higher plants, *J. Biol. Chem.* **44**, 27408–27415.

Zenk, M.H. (1996) Heavy metal detoxification in higher plants – a review, *Gene* **179**, 21–30.

Zoroddu, M.A., Kowalik-Jankowska, T., Kozlowski, H., Molinari, H., Salnikow, K., Broday, L., and Costa, M. (2000) Interaction of Ni (II) and Cu (II) with a metal binding sequence of histone H4: AKRHRK, a model of the H4 tail, *Biochim. Biophys. Acta* **1475**, 163–168.

CHAPTER 13

HEAVY METALS AND NITROGEN METABOLISM

GRAŻYNA KŁOBUS, MAREK BURZYŃSKI AND JÓZEF BUCZEK
Department of Plant Physiology, Institute of Botany
Wrocław University, PL 50–328 Wrocław, Poland

1. INTRODUCTION

Among macronutrients required by plants, nitrogen is consumed in the greatest abundance and most often limits growth and productivity. Its assimilation, like photosynthesis, is a life-dependent process that members of the animal kingdom are unable to perform for themselves. The only route by which plants can convert inorganic N into organic form is the uptake of nitrogen followed by its reduction and assimilation in plant cells. Available forms of nitrogen vary depending on the habitat of the plant and include inorganic (dinitrogen, nitrate and ammonium) and organic (urea and amino acids) compounds. In adapting to diverse habits, plants have developed multiple strategies for acquiring nitrogen, which range from the uptake of mineral nitrogen, to nitrogen fixation and even carnivory. Nitrogen taken up by plants serves as an important intracellular regulatory molecule altering N metabolism modulating genes in the nitrogen assimilation pathway. Also organic product(s) of N assimilation act as the feedback regulators of genes encoding cellular nitrogen transport systems and N assimilatory enzymes. Among these genes the most important are those encoding ammonium and nitrate carriers as well as the following enzymes: nitrate reductase (NR), nitrite reductase (NiR), glutamic synthetase (GS) and glutamate synthase (GOGAT). Resuming, the nitrogen uptake and nitrogen assimilation appear to control each other and thereby effectors altering one step of nitrogen acquisition totally disturb N the metabolism of a plant.

2. MODIFICATION OF INORGANIC NITROGEN UPTAKE UNDER STRESS

Although the most abundant source of nitrogen is the atmosphere, only a few plants establishing symbiotic interactions with bacteria genus fixing N_2 are capable of using it. Most of higher plants take up nitrogen from the soil as a nitrate or ammonium. The preference to the mineral nitrogen form varies from species to species (Glass and Siddigi, 1995, von Wiren et.al 1997). In general, absorption of ammonium and nitrate by most crop plants depends upon their concentration in the soil solution (Bloom 1988). It is know very well, that the level of ammonium and nitrate in the soil is determined mainly by the activity of ammonification (*Pseudomonas sp.*, *Bacillus cereus*) and nitrification (*Nitrosomonas* and *Nitrobacter*) processes. Heavy metals, through modulation of the bacteria activity influence the nitrate and ammonium concentration in the soil.

Alternatively, the N soil source may strongly influence the heavy metals absorption by plants according to changes in the rhizosphere pH. It has been shown that ammonium nutrition causes the net extrusion of protons and soil acidification (Dyhr-Jansen and Brix 1996), whereas NO_3^- nutrition leads to proton consumption increasing the pH of the soil (Glass 1988). The pH of the rhizosphere is one of the important factors affecting the heavy metals availability to plants. Generally, increasing of soil pH decreases the metal uptake by plants (Marschner 1995, Greger 1999). Thus, fertilization of plants with NO_3^- strongly reduces the uptake and accumulation of heavy metals whereas ammonium nutrition enhances their level in plant tissues. It was shown that in forest ecosystems nitrate fertilization effectively limited the uptake and toxicity of Pb. NO_3^- nutrition repressed also the uptake of Zn, Mn (Cox and Reisenauer (1977) and Cu (Tills and Alloway, 1981, Chesire et al 1982, Kumar et al. 1990, Weber et al. 1991). Compared with NO_3^-, ammonium ions used as an exclusive nitrogen source elevated the intracellular level of heavy metals.

2.1 Nitrate Uptake

In a typical aerobic agricultural soil, the amount of nitrate is much higher than ammonium and the major nitrogen source is NO_3^-. Thermodynamic evaluations have indicated that absorption of nitrate by plants is an active process, even at the highest nitrate concentrations found in soils (Glass et al. 1992). Physiological studies have provided evidences that NO_3^- influx into root cells is proton-coupled and therefore directly dependent on the H^+ pumping activity of plasmalemma H^+-ATPase (Kłobus and Buczek, 1995, Miller and Smith 1996). The uptake and transport of nitrate into the root cell across the plasma membrane is carried out by combined activities of a set of high– and low-affinity transport systems (Doddema and Telkamp, 1979, Hole et al. 1990, Glass et al. 1992). The high-affinity systems (HATS) are active at low NO_3^- concentrations (below 1mM), and the low-affinity systems (LATS) act at the concentration of nitrate over 1 mM (Crawford and Glass, 1998). Some of these NO_3^- transport systems are constitutively expressed, while the others are NO_3^- inducible and subject to the negative feedback regulation by the products of its assimilation (Shiddiqi et al. 1990, Aslam et al. 1992 and 1993). Molecular biology studies have shown that nitrate transporters in plants belong to two different families (*NRT1* and *NRT2*) represented by multiple genes encoding transporters with different regulatory and kinetic properties (for review see Crawford and Glass, 1998, Forde, 2000). Generally, the genes encoding the nitrate transport proteins involved in LATS are classified to the *NRT1* family and those involved in HATS belonged into *NRT2* family. Excess of heavy metal in the soil strongly represses the nitrate uptake in plants (Table 1).

2.1.1 Transport of nitrate into root cells.
Among heavy metals, the most effective inhibitor of the nitrate uptake in higher plants seems to be cadmium. Even a very low level of this ion (5 µmoles) present in the environment sharply restricted the NO_3^- absorption in many plant species (Burzyński 1988, Burzyński and Buczek 1994a, Shalaby et al. 1995, Chaoui et al. 1996, Hernandez et al. 1997, Boussama et al. 1999, Gouia et al. 2000). Cadmium repression of the nitrate absorption by plants was detected as a drop in NO_3^- loss from the external medium (Burzyński 1988, Burzyński and Buczek 1994, Hernandez et al. 1997, Boussama et al. 1999a and b, Gouia et al. 2000). Cd-induced alteration in the nitrate uptake was always parallel to the lowered concentration of NO_3^- in tissues of cucumber (Burzyński 1988), and maize (Hernandez et al. 1996). Contrary to other heavy metals, no distinct relationships between external concentration of cadmium and the nitrate uptake was observed (Burzyński 1988, Boussama et al. 1999, Hernandez et al. 1997, Gouia et al. 2000). The inhibition of NO_3^- absorption caused by 5 µM Cd and 25 µM Cd was almost the same (Burzyński 1988). Similar observations were made for bean and pea seedlings (Hernandez et al. 1997, Gouia et al. 2000). Furthermore, exclusion of Cd from the uptake solution did not reverse its inhibitory effect on the NO_3^- uptake up to 96 hours, suggesting a strong binding of metal ions in the plant cells (Burzyński 1988, Hernandez et al. 1997, Gouia et al. 2000). Lozano-Rodriguez et al. (1997) have found the largest, proportion of cadmium in the soluble fraction of pea plants, consisting mainly of vacuolar and cytosolic contents. Association of Cd with proteins or polypeptides inside of the cell might be the reason for the irreversible action of metal on the uptake.
Alteration in the nitrate absorption has also been demonstrated in cucumber seedlings treated with Cu and Ni (Burzyński and Buczek 1994a). After one-hour exposition of plants to 100 µM of metals the NO_3^- uptake dropped respectively about 50 and 80 per cent. Similarly to the action of cadmium, the Cu nor Ni effect on the nitrate uptake was irreversible up to eight hours. Weber et al (1991) found strong disturbances in the nitrate uptake by *Silene vulgaris* exposed even to as low Cu concentration as 4µM. However, it should be noted, that the duration of plant treatment with metal in this experiment was much longer (7 hours). Also tungstate ions were an effective inhibitor of the nitrate absorption in cucumber plants in short-term experiments (Kłobus et al. 1991). Its effect, like cadmium action, was not reversible up to 24 hours.
In short-time experiments, the nitrate uptake has been strongly affected at the presence of Pb in the environment, but the effective metal concentration was much higher and exceeds 50 µmoles (Burzyński and Grabowski 1984, Burzyński 1988, Burzyński and Buczek 1994). Inhibition of the nitrate absorption from the nutrient

Table 1. Inhibition of nitrate uptake under heavy metal stress

Taxon	Metal concentration (μmoles dm^{-3})	References
Anacystis nidulans	Cu (5)	Kashyap and Gupta (1982)
Cucumis sativus	Cd (5, 10, 25, 100)	Burzyński 1988, Burzyński and Grabowski, 1984,
C. sativus	Pb (25, 50, 100)	Burzyński 1988, Burzyński and Buczek 1994a,
	Cu (10)	Burzyński and Buczek 1994a,
	Ni (10)	Burzyński and Buczek 1994a,
	WO$_4^{2-}$ (100 μM)	Kłobus et al. 1991
C. sativus	Cd (50, 100)	Gouia et al., 2000
Phaseolus vulgaris v. Morgan	Cd (10, 50)	Hernandez et al., 1997
Pisum sativum	Cu (4,8,16)	Weber et al. 1991
Silena vulgaris	Cd	Hernandez et al., 1996
Zea mays		

solution decreased gradually with Pb concentration in the external medium and with duration of plant exposition to the metal (Burzyński 1988). Contrary to the Cd, Cu and Ni action on the nitrate uptake, the inhibitory effect of Pb was easily reversed. The nitrate uptake rate recovered the control level almost immediately after exclusion of Pb from the nutrient medium (Burzyński and Buczek 1994). The mechanism of the reversibility of Pb action is unclear. Positively charged lead ions, similarly to the other heavy metals, are attracted to the negative charges of the cell wall structures. In contrast to Cd, Cu and Zn, only a small part of lead taken up in the apoplast is later transported through the plasma membrane into the cytoplasm (Wierzbicka 1987, Kennedy and Gonsalves, 1987). Thus, the cell wall elements with negatively charged carboxylic groups form a specific barrier with a large capacity, which acts as a cation exchanger. When the extra-cellular concentration of Pb dropps, the ions accumulated in cell wall dissociated and could be easily released from plant tissues.

Aluminium, although not a "heavy" metal, is one of the major environmental causes for inhibition of the nitrate absorption in plants. Research on Al toxicity cleary showed that its effect depends on the several factors. Among them especially important are: the form of ion in the environment, its concentration and duration of plant exposition to the metal. Analysis of Archambault et al (1996) have predicted that the free activity of Al^{+3} was over tree times greater in AlCl$_3$ than in AlK(SO$_4$)$_2$ or Al$_2$(SO$_4$)$_3$. Low external concentration of aluminium applied as sulfate (up to 100 μmoles x dm^{-3}) in short-term experiments enhanced the nitrate uptake in barley (Nichol et al. 1993). Similar observations were made by Jerzykiewicz (2001) for cucumber seedlings exposed shortly (up to 3 hours) to the low AlCl$_3$. Lindberg et al. (1991) and Degehardt et al. (1998) have found increased net H$^+$ extrusion caused by Al^{3+} binding at the plasma membranes. Since nitrates are actively transported to plant cells (McClure et al. 1990), the elevation of the membrane electrochemical gradient could lead to the enhanced NO_3^+ absorption. Prolongation of Al-treatment over 3 hours, as well as higher external aluminium concentrations decreased dramatically the nitrate uptake and at least 5 mM $AlCl_3$ in nutrient solution initiated the efflux of NO_3^- from cucumber roots Jerzykiewicz (2001). The strong inhibition of the nitrate uptake was also shown in long-term studies with Al–treated white clover seedlings (Jarvis and Hatch, 1986).

2.1.2 Long-distance nitrate transport.
In most plant species only a portion of the absorbed NO_3^+ is assimilated in the roots. The remainder (seldom over 80%) is transported upwards through the xylem into the shoots. The data of experiments obtained by Hernandez et al. (1997) and Gouia at al (2000) evidenced that cadmium effectively restricted the nitrate transport from the roots to the shoots. Allocation of nitrate from roots to shoots was also inhibited in Pb-treated cucumber seedlings (Burzyński, personal communication). In plants, the NO_3^+ long distance transport via the xylem is mainly controlled by transpiration intensity (Barthes et al. 1996). Therefore, diminishing of the transpiration effectiveness after cadmium (Poschenrieder and Barceloy 1989, Barcelo and Poschenrieder 1990, Greger and Johansson 1992, Hernandez et al. 1997) or lead (Burzyński 1987) application to plants might partially explain the perturbations in nitrate long distance transport. According to De Filippis (1979) Pb^{2+} and, in much lesser degree, Cd^{2+} permanently bind to phosphates. Thus, both ions may form insoluble complexes in the plasma membrane, which together with a strong binding of ions to negatively charged carboxylic groups of the cell wall effectively block both, the apoplastic and symplastic way of long-distance nitrate transport.

In contrast to cadmium and lead, the copper ions seemed to have no effect on the transport of NO_3^- from root to shoot. Even at a relatively high Cu level in environment, practically all nitrogen taken up by plants appeared to be allocated into the shoots (Weber et al. 1991). To date, no experiments explaining the action of other heavy metals on the long-distance transport of nitrate were carried out.

2.2 Ammonium uptake
Similarly to the nitrate, ammonium concentration in the soil varies from micromolar to hundreds of milimolar depending on the microbial activity and environmental factors (Marschner 1996). For this reason plants have to evolve a repertoire of transporters that enable them to import efficiently ammonium over a wide range of environmental concentrations. Studies of ammonium uptake kinetics in roots indicated that the plant NH_4^+ transporters belong to at least two distinct systems (Wang et al. 1993). One of them, the passive, non-saturable, low-affinity transport system (LATS) operates at a high external concentration of ammonium (1 mM to 40 mM), whereas the second one, an active and saturable system with a high affinity to the substrate (HATS) is responsible for the transport of ions into the cells at a low NH_4^+ level in soil (below 1 mM) (Wang et al. 1993). Passive transport of ammonium into cells comprise both the NH_3 diffusion and the NH_4^+ channel transport. Molecular analyses have shown that the group of transport proteins encoded by *AMT1* genes (Ninneman et al. 1994), whose

expression is down regulated by extra-cellular level of N and intracellular glutamine concentration (Rawat et al. 1999) is responsible for the HATS activity in plants. Uncouplers of the plasma membrane proton motive force inhibited the activity of HATS suggesting its dependence on metabolism. On the other hand, changes in external pH had only little effect on HATS. Taken together, the results of the uncoupler and pH experiments indicate that active NH_4^+ transport facilitated by HATS is not directly coupled with H^+ transport (for review see Howitt and Udvardi 2000). Ammonium taken up by roots is generally not used for long-distance transport within the plants and most of it is assimilated locally in the cytoplasma and plastids of root cells.

Disturbances in ammonium uptake caused by heavy metals were studied less extensively than metal dependent alterations in the nitrate uptake (Table 2).

Weber et al (1991) have shown strong decrease of ammonium uptake in *Silene vulgaris* at the presence of copper. The toxic effect of Cu on the NH_4^+ absorption was obviously smaller then the Cu alteration in the uptake of nitrate, implying much less sensitivity of the NH_4^+ uptake system to copper. Discrepancy in the Cu action on the uptake of NH_4^+ and NO_3^- might be due to the differences in absorption mechanisms of both ions. As was mentioned above, the nitrate anions are transported through plasmalemma only in the energy-depending manner, whereas ammonium can pass the membrane both, actively and passively. Since experiments carried out with *Silene vulgaris* were made at a relatively high NH_4^+ concentration (4 mM), at least part of the ammonium-nitrogen could be taken up through passive diffusion of uncharged NH_3 and NH_4^+ cations. Thus, the Cu-dependent disturbances in the uptake of inorganic nitrogen forms can mainly reflect the toxic effect of copper on the ATPase system operating in the plasmalemma. Burzyński and Buczek (1998) have found similar repressive action of copper on the ammonium uptake in cucumber seedlings. Interestingly, Cu action was much stronger at acid pH, suggesting a better mobility and transport of copper in acid solutions. Also other heavy metals: Cd^{2+}, Pb^{2+} and Fe^{2+} significantly diminished NH_4^+ absorption by cucumbers (Burzyński and Buczek 1998). In contrast to copper, lead and cadmium-dependent alterations in the ammonium uptake were similar at acid and neutral pH. Cu, Cd, Ni, Zn and Mn strongly inhibited the ammonium uptake in young *Brassica napus* plants. Among the metal ions, Cu has been shown to affect the ammonium uptake to the greatest extent (Kubik-Dobosz et al. 2001).

Feeding of barley with aluminium has also been found to alter the ammonium in a different manner than nitrate absorption (Nichol et al. 1993). Concentrations of Al^{3+} stimulating the NO_3^- influx were repressive for the transport of NH_4^+. Moreover, the authors noted that Al^{3+} inhibited the influx of many different cations and stimulated the influx of other anions. According to Nichols and co-workers, Al^{3+} attaching to the negatively charged phospholipids and/or acidic amino acids residues of protein in the plasma membranes formed a positive charged layer that influenced ion movement to the binding sites of transporters. Generally, this positive charged layer might retard the movement of cations and increase the movement of anions. Such an effect, in turn, will inhibit the cation influx while stimulating the transport of anions into the cytoplasma (Nichols et al. 1993). On the other hand, Al^{3+} has been shown to block directly some cation channels (Schroeder 1988, Hung et al. 1992, Rengel and Elliot 1992b). Thus, it

Table 2. Inhibition of ammonium uptake under heavy metal stress

Taxon	Metal concentration (μmoles dm^{-3})	Literature
Silene vulgaris	Cu (4, 6, 18 μM)	Weber et al. 1991
Cucumis sativus	Cu, Cd, Pb (100 μM)	Burzyński and Buczek 1987, 1998
Brassica napa	Cd, Cu, Mn, Ni, Zn (100 μM)	Kubik-Dobosz et al. 2001

cannot be excluded that the aluminium inhibition of NH_4^+ absorption resulted from the direct action on NH_4^+ transport protein.

2.3 Mechanism of metal action on the uptake of mineral nitrogen forms

Heavy metal stress can induce a series of events, which lead to the decrease in inorganic nitrogen uptake but the mechanism of metal action on the nitrate and ammonium absorption by plants is still unclear. Apparently, the action of metals on ion uptake may be both direct and indirect.

Physiological studies clearly show that every metal repressed the nitrate absorption in uninduced as well as in NO_3^--induced plants, suggesting metal-failure of both, the constitutive and the inductive components of the nitrate transport systems. Burzyński and Buczek (1994a) working at low external nitrate (0.7 mM KNO_3) evidenced Cd, Pb, Cu and Ni dependent inhibition of the transporter(s) with high-affinity to the nitrate in cucumber seedlings. Metals also diminished the activity of high-affinity ammonium transporter(s) (Burzyński and Buczek 1998). Although, most of the experiments on heavy metals action were done with high external concentration of NO_3^- or NH_4^+ (over 1 mM), when both of the transport systems, with high affinity and with low affinity, were operated (Burzyński and Grabowski 1984, Burzyński 1988, Weber 1991, Boussama et al. 1999b, Hernandez et al. 1997, Gouia et al. 2000). Thus, the heavy metal repression of low-affinity nitrate and ammonium transport could not be excluded. A possible explanations of the lowered nitrogen absorption could be a direct interaction between heavy metals and the SH groups of the ammonium and nitrate transport proteins of both LATS and HATS. Every metal ion is easily bound to the sulfhydryl residues of amino acids (De Filippis 1979). The contribution of the thiol groups in the reactive center of the NO_3^- transporters operating in the plasma membranes was well documented for cucumber root cells (Kłobus et al. 1998).

Besides metal-SH interaction, the main reason of the heavy metal alteration in the inductive nitrate transporters and HATS-ammonium transporter could be due to the diminishing expression of the *NRT* and *AMT1* genes. Molecular investigations in *Nicotina* and *Arabidopsis* species pointed out two reasons of the lowered *NRT* gene expression in plant tissues: firstly, decrease of intracellular NO_3^- level and secondly, enhanced concentration of NH_4^+ and/or amino acids (Quesada et al, 1997, Krapp et al.

1998). Elevation of the amino acids content in the cell also repressed the expression of *AMT1* genes (Rawat et al. 1999). In cadmium-treated plants perturbations in the nitrate (Burzyński 1988, Hernandez et al. 1996) as well as ammonium and amino acids tissue content (Hernandez et al. 1997, Gouia et al. 2000) were found. Additional effects of heavy metals on the gene expression are also possible. For example, Cu^{2+}, Fe^{2+} and Ni^{2+} association with phosphate residues of nucleic acids caused DNA damage (Lloyd and Phillips 1999). Conformational changes of TFIIIA caused by the excess of Cu^{2+}, Mn^{2+} and Cd^{2+} induce the oxidative disorganization of the initial side of transcription (Stohs and Bagchi 1995). Copper additionally impairs the transcription process disrupting the nucleotide binding of DNA (Lloyd and Phillips 1999). Furthermore, elevated Cu^{2+}, Cd^{2+} and Ni^{2+} levels lead to an enhanced free radical production in cells, which oxidize a broad group of organic molecules, i.e. nucleic acids (Stohs and Bagchi, 1995). Recently, Kubik-Dobosz et al. (2001) have shown that the *AtAMT1;1* transcript level in *Brassica napus* decreased rapidly during Zn^{2+}, Cu^{2+} and Mn^{2+} treatment of plants. Diminished gene expression was parallel with the decrease in the NH_4^+ uptake. Unfortunately, to date, no molecular data are available on the effect of a heavy metals on the expression of particular *NRT* gene.

In addition to the direct effect of heavy metals on ions uptake, same of them, like Cu, Cd, Pb, Hg, Ni and Zn are also known to exert their effects indirectly by interaction with membrane components. Excess of heavy metals alters the membrane lipids, changing their total amount, qualitative composition and saturation (for review see Devi and Prasad, 1999). Metal induced failure in membrane lipids was often associated with its peroxidation (de Vos et al. 1991, Demidchik et al 1997, Hernandez et al. 1997). The consequence of the membrane lipid alteration is a general loss of its functionality and permeability (De Filipps 1979, Clarkson 1993, Metharg 1993). It has been well documented that heavy metals, especially Cu, Cd, Zn, Hg and Al induced the potassium leakage, which is a good parameter of permeability damage (De Filipps 1978, Kaltjens and Ulden 1987, Demidchik et al 1997, de Vos et al. 1991, Lindberg et al. 1993). Impairment of the nitrate and ammonium uptake in metal treated plants can thus be a consequence of alteration in membrane permeability caused by heavy metals. Additionally, another explanation for the inhibitory metal effect on the nitrogen uptake could be the repression by metals of the activity of the plasma membrane proton pump. As mentioned above, the transmembrane electrochemical gradient generated by H^+-ATPases is a motive force for transport of NO_3^- and NH_4^+ through the plasma membrane. According to Serrano (1990) Cu^{2+} ions are one of the most powerful inhibitors of the cell plasma membrane ATPase. Cadmium, zinc, mercury, lead, nickel and aluminium were shown to repress the enzyme activity in an almost equal degree (Kennedy and Gonsalves 1987, Fodore et al. 1995). Enzyme inactivation was probably due to the metal – SH binding in the active center of ATPase (Kennedy and Gonsalves 1987, Vara and Serrano 1982). Moreover, the activity of H^+-ATPase is also influenced by the lipid composition of membranes, especially the by sterols/phospholipids relationships (Kasamo and Nouchi 1987). Taken together, the altered plasmalemma H^+-ATPase function would partially explain the metal-induced perturbation of the uptake of both inorganic nitrate forms.

3. ASSIMILATION OF MINERAL NITROGEN FORMS

Nitrogen assimilation is one of the very important processes in plant metabolism, controlling the plant growth and development. Good conditions of enzyme activities involved in the conversion of inorganic nitrogen into organic form are crucial to the plant productivity, biomass and crop yield.

Table 3. Reaction catalysed by enzymes of primary nitrogen assimilation

NR NO_3^- + NADH or NADPH	→	NO_2^- + NAD or NADP
NiR NO_2^- + Fd_r	→	NH_4^+ + Fd_{ox}
GS NH_4^+ + ATP + glutamate	→	glutamine + ADP + P_i
GOGAT glutamine + 2-oxoglutarate + Fd_r	→	2 glutamate + Fd_{ox}
glutamine + 2-oxoglutarate + NAD(P)H	→	2 glutamate + NAD(P)
GDH 2-oxogltarate + NH_4^+ + NAD(p)H	↔	glutamate + NAD(P)
AspAT oxaloacetate + glutamate	↔	aspartate + 2-oxoglutarate
AlaAT enolopyruvate + glutamate	↔	alanine + 2-oxoglutarate
AS aspartate + glutamine + NH_4^+ +ATP	↔	asparagine+ glutamate +ADP +P_i

In the well aerated soils the nitrification bacteria are active and NO_3^- is the main nitrogen source for plants. Nitrate taken up by plants is reduced to ammonium by the sequential action of the enzymes nitrate reductase (NR) and nitrite reductase (NiR). Ammonium is the only reduced nitrogen form available to plants for assimilation into nitrogen carrying amino acids: glutamine, glutamate, asparagine and aspartate (Ireland and Rea 1999). The enzymes: glutamine synthetase (GS), glutamate synthase (GOGAT), glutamate dehydrogenase (GDH), aspartate amino transferase (AspAT), alanine amino transferase (AlaAT) and asparagine synthetase (AS) are responsible for the biosynthesis of these nitrogen-carrying amino acids.

Production of nitrite by NR is localised mainly in cytosol, whereas nitrite is afterwards reduced by NiR in the plastids. Incorporation of ammonium into carbon skeletons seems to proceed mainly via the glutamine synthetase-glutamate synthase cycle. Predominant forms of GS and GOGAT in the roots are cytosolic GS1 and NADH-GOGAT, and these isoenzymes have been recognized as being to be involved in primary nitrogen assimilation in this organ. In leaves, chloroplastic GS2 and ferredoxin-GOGAT (Fd-GOGAT) are predominant. These enzymes have been regarded as to functioning in the assimilation of nitrogen into glutamine and glutamate in the roots. GDH-NADH is less likely to be involved in primary nitrogen assimilation because of its high K_m for ammonia, but a nonredundant role for GDH is the assimilation of ammonia under conditions of NH_4^+ excess in the cells. Glutamate dehydrogenase is localised in mitochondria and, in smaller amounts, in chloroplasts.

Different sensitivity of enzymes to heavy metals, but also different localisation of enzymes in the cells and organs are the reasons for various influences of heavy metals on their activities. The action of metals on so differently localised enzymes depends also on the mobility of metals, their external concentration and the time of plant exposition to metals.

The influence of metals on nitrogen incorporation into amino acids may proceed in different ways:
- heavy metals can affect the activity of many enzymes, also the nitrogen assimilation one, by binding to important sulphydryl groups (Van Asche 1990). The order of capacity of heavy metals to SH groups is as follows: Hg>Ag>Cu>Pb>Cd>Ni>Co>Zn>Mn (De Filippis 1979). Changes in enzyme activities observed in the *in vitro* conditions (heavy metals added to the reaction mixture) are not always similar to the changes observed in the *in vivo* conditions (metals added to the growth nutrient solutions). The effect of heavy metals *in vivo* is very often stronger, and is visible when metal is used in a lower dose than in the *in vitro* experiments. The *in vivo* effect of metals is usually dependent on the duration of plants' exposition to metals.

The *in vivo* metals can interact with factors essential for gene expression and for enzyme activities such as transcriptional factors, proteins of kinases or phosphatases, and also may change the level of other metals, which are essential to enzyme activities.

The decrease of nitrogen assimilation process in heavy metals treated plants can also reflect the disturbing of the general homeostasis of plants metabolism. Heavy metals induce the changes in plant water relations (Poschenrieder and Barcelo 1999), and a change in the uptake and distribution of other than N essential nutrient elements (Siedlecka.1995, Burzyński 1987), as well as influence photosynthesis (Prasad and Strzałka 1999). Heavy metals like other stresses generate the toxic oxygen species, which can cause the breakdown of enzyme proteins directly by oxidative reaction or indirectly by increasing the proteolytic activity (Dietz et al. 1999).

3.1 Nitrate reductase activity

Reduction of nitrate to nitrite by the nitrate reductase is the first step in the assimilation of nitrate and is often considered to be the most regulated and rate-limiting step in the process. The amount of NR protein and NR activity is under control of transcriptional and posttranslational mechanisms. Nitrate, light, Gln/Glu balance, sucrose level resulting from photosynthesis rate and cytokinins are the main factors limiting NR genes expression. Posttranlational regulation of NR comprises the phosphorylation/ dephosphorylation modifications of the enzyme protein. Phosphorylation diminishes the activity of NR while dephosphorylation increases it. Thus, the activity of NR depends on protein kinases and protein phospatases. 14–3–3 binding protein and the level of divalent cations such as Ca^{2+} and Mg^{2+} are very important in posttranlational changes of the enzyme activity (for review see Campbell 1999 and Kaiser et al. 1999).

The mechanism of regulation in such a way decides about the wide sensitivity of the nitrate reductase to different environmental stresses. One of the factors mostly influencing NR activity is the level of heavy metals. Changes in activity after exposition of different plant species to heavy metals were observed many times (Table 4).

Table 4. Inhibition of nitrate reductase under heavy metals stress

Taxa	Metal	Reference
Albizia lebbek	Cr,Hg,Ni	Tripathi and Tripathi 1999
Chlorella	Cu, Ni	Rai et al. 1996
Cucumis sativus	Cd, Cu	Burzyński in press
Cucumis sativus	Cd, Pb	Burzyński 1988, 1990
Cucumis sativus	W, V	Buczek et al. 1980
Cucumis sativus	Al	Jerzykiewicz et al. in press
Glaucocystis nostochinearum	Cr	Rai et al. 1992
Hordeum vulgare	Cd	Boussama et al. 1999
Hydrilla verticilata	Hg	Gupta and Chandra 1996
Lycopersicum esculentum	V	Buczek 1973
Lycopersicum esculentum	Cd	Ouariti et al 1997
Nelumbo nucifera	Cr	Vajpayee et al. 1999
Nymphaea alba	Cr	Vajpayee et al. 2000
Phaseolus vulgaris	Cd	Gouia et al. 2000
Phasseolus vulgaris	Cd	Ouariti et al. 1997
Pisum sativum	Cd	Hernandez et al. 1997
Pisum sativum	Cd	Chugh et al. 1992
Sesamum indicum	Cd, Cu, Pb	Singh et al. 1994
Sesamum indicum	Cu+Cd, Pb+Cd	Bharti et al. 1996
Sorgum	Al.	Keltjens an Ulden 1987
Spirodela polyrrhiza	Cr	Tripathi and Smith 1996
Triticum aestivum	Cu	Luna et al. 1997
Valisneria spiralis	Hg	Gupta and Chandra 1998
Zea mays	Cd	Boussama et al 1999a
Zea mays	Hg	Pandey and Srivastava 1993

Nitrate reductase is a flavoprotein containing a hem-Fe and molybdenum complexed with unique pterin or molybdopterin. Campbell (1999) has proposed two important cysteine residues in the holoenzyme of nitrate reductase. The first is Cys 191, ligand of Mo, involved in binding to molibdenopterin, and the second is Cys 889. The -SH group of cysteine 889 directed the positioning of the NADH for optimum electron transfer to the FAD. It seems that these two cysteine residues are the most sensitive targets for heavy metals.

In the experiments of Luna et al. (1997) copper decreased nitrate reductase in segments of *Triticum aesetivum* leaves floated in a Cu solution. Inhibition of NR was visible before affecting of the active oxygen generation. Copper inhibition was reversed by subsequent incubation with EDTA indicating that the metal bonded to key -SH groups of the enzyme. Luna after Smarrelli and Campbell (1983) suggested that the Cys residue located in the Cyt b reductase fragment could be modified by copper. This conserved Cys residue, probably Cys 889 estimated by Cambell (1999) seemed to assist the electron transfer from NADH to FAD. The affinity of Cu to SH groups was stronger than Cd or Pb and Cu more drastically than Cd decreased the *in vivo* NR acivity in cucumber roots. Furthermore, Cu but not Cd, significantly decreased the *in vitro* NR activity (Burzyński 2001). The 30 minutes treatment of cucumber seedlings with 20 µM Cd decreased the activity of nitrate reductase due to the change of it phosphorylation level, but higher concentration (50 µM) diminished the enzyme activity in a different way (Reda – personal communication). Between Cd, Zn, Ni, Co, Cu and Mn copper was the strongest *in vitro* inhibitor of NR extracted from the leaves of *Silene cucubalus*. (Mathys 1975). For this reason, it was proposed that the NR activity might be used as an early indicator of excess accumulation of Cu^{2+} and plant tissue injury (Luna et al 1999 and Burzyński 2001). Nitrate reductase can display partial activities using artificial electron donors (such as reduced methyl viologen, rMV) to reduce nitrate to nitrite, or artificial electron acceptors (such as cytochrome c) with NADH as the electron donor. The first reaction requires an active Mo-pterin domain, the second one FAD and hem domains. Gouia et al. (2000) investigating the influence of Cd on partial activities of NR have shown that the rMV:NR reaction was more sensitive to Cd inhibition than the NADH:cytochrome C reductase activity. These results suggested that Cys191, which is important in bounding the NR holoenzym to Mo, is probably the most sensitive target for Cd.

A different correlation was obtained by Singh et al. (1994) and Bharti et al. (1996); *in vivo* nitrate reductase activity in the roots of *Seasamum indicum* seedlings was inhibited with Pb, Cu and Cd but the *in vitro* enzyme activity for this plant was increased due to the addition of metals. The authors suggested that the effect of the metals on NR was rather indirect, possibly through the inhibition of NADH and NO_3 supply. In cucumber seedlings indirect effects of Cd and Pb on NR activity were also observed (Burzyński 1988, Burzyński and Grabowski 1984). Accumulation of Cd as well as Pb in leaves after 24 or 48 hours of the plants' exposition to metals was negligible but inhibition of the enzyme activity was significant. Every lead and cadmium concentration, which inhibited NR activity, also inhibited NO_3^- uptake. Similarly, Boussama et al. (1999a) in experiments on *Zea mays* and Boussama et al. (1999b) on barley, as well as Ouariti et al. (1997) on bean observed that the NO_3^- content in plant tissue was sometimes much more affected by Cd treatments of plants, than the level of NR activity. Hernandez et al. (1997) investigating pea indicated that Cd almost completely inhibited the net NO_3^- uptake and dramatically decreased the shoot NR activity. Inhibition of nitrate uptake and NR activity was recovered after the transfer of plants to the Cd-free nutrient solution. NO_3^- anions are not only the substrate for nitrate reductase but also a factor inducing the synthesis of nitrate reductase protein.

The *in vivo* effect of Zn and *in vitro* effect of Cu, Zn, Cd, Ni, Co and Mn on nitrate reductase in *Silene cucubalus* was investigated by Mathys (1975). In zinc resistant population Zn in concentration up to 1 mM increased the enzyme activity. The activity of NR in leaves of copper resistant and Zn non-resistant cultivars was very sensitive. In the *in vitro* assays each metal added to the incubation mixture at the concentration below 1 mM distinctly decreased the activity of NR. In the Mathys experiments (1975) the metals have been arranged in following way with decreasing toxicity: Cu, Cd, Zn, Ni, Co, and Mn.

Hg bonded very easily with the sulphydryl groups and in concentration as low as 1 µM significantly decreased the *in vivo* NR activity in leaves of *Vallisneria spiralis*

(Gupta and Chandra 1998).The authors suggest that the inhibition of NR activity by Hg could be due to the mercury interaction with NR proteins and also to the reduced substrate availability. Gupta and Chandra (1996) observed the negative influence of Hg on nitrate level and nitrate reductase activity in *Hydrilla verticillata* and Pandey and Srivastava (1993) determined NR activity inhibition in maize leaf segments treated with Hg. Rai et al. (1996) found that Ni as well as Cu are the effective inhibitors of the nitrate reductase in chlorella cells.

Klejtens and Ulden (1987) have showed that Al-induced inhibition of NR activity was accompanied by a decrease of NO_3^- tissue concentrations and that enzyme repression was visible only in the shoots of sorgum. Klejtens and Ulden (1987) concluded that a direct repression of nitrate reductase activity by Al seems unlikely because NR activity in roots (cellular concentration of Al in roots should be higher than in shoots) was not affected in Al-treated plants. Inhibitory effect of aluminium on the soluble as well as the plasma membrane bound nitrate reductase activity in cucumber roots was also found by Jerzykiewicz (2001). However, the enzyme activity increased when a lower concentration of Al (0.5mM) was used. Positive influence of Al on NR activity in the *Triticum aestivum* root was found by Basu et al. (1994). Aluminium affected the cytoplasmic phosphate metabolism, competing with magnesium and this could be the reason of higher NR activity in Al treated plants, because Mg and NR phosphorylations are the natural inhibitors of nitrate reductase activity. In the *in vitro* experiment Al diminished NR activity (Jerzykiewicz 2001).

Vanadium in the in *vitro* and the *in vivo* experiments decreased nitrate reductase activity in tomato leaves and in *Cucumis sativus* seedlings (Buczek 1973, Buczek et al. 1980) An addition of sephadex or EDTA to the enzyme crude extracts completely restored nitrate reductase activity suggesting very labile bond between vanadium ions and enzyme molecule.

In roots and cotyledons of cucumber seedlings also WO_4^{-2} appeared to be an effective inhibitor of the nitrate absorption and nitrate reductase activity (Buczek et al. 1980, Kłobus et al. 1991). Data obtained in experiments clearly showed that tungsten primarily inhibited the synthesis and activity of NR and to a lesser extent affects the NO_3^- absorption.

Cr (VI), significantly repressed the NR activity and protein content of *Nymphae alba* L. (Vajpayee et al. 2000). Positive correlation between the decrease of chlorophyll content and NR activity was observed, so reduction in NR activity in chromium treated *Nymphae alba* was probably due to the disturbances in chlorophyll biosynthesis leading to lower levels of photosynthesis.

Mn, similarly as Ca^{2+} and Mg^{2+} inhibited NR activity through the binding to 14–3-3 protein. Ca, Mg or Mn determined association of 14–3-3 with NR protein and enzyme inhibition (Athwal et al. 1998). The regulation of NR activity by other divalent cations through their interactions with Ca^{2+} and Mg^{2+} cannot be excluded.

In the results presented above it was considered that binding of metals with important cysteine residues in the NR holoenzyme is the main reason of direct NR inhibition. However, restricted availability of nitrate in the plants treated with metals is the mechanism of indirect metals' influence on nitrate reductase activity. Long-term plant exposition to heavy metals represses NR activity by disturbance in plant water relations, disturbance in the production of sugars in photosynthesis, and by metal induced toxic oxygen generation.

3.2. Nitrite reductase activity.

Nitrite reductase (NiR) is usually considered to be a chloroplastic or, in non-photosynthetic tissue, a plastid enzyme. Reduced ferredoxin is the electron donor in green tissues, while in non-photosynthetic tissues, a ferredoxin-like protein is thought to be the reductant, which obtains its reducing power from NADPH. NiR contains one siroheme and one Fe_4S_4 cluster as prosthetic group. NR activity somehow determines the NiR gen expression. Metabolite from nitrate reduction and downstream from nitrite reduction repress the expression of genes encoding both, nitrate and nitrite reductase (Stitt 1999).

In general, nitrite reductase is more resistant than NR. It is mainly a constitutive enzyme, protein is more stable, and the enzyme is localised in plastids so that the metals' approach to it is not as easy as to the cytoplasmic nitrate reductase. The nitrite reductase is not an enzyme which limits nitrogen assimilation in plants, and probably this causes that works concerning the influence of metals on NiR activity are infrequent.

Inclusion of 100 µM cadmium into the incubation mixture decreased the NiR activity extracted from bean leaves (Gouia et al. 2000). At a similar concentration Cd did not change the enzyme activity of pea roots and leaves and of cucumber seedling roots (Chugh et al. 1992 and Burzyński 1990 respectively). The *in vivo* inhibiting effects of metals on NiR activity are more visible. After twelve days the activity of NiR was negatively affected by Cd treatment in *Zea mays* and barley leaves (Boussama et al. 1999a, Boussama et al. 1999b). The effect was more pronounced in the roots than in the shoots. Similarly, in cucumber seedlings 10 µM Cd and 50 µM Pb after 24 hours diminished the NiR roots activity to 50%, but activity in cotyledons did not change in comparison to the control.

As it was mentioned earlier, metals influence the NO_3^- uptake and nitrate homeostasis in plant cells. As to NR gene expression, nitrate also leads to the induction of NiR genes (Stitt 1999). The regulation of NiR genes' expression by nitrate could be the main mechanism of metals' inhibition of nitrite reductase *in vivo*.

3.3. Glutamine synthetase and glutamate synthase activities

Ammonium is the reduced nitrogen form available to plants for assimilation into amino acids and protein. In higher plants ammonium is mainly generated by:
a. reduction of nitrate by cytosolic nitrate reductase and localised in plastids nitrite reductase,
b. decarboxylation with deamination of glycine to serine during photorespiration,
c. amino acid deamination and nucleic acids catabolism.

Ammonium is incorporated into non-toxic glutamine and glutamate. This reaction is catalysed by two enzymes: glutamine synthetase (GS) and glutamate synthase (GOGAT). Both enzymes occur in multiple isoenzymic forms encoded by distinct genes (for review see Lam et al. 1996) and their participation in NH_4^+ assimilation is various. GS2 and ferredoxin-GOGAT (Fd-GOGAT) assimilate ammonium generated by NiR in the chloroplasts of leaves; Fd-GOGAT and those isoenzymes are the predominant forms in leaves. Since the predominant forms of GS and GOGAT in roots are cytosolic GS1 and NADH-GOGAT, these isoenzymes have been suggested as being involved in primary nitrogen assimilation in non-photosynthetic tissues. GS2 and Fd-GOGAT of chloroplasts also participate in reasimilation of photorespiratory ammonia. GS2 mutants lacking the ability to reasimilate ammonia may become lost during photorespiratory conditions (see Lam et al. 1996). The increased activities of cytosolic GS1 and NADH-GOGAT during germination and leaves' senescence have suggested the involvement of these particular isoenzymes in the assimilation of recycled nitrogen. GS1 isoenzymes play a role in the mobilisation of nitrogen for translocation and/or storage (Oaks 1993). The metal influence on glutamine synthetase and glutamate synthase is often presented as a mean of metal effect on all isoenzymes of GS or GOGAT localised in the roots or shoots. The excess of metals in nutrient solution may inhibit GS2 isoenzymes, but as a reason of senescence may stimulate GS1 and NADH-GOGAT isoenzymes. The decrease of one form of enzyme may be masked by the increase of an other one. So changes observed in the enzymes' activities after heavy metal treatment are different and sometimes hard to interpret.

As we mentioned earlier metals disturb NO_3^- absorption by plants. Intracellular NO_3^- level affects the expression of the genes encoding not only NR and NiR but also GS and GOGAT (Oaks 1993, Stitt 1999). Probably this dependence of GS and GOGAT genes on nitrate, besides the direct metal action on SH groups of enzyme proteins, is the reason of a general inhibitory effect of heavy metals on the activity of these enzymes.

Chien and Kao (2000) have showed that $CdCl_2$ induced ammonium accumulation in detached rice leaves. The authors considered that ammonium accumulation was not associated with senescence of detached leaves but rather with the decrease in glutamine synthetase activity. Similarly, Cd decreased the GS and GOGAT activities and simultaneously raised the level of NH_4^+ in *Zea mays* (Boussama et al. 1999a). The total protein content in both, shoots and roots was diminished in Cd-stressed plants. NADH-dependent isoform of GOGAT was less affected by Cd stress than the Fd-dependent one. The effect of cadmium on GS and GOGAT activities in pea seedlings was investigated by Chugh et al. (1992). The activity of glutamine synthetase in leaves was practically unchanged, whereas in the roots it declined markedly. Ferredoxin-dependent glutamate synthase activity in leaves dropped after 6 days of plant treatment by Cd but after 12 days it was comparable to the control. GS and both forms of GOGAT were repressed by Cd in leaves of been seedlings (Gouia et al. 2000) and in barley (Boussama et al. 1999b). 24 or 48 hours of cucumber seedlings treatment with Cd or Pb caused the decrease of GS activity in roots but metals did not change the GS activity in cotyledons (Burzyński et al. 1990). In the same experiment the GOGAT-NADH activity in cotyledons of seedlings treated with Pb or with Cd was higher than in the control plants. One hour of plant treatment with 100 µM Cu, Cd, Fe and 500µM Pb decreased the GS activity in the roots of cucumber seedlings to a great extent (Burzyński and Buczek

1998). In the *in vitro* experiments metals decreased the enzyme activity only when used in high concentrations: Cu at 1000 µM (Burzyński and Buczek 1997), Cd at 500 µM, Fe at 1000, only Pb was active at 10 µM concentration (Burzyński, unpublished data). The small effect of metals (except Pb) on GS activity suggested indirect action on the enzyme activity. Similarly, high concentrations (500- 1000µM) of Zn, Cu, Co, Ni, Cd, in a different degree, repressed both GS forms extracted from *Triticale* seedlings (Bielawski 1994). Only 10 µM Hg ions presented in the assay mixture distinctly diminished the GS activity. Orzechowski and Bielawski (1997) tested *in vivo* the toxicity of Cd, Zn and Pb on ammonium assimilation in *Triticale*. They found that the sensitivity of GS to metals was different in the roots and shoots. Cd and Zn decreased the specific GS activity in the shoots but in roots activity was even higher than in the control. Pb stimulated GS in the roots as well as in shoots. The authors suggested that the regulation mechanism of GS is different in the roots and in the leaves. Orzechowski and Bielawski (1994) implying that the inhibition of GS activity in leaves was the reason of lower Mg^{2+} concentration in this organ, observed after metals' treatment.

The GS-GOGAT pathway transfers ammonium to the carbon skeleton in the form of 2-oxoglutarate. To date, the exact enzyme origin of 2-oxoglutarate for plant ammonium assimilation remains unknown. Galves et al. (1999) proposed NADP-dependent isocitrate dehydrogenase (ICDH) and aspartate aminotransferase (AspAT) as enzymes responsible for 2-oxoglutarate production for ammonium assimilation. ICDH belongs to a multi-isoenzymes family whose members are located within cytosol and plastids, which GS and GOGAT forms operate. AspAT produces aspartate and 2-oxoglutarate. Heavy metals interrupting the photosynthesis and cells respiration can affect primary nitrogen assimilation through limitation of the carbon skeleton. Mathys (1975) observed the repression of ICDH activity after Zn, Ni and Cu, but not Cd, addition to the incubation medium. AspAT has shown similar cadmium resistance in both, *in vitro* and *in vivo* short–term (6 days) experiments (Chugh et al. 1992, and Burzyński, data unpublished). However, in long-term Cd treatment of plants (12 days) markedly decreased AspAT activities were observed by Chugh et al. (1992).

3.4. Glutamate dehydrogenase (GDH) activity

GDH catalyses the reductive amination of 2-oxoglutarate in the presence of NADH or NADPH, or oxidative deamination of glutamate in the presence NAD or NADP. There are two major types of GDH, an NADH-requiring form localised in the mitochondria, and an NADPH form found in the chloroplast. The role of GDH in nitrogen metabolism of plants is not well established. Higher affinity of GS to NH_4^+ (higher K_m for GS compare to GDH) pointed out the dominance of the GS-GOGAT system in glutamate synthesis. GDH is involved rather in its oxidation than in reduction. (Miflin and Lea 1976). GS and GDH are spatially separated within the cell. GS is restricted to the cytoplasm and plastids whereas GDH is localised in mitochondria. Because of photorespiration, NH_4^+ level within mitochondria is much higher than in cytosol, high enough to support a GDH action in the direction of glutamate synthesis in the primary nitrogen assimilation (Oaks 1993). The GDH enzyme activity can be induced in plants exposed to a high external level of ammonia. For this reason GDH has been proposed to be specifically important for ammonia-detoxification purposes.

The catabolic role of GDH has been observed during germination and senescence, the two periods when visible amino acid catabolism occurs. Oxidative deamination activity of GDH increase in response to carbon limitation suggesting the regulation of the GDH gen expression by light and sucrose (Lam et al. 1996).

The amination reaction of GDH is activated by Ca^{2+}, Zn^{2+} and Mn^{2+} ions and amination GDH activity is always on a high level when ammonium exceeds the normal cell concentrations.

Despite of the lower mineral nitrogen uptake in plants treated with heavy metals a higher concentration of ammonium in plant tissues has been usually observed. Accumulation of ammonium after plant exposition to metals could be attributed to a decrease in the GS-GOGAT cycle activity but also might be associated with tissues senescence and higher proteolysis, especially in long-term experiments. Intracellular ammonium content markedly increased upon Cd exposure, and GDH activity was four to six-fold higher than in the control beans plant (Gouia et al. 2000). Boussama et al (1999a) have observed a similar correlation between the NH_4^+ content and GDH activity in barley. The aminating form of GDH (NADH-GDH) increased but the oxidative deaminating form (NAD-GDH) decreased in Cd treated roots and shoots of Zea mays. (Boussama et al.1999a). Cd enhanced the level of glutamate dehydrogenase in pea (Chugh et al. (1992). Higher GDH-NADH activity in roots and shoots in triticale after Cd, Zn and Pb treatment was also observed by Orzechowski et al. (1997). In experiments mentioned above metals' action on plants was long: several days to a month.

In short-term experiments (one hour plant exposition to Cu, Cd, Pb, Fe), when an excess of NH_4^+ in tissues was not so visible, the inhibition of NADH – GDH activity in cucumber seedling roots was observed (Burzyński and Buczek 1997 and 1998). But after 24 and 48 hrs of cucumber seedling treatment with Cd or Pb, the activity of NADH-GDH in roots was significantly higher than in control roots, free from Cd and Pb.

A dual physiological role is suggested for the induction of aminating NADH-GDH in heavy metals' treated plants:
- a detoxification role for the recycling of the high ammonium content originating both, from a decrease in GS/GOGAT activities and from the degradation of the organic nitrogen compounds (Srivastava and Singh 1987),
- a role in the replenishment of the glutamate pool, needed for the synthesis of Cd-binding peptides as for instance phytochelatins. It was well established that heavy metals and especially Cd in plant cells induce the activity of γ-glutamylcysteine synthetase and the formation of class III metallothioneines named phytochelatins (for review see Prasad 1999). These phytochelatins (PC) are glutamate and cysteine-rich peptides. They bind metals and this is one of the defence mechanisms of plants against the toxic properties of heavy metals. Gouia et al. (2000), Orzechowski et al. (1997), Boussama et al. (1999) suggested that NADH- GDH is an enzyme responsible for the production of glutamate, necessary to PC synthesis, when activity of the GS-GOGAT system is repressed by metal. Ju et al. (1997) has examined the response to Cd of

enzymes involved either directly or indirectly in the synthesis of glutamate in *Zea mays* seedlings. The enzyme level was estimated on the basis of its polypeptide amount using SDS-PAGE and western blotting methods. Simultaneously with the increase of γ-glutamylcysteine the authors observed an increase of GS polypeptide, and a marked decrease of the NR protein in the roots. The amount of NiR polypeptides also diminished. Cd did not affect the Fd-GOGAT, NAD(P)H-GOGAT or GDH polypeptides. Although the activity of aminating, as well as deaminating forms of GDH decreased in the Cd treated plant also. So, after these results the role of NADH-GDH in the production of glutamate for PC synthesis is doubtful. According to Brunold (1993) the synthesis of γ-glutamylcysteine in metal-treated plant cells is rather limited by the cysteine synthesis and not by the activity of enzymes which are responsible for glutamate synthesis.

3.5 Proline synthesis

In plant cells after heavy metals' treatment, as a defence mechanism, the production of phytochelatins and peptides like metallotioneins II, rich in cysteine residues, is observed. Proline, is another nitrogen compound whose accumulation in heavy metal resistant plants is detected (Schat et al. 1999 and Sharma et al. 1999). Cu was the most effective inducer of proline accumulation in tissues, followed by Cd and Zn, respectively. The results demonstrated that metal-induced proline accumulation is indirect and depened on the development of metal-induced water deficit in leaves. In Sharma et al. (1999) experiments, proline appeared to protect *in vitro* glucose-6-phosphate dehydrogenase and nitrate reductase against Zn and Cd. The authors concluded that this protection was based on a reduction of the free metal ion activity in the assay buffer, due to the formation of metal-proline complex. A similar proline role in metal sequestration in the *in vivo* could not be excluded, but it seems that the main function of metal-induced proline accumulation may be associated with osmoregulation and enzyme protection against dehydration.

In the green tissue of bean As and heavy metals such as Hg, Cu, Zn, similarly as other abiotic stresses induced the expression of genes encoding proline-rich proteins (Chai et al. 1998). Proline-rich proteins, glycine-rich proteins and hydroxyproline-rich glycoproteins are the three major classes of structural proteins of plant cell walls. The authors suggest that proline-rich proteins may have important roles in the lifting of plant resistance to heavy metals actions.

4. NITROGEN FIXATION IN METAL STRESS

Heavy metals affect N_2–fixation in legume plants and free-living bacteria that are capable of fixing molecular nitrogen in various pathways of the symbiotic interactions. It is well known that some metals at relatively low concentrations are essential for the growth of different *Rhizobium* and *Bradyrhizobium* strains. For example, every N_2-fixing bacteria has obligatory requirement for Mo, because molybdenum is a metal component of nitrogenase (Thorneley 1992). Also, the introduction of zinc into a zinc-deficient nutrient stimulated the growth of *Rhizobium* (Wilson and Reisenauer 1970), while growth and hydrogenase activity in free-living *Bradyrhizobium japonicum* (Klucas et al. 1983) and *Azotobacter chroococcum* (Portridge and Yates 1982) required Ni in the environment. Copper nutrition of *Lupinus luteus* inoculated with *Bradyrhizobium lupinus* markedly increased the yield, total nitrogen content and dry weight of nodulas (Seliga 1993). Moreover, the lack of Cu in the soil limited the iron uptake and its translocation to the nodules and significantly decreased the concentration of nodule leghemoglobin (Seliga 1993). Iron, like molybdenum is an essential constituent of the nitrogenase complex and for this reason it is very important for the initiation of nodule formation (Tang et al. 1990), as well as for nodule development (O'Hara et al. 1988). The bacteria requirement for Mo, Cu, Fe and Co is generally at trace level. Higher concentration of heavy metals in the environment can limit the infection process, nodule formation, their function and growth of the host plant.

It has been found that symbiotic N_2-fixing systems were especially sensitive to some heavy metals introduced into soils as a result of past sewage sludge application (McGrath et al. 1988). Industrial and communal sludge are rich in inorganic nutrients such as N, P, Ca and Mg, but this sludge (principally from industrial areas) often contain considerable amounts of potentially toxic metals such as Cd, Zn, Cu, Pb, Cr, Hg, Al and others. A long-time application of sewage sludge greatly elevates the concentration of heavy metals, preferentially in the topsoil, indicating that application of different sludge on agricultural soils is very hazardous to crop production and microbial activity, including symbiotic N_2–fixation. The nitrogen fixation by free-living, non-symbiotic bacteria has been found to be sensitive to small concentrations of heavy metals added experimentally to the soils (Letunova et al. 1985, Martensson 1993), or after introduction of some metals with sludge to the soil (Martenssen and Witter 1990, Lorenz et al. 1992). In both cases the reduction of N_2–fixation in heterotrophic biological nitrogen fixing microorganisms have been observed (Martensson and Torstensson 1996). The possible mechanism by which the elevated metal accumulation in the soil diminished the nitrogen fixation in symbiotic systems could involve either the prevention of nodule formation on white clover roots, or elimination of effective *Rhizobium* strains (McGrath et al. 1988).

The experiments of Giller et al. (1989) have shown that *Rhizobium leguminosarum* strains isolated from the nodules of white clover growing on metal contaminated soils were incapable of N_2–fixation. When the metal contaminated soils were inoculated with standard *R. leguminosarum* bv. *trifolii* strains and subsequently incubated for two month before sowing the white clovers, no effective nodules were formed on plant roots and no nitrogen fixation was detected, unless the initial inoculum was applied at high concentration (10^{10} cells g soil^{-1}). On the other hand, when metal contaminated soil was inoculated with the standard effective *Rhizobium* strains and white clover was sown immediately, the plant formed effective nodules and grew normally. Giller and co-workers (1989) suggested that heavy metals present in the soils, contaminated with sewage sludge or with the other industrial wastes, were sufficient to eliminate the free-living clover *Rhizobium* present in soils at low concentrations (below 10^{10} cells g soil^{-1}). In other words, the effective clover *Rhizobium* strains were unable to survive in the free-living state, until the root nodules were developed. Formation of nodules before the metal application perfectly protected rhizobia from the toxic effect of heavy metals. Since the sewage sludge treated soils contained mixture of various toxic metals, it was impossible to determine which metal or metals were responsible for the toxic effect on *Rhizobium leguminosarum*. To answer this question Chaudri et al. (1992b) made experiments with soils contaminated only with one metal (Cd, Zn, Cu or Ni) and found that the most toxic to *R. Leguminosarum* bv. *trifolii* was cadmium followed by zinc and copper. Cd and Zn treatment at the concentration of about 7 mg and 389 mg per kg of soil declined significantly the number of rhizobia. The effect of Cu was much more weaker and the cells of rhizobia survived even at a relatively high concentration of metal. Surprisingly, nickel had no effect on the population of *R. Leguminosarum* bv. *trifolii*. It should be noted, that the inhibition of nodule formation was detected in laboratory as well as in field conditions (Chaudri et al. 1993).

Huang et al. (1974) have shown the direct action of cadmium and lead on the nitrogenase activity in *Glycine max*. inoculated with standard strains of *Rhizobium japonicum*. After 59 days of soybean growth the enzyme activity measured as an acetylene reduction was inhibited about 71% in the presence of 15 μM Cd, whereas Pb affected acetylene reduction only in high concentration (50 μm Pb) by about 56%. Vesper and Weidensaul (1978) have examined the influence of Cd, Ni, Zn and Cu on *Glycine max* bv. Weine cultivated in sand culture, and have found some differences in metal action on the nodule formation and N_2–fixation. Cadmium even at such low concentrations as 1ppm drastically reduced the nodule number, N_2–fixation intensity and total dry weight of plants. Nickel did not influence the nodule formation in soybean roots, whereas the nitrogen fixation was clearly diminished. On the contrary, zinc inhibited formation of nodule but only slightly diminished the N_2 – fixation, while copper reduced both the nodulation and N_2–fixation. Furthermore, Cd and Ni significantly decreased the level of leghemoglobine in the nodules while Zn and Cu had no effect on it. According to the above data the following order for decreasing metal toxicity was proposed: Cd>Ni>Cu>Zn (Vesper and Weidensaul, 1978).

The inhibitory effect of cadmium on the nitrogen fixation and activities of some nitrogen assimilatory enzymes in *Pisum sativum* cv. Bonnerville have been reported by Chough et al. (1992). After supplying of pea plants with Cd in different concentrations the N_2-fixation and some nitrogen metabolism enzymes were determined. A higher level of cadmium, significantly depressed the N_2–fixation per plant, while acetylene reduction per g of nodules was decreased only negligibly, suggesting that Cd^{2+} reduced the total nitrogen fixation through its inhibitory effect on nodule development. These results agreed with the conclusions of other workers (Huang et al. 1974) that the relatively low amount of cadmium presence in the environment limited the capacity of plants to nitrogen fixation and inhibited some enzymatic reactions involved in the biological nitrogen fixation. The inhibitory effect of cadmium was more pronounced in roots than in shoots (Chugh et al. 1992, Van Asche et al. 1988). This appears convincing, because Cd^{2+} is accumulated in much larger quantities in the cytosol of root cells (Grill et al. 1985, Grill et al. 1986) mainly influencing the nitrogen metabolism in roots and nodules.

It has been well documented that metals like cadmium and lead are readily taken up by plants and interfere with plant metabolism causing inhibition of photosynthesis (Baszyński 1986, Skórzyńska- Polit and Baszyński 1995) and plant growth (Huang et al.

1974). Since N_2–fixation is a highly energy consuming process, the reduction of photosynthesis and production of photosynthetate could also be indirectly responsible for inhibition of nitrogenase activity. Although, to date, no information on the metal-depending alteration in the nitrogen fixation due to the metal induced photosynthesis disturbances are available.

Purchase et al. (1997) have found that *R. leguminosarum* bv. *trifolii* strains isolated from the nodules of *T. repens* growing on experimental plots treated with heavy metal contaminated sludge, formed mucoid and non-mucoid colonies on agar enriched with 4 mM $ZnCl_2$. These non-mucoid strains compared to mucoid ones produced more extracellular carbohydrate and were more resistant to Cd, Cu, Ni and Zn, and effectively fixed the nitrogen in white clover. The mechanism of its resistance remains however unknown.

In the acid soil aluminium and manganese are main toxic metals which limit the productivity of plants. At pH below 5.0, aluminium salts are hydrolysed to trivalent species, which are mobilised into the soil solution impairing the growth and modifying many metabolic and development processes (Foy at al. 1978, Kochian 1995). In neutral and weakly acidic soils Al formed immobile or poorly soluble species.

The primary symptom of aluminium toxicity is the reduction of root growth (Delhaize and Ryan 1995) as a result of Al^{3+} absorption by root apex (Ryan et al. 1993, Bennet and Breen 1991). Reduced growth of roots via the action of Al^{3+} on the root apex limits the nodule formation and N_2-fixation. Howieson et al. (1988) has suggested that poor nodulation in acid soil could also be due to the low survival of *Rhizobium* or *Bradyrhizobium*. White clover (*Trifolium repens* L.) is very sensitive to aluminium in low pH, and most likely this is why white clover is rarely found in acid soils. Some clover cultivars are, however, relatively tolerant to Al^{3+} added to the soil as aluminium sulphate (Crush and Caradus 1992). The concentration of $Al_2(SO_4)_3$ increased with zone depth in the experimental soil profiles. Despite of this, the Al^{3+} tolerant genotype (G-77) had a greater root mass deeper in the soil profiles than Al susceptible genotype (G-129). Also nitrogenase activity measured as acetylene reduction was greater in nodules isolated from roots of Al-tolerant genotypes compared to Al-susceptible genotype of white clover at every tested depth of soil profiles. Crush and Caradus (1992) supposed that the Al-tolerant genotype of white clover either prevents absorption of Al^{3+} or excretes aluminium from roots to the rhizosphere.

Aluminium, like other metals, has been shown to directly affect the growth and survival of different *Rhizobia* and *Bradyrhizobia* strains (Kim et al. 1985, Flis et al. 1993). On the other hand, the infection with *Rhizobium* strains of legume plants, as well as nodule growth were strongly sensitive to relatively low concentration of Al^{3+} (ranging from 7 to 16 µM) in nutrient solution (Alva et al. 1987 and Hoheneberg and Munns 1984)

Individual *Bradyrhizobium japonicum* strains might differ not only in the sensitivity to aluminium but also in their ability to grow in acid soils or media (Munns and Keyser 1981, Keyser and Munns 1979). Thus, the growth of bacteria strains in acid soils contaminated with aluminium can be repressed by the high H^+ concentration as well as better Al mobility. Apart from Al^{3+} whose availability raises in acid soils, phosphorus insufficiency in these conditions may additionally reduce the multiplication rate of *Rhizobium* strains (Keyser and Munns 1979). This is very important because the number of active symbiotic microorganisms in the soil is significant for colonisation of the soil and inoculation of the host roots (Munns and Keyser 1981). Taylor et al. (1991) examined the efficiency of N_2 fixation in *Glycine max* cultivars inoculated with five *Bradyrhizobium japonicum* strains with different tolerance to acidity and Al^{3+} toxicity. They found that more tolerant strains of *B. japonicum* formed a symbiosis which fixed nitrogen much more effectively because of the larger nitrogenase activity.

Considering the fact that the root apex is the primary and main site of Al-toxicity and the primary effect of aluminium is due to the inhibition of the uptake of some cations (Ca^{2+}, Mg^{2+} and K^+), Huang et al. (1992a and b) and Ryan et al. (1992) investigated the ion transport in apical roots. The results of their experiments implicated that Al^{3+} blocked the Ca^{2+} channels in root cell plasma membranes. It would thus appear that the inhibition of root growth generated by blockage of the Ca^{2+} channels in the plasma membrane leads to various disturbances of many metabolic processes depending on an adequate Ca^{2+} level in the cytosol.

Investigations on the interaction of Al^{3+} and Ca^{2+} on the growth of leguminous plant, nodulation and nitrogen fixation indicated that calcium deficiency in the apical meristem of roots was the result of Al^{3+} toxicity (Foy et al. 1969, Foy et al. 1972). Alva et al. (1986) reported a 50% decrease in the root length of *Medicago sativa*, *Trifolium subterraneum* and *Glycine max* treated with Al^{3+} ranging from 7 to 16 µM. The toxic effect of aluminium was significantly decreased after elevation of Ca^{2+} concentration in the solutions. In contrast to the experiments of Alva and co-workers (1986), Shamsuddin et al. (1992) have noted that increased Ca^{2+} concentration in ambient solution only slightly diminished the aluminium toxicity in *Arachis hypogea* exposed to 35 µM Al^{3+}. Furthermore, low intracellular concentration of magnesium found at a high level of Ca^{2+} and Al^{3+} in nutrient solution, suggested that the lack of alleviating effect of Ca^{2+} on Al^{3+} toxicity resulted from interference with Mg^{2+} nutrition. Some experimental evidences confirming this hypothesis have been elucidated by Bennet et al. (1986) and Grimme (1983).

Igual et al. (1997) investigated the effect of Al^{3+} on nodule formation and symbiotic nitrogen fixation in *Casaurina conninghamiana*, inoculated with nitrogen fixing actinomycetes *Frankia*. They discovered that only 50% of the plants were nodulated, while the weight of nodules and the nitrogenase activity was not affected by Al^{3+}. Such results suggested that aluminium affected mainly the nodulation of roots without any change of the activity of already formed nodules. This is why some nodulated actinorhizal plants could grow on acid soils (Dixon and Wheeler 1983). But the extreme Al concentration (880 µM Al) decreased nitrogen fixation efficiency about twice. Igual and co-workers suggested that high Al^{3+} concentration inhibited the formation of root hairs and in this way decreased the nodulation process. A similar explanation of the Al toxicity at high concentrations was given for *Glycine max* (Brady et al. 1993) and *Arachis hypogea* (Shamsuddin et al. 1992).

The high level of Mn^{2+} in the soil more distinctly affects the nitrogen content in plants using atmospheric N_2 as an exclusive nitrogen source than in plants taking up the mineral forms of nitrogen (Dobereiner 1966). Manganese may exert a negative effect on the nodulation of some legume species (Foy et al. 1978). For example, the excess of

Mn^{2+} in the soil decreased the N_2–fixation and nodulation in *Phaseolus vulgaris* L. (Dobereiner 1991). A similar negative effect of high manganese level on the nodulation has been observed in *Trifolium subterraneum* by Evans et al. (1987). Manganese toxicity was also found as the main toxic factor limiting the growth of soybean (Reddy et al. 1991).

As mentioned above, the roots are the crucial plant organ accumulating heavy metals (Pb, Cu, Cd). It thus seems that plants which derive nitrogen mainly by symbiotic nitrogen fixation are especially exposed to toxic effect of heavy metals. The synthesis of phytochelatins and metallothioneine-like proteins as well as antioxidant enzymes is a very important defence system against the metal toxicity in nodule formation and nitrogen fixation.

The growth and nodulation of many legumes are affected by various metals (Cd, Zn, Ni, Cu, Cr, Al and others) if their concentrations in the soil are relatively high. The negative effect of metals in the soil on the survival and behaviour of different *Rhizobium* and *Bradyrhizobium* strains is well known, however, the mechanisms of heavy metals' toxicity as well as the resistance of symbiotic microorganisms and host plants, are to date poorly explained. Nevertheless, the heavy metals' resistance of legumes, *Rhizobium* and *Bradyrhizobium* strains, undoubtedly may involve the following: precipitation of metals as insoluble salts, production of chelating agents, alteration in membrane permeability, release of organic acids from the root apex to the rhizosphere, alterations in the rhizosphere pH and others. Unfortunately, the effect of heavy metals on metabolic processes during symbiotic nitrogen fixation in nodules was poorly investigated. It thus appears that further experiments in this field are essential.

5. CONCLUSION

Heavy metals can affect the nitrogen metabolism at multiple levels. Potential sides of their toxicity are presented in Fig. 1. The first place of metal action which disturb the nitrogen acquisition by plants is undoubtedly connected with plasma membrane. At this level metals can alter the nitrogen transport directly, through the action on the specific ammonium and nitrate transport proteins or indirectly, changing the membrane composition, its fluidity and H^+ pumping. It is becoming clear that changes in the intracellular concentration of nitrogen could initiate a signal transduction pathway leading to the modification of the expression of genes encoding enzymes involved in its assimilation. However, the direct binding of metal to the enzyme protein responds for its inhibition also. Assuming that nitrogen uptake is downregulated by the products of nitrogen assimilation, it is obvious that the heavy metal inhibition of particular enzymes of the nitrogen assimilatory pathway can alter the nitrogen uptake. Until recently, most experiments concerned only with the metal influence on the uptake processes or on the enzyme activities.

Future experiments should be address the possible interferences between the nitrogen transport and its acquisition in heavy metal stress. New genomic tools such as

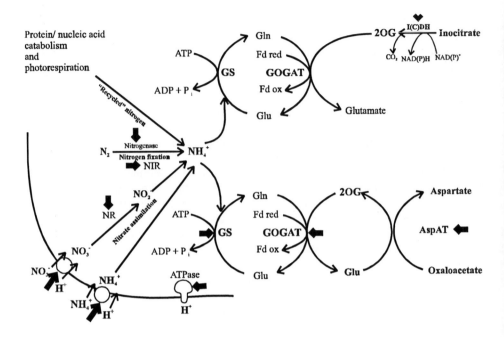

Figure 1. A simplified scheme showing the heavy metals actions (black arrow) on the particular steps of the nitrogen uptake, its assimilation by the nitrogenase, nitrate reductase (NR), nitrite reductase (NiR), glutamine synthetase – glutamate synthase (GS-GOGAT) pathway and the production of the organic acid 2-oxoglutarate (2OG) either by an isocitrate dehydrogenase (ICDH) or an aspartate aminotransferase (AspAT).

expression of specific genes encoding the nitrogen transporters and nitrogen assimilatory enzyme proteins will allow researchers to dissect the molecular complexities of metal action. This will certainly lead to a better understanding of the complex biochemical and physiological responses observed in the metal-stressed plants, and ultimately to the development of molecular strategies which improve plant resistance to the heavy metals.

REFERENCES

Alva, A.K., Asher, C.J. and Edwards, D.G. (1986) The role of calcium in alleviating aluminium toxicity, *Aust. J. Agric. Res.* **37**, 375–382

Alva, A.K., Edwards, D.G., Asher, C.J. and Suthipradit, S. (1987) Effects of acid soil in fertility factors on growth and nodulation of soybean, *Agron. J.* **79**, 302–306.

Archambault, D.J., Zhang, G.M., and Taylor G.J. (1996) A comparison of the kinetics of aluminium (Al) uptake and distribution in roots of wheat (*Triticum aestivum*) using different aluminium sources. A revision of the operational definition of symplastic, *Physiol. Plant.* **98**, 578–586.

Aslam, M., Travis, R.L., and Huffaker, R.C. (1992) Comparative kinetics and reciprocal inhibition of nitrate and nitrite uptake in roots of uninduced and induced barley seedlings, *Plant Physiol.* **99**, 1124–1133.

Aslam, M., Travis, R.L., and Huffaker, R.C. (1993) Comparative induction of nitrate and nitrite uptake and reduction systems by ambient nitrate and nitrite in intact roots of barley seedlings, *Plant Physiol.* **102**, 811–819.

Athwal, G.S., Huber, J.I., and Huber, S.C. (1998) Biological significance of divalent metal ion binding to 14-3-3 proteins in relationship to nitrate reductase inactivation, *Plant Cell Physiol.* **39**, 1065–1072.

Barcelo, J. and Poschenrieder, C. (1990) Plant water relations as affected by heavy metal stress: a review, *J. Plant Nutr.* **13**, 1–37.

Barthes, L., Deleens, E., Bousser, A., Hoarau, J., and Prioul, J.L. (1966) Xylem exudation is related to nitrate assimilation pathway in detoped maize seedlings: use of nitrate reductase and glutamine synthetase inhibitors as tools, *J. Exp. Bot.* **47**, 485–495.

Basu, A., Basu, U., and Taylor, G.J. (1994) Induction of microsomal membrane proteins in roots of an aluminum-resistant cultivar of *Triticum aestivum* L. under condition of aluminium stress, *Plant Physiol.* **104**, 1007–1013.

Baszyński, T. (1986) Interference of Cd^{2+} in functioning of the photosynthetic apparatus of higher plants, *Acta Soc. Bot. Polon.* **55**, 292–302.

Bennet, R.J. and Breen, C.M. (1991) The aluminium signal: new dimensions to mechanism of aluminium tolerance, *Plant Soil*, **134**, 153–166.

Bennet, R.J., Breen, C.M., and Foy, M.V. (1986) Aluminium toxicity and induced nutrient disorders involving the uptake and transport of P, K, Ca and Mg in *Zea mays* L., *Afr. J. Plant Soil* **3**, 11–17.

Bharti, N., Singh, R.P, Sinha, S.K. (1996) Effect of calcium chloride on heavy metal induced alteration in growth and nitrate assimilation of *Sesamum indicum* seedlings, *Phytochemistry* **41**, 105–109.

Bielawski, W. (1994) Effect of some compounds on glutamine synthetase isoform activity from *Triticale* seedling leaves, *Acta Physiol. Plant.* **16**, 303–308.

Bloom, A.J. (1988) Ammonium and nitrate as nitrogen sources for plant growth, *ISI Atl. Sci.* **1**, 55–59.

Boussama, N, Ouariti, O., and Ghorbal, M.H. (1999b) Changes in growth and nitrogen assimilation in barley seedlings under cadmium stress, *J. Plant Nutr.* **22**, 731–752.

Boussama, N., Ouariti O., Suzuki A., and Ghorbal, M.H. (1999a) Cd-stress on nitrogen assimilation, *J. Plant Physiol.* **159**, 310–317.

Brady, D.J., Edwards, D.G., Asher C.J., and Blamey, F.P.C. (1993) Calcium amelioration of aluminium toxicity effects on root hair development in soybean (*Glycine max* L.), *New Phytol.* **123**, 531–538

Brunold, C. (1993) Regulatory interactions between sulfate and nitrate assimilation, in L.J. De Kok, I. Stulen, H. Rennenberg, C. Brunold and W.E. Rauser (eds.), *Sulfur nutrition and assimilation in higher plants*, SPB Acad. Publishing, The Hague, pp. 62–78.

Buczek, J. (1973) Effect of vanadium on nitrate reductase activity in tomato leaves, *Acta Soc. Bot. Polon.* **42**, 223–232.

Buczek, J., Kowalińska, E., and Kuczera, K. (1980) Reduction of nitrates in *Cucumis sativus* L. seedlings, *Acta Soc. Bot. Polon.* **49**, 259–267.

Burzyński, M. (1987) The influence of lead and cadmium on the absorption and distribution of potassium, calcium, magnesium and iron in cucumber seedlings, *Acta Physiol. Plant.* **9**, 229–238.

Burzyński, M. (1988) The uptake and accumulation of phosphorus and nitrates and the activity of nitrate reductase in cucumber seedlings treated with $PbCl_2$ or $CdCl_2$, *Acta Soc. Bot. Polon.* **3**, 349–359.

Burzyński, M. (1990) Activity of some enzymes involved in NO_3^- assimilation in cucumber seedlings treated with lead or cadmium, *Acta Physiol. Plant.* **12**, 105–110.

Burzyński, M. (2001) Influence of pH on Cd and Cu upt, distribution, and their effect on nitrate reductase activity in cucumber *(Cucumis sativus* L.) seedlings roots, *Acta Physiol. Plant.* **23**, 201-206.
Burzyński, M. and Buczek, J. (1994a) The influence of Cd, Pb, Cu and Ni on NO_3^- uptake by cucumber seedlings. II. *In vivo* and *in vitro* effects of Cd, Pb, Cu and Ni on the plasmalemma ATPase and oxidoreductase from cucumber seedling roots, *Acta Physiol. Plant.* **16**, 297–302.
Burzyński, M. and Buczek, J. (1994b) The influence of Cd, Pb, Cu and Ni on NO_3^- uptake by cucumber seedlings. I. Nitrate uptake and respiration of cucumber seedling roots treated with Cd, Pb, Cu and Ni, *Acta Physiol. Plant.* **16**, 291–296.
Burzyński, M. and Buczek, J. (1997) The effect of Cu^{2+} on uptake and assimilation of ammonium by cucumber seedlings, *Acta Physiol. Plant.* **19**, 3–8.
Burzyński, M. and Buczek, J. (1998) Uptake and assimilation of ammonium ions by cucumber seedlings from solutions with different pH and addition of heavy metals, *Acta Soc. Bot. Polon.* **67**, 197–200.
Burzyński, M. and Grabowski, A. (1984) Influence of lead on NO_3^- uptake and reduction in cucumber seedlings, *Acta Soc. Bot. Polon.* **53**, 77–86.
Campbell, W.H. (1999) Nitrate reductase structure, function and regulation: bridging the gap between biochemistry and physiology, *Annu. Rev. Plant Physiol. Plant Mol. Biol.* **50**, 277–303.
Chai, T.Y., Didierjean, L., Burkard, G., and Genot, G. (1998) Expression of green tissue-specific 11 kDa proline-rich protein gene in bean in response to heavy metals, *Plant Sci.* **133**, 47–56.
Chaoui, A., Ghorbal, M.H., Ghorbal, E.L., and Ferjani, E. (1996) Effect of cadmium-zink interactions on hydroponically grown bean *(Phaseolus vulgaris), Plant Sci.* **127**, 139–147.
Chaudri, A.M., McGrath, S.P., Giller, K.E., Rietz, E., and Sauerbeck, D.R. (1993) Enueration of indigenous *Rhizobium leguminosarum* biovar. Sludge, *Soil Biol. Biochem.* **25**, 301–309.
Chaudri, A.M., McGrath, S.P., and Giller, K.E. (1992) Survival of indigenous population *of Rhizobium leguminosarum* biovar. *trifolii* in soil spiked with Cd, Zn, Cu and Ni salts, *Soil Biol. Biochem.* **24**, 625–632.
Chesire, M.V., Bick, W., De Kok, P.C., and Inkson, R.H.E. (1982) The effect of copper and nitrogen on amino acid composition of oat strow, *Plant Soil* **66**, 139–148.
Chien, H-F. and Kao, C.H. (2000) Accumulation of ammonium in rice leaves in response to excess cadmium, *Plant Sci.* **156**, 111–115.
Chugh, L.K., Gupta, V.K., and Sawhney, S.K. (1992) Effect of cadmium on enzymes of nitrogen metabolism in pea seedlings, *Phytochemistry* **31**, 395–400.
Clarkson, T.W. (1993). Molecular and ionic mimicry of toxic metals, *Annu. Rev. Pharmacol. Toicol.* **32**, 545–571.
Cox, W.J. and Reisenauer, H.M. (1977) Ammonium effects on nutrient cation absorption by wheat, *Agron. J.* **69**, 868–871.
Crawford, N.M. and Glass, A.D.M (1998) Molecular and physiological aspects of nitrate uptake in plants, *Trends Plant Sci.* **3**, 389–395.
Crush, J.R. and Caradus, J.R. (1992) Response to soil aluminium of two white clover *(Trifolium repens* L.) genotypes, *Plant Soil* **146**, 39–43.
De Filippis, L.F. (1979) The effect of heavy metal compounds on the permeability of *Chlorella* cells, *Z. Pflanzenphysiol.* **92**, 39–49.
De Vos, C.H., Schat, H., De Waal, M.A.M., Voijs, R., and Ernst, H.O. (1991) Increased resistance to copper-induced damage of the root cell plasmalemma in copper tolerant *Silene cucubalus*, *Physiol. Plant.* **82**, 523–528.
Degenhardt, J., Larsen, P.B., Howell, S.H., and Kochian, L.V. (1998) Aluminum resistance in the *Arabidopsis* mutant alr-104 is caused by aluminum-induced increase in rhizosphere pH, *Plant Physiol.* **117**, 19–25.
Delhaize, E. and Ryan, P.R. (1995) Aluminium toxicity and tolerance in plants, *Plant Physiol.* **107**, 315–321.
Demidchik, V., Sokolik A., and Yurin, V. (1997) The effect of Cu^{2+} on ion transport systems of the plant cell plasmalemma, *Plant Physiol.* **114**, 1313–1325.
Devi, S.R. and Prasad, M.N.V. (1999) Membrane lipid alterations in heavy metal exposed plants, in M.N.V. Prasad and J. Hagemeyer (eds.), *Heavy Metal Stress in Plants. From Molecules to Ecosystemes*, Springer-Verlag, pp. 99–116.
Dietz, K.J., Baier, M., and Krämer, U. (1999) Free radicals and reactive oxygen species as mediators of heavy metals toxicity in plants, in M.N.V. Prasad and J. Hagemeyer (eds.), *Heavy Metal Stress in Plants. From Molecules to Ecosystemes*, Springer-Verlag, pp. 73–98.

Dixon, R.O.D. and Wheeler, C.T. (1983) Biochemical, physiological and environmental aspects of symbiotic nitrogen fixation, in J.C. Gordon and C.T. Wheeler (eds.), *Biological Nitrogen Fixaation in Forest Ecosystems: Foundation and Aplocations*, Nijhoff/dr W. Junk, The Hague, pp. 107–171.

Dobereiner, J. (1966) Manganase toxicity effects on nodulation and nitrogen fixation of beans (*Phaseolus vulgaris* L.) in acid soils, *Plant Soil* **24**, 153–166.

Doddema, H. and Telkamp, G.P. (1979) Uptake of nitrate by mutants of *Arabidopsis thaliana*, distributed in uptake or reduction of nitrate. II. Kinetics, *Physiol. Plant.* **45**, 332–338.

Dyhr-Jensen, K. and Brix, H. (1996) Effect of pH on ammonium uptake by *Typha latifolia* L., *Plant Cell Environ.* **19**, 1431–1436

Evans, P.T., Scott, B.J., and Lill, V.J. (1987) Manganase tolerance in subterraneum clover (*Trifolium subterraneum* L.) genotypes grown with amonium nitrate or symbiotic nitrogen, *Plant Soil* **97**, 207–215.

Flis, S.E., Glenn, A.R., and Dilworth, M.J. (1993) The interaction between aluminium and root nodule bacteria, *Soil Biol. Biochem.* **25**, 403–414.

Fodor, E., Szabó-Nagy, A., and Erdei L. (1995) The effects of cadmium on the fluidity and H^+-ATPase activity of plasma membrane from sunflower and wheat roots, *J. Plant Physiol.* **147**, 87–92.

Forde, B.G. (2000) Nitrate transporters in plants: structure, function and regulation, *Biochem. Biophys. Acta* **1465**, 219–235.

Foy, C.D., Chaney, R.L., and White, M.C. (1978) The physiology of metal toxicity in plants, *Annu. Rev. Plant Physiol.* **29**, 511–566.

Foy, C.D., Fleming, A.L., and Armiger, W.H. (1969) Aluminium tolerance of soybean varietes in relation to calcium nutrition, *Agron. J.* **61**, 505–511.

Foy, C.D., Fleming, A.L., and Gerloff, G.C. (1972) Differential aluminium tolerance on two snap bean varietas, *Agron. J.* **64**, 815–818.

Galves, S., Lancien, M., and Hodges, M. (1999) Are isocitrate dehydrogenases and 2-oxogltarate involved in the regulation of glutamate synthesis, *Trends Plant Sci.* **4**, 484–489.

Giller, K.E., McGrath, S.P., and Hirsch, P.R. (1989) Absence of nitrogen fixation in clover grown on soil subject to long-term contamination with heavy metals is due to survival of only infective *Rhizobium*, *Soil Biol. Biochem.* **21**, 841–848.

Glass, A.D.M. (1988) Nitrogen uptake by plant roots, *ISI Atl. Sci.* **1**, 151–156

Glass, A.D.M., Shaff, J.E., and Kochian, L.V. (1992) Studies of the uptake of nitrate in barley. IV. Electrophysiology, *Plant Physiol.* **99**, 456–463.

Glass, A.D.M. and Siddiqi, M.Y. (1995) Nitrogen absorption by plant roots, in H.S. Srivastava and R.P Singh (eds.), *Nitrogen, Nutrition in Higher Plants*, Assoc. Publishing Co., pp. 21–56.

Gouia, H., Ghorbal, M.H., and Meyer, C. (2000) Effects of cadmium on activity of nitrate reductase and on other enzymes of the nitrate assimilation pathway in bean, *Plant Physiol. Biochem.* **38**, 629–638.

Greger, M. (1999) Metal Availability and Bioconcentration in Plants, in M.N.V. Prasad and J. Hagemayer (eds.), *Heavy Metal Stress in Plants. From Molecules to Ecosystems,* Springer-Verlag, pp. 1–28.

Greger, M. and Johansson, M. (1992) Cadmium effects on leaf transpiration of sugar beet (*Beta vulgaris*), *Physiol. Plant* **86**, 465–473.

Grill, E., Winnacker, E.L., and Zenk, M.H. (1985) Phytochelatins: the principal heavy metal complexing peptides of higher plants, *Science* **230**, 674–676.

Grill, E., Winnacker, E.L., and Zenk, M.H. (1986) Homo-phytochelatins are heavy metal binding peptides of homo-glutatione containing *Fabales*, *FEBS Lett.* **205**, 47–50.

Grimme, H. (1983) Aluminium induced magnesium deficiency in oat, *Z. Pflanzenernaehr. Bodenk.* **146**, 666–676.

Gupta, M. and Chandra, P. (1998) Bioaccumulation and toxicity of mercury in rooted-submerged macrophyte *Vallisneria spiralis, Environ. Pollut.* **103**, 327–332.

Gupta, M. and Chandra, P. (1996) Bioaccumulation and physiological changes in *Hydrilla verticillata* (l.f.) royle in response to mercury, *Bull. Environ. Contam. Toxicol.* **56**, 319–326.

Hernandez, L.E, Gárate, A., and Carpenta-Ruiz, R. (1997) Effect of cadmium on the uptake, distribution and assimilation of nitrate in *Pisum sativum*, *Plant Soil* **189**, 97–106.

Hernandez, L.E., Gárate, A., and Carpenta-Ruiz, R. (1996) Influence of cadmium on the assimilation of two cultivars of *Zea mays*, in O. Van Cleemput, G. Hofman and A. Vermoesen, *Progress in Nitrogen Cycling Studise*, Kluwer Acad. Publ., Dordrecht, pp. 115–132.

Hohenberg, J.S. and Munns, D.N. (1984) Effect of soil acidity factors on nodulation and growth of *Vigna unguiculata* in solution culture, *Agron. J.* **76**, 477–481.

Hole, D.J., Emran, A.M., Fares, Y. and Drew, M.C. (1990) Induction of nitrate transport in maize roots, and kinetics of influx, measured with nitrogen-13, *Plant Physiol.* **93**, 642–647.
Howieson, J.G., Ewing, M.A., and D'Antuono, M.F. (1988) Selection for acid tolerance in *Rhizobium meliloti, Aust. J. Plant Physiol.* **19**, 287–296.
Howitt, S.M. and Udvardi, M.K. (2000) Structure, function and regulation of ammonium transporters in plants, *Biochem. Biophys. Acta* **1465**, 152–170.
Huang, C.Y., Bazzaz, F.A., and Vanderhoef, L.N. (1974) The inhibition of soybean metabolism by Cd and Pb, *Plant Physiol.* **54**, 122–124.
Huang, J.W., Grunes, D.L., and Kochian, L.V. (1992a) Aluminium effects on the kinetics of calcium uptake into cells of the wheat root apex. Quantification of calcium fluxes using a calcium-selective vibrating microelectrode, *Planta* **188**, 114–121.
Huang, J.W., Shaff, J.E., Grunes, D.L., and Kochian, L.V. (1992b) Aluminium effect on calcium fluxes at the root apex of aluminium-tolerant and aluminium-sensitive wheat cultivars, *Plant Physiol.* **98**, 230–237.
Igual, I.M., Rodriguez-Borrucio, C., and Cervantes, E. (1997) The effect of aluminium on nodulation and symbiotic nitrogen fixation in *Casaurina cunninghamiana* Miq., *Plant Soil* **190**, 41–46.
Ireland, R.J. and Lea, P.J. (1999) The enzymes of glutamine, glutamate, asparagine and aspartate metabolism, in B.K. Singh (ed.), *Plant Amino Acids. Biochemistry and Biotechnology*, Dekker, New York, pp. 49–109.
Jarvis, S.C. and Hatch, D.J. (1986) The effect of the low concentrations of aluminum on the growth and uptake of nitrate-N by white clover, *Plant Soil* **95**, 43–55.
Jerzykiewicz, J. (2001) Aluminium effect on the nitrate assimilation in cucumber (*Cucumis sativus* L.) roots, *Acta Physiol. Plant..* **23**, 213-219
Ju, G.C., Li, X., Rauser, W.E., and Oaks, A. (1997) Influence of cadmium on the production of γ-glutamylcysteine peptides and enzymes of nitrogen assimilation in *Zea mays* seedlings, *Physiol. Plant* **101**, 793–799.
Kaiser, W.M., Weiner, H., and Huber, S.C. (1999) Nitrate reductase in higher plants: A case for transduction of environmetal stimuli into control of catalytic activity, *Physiol. Plant.* 105, 385–390.
Kasamo, K. and Nouchi, I. (1987) The role of phospholipids in plasma membrane ATPase activity in *Vigna radiata* (mung bean) roots and hypocotyls, *Plant Physiol.* **83**, 823–828.
Kashyap, A.K. and Gupta, S.L. (1982) Effect of lethal copper concentration on nitrate uptake, reduction and nitrite release in *Anacystis nidulans, Z. Pflanzenphisiol.* **107**, 289–294.
Keltjens, W.G. and Ulden, P.S.R. (1987) Effects of Al on nitrogen (NH_4^+ and NO_3^-) uptake, nitrate reductase activity and proton release in two sorghum cultivars differing in Al tolerance, *Plant Soil* **104**, 227–234.
Kennedy, C.D. and Gonsalves, F.A.N. (1987) The action of divalent Zn, Cd, Hg, Cu nad Pb on the ATPase activity of a plasma membrane fraction isolated from roots of *Zea mays, Plant Soil* **117**, 167–175.
Keyser, H.H. and Munns, D.N. (1979) Tolerance of *Rhizobia* to acidity, aluminium and phosphate, *Soil Sci. Soc. Am. J.* **43**, 519–523.
Kim, M.K., Asher, C.J., Edwards, D.G., and Date, R.D. (1985) Aluminium toxicity. Effects on growth and nodulation of subterranean clover, *Proc. 15th Int. Grassland Congr.*, Kyoto, Japan, 501–503.
Kłobus, G. and Buczek, J. (1995) The role of plasma membrane oxidoreductase activity in proton transport, *J. Plant Physiol.* **146**, 103–107.
Kłobus, G., Jerzykiewicz, J. and Buczek, J. (1998) Characterization of the nitrate transporter in root plasma membranes of *Cucumis sativus* L., *Acta Physiol. Plant.* **20**, 323–328.
Kłobus, G., Łubocka, J. and Buczek, J. (1991) Effect of sodium tungstate on NO_3^- uptake by cucumber seedlings, *Acta Phys. Plant.* **13**, 227–233.
Klucas, R.V., Hanus, F.J., Russell, S.A., and Evans, H.J., (1983) Nickel: a micronutrient element for hydrogen-dependent growth of *Rhizobium japonicum* and for expression of urease activity in soybean leaves, *Proc. Natl. Acad. Sci. USA.* **80**, 2253–2257.
Kochian, L.V. (1995) Cellular mechanisms of aluminium toxicity and resistance in plants, *Annu. Rev. Plant Physiol. Plant Mol. Biol.* **46**, 237–260.
Krapp, A., Fraisier, V., Schleible, W-R., Quesada, A., Gojon, A., Stitt, M., Caboche, M., and Daniel-Vedale, F. (1998) Expression studies of Nrt2:1Np, a putative high affinity nitrate transporter: evidence for its role in nitrate uptake, *Plant J.* **6**, 723–732.
Kubik-Dobosz, G., Hałajko, T., and Górska, A. (in press) Expression of *amt1* gene in the presence of some toxic ions, *Acta Physiol. Plant.***23**, 187-192

Kumar, V., Yadav, D.V., and Yadav D.S. (1990) Effects of nitrogen sources and copper levels on yield, nitrogen and copper contents of wheat (*Triticum aestivum* L.), *Plant Soil* **126**, 79–83.

Lam, H.M., Coschigano, K.T., Oliveira, Melo-Oliveira, I.C., and Coruzzi, G.M. (1996) The molecular-genetics of nitrogen assimilation into amino acids in higher plants, *Annu. Rev. Plant Physiol. Plant Mol. Biol.* **47**, 569–593.

Letunova, S.V., Umarow, M.M., Niyazova, G.A., and Melekhin, Y.I. (1985) Nitrogen fixation activity as a possible critrion for determining permissible concentrations of heavy metals in soil, *Soviet Soil Sci.* **17**, 88–92.

Lindberg, S. and Griffiths, G. (1993) Alumnium effects on ATPase activity and lipid composition of plasma membrane in sugar beet roots, *J. Exp. Bot.* **44**, 1543–1550.

Lloyd, D.R. and Phillips, D.H. (1999) Oxidative DNA damage mediated by copper (II), iron (II) and nickel (II). Fenton reactions: evidence for site-specific mechanisms in formation of double-strand breaks, 8-hydroksydeoxyguanosine and putative intrastrand cross-links, *Mut. Res. Mar.* **424**, 23–36.

Lorenz, S.F., McGrath, S.P., and Giller, K.E. (1992) Assessment of free-living nitrogen fixation activity as a biological indicator of heavy metal toxicity in soil, *Soil Biol. Biochem.* **24**, 601–606.

Lozano-Rodrigez, E., Hernandez, L.E., Bonay, P., and Carpena-Ruiz, R.O. (1997) Distribution of cadmium in shoot and root tissues of maize and pea plants. Physiological disturbances, *J. Exp. Bot.* **48**, 123–128.

Luna, C.M., Casano, L.M., and Trippi, V.S. (1997) Nitrate reductase is inhibited in leaves of *Triticum aestivum* treated with high level of copper, *Physiol. Plant.* **101**, 103–108.

Marschner, H. (1985) *Mineral Nutrition of Higher Plants*, Acad. Press, London, pp. 6–73.

Martensson, A.M. (1993) Use of heterotrophic and cyanobacterial nitrogen fixation to study the impact of anthropogenic substances on soil biological processes, *Bull. Environ. Contam. Toxicol.* **50**, 466–473.

Martensson, A.M. and Torstensson, L. (1996) Monitoring sewage sludge using heterotrophic nitrogen fixing microorganisms, *Soil Biol. Biochem.* **28**, 1621–1630.

Martensson, A.M. and Witter, E. (1990) Influence of various soil amedments on nitrogen-fixing soil microorganisms in a long-term field experiments, with special reference to sewage, *Soil Biol. Biochem.* **22**, 977–982.

Mathys, W. (1975) Enzymes of heavy metals resistant and non resistant populations of *Silene cucubalius* and their interactions with some heavy metals *in vitro* and *in vivo*, *Physiol. Plant.* **33**, 161–165.

McClure, P.R., Kochian, L.V., Spanswick, R.M., and Shaff, J.E. (1990) Evidences for cotransport of nitrate and protons in maize roots. I. Effect of nitrate on membrane potential, *Plant Physiol.* **93**, 281–289.

McGrath, S.P., Brookes, P.C., and Giller, K.E. (1988) Effects of potentially toxic elements in soil derived from past applications of sewage sludge on nitrogen fixation by *Trifolium repens* L., *Soil Biol. Biochem.* **20**, 415–425.

Meharg, A.A. (1993) The role of plasmalemma in metal tolerance in angiosperms, *Physiol. Plant.* **88**, 191–198.

Miflin, B.J. and Lea, P.J. (1976) The pathway of nitrogen assimilation in plants, *Phytochemistry* **15**, 873–885.

Miller, A.J. and Smith, S.J. (1996) Nitrate transport and compartmentation in cereal root cells, *J. Exp. Bot.* **47**, 1455–1463.

Munns, D.N. and Keyser, H.H. (1981) responses of *Rhizobium* strains to acid and aluminium stress, *Soil Biol. Biochem.* **13**, 115–118.

Nichol, B.E., Oliveira, L.A., Glass, A.D.M., and Siddiqi, M.Y. (1993) The effects of aluminum on the influx of calcium, potassium, ammonium, nitrate and phosphate in an aluminum-sensitive cultivar of barley (*Hordeum vulgare* L.), *Plant Physiol.* **101**, 1263–1266.

Ninneman, O., Jauniaux, J.C., and Frommer, W.B. (1994) Identification of a high affinity NH_4^+ transporter from plants, *EMBO J.* **13**, 3464–3471.

Oaks, A. (1994) Primary nitrogen assimilation in higher plants and its regulation, *Can. J. Bot.* **72**, 739–750.

O'Hara, G.W., Boonkerd, N., and Dilworth, M.J. (1988). Mineral constrainnts to nitrogen fixation, *Plant Soil* **108**, 93–110.

Orzechowski, S., Kwinta, J., Gworek, B., and Bielawski, W. (1997) Biochemical indicators of environmental contamination with heavy metals, *Pol. J. Environ. Stud.* **6**, 47–50.

Ouariti, O., Gouia, H., and Ghorbal, M.H. (1997) Responses of bean and tomato plants to cadmium: Growth, mineral nutrition, and nitrate reduction, *Plant Physiol. Biochem.* **35**, 347–354.

Pandey, M. and Srivastava, H.S. (1993) Inhibition of nitrate reductase activity and nitrate accumulation by mercury in maize leaf segment, *J. Environ. Health* **35**, 110–114.

Portridge, C.D.P. and Yates, M.G. (1982) Effect of chelating agents hydrogenase in *Azotobacter chroococum*, *Biochem. J.* **204**, 339–344.

Poschenrieder, C. and Barcelo, J. (1999) Water relation in heavy metals stressed plants, in M.N.V. Prasad and J. Hagemeyer (eds.), *Heavy Metal Stress in Plants. From Molecules to Ecosystemes*, Springer-Verlag, pp. 207–231.

Prasad, M.N.V. (1999) Metalothioneins and metal binding complexes in plants. in M.N.V. Prasad and J. Hagemeyer (eds.), *Heavy Metal Stress in Plants. From Molecules to Ecosystemes*, Springer-Verlag, pp. 51–73.

Prasad, M.N.V. and Strzałka K. (1999) Impact of heavy metals on photosynthesis, in M.N.V. Prasad and J. Hagemeyer (eds.), *Heavy Metal Stress in Plants. From Molecules to Ecosystemes*, Springer-Verlag, pp. 117–139.

Purchase, D., Miles, R.J., and Young, T.W.K. (1997) Cadmium uptake and nitrogen fixing ability in heavy-metal-resistent laboratory and field strains of *Rhizobium leguminosarum* biovar. *trifolii, FEMS Microbiol. Ecol.* **22**, 85–93.

Quesada, A., Hidalgo, J., and Fernandez, E. (1997) Three *nrt* genes are differentially regulated in *Chlamydomonas*, *Mol. Gen.* **258**, 373–377.

Rai, L.C., Rai, P.K., and Mallick, N. (1996) Regulation of heavy metal toxicity in acid-tolerant *Chlorella*: physiological and biochemical approaches, *Eniviron. Exp. Bot.* **36**, 99–109.

Rai, U.N., Tripathi, R.D., and Kumar, N. (1992) Bioaccumulation of chromium and toxicity on growth, photosynthetic pigments, photosynthesis *in vivo* nitrate reductase activitry and protein content in a chlorococcalean green alga *Glaucocystis nostochinearum* Itzigsohn., *Chemosphere* **25**, 721–732.

Rawat, S.R., Silim, S.N., Kronzucker, H.J., Siddiqi, M.Y., and Glass, D.M. (1999) AtAMT gene expression and NH_4^+ uptake in roots of *Arabidopsis thaliana*: evidence for regulation by root glutamine levels, *Plant J.* **19**, 143–152.

Reddy, M.R., Ronaghi, A., and Bryant, J.A. (1991) Differential responses of soybean genotype to excess manganese in an acid soil, *Plant Soil* **134**, 221-226.

Rengel, Z. and Elliott, D.C. (1992a) Aluminum inhibits net $^{45}Ca^{2+}$ uptake by *Amaranthus* protoplasts, *Biochem. Physiol. Pflanzen* **188**, 177–186.

Rengel, Z. and Elliott, D.C. (1992b) Mechanism of aluminum inhibition of net $^{+45}Ca^{2+}$ uptake by *Amaranthus* protoplasts, *Plant Physiol.* **98**, 632–638.

Ryan, P.R., Ditomaso, J.M., and Kochian, L.V. (1993) Aluminium toxicity in roots: An investigation of special sensitivity and the role of the root cap, *J. Exp. Bot.* **44**, 437–446.

Ryan, P.R., Shaff, J.E., and Kochian, L.V. (1992) Aluminium toxicity in roots. Corelation among ionic currents, ion fluxes and root elongation in aluminium-sensitive and aluminium tolerant wheat cultivars, *Plant Physiol.* **99**, 1193–2000.

Schat, H, Sharma, S.S., and Vooijs, R. (1997) Heavy metal-induced accumulation of free proline in a metal-tolerant and a nontolerant ecotype of *Silene vulgaris*, *Physiol. Plant.* **101**, 477–482.

Schroeder, J.I. (1988) K^+ transport properties of K^+ channels in the plasma membrane of *Vicia faba* guard cells, *J. Gen. Physiol.* **92**, 667–683.

Seliga, H. (1993) The role of copper in nitrogen fixation in *Lupinus luteus* L., *Plant Soil* **155/156**, 349–352.

Serrano, R. (1990) Plasma membrane ATPase, in L. Larsson and I.M. Moller (eds.), *The Plant Plasma Membrane*, Springer-Verlag, Berlin, pp. 127–154.

Shalaby, A. M. and Al-Wakeel, S.A.M. (1995) Changes in nitrogen metabolism enzyme activities of *Vicia faba* in response to aluminum and cadmium, *Biol. Plant.* **37**, 101–106.

Shamsuddin, Z.H., Kasran, R., Edwards, D.G., and Blamey, F.P.C. (1992) Effects of calcium and aluminium on nodulation, nitrogen fixation and growth of groundnut in solution culture, *Plant Soil* **144**, 273–279.

Sharma, S.S, Schat, H.A., and Vooijs, R. (1998) *In vitro* alleviation of heavy metal-induced enzyme inhibition by proline, *Phytochemistry* **49**, 1531–1535.

Siddiqi, M.Y., Glass, A.D., Ruth, T.J., and Rufty, T.W. (1990) Studies on the uptake of nitrate in barley, *Plant Physiol.* **93**, 1426–1432.

Siedlecka, A. (1995) Some aspects of interactions between heavy metals and plant minerals, *Acta Soc. Bot. Polon.* **64**, 265–272.

Singh, R.P, Bharti, N., and Kumar, G. (1994) Differential toxicity of heavy metals to growth and nitrate reductase activity of *Sesamum indicum* seedlings, *Phytochemistry* **35**, 1153–1156.

Skórzyńska-Polit, E. and Baszyński, T. (1995) Photochemical activity of primary leaves inb cadmium stressed *Phaseolus coccineus* depends on their growth stages, *Acta Soc. Bot. Polon.* **64**, 273–279.

Smarrelli, J. Jr. and Campbell, W.H. (1983) Heavy metal inactivation and chelator stimulation of higher plant nitrate reductase, *Biochem. Biophys. Acta* **742**, 435–445.

Srivastava, H.S. and Singh, R.P. (1987) Role and regulation of L-glutamate dehydrogenase activity in higher plants, *Phytochemistry* **26**, 597–610.
Stitt, M. (1999) Nitrate regulation of metabolism and growth, *Curr. Op. Plant Biol.* **2**, 178–186.
Stohs, S.J. and Bagchi, D. (1995) Oxidative mechanisms in the toxicity of metal ions, *Free Rad. Biol. Med.* **41**, 553–575.
Tang, C., Robson, A.D., and Dilworth, M.J. (1990) A split-root experiment shows that iron is required for nodule initiation in *Lupinus angustifolius* L., *New Phytol.* **115**, 61–67.
Taylor, R.W., Williams, M.I., and Sistani, K.R. (1991) N_2 fixation by soybean-*Bradyrhizobium* combinations under acidity, low P and high Al stress, *Plant Soil* **131**, 293–300.
Thorneley, R.N.F. (1992) Nitrogen fixation: a new light on nitrogenase, *Nature* **360**, 532–533.
Tills, A.R. and Alloway, B.J. (1981) The effect of ammonium and nitrate nitrogen source on copper uptake and amino acid status of cereals, *Plant Soil* **62**, 279–290.
Tripathi, R.D. and Smith, S. (1996) Effect of chromium (IV) on growth, pigment content, photosynthesis, nitrate reductase activity, metabolic nitrate pool and protein content in duckweed *Spirodela polyrrhiza* L., in M. Yunus, (ed.), ICPEP, 96, Book of abstracts, pp. 159.
Tripathi, A.K. and Tripathi, S. (1999) Changes in same physiological and biochemical characters in *Albizia lebbek* as bioindicators of heavy metal toxicity, *J. Environ. Biol.* **20**, 93–98.
Vajpayee, P., Sharma, S.C., Tripathi, R.D., and Yunus, M. (1999) Bioaccumulation of chromium and toxicity to photosynthetic pigments nitrate reductase activity and protein content of *Nelumbo nucifera* Gaertn., *Chemosphere* **39**, 2159–2169.
Vajpayee, P., Tripathi, R.D., Rai, U.N., Ali, M.B., and Singh, S.N. (2000) Chromium (VI) accumulation reduces chlorophyll biosynthesis, nitrate reductase activity and protein content in *Nymphaea alba* L., *Chemosphere* **41**, 1075–1082.
Van Assche, F., Cardinaels, C., and Clijsters, H. (1988) Induction of enzyme capacity in plants as a result of heavy metal toxicity: dose response relations in *Phaseolus vulgaris* L., treated with zinc and cadmium, *Environ. Poll.* **52**, 103–115.
Van Assche, F. and Clijsters, H. (1990) Effects of metals on enzyme activity in plants, *Plant Cell Environ.* **13**, 195–206.
Vara, F. and Serrano, R. (1982) Partial purification of the proton-translocating ATPase of plasma membranes, *J. Biol. Chem.* **257**, 12826–12830.
Vesper, S.J. and Weidensaul, T.C. (1978) Effects of cadmium, nickel, copper and zinc on nitrogen fixation by soybean, *Water Air Soil Pollut.* **9**, 413–422.
Von Wiren, N., Gazzarini, S., and Frommer, W.B. (1997) Regulation of mineral nitrogen uptake in plants, *Plant Soil* **196**, 191–199.
Wang, M.Y., Siddiqi, M.Y., Ruth, T.J., and Glass, A.D.M. (1993). Ammonium uptake by rice roots, *Plant Physiol.* **103**, 1259–1267.
Weber, M.B., Schat, H., Ten Bookum-Van Der Maarel, W.M. (1991). The effect of copper toxicity on the contens of nitrogen compounds in *Silvena vulgaris* (Moench) Garcke., *Plant Soil* **133**, 101–109.
Wierzbicka, M. (1987) Lead accumulation and its translocation barriers in roots of *Allium cepa* L. – autographic and ultrastructural studies, *Plant Cell Environ.* **10**, 345–351.
Wilson, D.O. and Reisenauer, H.M. (1979) Effect of manganese and zinc ions on growth of *Rhizobium*, *J. Bacteriol.* **103**, 729–732.

CHAPTER 14

PLANT GENOTYPIC DIFFERENCES UNDER METAL DEFICIENT AND ENRICHED CONDITIONS

S. LINDBERG[a] AND M. GREGER[b]

[a]*Department of Plant Biology, Swedish University of Agricultural Sciences SE-750 07 Uppsala, Sweden*
[b]*Department of Botany, Stockholm University, SE-106 91 Stockholm, Sweden*

1. INTRODUCTION

Plants are able to grow under various conditions and different genotypes have developed which are able to cope with extreme environments. Since plants are stationary they have to develop different strategies for surviving in e.g. soils which are deficient in nutrients or in soil with toxic elements. To be able to grow under extreme conditions such species will be competitive with other species.

Some heavy metals are micro nutrients (Cu, Fe, Mn, Mo, Ni and Zn), and are thus essential to plant growth and development, usually at very low (μmolar) concentrations. However, elevated concentrations of micro nutrients, or the presence of heavy metals without nutritional functions, cause toxicity to most plant species and develop stress symptoms in the plant. Within different species, or cultivars, there are genotypes that are tolerant to toxic metals at different degree. Tolerance to one metal may sometimes also be associated with tolerance to another metal, or to other stress conditions. This chapter will focus on genotypic differences in tolerance both to low concentrations of micronutrients and to high concentrations of toxic metals such as heavy metals and aluminium. Aluminium is a light metal that causes serious toxicity to plants on acidic soils. The chapter will also deal with some applications for highly metal-tolerant genotypes, such as their use in phytoremediation and mining.

2. GENOTYPIC DIFFERENCES UNDER METAL DEFICIENT CONDITIONS

Plant suffer from nutrient deficiency stress when the nutrient availability is below that required for optimal growth. There are different reasons for mineral deficiency such as low nutrient concentration in the soil, low mobility of nutrients, that the nutrients are present in a chemical form which can not be taken up by the roots or has a low availability in the tissue. The mobility of nutrients depends on the mass flow of water, adsorption capacity of the soil and soil pH (Pearson and Rengel 1997). The chemical form of the element within the soil determines its availability; e.g. of the two forms of iron; Fe(II) is available for plant cell uptake, but Fe(III) is not.

Different genotypes of a plant have different genetic constitution, that is a special set of alleles present in each cell of the organism (Rengel 1999a). Some plant species and genotypes have a capacity to grow and develop on soils of mineral deficiency. They are tolerant to nutrient deficiency by specific mechanisms either for efficient uptake of nutrients or for more efficient utilization of nutrients (Sattelmacher et al. 1994). Efficient genotypes may have the possibility to convert non-available nutrient forms in the soil into plant-available ones, or they may easily transport nutrients across the plasma membrane (Rengel 1999a) or within the plant. Plants may also be able to change the morphology of the roots, and form symbiotic associations to increase the uptake of nutrients (Marschner 1995). Moreover, plants are able to extrude chemical compounds into the soil to increase the efficiency at which nutrients are taken up. These characteristics are especially important under conditions when the level of micro nutrients is lower than required. Both mineral nutrients and toxic metals such as heavy metals and Al may interfere with the uptake and utilization of micro nutrients and make them more deficient. In Table 1 are shown some different resistance mechanisms which plants use to increase micronutrient uptake under deficient conditions.

Table 1. Resistance mechanisms in the roots under micronutrient stress

Mechanism	Micronutrient	References
cluster roots	Fe	White and Robson (1989)
	Zn	Dinkelaker et al. (1995)
exudation of organic acids	Fe	Marschner (1995)
	Zn	Pearson and Rengel (1997)
	Mn	”
mucilage	Fe	Marschner (1995)
	Zn	Pearson and Rengel (1997)
phytometallophore release	Cu	Kinnersley (1993)
	Zn	Cakmak et al. (1994)
phytosiderophore release	Fe	Römheld and Marschner (1995)
		Mori (1994)
reductase activity/H^+-release	Fe	Rengel (1999a,b)
retranslocation	Zn	Pearson and Rengel (1994)
	Ni	Gerendas et al. (1999)
symbiotic associations	Cu	Li et al. (1991)
	Zn	Bürkert and Robson (1994)
	Mn	Rengel et al. (1998)
transfer cell formation	Fe	Bienfait (1989)
V_{max} increase	Fe	von Wiren et al. (1995)

2.1. Copper as a micronutrient

Copper (Cu) is required for several important biochemical and physiological processes (Bussler 1981). This element is an essential component of the plant enzymes diamine oxidase, ascorbate oxidase, o-diphenol oxidase, cytochrome c oxidase, superoxide dismutase, plastocyanin oxidase and quinol oxidase (Delhaize et al. 1985). In the absence of Cu these enzymes are inactivated. Like iron, copper is involved in redox reactions ($Cu^{2+}+e^- <=> Cu^+$) in the mitochondria and in the light reactions of photosynthesis (Taiz and Zeiger 1998). In *legumes* receiving a low copper supply, nodulation and N_2 fixation are depressed. Under condition of Cu deficiency, there is also a decrease in phenolase activity resulting in accumulation of phenolics (Judel 1972) and a decrease in the formation of melanotic substances. The decline in phenolase activity may also indirectly be responsible for a delay in flowering and maturation (Reuter et al. 1981).

The initial symptom of copper deficiency is the production of dark green leaves, sometimes with necrotic spots. However, the demand for Cu is within a very small concentration range - plant Cu concentrations are usually between 2-20 ppm and, therefore, there is a need for mechanisms that secure Cu homeostasis. Many investigations have shown different tolerance strategies in plants for avoiding toxic Cu concentrations (See below), but there is scarce information on plant strategies to cope with deficient levels of Cu. Among the cereals, rye is more tolerant to Cu deficiency than wheat and oats (Pearson and Rengel (1997). Differences in Cu efficiency also exist between different wheat cultivars (Marschner 1995). The Cu uptake and concentrations in Canadian spring wheat may be affected by nitrogen fertilizers (Soon et al. 1997). The Cu concentration in the plant was correlated with grain yield, suggesting that either varietal differences in Cu uptake may affect yield, or a prior selection for yield may have resulted in the selection for Cu-efficient plants as well.

Triticale, a hybrid of wheat and rye has a high copper efficiency similar to rye, indicating that the uptake mechanism for Cu is genetically controlled. Graham et al. (1987) was able to isolate the tolerant gene found in the long arm of chromosome 5R of rye and introduce it into wheat and oats. Thereby they could double the yield of wheat and oats on Cu-deficient soils.

2.2. Manganese as a micronutrient

Manganese (Mn), like copper, is a necessary element for plants. At a high soil pH manganese forms relatively insoluble compounds not available to the plant roots. Availability of Mn also depends on its oxidation state; the oxidized form Mn^{4+} is not available to plants, while the reduced form Mn^{2+} is (Rengel 1999a,b). Therefore, under anoxia Mn and Fe serve as alternate electron acceptors for microbial respiration and are transformed to the reduced form which increases the solubility and availability for plant uptake. Manganese availability may also be affected by Mn-oxidizing and Mn-reducing microorganisms that colonize the roots (Rengel 1997).

Manganese ions (Mn^{2+}) activate several enzymes, such as decarboxylases and dehydrogenases involved in Krebs cycle (Taiz and Zeiger 1998) and this element is necessary for the photosynthetic oxygen evolution (Marschner 1995). Manganese is a cofactor for both phenylalanine Ammonoia-lyase, that mediates production of cinnamic acid and other phenolic compounds, and for peroxidase involved in polymerization of cinnamyl alcohols into lignin (Burnell 1988). Deficiency of Mn, therefore, inhibits the building of phenolic substances and lignin which is considered important for the defense of fungal infections (Matern and Kneusel 1988, Rengel 1997). Manganese deficiency causes intravenous chlorosis and small necrotic spots. As under Fe deficiency, root growth is more inhibited than shoot growth.

In different genotypes of barley (*Hordeum vulgare* L.) manganese efficiency was demonstrated in soil culture in terms of vegetative growth and relative grain yield, and under controlled conditions in a nutrient solution it was expressed in terms of dry matter production and Mn concentration in the plant tissue (Huang et al. 1994).

However, when manganese was supplied as Mn^{2+} buffered by HEDTA (N-(2-hydroxyethyl)ethylenedinitrilotriacetic acid) Mn efficiency was not expressed, either in terms of dry matter production, Mn concentration, or Mn accumulation. Therefore, it is likely that Mn efficiency in barley is not related to differences in uptake rate of Mn, but may involve some process which increases the availability of soil Mn to plant roots, e.g. colonization of root bacteria. It was shown that Zn and Mn availability modify the rhizoflora of *wheat* roots and that the effect was strongly dependent on the genotype (Rengel et al. 1998). The most Mn-efficient wheat genotype C8MM (which grows better and yields more than Mn-inefficient genotypes under Mn-deficiency conditions) had roots colonized with more pseudomonads than other genotypes. Zn-efficient Aroona sustained colonization of many non-pseudomonads under Zn deficiency compared with control conditions. Other genotypes showed the same tendency. Also in dicots the Mn-efficiency may depend on bacteria. In Mn-deficient environment of soybean roots, up to ten-fold greater numbers of Mn-oxidizing bacteria have been found, but fewer Mn-reducers (Huber and McCay-Buis 1993).

2.3. Molybdenum as a micronutrient

In solution molybdenum (Mo) mainly occurs as molybdate oxyanion, MoO^{2-}_4 (Mo VI). In mineral soils of low pH both phosphate and molybdate strongly absorb to iron oxide hydrate and in the uptake by roots these anions compete (Marschner 1986). Molybdenum ions are cofactors of several enzymes, such as nitrate reductase, nitrogenase, xanthine oxidase/dehydrogenase, aldehyde oxidase and sulfite oxidase. Since Mo is involved both in nitrate assimilation and nitrogen fixation, a molybdenum deficiency may also cause nitrogen deficiency, if the nitrogen source is mainly nitrate, or if the plants depend on symbiotic nitrogen fixation (Taiz and Zeiger 1998). The molybdenum requirement of root nodules in legumes and non-legumes, e.g. *alder*, is relatively high. Under Mo deficiency older leaves will get chlorosis between veins and necrosis. In broccoli and cauliflower the leaves do not show necrosis but become twisted and even die due to insufficient differentiation of vascular bundles (whiptail desease). Flower formation may be prevented or the flowers abscise before they are fully developed. Molybdenum deficiency also affects the pollen formation in maize (Gubler et al. 1982).

2.4. Nickel as a micronutrient

Nickel (Ni) is an essential element for plant growth and is probably involved in different physiological processes. Deficiency of Ni is not very common. This element is a component of the enzyme urease and is essential for its function (Dixon et al. 1975, Gerendas et al. 1999). The incorporation of Ni into urease apo-protein requires an active participation of several accessory proteins (Gerendas et al. 1999). When nitrogen is supplied as urea, Ni increases plant growth of many different species, such as rye, wheat, soybean, rape, zucchini and sunflower (Gerendas et al.1999). Nickel is essential for plant species that use ureides in their metabolism (Marschner 1986). Nickel-deficient plants accumulate urea in their leaves and show leaf tip necrosis (Taiz and Zeiger 1998). Nitrogen-fixing microorganisms require nickel for hydrogen uptake hydrogenase and Ni-deficiency may, therefore, cause nitrogen-deficiency in the plant. Compared with other heavy metals the mobility of Ni within the plant is rather high (Gerendas et al 1999).

2.5. Zinc as a micronutrient

Zinc deficiency is common both in highly weathered acid soils and in calcareous soils of high pH. In the latter case also iron deficiency is common. Zinc is a co-factor of more than 200 enzymes, such as oxidoreductases, hydrolases, transferases, lyases, isomerases, and ligases (Jackson et al 1990). Many of the metalloenzymes are involved in the synthesis of DNA and RNA and protein synthesis and metabolism. Zinc is important for nitrogen metabolism in plants (Mengel and Kirkby 1987).

The most characteristic symptoms of zinc deficiency in dicots are stunted growth due to shortening of internodes, "rosetting" and a decreases in leaf size (Marschner 1986). The shoot/root ratio usually decrease under Zn deficiency, because root growth is less affected than shoot growth. Deficiency also causes interveinal chlorosis in leaves and reduction in crop production. In cereals chlorotic band along the midrib and red necrotic spots often occur on the leaves and are likely the result of phosphorus toxicity (Rahimi and Bussler 1979). Usually grain and seed yield are more depressed by zinc deficiency than total dry matter production. During the fertilization most of the zinc is incorporated into the developing seed and high amount of zinc is also found in the pollen grains (Polar 1975)

Cereals are differently responding to zinc deficiency. Three wheat genotypes (*Triticum aestivum* and *T. turgidum*) differed in their root-growth response to low zinc levels (Dong et al. 1995). The zinc-efficient genotype increased root and shoot dry matter and developed longer and thinner roots (a greater proportion of fine roots with diameter = 0.2 mm) compared with the less efficient genotype. Due to a larger root surface area the efficiency of zinc uptake increased. In wheat Zn can be remobilized from leaves under Zn deficiency (Pearson and Rengel 1997). Also spinach, potato, navy bean, tomato, sorghum and maize show great variations in Zn efficiency (Graham and Rengel 1993).

Compared to Zn-sufficient roots, wheat under Zn-deficiency develops thicker roots on average, with a shorter length of the fine roots (diameter = 0.2 mm). Treatment with the herbicide chlorsulfuron decreased the growth of these fine roots during the first 4 d of exposure of the herbicide and then completely inhibited growth of these roots (Rengel 1999a). The Zn-efficient genotype Excalibur (*Triticum aestivum*) maintained a longer length of fine roots and had a greater surface area than the non-efficient genotypes Gatcher (*Triticum aestivum*) and Durati (*Triticum turgidum* conv. durum) regardless of plant age or Zn nutrition. Chlorsulfuron also decreased net uptake of Cu and Mn in all three genotypes and net uptake of Zn in Gatcher and Durati genotypes. For the Zn-efficient Excalibur a longer time of exposure to chloroform was needed before Zn uptake was reduced.

The capacity for root exudation of phytosiderophores (PSs) differs in different graminaceous species under Zn deficiency (Cakmak et al 1994, 1996, Rengel 1997, Rengel et al. 1998. When six bread wheat genotypes (*Triticum aestivum* L.) were compared with one durum wheat genotype (*Triticum durum* L. cv. Kunduru-1149). Erenoglu et al (1996) found that all bread wheat genotypes contained similar Zn concentration in dry matter despite quite different Zn efficiency. In all genotypes supplied adequately with Zn, the rate of phytosiderophore release was very low and did not exceed 0.5 µmol/48 plants/3 h. Under Zn deficiency the release of PS increased in all bread wheat genotypes, but not in the durum wheat genotype. However, Dagdas-94, the most efficient genotype, did not show the highest rate of PS release. It is not yet proved that the PSs have an important role in zinc uptake and transport. For instance the PSs have a two-fold higher affinity for Fe than for Zn. In experiments with different wheat genotypes it was shown that Zn deficiency, in comparison with Zn sufficiency, caused an increase in Cu/ZnSOD concentration in tolerant genotypes, but not in sensitive ones (Yu et al. 1998). It was concluded that Cu/ZnSOD might be involved in reducing Zn-deficiency-related oxidative damage in genotypes tolerant to Zn deficiency.

In *maize* genotypes zinc accumulation was found to be genetically controlled and affected by additive genes (El-Bendary et al. 1993). When two high accumulators (Rg-5 and Rd-2) and two low Zn accumulators (Rg-8 and G-307) and a moderate one (K-64) were used to obtain all possible ten F1 hybrids, the high Zn accumulating parents were the best general combines. Four genes were found to be the minimum segregation factors in the (high X low) crosses for Zn accumulation. Zinc deficiency also increases root exudation of amino acids, sugars and phenolic substances at different degrees in different species (Zhang 1993).

2.6. Compatibility with Al^{3+}

Although plants usually have a possibility to take up aluminium, this element is not necessary for plant growth and development. Beneficial effects of Al in the growth medium are also seldom reported but may exist at very low soil pH, when toxic protons can be exchanged for less toxic aluminium ions (Al^{3+}; Kinraide 1993). Aluminium may also alleviate toxic effects by copper on *Triticum aestivum* (Hiatt et al. 1963) and by cadmium on *Holcus lanatus* (McGrath et al. 1980).

2.7. Other compatibilities in micronutrient uptake

If one micronutrient is deficient, uptake of other micronutrients may be facilitated (Kochian 1991). Such compatibility was reported for manganese and copper or zinc (Del Rio et al. 1978) and between zinc and iron, manganese or copper (Rengel and Graham 1995, 1996; Rengel 1999a) and also between iron and manganese (Iturbe-Ormaetxe et al. 1995). Manganese interaction with nitrogen has also been reported for barley genotypes differing in manganese efficiency (Tong et al. 1997).

Often zinc (Zn) uptake in plants is affected by the presence of phosphate. Two oilseed rape genotypes (*Brassica napus* L.) responded differently to Zn application (Lu et al. 1998). Both shoot and root dry matter production at different Zn concentrations was significantly greater under high phosphate (P) supply compared with low phosphate supply. A high P supply did not accentuate Zn deficiency symptoms compared with a low supply, or even absence of Zn. On the other hand, P concentration in plant parts was higher in the absence of Zn than in its presence. An increase in P supply had no effect on Zn concentration in the genotypes, but increased P concentration and P uptake in shoots.

In the presence of high cobalt level the concentrations of manganese (Mn) and iron (Fe) in the shoot may decline, causing clorosis in the young leaves. In mung beans treated with 5 µM cobalt (Co), the concentrations of Mn and Fe decreased with 55% and 80%, respectively (Liu et al. 2000). Co did not affect the uptake of iron (Fe) into the roots, but the transport of Fe to the shoots was greatly reduced. In mung beans Co increased the uptake of sulphur into the plant and its transport to the shoots, where as the S concentration in the leaves was increased 2-fold. In oilseed *rape* also boron (B) and zinc may interact. Grewal et al. (1998) found that Zn deficiency enhanced boron concentration in younger and older leaves. Boron concentration was higher in older leaves than in younger leaves, irrespective of B deficiency, and indicated immobility. Zinc concentration was higher in younger leaves than in older ones, indicating mobility. An increased supply of Zn increased B uptake under high Zn supply in the genotype that was most tolerant to low B and low Zn.

2.8. Strategies I and II for uptake of Fe in different species and genotypes

Plants are unable to take up iron in the Fe(III)-form and have developed different mechanisms to cope with this problem. Within strategy I, adopted by non-graminaceous monocots and dicots, the plant roots exclude protons and are able to reduce Fe(III) to Fe(II) in their plasma membranes (Figure 1). Within Strategy II, adopted by grasses, the roots are able to exclude phytosiderophores that form complex with Fe(III). The chelated Fe can then be transported across the plasma membranes (Rengel 1999 a,b). It was suggested that in soybean roots the Fe reduction shows diurnal rhythmicity, when the plants are grown in a normal 16/8 h light/dark cycle (Stevens et al. 1994), similar to the phytosiderophore release in some grasses. Under continuous illumination no such diurnally rhythmic pattern was obtained.

Within both Strategy I and Strategy II various species and genotypes within species differ in Fe efficiency (Jolley and Brown 1994). A Fe-efficient soybean genotype accumulated more apoplastic Fe in the roots than a Fe-inefficient genotype, and the former then easier could increase the Fe-pool in the shoot (Longnecker and Welch 1990). Wei et al (1997) showed that Fe-efficient *clover* showed exudation of protons at a three-fold higher rate than the Fe-inefficient genotype. On the other hand, no difference in Fe^{3+}-reduction rate or exudation of complexing agents was found in the genotypes. Within woody plants with Strategy I, *Vitis berlandieri* and *V. vinifera* also showed greater rhizosphere acidification, but excreted more organic acids and had a better capacity to reduce Fe(III) chelates in comparison with Fe-inefficient *V. riparia* (Brancadoro 1995).

2.8.1. Strategy I genotypes
Both iron-efficient and iron-inefficient genotypes of apple have been found. The roots and rhizosphere of *Malus xiaojinensis* (Fe-efficient) and *Malus baccata* (Fe-inefficient) responded differently to four different pH-levels tested (Han et al. 1997). The efficient genotype had a higher weight per seedling than the inefficient one at pH's from 7.4 to 8.4. The reducing abilities of root exudates were decreased with increasing of the solution pH from 5.4 to 8.4 and were more than twice as high in *M. xiaojinensis* than in *M. baccata*. The former genotype had a significantly higher respiration rate than the latter genotype at high pH and more reduced the solution pH in distances of 0-4 mm to root surface or in distances of 5-10 mm along the root from the root tip. Although Fe-reducing capacity of +Fe-treated plants proved to be linearly and positively correlated with Fe uptake in *pear* and *quince* genotypes, pear and quince roots do not increase their ability to reduce Fe (III) under Fe-deficiency stress (Tagliavini et al 1995)

When a Fe-hyperaccumulation mutant (dgl) of *pea* (*Pisum sativum*) and a parental (cv Dippes Gelbe Viktoria (DGV) genotype were raised in the presence and absence of Fe added as Fe(III)-N,N'-ethylene bis[2-(2-hydroxyphenyl)-glycine, the reductase activity of both initially increased with low rates followed by a nearly 5-fold stimulation in rates by day 15 (Grusak and Pezeshgi 1996). In DGV, the root Fe(III)reductase activity increased only a little by day 20 in +Fe-treated plants and about 3-fold in -Fe-treated plants. The net proton efflux was enhanced in roots of -Fe-treated DGV and both dgl types, relative to +Fe-treated DGV. In the dgl mutant, the enhanced proton efflux occurred prior to the increase in reductase activity. Moreover, reductase studies using reciprocal shoot:root grafts demonstrated that shoot expression of the dgl gene leads to the generation of a signal that enhances the reductase activity in roots. The dgl gene product may alter or interfere with a normal component of a signal transduction mechanism regulating the Fe homeostasis in plants.

2.8.2. Strategy II genotypes
In graminaceous species the rate of phytosiderophore (PS) release was related to Fe deficiency, which decreased in the order barley > maize > sorghum (Römheld and Marschner 1990) and oat > maize (Mori 1994). On the other hand, there was no difference in PS released in the investigated maize genotypes under Fe-deficiency. Moreover, a Fe-inefficient genotype had a 10 times lower V_{max} of the saturable component of uptake than the Fe-efficient one.

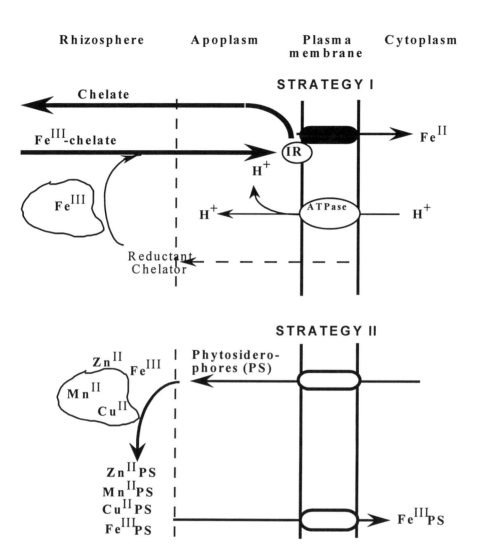

Figure 1. Strategy I and II for Fe^{III} uptake by plant roots (Marschner 1995). IR, inducible reductase.

In barley genotypes with differing manganese (Mn) efficiency, iron deficiency stress specifically induced barley genes *Ids1* and *Ids2* (Huang et al. 1996). Ids1 was equally expressed in the roots of both genotypes investigated, and this expression was not affected by Mn supply level and does not contribute to Mn efficiency. On the other hand, the gene *Ids2* was expressed at a higher level in roots of the Mn-efficient genotype than in the Mn-inefficient genotype, but the expression levels were not affected by Mn supply. Both genes had higher expressions in three-week-old plants than in two-week-old ones. The different expression of *Ids2* may indicate that this gene plays a role both in the Fe deficiency response and in the Mn efficiency mechanism.

Under both iron and zinc deficiency phytosiderophores are released by grass roots. In an investigation with two wheat cultivars; Zn-efficient Aroona and Zn-inefficient Durati, it was found that the zinc-efficient cultivar responded to Zn deficiency by increasing phytosiderophore release usually after 10 days, whereas the release in Zn-inefficient cultivar was very low during 25 days' growth (Cakmak et al. 1994). In contrast, under Fe deficiency, and also under both Fe and Zn deficiency, Aroona and Durati released similar amounts of phytosiderophores. Moreover, both under Zn and Fe deficiency the same phytosiderophores were released. The results show that phytosiderophore release may be involved in Zn efficiency in graminous species.

Deficiency of iron often causes chlorosis in leaves but is not necessarily depending on a low shoot content of iron. A high accumulation of phosphorus, aluminium, copper, manganese and zinc may interfere with Fe metabolism within the plant (Foy et al. 1998). In acid soils (pH 5.0) six *Nilegrass* (*Acroceras macrum,* Stapf.) clones differing in ploidy levels showed different symptoms of chlorosis. When they were grown in nutrient solution chlorosis was caused both by low Fe and high Mn levels. More often Fe deficiency was obtained in tetraploids, than in hexaploid or pentaploids clones.

3. MECHANISMS BEHIND TOLERANCE TO METAL DEFICIENCY

Plants cope with nutrient deficiency stress by adapting different mechanisms, e.g. mechanisms that increase the ability of the root to take up nutrients, such as development of cluster roots or by increasing the root uptake area (Pearson and Rengel 1997). Another mechanism is to release chemical compounds into the soil in order to increase the solubility or availability of nutrients. For uptake of Fe and Zn the mucilage at the apical part of the root, mostly containing polysaccharides, is important. Also symbiotic associations with microorganisms, such as legumes with *Rhizobium* or terrestrial plants with mycorrhizal fungi increase the uptake of nutrients.

It is likely that many mechanisms are present in plants showing mineral efficiency, i.e. effective uptake and utilization of nutrients. Rengel (1999a) suggested that 1. environmental factors are important in the phenotypic expression of a certain efficiency mechanism as well as in the effectiveness of different mechanisms; 2. more than one mechanism may be responsible for the efficiency level in a particular genotype, 3. increased efficiency found in one genotype compared with another involves additional mechanisms not present in the less efficient genotype and 4. the effects of different mechanisms in a particular genotype may be additive.

3.1. Chelating agents increase the uptake of Fe, Cu, Mn, Ni and Zn

Plants require minimal amounts of certain metals, such as Cu, Fe and Zn for optimal growth and productivity, but excess of these metals is toxic to plants. Both at metal excess and metal deficient condition plants respond by synthesizing chelating substances (Kinnersley 1993).

Non-graminaceous monocots and dicots have a special strategy (Strategy I) to facilitate uptake of Fe, such as enhancement of Fe^{3+} reduction, exudation of protons or reducing/chelating compounds into the rhizosphere (Rengel 1999a). However, when grasses and other graminaceous monocots are subjected to Fe deficiency, the roots are secreting phytosiderophores (PS) that chelate with Fe^{3+}(Strategy II). The PS-Fe complex is transferred actively across the plasma membrane by a specific uptake system (Marschner and Römheld 1994, Welch 1995). It was shown that high temperature increases the exudation of PS (Mori 1994). Both phytosiderophores and phytochelatins, the latter induced in the plant at excess of Cu, Zn and Cd, have properties in common with biostimulants, such as humic substances and oligomers of lactic acid (Kinnersley 1993).

Plants also produce a variety of other compounds, such as organic acids, that are secreted from roots and capable of forming complexes with metal ions in the soil. The composition of the organic acids varies with different species and soil conditions. For example roots of *Banksia integrifolia* excrete mostly citric acid followed by malic acid and aconitic acid (Grierson 1992).

Besides organic acids, amino acids, such as histidine, can form complexes with metal ions. Krämer et al (1997) found that histidine could bind to nickel in the nickel hyperaccumulator *Alyssum lesbiacum*.

3.2. Resistance to copper, iron and manganese deficiencies

Mechanisms of resistance to copper and manganese deficiencies include exudation of organic acids from the apical part of the root and uptake of these elements by phytosiderophores. These mechanisms are also present for iron uptake (Marschner 1995). When malic acid is present at the surface of MnO_2, (Mn^{4+}), Mn^{2+} is released and chelated and will become more mobile in the rhizosphere (Pearson and Rengel 1997).

In addition to phosphorus deficiency, also iron and zinc deficiencies induce development of cluster roots (Marschner et al. 1987, Dinkelaker et al. 1995). Within the family Proteaceae so called "proteoid" roots are formed showing dense clusters of rootlets along the lateral roots (Pearson and Rengel 1997). Proteoid roots have been associated with local high concentrations of organic acids, reducing agents and hydrogen ions (Gardner et al. 1983). The efficiency of cluster roots was thought to be due to a high exudation rate of organic acids and coupled to H^+ATPase activity (Dinkelaker et al. 1995). Citrate excreted by proteoid roots could react with Fe^{3+} in the soil forming Fe-citrate complex (Gardner et al. 1983). In *Cyperaceae, Restionaceae* and eucalypts "non-proteoid-like" root clusters are formed including dauciform, capillaroid and stalagmiform roots.

3.3. Resistance to nickel and zinc deficiency

The Zn-efficient plants use many different strategies to increase Zn uptake and utilisation. Such mechanisms include development of longer and thinner roots (Dong et al 1995), changes in the root environments by release of Zn-chelating phytosiderophores (Cakmak et al. 1994, Rengel et al. 1998), increase of net Zn accumulation (Rengel et al. 1998), efficient utilization and compartmentalization of Zn within the cells, tissues and organs (Graham and Rengel 1993). Also changes in the plasma membranes such as maintaining the sulphydryl groups in a reduced state (Rengel 1999a) and increased synthesis of specific root-cell polypeptides have been found. Other mechanisms such as symbiotic associations with mycorrhizal fungi can increase the uptake of both Zn and Cu in plant roots by increasing the absorptive area (Bürkert and Robson 1994.

The genotypic differences in tolerance to iron-, zinc- and manganese-deficient soils were further described by Marschner (1986), by Prasad and Hagemeyer (1999) and by Rengel (1999a,b).

4. GENOTYPIC DIFFERENCES UNDER METAL ENRICHED CONDITIONS

In nature, trace element toxicity may be present in all living organisms. The toxic metal ions can be divided into two categories based on their capacity to bind to different ligands of organic molecules. Class A metals include Al^{3+} and form stable complexes with ligands containing oxygen, while class B metals, e.g. heavy metals such as Cu^{2+} and Cd^{2+}, form more stable complexes with ligands containing sulfur and nitrogen. Often the organic molecules produced by plants contain several of the specific ligands for a certain metal. The toxicity of the metals may be related to the interaction with these critical molecules, but also to the presence of other metal-binding organic substances (Jackson et al. 1990).

Aluminium is the third most abundant element in the Earth's crust, but most of the Al is present in chelated or precipitated forms or in minerals and it is therefore not available to plant roots. Heavy metals occur in comparatively small amounts under natural condition, but may be introduced into the environment in large amounts as pollutants and retained in soils in an irreversible way (Tiller 1989).

When present together various toxic metal ions interact to produce a range of additive, multiplicative, synergistic and antagonistic responses (Taylor 1989). Aluminium may act synergistically with ferrous iron and decrease growth of suspension-cultured cells and also alleviate the toxicity of Cu (Yamamoto et al. 1994).

Some plant species show a large variation in the reaction to toxic levels of metals, e.g. in Salix viminalis the response of different clones to Cd, Cu and Zn varies (Greger and Landberg 1999). It has been thought that the degree of tolerance to the

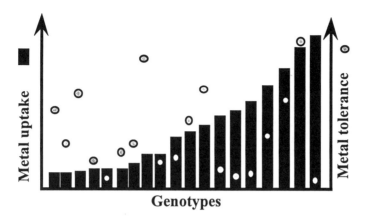

Figure 2 Metal tolerance (grey circle) and metal accumulation (black bars) of various genotypes of one and the same species. Each bar and circle represent a single genotype.

metals is related to the amount taken up, however, there is no correlation between uptake and tolerance (Greger and Landberg 1999; Figure 2).

4.1. Toxicity of heavy metals

In general, the symptoms of toxicity to heavy metals occur as chlorosis, yellowing of leaves, and as growth inhibition and burning. Toxicity levels of trace elements are extremely different for different species, and within different species, and range from 20 to 50 µg for copper and boron to several hundred µg for manganese, molybdenum and zinc.

4.1.1. Toxicity of cadmium
Cadmium has a high mobility in the soil and is easily taken up by plant roots. The uptake form is Cd^{2+}. Cadmium uptake in roots of sugar beet depends on external concentration, the supply method for cadmium and supply rate of mineral nutrients (Greger et al. 1991). Most of the cadmium is bound to the cell walls but some cadmium enters the plasma membrane. Recent results with wheat protoplasts show that cadmium can be taken up by calcium and potassium channels in the plasma membrane (S. Lindberg, and M. Greger, unpublished). Cadmium is probably also transported in free form within the plant (Greger and Lindberg (1986). Cadmium interferes with many essential metabolic reactions in the plant such as the carbohydrate metabolism, root tip respiration, formation of chlorophyll, and photosynthetic CO_2-fixation (Greger and Ögren 1991, Greger and Bertell 1992, Greger et al. 1991). It also irreversibly substitutes for micronutrients, such as copper and zinc, used as co-factors of different enzymes in RNA, DNA and protein metabolism (Jackson et al. 1990). Cadmium inhibits the uptake of potassium, magnesium and iron into roots; the transport of Mg and Fe to the shoots,

and increases the transport of calcium to the shoots (Lindberg and Wingstrand 1985, Greger and Lindberg 1986, Ouzounidou et al. 1997).

Toxicity symptoms are similar to Fe chlorosis, necroses, wilting, red-orange leaves and growth reduction (Greger and Lindberg 1986, Inouhe et al. 1994).

4.1.2. Genotypic differences in toxicity to cadmium
The growth of dicots is usually more severely inhibited than monocots but great differences occur also between species and genotypes. Quariti et al. (1997) found that 2 µM Cd caused 50% root growth reduction of bean (*Physeolus vulgaris* L.), but 50 µM Cd was needed for 50% root growth reduction of tomato (*Lycopersicon esculentum*) after 7 days. Also for 50% shoot growth reduction more Cd was needed for tomato than for bean.

Deciduous trees such as birch, are less sensitive to cadmium than spruce and pine (Österås et al. 2000). The sensitivity is also different between ecotypes within each species.

A great deal of cadmium is bound to the cell walls, e.g. in *Azolla* (Sela et al. 1988), but in many plants it is found in the cytoplasmic fractions of roots and leaves (Jackson et al. 1990). Cytosolic cadmium is bound to poly (gamma-glutamylcysteinyl) glycines, such as cadystins, phytochelatins, class III metallothioneins or gamma-glutamyl binding peptides (Prasad 1995, 1999). These thiol-rich compounds are induced in plants by the presence of cadmium, but tolerance mechanisms are not directly correlated with the ability to synthesize these substances, but rather to effectively form cadmium-polypeptide-complexes.

4.1.3. Genotypic differences in toxicity to copper
Copper and cadmium toxicity is primary the result of these metals high affinities for sulhydryl-containing molecules, e.g. enzymes and proteins. By binding to Cu and Cd the protein conformation is changed and thus also activities or catalytic specificities of the enzymes and proteins. Both phytochelatins (PCs) and metallothioneins (MTs) have been found which form complexes with Cu and Cd at different degree due to plant species (Jackson 1990, Reddy and Prasad 1990, Steffens 1990, Rengel 1997, Prasad 1999). The PCs are synthesized from glutathione by the enzyme phytochelatin synthase. The synthesis of the MTs seems to be constitutive and are not greatly elevated in plants exposed to high concentration of either metal ion. The thiol-rich copper-binding proteins occur in both copper-tolerant and copper-sensitive plants and are not necessary correlated with tolerance. Copper induces a poly (gamma-glutamylcysteinyl) glycine protein that complexes with Cu and has a higher affinity for Cu^{2+} than Cd^{2+}. In maize, 60% of the Cu was bound to MT-like proteins and about 25% was bound to PC peptides [possibly gamma-$EC)_n G$]. All plants analyzed seem to have the enzymes required for the polypeptide synthesis. In Cu-sensitive and Cu-tolerant *Silene vulgaris* both populations produced PCs when exposed to Cu. However, the sulphhydril content of roots was different and related to root growth. Also in Cu-sensitive and Cu-tolerant *Mimulus guttatus* polypeptides are induced by Cu in similar amounts.

4.1.4. Genotypic differences in toxicity to manganese
Nutrient toxicities in crops are more frequent for manganese and boron than for other nutrients. Manganese toxicity is present on acid soils in many parts of the world (Gupta and Gupta 1998). The problem is elevated when acid-forming nitrogenous fertilizers are applied. Boron toxicity occurs in irrigated regions where the irrigation water is high in B.

4.1.5. Genotypic differences in toxicity to nickel
When nickel is taken up into root cells, it may interact with different metabolic processes. Therefore, plants have developed mechanisms that prevent such interaction, e.g. Ni binding to polypeptides or amino acids in the xylem (Jackson et al. 1990). Nickel accumulates in leaves of plants that contain high concentrations of citrate and citric acid may be involved in detoxification of Ni (Lee et al. 1978).

In most plant species excess of Ni causes chlorosis. Dicots obtain interveinal chlorosis, while monocots obtain yellow stripes along the leaves and sometimes necrotic leaf margins (Mengel and Kirkby 1987). Many different methods were used for screening of Ni-tolerance among species and genotypes, such as measurements of fresh weight and root growth as well as histological examination of nucleoli (Liu et al. 1994), root length, shoot length, root/shoot dry biomass production and root/shoot tolerance index (Samantaray et al. 1998). For mung bean (*Vigna radiata* L) the root tolerance index (RTI) and shoot tolerance index (STI) were the best parameters for Ni tolerance. Eight cultivars of *V. radiata* were ranked with respect to their tolerance to Ni: Dhauli > PDM-116 > LGG-407 > K-851 > TARM-22 > TARM-1 > TARM-21 > TARM-26.

4.1.6. Beneficial effects by high uptake of nickel
Tolerance to high nickel concentration may be correlated to other types of stress tolerance, e.g. drought stress. When the effects of nickel were studied in two cultivars of durum wheat with different sensitivities to drought stress, *Triticum durum* cv. *Ofanto* (drought tolerant, DT) and Adamello (drought sensitive, DS), it was found that the DT cultivar was taking up approx. 3.5 times more Ni in the roots than the DS one did when grown at 35 µM Ni (Pandolfino et al. 1996). Moreover, the DT cultivar exhibited both a better growth and nutritional status when compared to the DS cultivar. Only in the latter cultivar the content of chlorophyll b decreased with Ni treatment, and the content of chlorophyll a was more decreased than in the DT cultivar. Nickel caused a less decrease in water potential and relative water content in the DT, than in the DS cultivar. Moreover, the antioxidative defence enzymes, guaiacol peroxidase, ascorbate peroxidase and glutathione reductase showed increased activity in Ni-treated DS seedlings; this increase in activity was not observed in the DT seedlings. Thus, in different wheat genotypes an increase capacity to take up nickel and to counteract Ni stress may be associated with drought resistance.

4.1.7. Effects of Pb on photosynthesis and respiration
Except for in some grasses and mosses, the transport of Pb^{2+} from roots to shoots is usually negligible in most plants. In tall fescue genotypes Pb supplied as $Pb(NO3)_2$ can reach the leaves by the transpiration stream and within minutes inhibit the photorespiration and dark respiration (Poskuta and Waclawczyk 1995). The inhibition of apparent photosynthesis and the activity of PSII was obtained within hours, required higher amounts of Pb, and was less pronounced. It was concluded that the primary targets of Pb toxicity in leaves of tall fescue were photosynthetic CO_2-fixation and photorespiratory CO_2 evolution, and that the resistance of PSII to Pb toxicity decreased with ploidy. The resistance of apparent photosynthesis and photorespiration instead increased with ploidy, when hexaploid, decaploid and their progeny an octaploid genotype were compared.

4.1.8. Genotypic differences in toxicity of zinc
General symptoms of Zn toxicity are turgor loss, necroses on older leaves and reduced growth (Wallnöfer and Engelhardt 1984). Some metallophytes, such as *Thlaspi caerulescens* have very efficient adaptations to high Zn concentration in the soil and can accumulate up to 4% Zn in the leaf dry matter (Tolra et al. 1996). The growth of the plant is stimulated by Zn concentrations which are toxic to most other plants.

Resistant genotypes sometimes contain larger amounts of metal in a detoxified form than sensitive genotypes, e.g. globular deposits of Zn-phytate are more abundant in Zn-resistant than Zn-sensitive *Deschampsia caespitosa* (Van -Steveninck et al. 1987). For *Betula* no difference was found (Denny and Wilkins 1987).

4.1.9. Favorable high Cu, Fe, Zn and Mn contents in plants
High contents of micronutrients may be favorable for crop plants used for food. Yang et al (1998) investigated different rice genotypes (*Oryza sativa* L.) and found sixteen different genotypes that contained higher grain Zn, Fe, Cu and Mn concentrations. Differences were found in the concentrations of these elements in polished rice and a fairly normal distribution among rice genotypes. In Indica rice the concentrations of Cu and Zn were about 2-fold higher than in Japonica rice, while Fe concentrations of Japonica rice were slightly higher than in Indica rice. Among Indica rice genotypes, red rice contained higher Zn than did white rice. Also protein and lysine concentrations differed considerably among the genotypes, but no close relationship between micro nutrients and protein and lysine concentrations were observed.

The copper content as well as N, P, and K contents in third whole leaf of sugarcane were shown to positively correlate with cane yield and juice quality parameters in different genotypes (Kumar and Verma 1997). Among 11 genotypes investigated, CoH 35 and CoH 108 showed the highest cane yield and contents of nutrients, while CoJ 64 had the lowest values.

4.2. Toxicity to aluminum

Aluminum toxicity is a major problem in naturally occurring acid soils and in soils with acidic precipitation. Toxicity occurs in soils with a pH below 5.5 and becomes more severe when pH drops below 5.0 (Foy et al. 1978). At a high pH Al is precipitated in insoluble forms. Aluminium speciation differs with pH and other ion present (Lindsay 1979). Table 2 shows the Al speciation at different pH in the presence of $AlCl_3$ and $CaSO_4$. Several monomeric species of Al, such as Al^{3+} are suggested to be toxic, but also the sum of all monomeric species may contribute to phytotoxicity (Blamey et al. 1983), as well as aluminum polymers (Wagatsuma and Ezoe 1985, Parker et al. 1989). Toxicity symptoms include inhibition of root and root hair growth and mineral uptake, especially uptake of calcium, phosphorus and potassium (Lindberg 1990, Rengel 1992a,b, Huang et al. 1995). Both drought effects caused by root growth inhibition and nutrient deficiency will cause a low crop production (Kochian 1995). Aluminium also causes accumulation of callose, exudation of malate, increased synthesis of phytochelatins and synthesis of heat-shock proteins (Rengel 1992b, Delhaize and Ryan 1995).

Aluminium can form cross-links between proteins and pectins within the cell walls (Chang et al. 1999) and can bind to membrane proteins and lipids (Lindberg and Griffiths 1993, Akeson et al. 1989). In tobacco cells Al sensitizes the membrane to the Fe(II)-mediated peroxidation of lipids (Ikegawa et al. 2000). Such peroxidation of lipids is probably a cause for loss of integrity of the plasma membrane and for cell death. Most of the available aluminum is bound into the cell wall but some Al can enter the plasma membrane and concentrate in the nucleus and interfere with DNA replication (Foy et al. 1978). Kinetic studies showed a biphasic uptake of Al in wheat (Pettersson and Strid 1989, Zhang and Taylor 1990). Recently it was shown by Sivaguru et al. (1999) that Al causes changes in microtubules (MTs) organization of suspension-cultured tobacco cells (*Nicotiana tabacum* L. cv.Samsun). The Al-induced changes both caused depolymerisation and stabilization of MTs depending on the phase of plant cell growth.

Table 2. Al speciation at different pH values in the external root media. In the presence of 50 μM $AlCl_3$ and 0.5 mM $CaSO_4$. $Al(OH)_3(s)$ corresponds to $10^{-4.4}$ M at pH 6.5.

pH in root medium	Log concentration (M)						
	$[Al^{3+}]$	$[AlOH^{2+}]$	$[Al(OH)_2^+]$	$[Al(OH)_3]$	$[Al(OH)_4^-]$	$[AlSO_4^+]$	$[Al(SO4)_2^-]$
4.0	-4.5	-5.5	-5.9	-7.5	—	-4.6	-6.1
5.0	-5.2	—	-4.5	-5.3	-8.6	-5.2	-6.9
6.5	—	-8.3	-6.0	-5.2	-7.1	—	—

4.3. Genotypic differences in toxicity to aluminum

In different *rice* cultivars Al is mainly confined to the roots and results in depressed root growth. In rice cultivars a better measure of Al toxicity is the relative shoot growth (Jan 1992). Tolerant cultivars transport only small amounts of Al to the shoots. When six different upland rice from Bangladesh were compared with three Philipinian cultivars from International Research Institute (IRRI), it was suggested that the local rice cultivars had less Al sensitivity than high yielding cultivars from IRRI. Aluminium sensitivity could be correlated with a low uptake and distribution of Ca and P, and not to the generally depressed internal Mg or Mn status of the plants (Jan 1992). Increased sensitivity to Al was also correlated with a high accumulation of Fe, Zn and Cu in the roots. Pregrowth of rice seadlings in the absence of Al for three weeks improved Al tolerance during subsequent growth in the presence of Al. In the pregrowth plants, high Al sensitivity was related to a decrease in mineral uptake, particularly of P and Mg.

In *wheat* cultivars (*Triticum aestivum* L) a visual symptom of Al toxicity is a reduction of both root and shoot growth. Little Al is usually transported to the shoots. Roots are more affected by Al and become short and thick and root tips turn brown (Strid 1996). Aluminium-treated cultivars cv. Kadett (Al-tolerant) and cv. WW20299 (Al-sensitive) took up more K and P than the control, whereas the concentrations of Mg and Ca were similar. In the shoots, concentrations of K, P and Mg were unaffected by Al, but the concentration of Ca was very much reduced. Experiments at 10°C and 25°C root zone temperature showed that Al was differently distributed in roots of Kadett at the two temperatures. At 25°C metabolic processes in Kadett delayed the Al transport to the site of toxicity but not in WW20299.

Durum wheat (*Triticum durum* Desf) was shown to be more sensitive to Al-toxicity in acid soils than the hexaploid wheat (*Triticum aestivum* L. em. Thell) but the former has not been adequately screened for Al tolerance (Foy 1996). When 15 lines of durum wheat were grown for 28 days at pH 4.5 and 6.0, a great difference in tolerance to acid soil was obtained based on relative shoot and root dry weight (wt at pH 4/wt at pH 6.0 times 100). Relative shoot dry weight alone was an acceptable indicator of acid soil tolerance. Durum lines PI 195726 (Ethiopia) and PI 193922 (Brazil) were significantly more tolerant than all other entries, even the Al-tolerant, hexaploid Atlas 66, used as a standard. The most sensitive durum lines (PI 322716, Mexico; PI 264991, Greece, PI 478306, Washington State, USA, and PI 345040, Yugoslavia, were as sensitive to acid soil as the Al-sensitive Scout 66 (*Triticum aestivum*). Concentrations of Al and phosphorus were significantly higher in durum wheat shoots of acid soil sensitive, than in those of tolerant lines, and exceeded those reported to cause Al and phosphorus toxicities in wheat and barley (Foy 1996).

4.3.1. Aluminum effects on root growth
The root hair development in different plants is often reduced by Al^{3+} activities greater than 2.5 µM. For different *white clover* genotypes the root hair length may differ three times. It was reported by Care (1995) that the root hairs of the long-haired population of the cultivar Tamar was more decreased by 2.2 µM Al^{3+} than the short-haired one. However, the root hair numbers decreased in a similar number for both populations.

4.3.2. Aluminium interference with mineral nutrients

Two important reasons for Al toxicity to plants are the inhibition of mineral uptake by Al and the impair of Al on the plasmalemma proton pump activity. Aluminium e.g. partly blocks the uptake of macro nutrients such as calcium, phosphate, potassium and magnesium and micro nutrients such as manganese. In sugar beet *(Beta vulgaris* L) cultivars with different sensitivity to Al in the order Primahill > Monohill > Regina, it was shown that in the less sensitive cv. Regina, two Al ions are probably bound per Ca^{2+} uptake site, but in the more sensitive cultivars only one Al ion is bound (Lindberg 1990. However, Al competitively inhibited the metabolic efflux of calcium mainly in the same way in the three cultivars. Only the less sensitive cv. Regina had a capacity to actively exclude phosphate from the roots and thus to prevent Al-phosphate precipitation within the plant roots. Al did not influence the non-metabolic (passive) transport of phosphate in sugar beets. The cv. Regina also more effectively transported Al to the shoot compared with the other cultivars.

Opposite effects of Al were recorded on K^+ ($^{86}Rb^+$) influx in absence (non-metabolic) and presence (metabolic) of dinitrophenol (DNP) (Lindberg 1990). With DNP Al inhibited the influx of K^+ ($^{86}Rb^+$) in roots in a competitive way in all three cultivars, but the inhibition constants were different. Therefore, the inhibition of non-metabolic influx was greater in Primahill (more sensitive) than in the other cultivars. In sorghum *(Sorghum bicolor* L. Moench) differing in Al-sensitivity, potassium concentration in roots and Mn concentration in shoots were higher in the tolerant than in the sensitive genotypes (Bernal and Clark 1997). However, the tolerant genotypes contained less calcium in both roots and shoots, and lower concentrations of P, Mg and Mn in the roots than the sensitive genotypes, and the mineral acquisition in shoots and roots of these genotypes was not a good indicator for distinguishing tolerance to Al toxicity. At moderate soil acidity, Mg deficiency limited growth of sorghum, whilst at high acidity root damage overruled the effect of Mg deficiency (Tan and Keltjens 1995).

In *Triticum aestivum* Al also blocks the influx of calcium by calcium channels in long-term experiments (Huang et al. 1995). The immediate reaction by Al addition is a transient increase of calcium in the cytosol (Lindberg and Strid 1997). Both these calcium reactions were found in sensitive and tolerant cultivars of wheat.

Besides calcium and magnesium (Keltjens and Tan 1993) silicon also may ameliorate Al toxicity (Hodson and Evans 1995). The positive effect by silicon has been shown in sorghum, barley, teosinte and soybean, but not in rice, wheat, cotton and pea. Silicon and Al may co-deposit within the cell walls as hydroxyaluminosilicate and/or aluminosilicate (Cocker et al. 1998). Calcium more than magnesium generally increases growth in dicots and magnesium more than calcium increases the growth of monocots under acid conditions (Keltjens and Tan 1993). The reason for the differing effect by Ca and Mg in dicots and monocots may be that Al has different capacity to bind to uptake sites for Ca and Mg in different plants.

4.3.3. Aluminum interference with plasma membrane $H^+ATPases$ and membrane potentials

By use of microelectrode technique it was shown that aluminum similarly affects the transmembrane potential (PD) of root cells in various cultivars and plant species (Lindberg 2000). The immediate effect depends on Al concentration, external pH, experimental and pretreatment temperatures, and other ions present (Lindberg et al. 1991, 1998). At external pH 4-5, Al usually causes a hyperpolarisation of PDv and PDc, the PD between the vacuole and cytosol of the impaled cell and the surrounding medium. At least in wheat cells the hyperpolarisation depends on an efflux of potassium (Lindberg and Strid 1997). However, at external pH 6.5, Al depolarizes the PDc, probably as a consequence of other Al species present at that pH, than at pH 4 and 5. Both K^+ and Ca^{2+} diminish the effects of Al on PD (Lindberg 2000). Table 3 and 4 show the Al effects on membrane potentials of sugar beet and spruce root cells.

Table 3. Effects of $AlCl_3$ on PDs in root cells of sugar beet. PD between the external medium and cell vacuoles (ΔPD_v) or cell cytoplasm (ΔPD_c). ND, no data. * Significantly different from Cl^- effect (0.01 < P < 0.05)

External root medium			Addition of $AlCl_3$			
			10 or 50 µM		100 µM	
KCl MM	CaSO4 mM	pH	ΔPDv MV	ΔPDc mV	ΔPDv mV	ΔPDc mV
0	0.5	4.0	-12±3 (5)*	-21±1 (4)*	+10±3 (4)	+7±2 (4)
0	0.5	5.0	-4±1 (4)*	-17±2 (4)*	+6±1 (5)	+4±2 (4)
0	0.5	6.5	-18±3 (8)*	+17±4 (5)*	-15±4 (6)*	+11±2 (3)*
0	1.0	4.0	-12±2 (6)*	+6±2 (4)	+9±6 (4)	-6±2 (5)
0	1.0	5.0	+3±2 (4)	+3±1 (3)	+1±1 (3)	+1±2 (3)
0	1.0	6.5	+17±2 (5)*	ND	+5±5 (3)	ND
2.0	1.0	6.5	+2±1 (6)	+3±3 (3)	+2±3 (4)	ND

Table 4. Effects of $AlCl_3$ on Pds in root cells of spruce. PD between the external medium and cell vacuoles (ΔPD_v) or cell cytoplasm (ΔPD_c) of intact roots of spruce after cultivation at 17/21°C (night/day). Electrical measurements at 21°C. ND, no data

pH of external root medium	Addition of $AlCl_3$	
	ΔPD_v, mV	ΔPD_c, mV
4.0	-10±3 (3)	ND
5.0	-5±3 (6)	-9±2 (3)
6.5	-17±4 (5)	+19±3 (6)

Moreover, upon several additions and removals to and from the root medium of Al-sensitive cultivars of wheat and sugar beet, Al gradually causes a depolarisation of PD of the root cells. The depolarisation probably depends on Al inhibition of the H^+ATPase activity, reflected as a decrease in cytosolic pH (Lindberg and Strid 1997). Also in other sensitive wheat cultivars, and in snapbean, Al induces depolarization (Miyasaka et al. 1989, Olivetti et al. 1995).

When sugar beets are raised in the presence of $AlCl_3$, the PDs largely depolarize both at low and high pH (Lindberg et al. 1991) and are likely reflecting an inhibition of the metabolic component of the PD by Al. This was confirmed by *in vitro* experiments with plasma membrane (PM) vesicles from sugar beet roots (Lindberg and Griffiths 1993), which showed that Al inhibits the PM ATPase in an uncompetitive way. Matsumoto and Yamaya (1986) also showed that Al inhibits the ATPase activity in membranes from *pea* roots.

5. MECHANISMS BEHIND TOLERANCE TO METAL ENRICHED CONDITIONS

Plants are able to cope with high metal levels in the surroundings either by a low net uptake or accumulation. Thereby they are using specific tolerance mechanisms to detoxify the metals by making them unavailable for toxic action. Two different strategies are thus found; excluders which avoid uptake also at quite high metal levels in the soil, and accumulators which accumulate high metal levels also at low metal concentrations in the soil (Figure 3). Furthermore, some plants tolerate metals by having a low metal translocation rate to the shoot, and in this way protecting the photosynthesis from damages.

Metal tolerance can be achieved by many different mechanisms (Figure 4), e.g. by binding to the cell wall, by metal chelation or precipitation with high-affinity ligands outside and inside the cytoplasm, by sequestration of the metal in the vacuole (Lichtenberger and Neumann 1997), in leaf epidermis and trichomes (Neumann et al. 1995, Krämer et al. 1997), or by exclusion or active extrusion of the metal (Zhang and Taylor 1989, Lindberg 1990). The toxic element is then hindered from interfering with sensitive metabolic reactions within the plant. In the presence of toxic metals, the metal tolerance can also be achieved by a capacity to keep, or to fast re-establish the internal Ca^{2+}, K^+ or H^+-homeostases. By the latter mechanism the toxic element little interferes with uptake and usage of essential nutrients or not directly affects the proton pumps. In addition the resistance can also be obtained by avoidance of contaminated areas, and formation of metal-resistant enzymes.

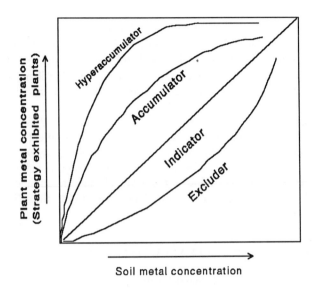

Figure 3. Different plant strategies to cope with high levels of toxic metals.

Tolerance is often specific for a certain metal, but the above mentioned tolerance mechanisms are general for more than one metal. It is therefore suggested that the reason for tolerance is not the existence of a single tolerance mechanism. Various sites of toxic actions are found in different plant genotypes and therefore different tolerance mechanisms, therefore, are needed (Landberg and Greger unpublished results).

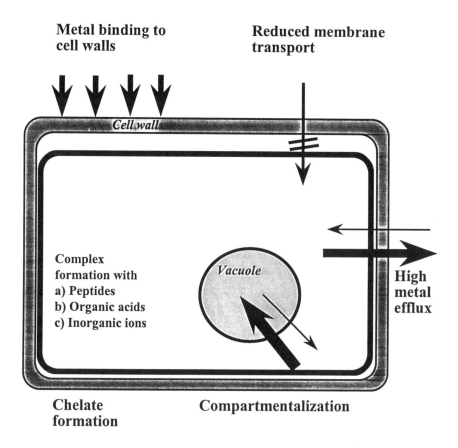

Figure 4. Cellular metal-tolerance mechanism

5.1 Cadmium tolerance mechanisms

Since cadmium has several sites of action within the plant, it is likely that tolerance is associated with mechanisms that either exclude it from the plant, or sequesters it in a less toxic form within the plant (Jackson et al. 1990). In birch (*Betula pendula*) cadmium was accumulated particularly in the fine roots and this accumulation together with a preference for root growth may be part of a mechanism for Cd tolerance (Gussarsson 1994). It was also suggested that tolerance of birch to Cd was due to a very low translocation rate of Cd to the shoot (Österås et al. 2000). However, root and shoot could be differently sensitive to a metal (Landberg and Greger unpublished results). This was also shown for birch which had much more sensitive shoots, than spruce and pine had, while the latter had much more sensitive roots (Österås et al. 2000).

Tolerance may also be due to a high root efflux of the metal. Cadmium-tolerant *Salix* clones were shown to have twice as high Cd efflux from roots as Cd-sensitive clones had (Landberg, Jensén and Greger unpublished results). The species *Salix* seem to have a low net uptake of Cd as a strategy to tolerate this metal (Landberg and Greger unpublished results). It is possible that the first Cd ions, which enter the roots, function as a signal to onset an efflux of Cd, in order to decrease the net uptake (Landberg, Jensén and Greger unpublished results).

5.2 Copper and nickel tolerance mechanisms

Large amounts of copper and nickel usually are found at copper mine areas as well as in serpentine soils. Previous investigations on *Mimulus guttatus* suggested that the same genes could be involved in tolerance to both these elements. Copper tolerance in *M. guttatus* was correlated with one major gene and a number of minor genes or modifiers. In order to study the effects of the major Cu tolerance gene on Cu - Ni co-tolerance, homozygous copper tolerant and non-tolerant lines were screened against nickel (Tilstone and Macnair 1997). There were significant differences between these lines for copper, but the same was not found for nickel, indicating that the major gene for copper tolerance does not govern Cu-Ni tolerance. By the screening of modifier lines, in which modifiers for different degrees of Cu tolerance were inserted into a non-tolerant background, it was found that genotypes possessing fewer copper modifiers yielded higher nickel tolerance than those genotypes with greater number of copper modifiers. Nickel tolerance, at least in this species, is thus heritable and under the control of different genes to those producing copper tolerance.

In some species Cu-tolerance involves the synthesis of thiol-rich polypeptides, but in others, such as spinach this is not the case (Turkendorf et al. 1984). Copper-tolerant *Deschampsia caespitosa* even produced less of thiol-rich Cu-complex than non-tolerant plants of the same species. Even if sulfur deficiency caused a reduced amount of thiol-rich compounds in *Deschampsia,* it did not interfere with Cu-tolerance. It was found that copper could be stored in the cell walls of some species and that Cu was more slowly taken up in some species. They also reported that Cu was excluded from the chloroplasts in some species. and suggested that these mechanisms were related to tolerance.

Tolerant species sometimes also are able to excrete chelating compounds, such as citrate and malate into the rhizospere and thus preventing the cells from Cu-toxicity

Plants with a high tolerance to nickel can be found in serpentine soils. These soils are present on all continents (Roberts and Proctor 1992) and can contain very high Ni concentration (up to 8000 ppm). They also contain high amounts of Cr and Co as well as a low Ca/Mg ratio. Many endemic species are adapted to serpentine areas and some of the species are hyperaccumulators, they can accumulate extremely high concentrations of trace elements. They also can stand low concentrations of calcium. Usually growth reduction on serpentine soils is not only due to toxic metals, these plants suffer from macronutrient deficiency (Nagy and Proctor (1997).

5.3. Zinc tolerance mechanisms

Some metallophytes can cope with extremely high concentration of Zn in the soil. It was suggested by Tolra et al. (1996) that only a part of the absorbed Zn in *Thlaspi* was involved in metabolic processes and that most of the Zn was either transported into the vacuoles or detoxified in chelated form. Another metallophyte that is frequent on metal rich soil is *Armeria maritima*. From experiments with this species Köhl (1997) concluded that plants from non-metalliferous soil populations were more sensitive to 1 mM Zn than plants from metalliferous soil. The former plants could, however, adopt to such high Zn concentration. It was suggested that genes for metal resistance are present in all genotypes. In plants from metalliferous soils these genes are constitutively active, while in plants from non-metalliferous soils the genes for metal resistance are activated by metal stress.

5.4. Aluminum tolerance mechanisms

Plants have adopted several tolerance mechanisms to cope with Al stress. Two classes of mechanisms have been proposed: those that allow plant to tolerate Al accumulation in the symplasm and those, which exclude Al from the root apex (Kochian and Jones 1997). The tolerance mechanisms include binding of Al to the cell wall, precipitation of Al with phosphate, secretion of citrate for chelation outside and inside the roots, secretion of proteins for Al binding, little interference with cytosolic calcium and pH homeostasis, and active exclusion of Al (Wagatsuma 1983, Koyama et al. 1988, Zhang and Taylor 1989, Lindberg 1990, Lindberg and Strid 1997). Usually an avoidance/exclusion mechanism is more important for Al tolerance than for heavy metal tolerance. Bennet and Breen (1991) proposed a signal transduction mechanism, where mucilaginous cap secretion and release of calcium may be mediated by endogenous growth regulators. In higher plants some of the tolerance mechanisms towards aluminum is specific for this element and associated with a dominant gene (Conner and Meredith 1985).

5.4.1. Aluminum tolerance mechanisms in sugar beets

From an investigation of three cultivars of sugar beet with different sensitivity to Al in the order Primahill > Monohill > Regina, it was concluded that the less sensitive Regina showed less growth reduction by Al due to less inhibition of DNP-independent Ca^{2+}influx, and to DNP-dependent extrusion of phosphate (Lindberg 1990). Also the more generally existing enhancement of DNP-dependent K^+ ($^{86}Rb^+$) influx by Al could serve as a defense mechanism against Al.

Moreover, Al differently affected the plasma membrane ATPase activity of sugar beet roots. Although Al *in vitro* inhibited the MgATPase (H^+ATPase) activity in an uncompetitive way for both Monohill and Regina, the specific K^+-activation of the ATPase was only inhibited in the more sensitive cv. Monohill (Lindberg 1986, Lindberg and Griffiths 1993). One reason for this could be a different composition or distribution of surface charges of the plasma membrane (PM), since both the binding of Al and interference with cation fluxes are different in the two cultivars (Lindberg 1990). In tolerant cultivars with less binding of Al to the PM and thus less inhibition of ATPase activity (Lindberg and Griffiths 1993), less interference with ion fluxes should be expected.

5.4.2. Aluminum tolerance mechanisms in wheat

Aluminum-tolerant cultivars of wheat (*Triticum aestivum*) may take up less ammonium ions than sensitive cultivars do, resulting in a higher solution pH and less solubility of the toxic ion (Taylor and Foy 1985). In aluminum-sensitive cultivars of wheat Al inhibits root growth within minutes at low pH, and in sensitive maize within 30 min (Rengel 1992, Ryan et al. 1992), but the main reason for the toxicity reaction has been debated.

In two wheat cultivars, Kadett (Al-tolerant) and WW 20299 (Al-sensitive) Al induced an increase of cytosolic calcium, and a decrease in potassium concentration but since the increase and decrease, respectively, were found in both cultivars, the reactions were not correlated with Al-tolerance (Lindberg and Strid 1997). By addition of $AlCl_3$ to root protoplasts isolated from both cultivars, it was found that that addition of 5 µM Al to the sensitive cultivar caused an acidification of the cytosol and this was gradually increasing by further addition. On the other hand, addition of Al to the tolerant cultivar caused only little acidification. At the same time the membrane potential of root cells were gradually depolarized in the sensitive cultivar, but was stable in the tolerant one. It was therefore concluded that Al inhibited the H^+ATPase activity in the sensitive cv, reflected by the depolarization, but little affected the H^+ATPase activity in the tolerant one, and that these mechanisms are important for Al-tolerance in wheat.

Delhaize (1993) showed that in near-isogen wheat lines differing in Al tolerance only the tolerant isoline released malate from the root apices. Also in other wheat genotypes malate exudation into the rhizosphere is connected with exclusion of Al from the root tips (Basu et al. 1994, Ryan et al. 1995)

5.4.3. Aluminum tolerance mechanisms in rice

From investigations with different rice cultivars it was concluded that aluminum-tolerant species accumulate Al rather in the roots than in the shoots (Thornton et al. 1986, Jan 1992).

5.4.4. Other aluminum tolerance mechanisms

Meredith (1978) selected aluminum-tolerant cultivars of tomato. Cell cultures do not exclude aluminum and should therefore be able to chelate the ion in a less toxic form within the cells. In barley and maize, tolerance is correlated with the ability to maintain high concentrations of organic acids within the roots in the presence of Al. Chelation with organic acids may occur both within the root cells, shoot cells or in the rhizospere of tolerant genotypes. Buck wheat (*Fagopyrum esculentum* Moench. cv. Jianxi) which shows high Al-tolerance accumulates Al in the leaves. About 90% of Al found in the leaves is present in the cell sap and complexes with oxalic acid (1:3). When the Al-oxalate complex was supplied to maize roots no phytotoxic reaction occurred. Aluminum also induces release of oxalic acid from root tips of buck wheat (Zheng et al. 1998). However, an overproduction of organic acids, will also affect the uptake and chelation of micronutrients and cost energy. It is, therefore, not clear if such a mechanism is sufficient to explain Al-tolerance in most species (Jackson et al. 1990).

Many Al-tolerant genotypes accumulate similar amounts of the toxic element in the root apoplasm as the sensitive genotypes do. Cell suspension of Al-sensitive *Phaseolus vulgaris* accumulates initially the same amount of Al as a Al-tolerant one, but after desorption of the roots in citric acid, the Al-sensitive genotype retained a greater amount of Al, than the Al-resistant one did (McDonald-Stephens and Taylor 1995).

6. GENOTYPIC VARIATIONS OCCURRING AT DIFFERENT HABITATS

Plants also present in environments with very high levels of metals in the ground. Most of these plants have an excluding strategy to avoid metal intoxication. It is common to found plants which either have low metal concentrations in the whole plant and exclude the metals at uptake, or they have a restricted metal translocation to the shoot and thus a high root accumulation (Coughtrey and Martin 1978, Landberg and Greger 1996, Stoltz and Greger unpublished). The reason for this has been discussed and it is possible that plants with these manners may protect the photosynthesis from damage.

On highly contaminated are as there also exist hyperaccumulators which are taking up very high amounts of metals. Thus a high accumulation factor exists (Figure 5). The translocation of metals to the shoot in such plants is high, and the concentration of the metal in the shoot often overcomes the concentration in the roots (Rascio 1977, Homer et al. 1991), which is not the case in a normal plant. The reason for this could be to protect the perennial organs of these plants and eliminating the metals through leaf-fall. Hyperaccumulators may also have the same high accumulation of metals also from low contaminated soils, as well as different populations of the same species may have a variability in hyperaccumulation (Lloyd-Thomas 1995). Taxonomically, hyperaccumulators of metals identified so far, account for less than 0.2% of all angiosperms. Many of the known hyperaccumulators are biennial or short-lived perennial herbs, or are shrubs or small trees.

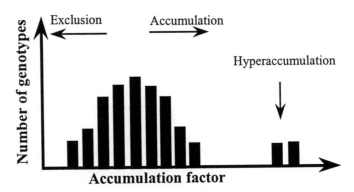

Figure 5. Genotype distribution in relation to response strategies of metal uptake

The reason for the adapted value of hyperaccumulation has been discussed and was concluded that hyperaccumulation benefits plants as a defense against herbivores or pathogens. To be able to resist these high concentrations of metals, tolerance mechanisms are needed and different kinds of sequestration or precipitation are likely to occur. One has thought that the mechanisms of tolerance and accumulation are the same. However, these two characters might be independent and in general, tolerance and accumulation may not be related characters (Ingrouille and Smirnoff 1986).

Plants with a high accumulation of e.g. Zn might also need higher levels of this element than other plants, not to suffer from deficiency. McGrath (1998) showed this phenomena for *Thlaspi caerulescens*. The reason for its high requirement may be related to the operation of strong constitutive mechanisms for sequestration and tolerance of Zn (Lloyd-Thomas 1995). An effective intrinsic mechanism may be able to make the metal unavailable in the cell, and could lead to deficiency even though the tissue concentration of the element is high.

On metalliferous outcrops or on antropogenically metal-contaminated substrates, hyperaccumulators are able to survive the strong selection pressure exerted upon other species and become dominant. The most common hyperaccumulators are those which accumulate Ni (Table 5). Many hyperaccumulators are found on serpentine (ultramafic) soils, e.g. at New Caledonia and Cuba (Brooks et al. 1979, Reeves et al 1996).

Table.5. Metal concentrations in plants with different properties to accumulate metals (Reeves and Baker 2000) and numbers of hyperaccumulator plants and their metal concentration (Baker et al. 2000).

Metal	Conc. in plant leaf, $\mu g(gDW)^{-1}$			Hyperaccumulators		
	Excluders	Normal plants	Accumulators	Conc. in leaf $\mu g(gDW)^{-1}$	No of Taxa	No of Families
Cadmium	0.03	0.1 - 3	20	> 100	1	1
Cobalt	0.01	0.03 - 2	20	> 1000	28	11
Copper	1	5 - 25	100	> 1000	37	15
Lead	0.01	0.1 - 5	100	> 1000	14	6
Manganese	5	20 - 400	2000	> 10000	9	5
Nickel	0.2	1 - 10	100	> 1000	317	37
Zinc	5	20 - 400	2000	> 10000	11	5

7. THE USE OF DIFFERENT TRAITS IN PLANT ACTIVITIES

Plants, tolerant to metals, can be used to remediate polluted sites. This technique is called phytoremediation and includes both the removals of metals from the site, so called phytoextraction, as well as immobilization of the metals in the rhizosphere, so called phytostabilization.

7.1 Phytoextraction

Plant genotypes with high uptake and high translocation to the shoot of the element in question can be used. These plants work as pumps, and use of perennial plants makes it possible to gradually harvest the above ground parts and thereby both remove the metals from the system and decrease the concentration of them in the soil.

Hyperaccumulators have been focused on a great deal, since they have the property to concentrate a very high metal level in the leaves (Table 5). *Thlaspi caerulescens* and *Alyssum murale* are examples on well-studied plant species with extraction of Zn, Cd and Ni. It has been shown, that *T. caerulescens* has a preference to grow roots into hot spots of Zn in the soil, which facilitates the uptake of this metal (Schwartz et al. 1999). The extreme high concentrations in hyperaccumulator plants at high contaminated sites such as mining areas make such plants serve as phytominers, which means that the metals in the plant tissues are high enough to be used in the metal industry. However, to purify sites with elevated metal levels by use of hyperaccumulators will take too long time (about a couple of 100 years).

When soils are to be purified from metals by plants, the metal removal can be calculated per area. To be a good phytoextractor the properties needed are not only a high tolerance and a specific uptake and translocation of the metal to the shoot, as well as a deep root system. The plant should also have a high biomass production, and this is not common among the hyperaccumulators. Thus, accumulator plants could better be used in phytoextraction, since they may have high biomass production. Moreover, phytoextraction is best suited at low-medium contaminated soil, such as metal-elevated arable soils. Cadmium accumulator clones of *Salix* have been shown to work well as phytoextractors on Cd-elevated arable land, due to its high biomass production and nearly similar Cd accumulation as some hyperaccumulators. *Salix* had a higher removal time of Cd than *Thlaspi caerulescence* and *Allysum murale* (Greger and Landberg 1999).

7.2 Phytostabilisation

The metals in soil at highly metal-contaminated sites should be immobilized to prevent the metals to be spread in the environment. Plant genotypes used in phytostabilisation have to be metal tolerant and to have a dense root system to prevent erosion. The most important function of these plants is to be able to colonize, or easily be established, on metal-contaminated land to diminish wind erosion. Furthermore, to decrease the mobilization of metals in the soil, plants have high metal accumulation in the rhizosphere and an extremely low translocation of the metals to the shoot. Plant roots also help to minimize water percolation through the soil, further reducing contaminant leaching. Plant roots also provide surfaces for sorption or precipitation of metals (Laperche et al. 1997). It was shown that in submersed *Eriophorum angustifolium*, *E. scheuchzeri* and mine tailings are able to reduce the leakage of metals from the tailings (Stoltz and Greger 2001).

REFERENCES

Akeson, M.A., Munns, D.N. and Burau, R.G. (1989) Adsorption of Al^{3+} to phosphatidylcholine vesicles. *Biochem. Biophys. Acta* **986**, 33-40.

Baker, A. J. M., McGrath, S. P., Reeves, R.D., and Smith, J. A. C. (2000). Metal hyperaccumulator plants: A Review of the ecology and physiology of a biological resource for phytoremediation of metal-polluted soils. In: (N. Terry, G. Banuelos, eds.) *Phytoremediation of contaminated soil and water.*, Lewis Publ., Boca Raton, pp. 85-107.

Basu, U., Godbold, D. and Taylor, G.J. (1994) Aluminum resistance in *Triticum aestivum* associated with enhanced exudation of malate. *J. of Plant Physiol.* **144**, 747-753.

Bennet, R.J. and Breen, C.M. (1991) The aluminum signal: new dimensions to mechanisms of aluminium tolerance. *Dev. Plant Soil Sci.* **45**, 703-716.

Bernal, J.-H. and Clark, R.B. (1997) Mineral acquisition of aluminum-tolerant and –sensitive sorghum genotypes grown with varied aluminum. *Communications in Soil Science and Plant Analysis* **28**, 49-62.

Blamey, F.P.C., Edwards, D.G. and Asher, C.J. (1983) Effects of aluminium, OH: Al and P:Al molar ration, and ionic strength on soybean root elongation in solution culture. *Soil Sci,* **136**, 197-207

Brancadoro, L. Rabotti, G., Scienza, A. and Zocchi, G. (1995) Mechanisms of Fe-efficiency in roots of *Vitis* spp. In response to iron deficiency stress. *Plant and Soil* **171**, 229-234.

Brooks, R.R., Morrison, R.S., Reeves, R.D., Dudley, T.R. and Akman, Y. (1979) Hyperaccumulation of nickel by *Alyssum* Linnaeus (Cruciferae). *Proc. R. Soc. London, Ser. B.* **203**, 387-403.

Bürkert, B. and Robson, A.D. (1994) ^{65}Zinc uptake in subterranean clover (*Trifolium subterraneum* L.) by three vesicular-arbuscular mycorrhizal fungi in root-free sandy soil. *Soil Biol. And Biochem.* **26**, 1117-1124.

Burnell, J.N. (1988) The biochemistry of manganese in plants. In: R.D. Graham, R.J. Hannam, N.C. Uren (eds) Manganese in soils and plants. Kluwer Academic Publishers, Dordrecht, pp. 125-137.

Bussler, W. (1981) *Copper in soils and plants.* J.F. Lonergan, A.D. Robson and R.D. Graham (eds.) pp. 213-234, Academic press, New York

Cakmak, S., Gulut, K.Y., Marschner, H. and Graham, R.D. (1994) Effect of zinc and iron deficiency on phytosiderophore release in wheat genotypes differing in zinc efficiency. *J. of Plant Nutr.* **17**, 1-17.

Care, D.-A. (1995) The effect of aluminium on root hairs in white clover (*Trifolium repens* L.). *Plant Soil* **171**, 159-162.

Chang, Y.C., Yamamoto, Y. and Matsumoto, H. (1999) Accumulation of aluminium in the cell wall pectin in cultured tobacco (*Nicotiana tabacum* L.) cells treated with a combination of aluminium and iron. *Plant Cell and Environ.* **22**, 1009-1017.

Cocker, K.M., Evans, D.E. and Hodson, M.J. (1998) The amelioration of aluminium toxicity by silicon in higher plants: solution chemistry or an in planta mechanisms? Physiol. Plant. **104**, 608-614.

Conner, A.J. and Meredith, C.P. (1985) Strategies for selection and characterization of aluminium-resistant variants from cell cultures of *Nicotiana plumbaginifolia*. *Planta* **166**, 466-473.

Coughtrey, P.J. and Martin, M.H. (1978). Cadmium uptake and distribution in tolerant and nontolerant populations of *Holcus lanatus* grown in solution culture. *Oikos* **30**, 555-560.

Del Rio, L.A., Sevilla, F., Go'mez, M., Yanez, J., Lopéz-Gorgé, J. (1978) Superoxide dismutase: an enzyme system for the study of micronutrient interactions in plants. *Planta* **140**, 221-225.

Delhaize, E. and Ryan, P.R. (1995) Aluminum toxicity and tolerance in plants. *Plant Physiol.* **107**, 515-321.

Delhaize, E., Lonergan, J.F. and Webb, J. (1985) Development of three copper metalloenzymes in clover leaves. *Plant Physiology* **78**, 4-7

Denny, H.J. and Wilkins, D.A. (1987) Zinc tolerance in *Betula* ssp. II Microanalytical studies of zinc uptake into root tissues. *New Phytologist* **106**, 525-534.

Dinkelaker, B., Hengeler, C. and Marschner, H. (1995) Distribution and function of proteoid roots and other root cluster. *Botan. Acta* **108**, 183-200.

Dixon, N.E., Gazola, C., Blakeley, R.L. and Zerner, B. (1975) Jack bean urease (EC 3.5.1.5) a metalloenzyme. A simple biological role for nickel? *J. Am. Chem. Soc.* **97**, 4131-4133.

Dong, B., Rengel, Z. and Graham, R.D. (1995) Root morphology of wheat genotypes differing in zinc efficiency. *J. of Plant Nutr.* **18**, 2761-2773.

El-Bendary, A.A., El-Fouly, M.M., Rakha, F.A., Omar, A.A. and Abou-Youssef, A.Y. (1993) Mode of inheritance of zinc accumulation in maize. *J. of Plant Nutr.* **16**, 2043-2053.

Erenoglu, B., Cakmak, I., Marschner, H., Römheld, V., Eker, S., Daghan, H., Kalayci, M. and Ekiz, H. (1996) Phytosiderophore release does not relate well with zinc efficiency in different bread wheat genotypes. *J. of Plant Nutr.* **19**, 1569-1580.

Foy, C.D. (1996) Tolerance of durum wheat lines to an acid, aluminum-toxic subsoil. *J. Plant Nutr.* **19**, 1381-1394.

Foy, C.D., Chaney, R.L. and White, M.C. (1978) The physiology of metal toxicity in plants. *Annu, Rev. Plant Physiol.* **29**, 511-566.

Foy, C.D., Farina, M.P.W. and Oakes, A.J. (1998) Iron-manganese interactions among clones of nilegrass. *J. of Plant Nutr.* **21**, 987-1009.

Gardner, W.K., Barber, D.A. and Parbery, D.G. (1983) The acquisition of phosphorus by *Lupinus albus* L. III. The probable mechanism by which phosphorus movement in the soil/root interface is enhanced. *Plant and Soil* **70**, 107-124

Gerendas, J., Polacco, J.C., Freyermuth, S.K. and Sattelmacher, B. (1999) Significance of nickel for plant growth and metabolism *J. of Plant Nutr.* **162**, 241-256.

Graham, R.D. and Rengel, Z. (1993) Genotypic variation in zinc uptake and utilization by plants. In: Zinc in soils and plants. A:D. Robson (ed) Kluwer Academic Publishers, Dordrecht, pp. 107-118

Graham, R.D., Ascher, J.S., Ellis, P.A.E. and Shephard, K.W. (1987) Transfer to wheat of the copper efficiency factor carried on rye chromosome arm 5RL. *Plant and Soil* **99**, 107-114

Greger, M and Ögren, E. (1991) Direct and indirect effects of Cd2+ on the photosynthesis and CO2-assimilation in sugar beets *(Beta vulgaris)*. *Physiol. Plant.* **83**, 129-135.

Greger, M. and Bertell, G (1992) Effects of Ca^{2+} and Cd^{2+} on the carbohydrate metabolism in sugar beet *(Beta vulgaris). J. of Exp. Bot.* **43**, 167-173.

Greger, M. and Landberg, T. (1999). Use of willow in phytoextraction. *Int. J. Phytorem.* **1**, 115-124.

Greger, M. and Lindberg, S. (1986) Effects of Cd^{2+} and EDTA on young sugar beets (*Beta vulgaris*). I. Cd^{2+} uptake and sugar accumulation. *Physiol. Plant.* **66**, 69-74.

Greger, M., Brammer, E., Lindberg, S., Larsson, G. and Idestam-Almquist, J. (1991) Uptake and physiological effects of cadmium in sugar beet (*Beta vulgaris*) related to mineral provision. *J. Exp. Bot.* **42**, 729-737.

Grewal, H.S., Graham, R.D. and Stangoulis, J. (1998) Zinc-boron interaction effects in oilseed rape. *J. of Plant Nutr.* **21**, 2231-2243.

Grierson, P.F. (1992) Organic acids in the rhizosphere of *Banksia integrifolia* L. Plant and Soil 144, 259-265.

Grusak, M.A. and Pezeshgi, S. (1996) Shoot-to-root signal transmission regulates root Fe(III) reductase activity in the dgl mutant of pea. *Plant Physiol.* **110**, 329-334.

Gubler, W.D., Gorgan, R.G. and Osterli, P.P. (1982) Yellows of melons caused by molybdenum deficiency in acid soil. *Plant Dis.* **66**, 440-451.

Gupta, U.C. and Gupta, S.C. (1998) Trace element toxicity relationship to crop production and livestock and human health. *Comm. Soil Sci. Plant Anal.* **29**, 1491-1522.

Gussarsson, M. (1994) Cadmium-induced alterations in nutrient composition and growth of *Betula pendula* seedlings: the significance of fine roots as a primary target for cadmium toxicity. *J. Plant Nutr.* **17**, 2151-2163

Han, Z.H., Wang, Q. and Chen, L. (1997) Root and rhizospere responses of iron-efficient or -inefficient apple genotypes to solution pH. *J. of Plant Nutr.* **20**, 1517-1525.

Hiatt, A.J., Amos, D.F. and Massey, H.F. (1963) Effects of aluminum on copper sorption by wheat. *Agron. J.* **55**, 284-287.

Hodson, M.J. and Evans, D.E. (1995) Aluminium/silicon interactions in higher plants. *J. Exp. Bot.* **46**, 161-171

Homer, F.A., Morrison, R.S., Brooks, R.R., Clemens, J. and Reeves, R.D. (1991). Comparative studies of nickel, cobalt and copper uptake by some nickel hyperaccumulators of the genus *Alyssum*. *Plant Soil* **138**, 195-205.

Huang, C., Graham, R.D., Barker, S.J. and Mori, S. (1996) Diffrential expression of iron deficiency-induced genes in barley genotypes with differing manganese efficiency. *J. Plant Nutr.* **19**, 407-420.

Huang, C., Grunes, D.L. and Kochian, L.V. (1995) Aluminium and calcium transport interactions in intact roots and root plasmalemma vesicles from aluminum-sensitive and tolerant wheat cultivars. *Plant and Soil* **171**, 131-135

Huang, C., Webb, M.J. and Graham, R.D. (1994) Manganese efficiency is expressed in barley growing in soil system but not in solution culture. *J. of Plant Nutr.* **17**, 83-95.

Huber, D.M. and McCay-Buis, T.S. (1993) A multiple component analysis of the take-all desease of cereals. *Plant Dis.* **77**, 437-447.

Ikegawa, H., Yamamoto, Y. and Matsumoto, H. (2000). Responses to aluminium of suspension-cultured tobacco cells in simple calcium solution. *Soil Sci. and Plant Nutr.* **46**, 503-514.

Ingrouille, M.J. and Smirnoff, N. (1986). *Thlaspi caerulescens* J. & C. Presl (*T. alpestre* L.) in Brittain. *New Phytol.* **102**, 219-233.

Inouhe, M., Ninomiya, S., Tohoyama, H., Joho, M. and Murayama, T. (1994) Different characteristics of roots in cadmium-tolerance and Cd-bindning complex formation between mono- and dicotyledonous plants. *J. Plant Res.* **107**. 201-207.

Iturbe-Ormaetxe, I, Moran, J.F., Arrese-Igor, C., Gogoreena, Y., Klucas, R.V. and Becana, M. (1995) Activated oxygen and antioxidant defences in iron-deficient pea plant. *Plant Cell Environ* **18**, 421-429

Jackson, P.J., Unkefer, P.J., Delhaize, E. and Robinson, N.J. (1990) Mechanisms of trace metal tolerance in plants. In: F. Katterman (ed) *Environmental injury to plants*, Academic Press, Inc. San Diego, New York.

Jan, F. (1992) Varietal diversity of upland rice in tolerance to aluminium. Doctoral thesis, Department of Plant Physiology, SLU, Box 7047, ISBN 91-576-4636-8.

Jolley, V.D. and Brown, J.C. (1994) Genetically controlled uptake and use of iron by plants. In: Biochemistry of metal micronutrients in the rhizospere J.A. Manthey, D.E. Crowley and D.G. Luster, Boca Raton (eds), Lewis Publishers, pp. 251-266.

Judel, G.K. (1972) Änderungen in der Aktivität der Peroxidase und der Katalase und im Gehalt an Gesamtphenolen in den Blättern der Sonnenblume unter dem Einfluss von Kupfer- und Stickstoffmangel. *Z. Pflanzenernaehr. Bodenkkd.* **133**, 81-92.

Keltjens, W.G. and Tan, K. (1993) Interactions between aluminium, magnesium and calcium with different monocotyledonous and dicotyledonous plant species. *Plant Soil* **156**, 485-488.

Kinnersley, A.M. (1993) The role of phytochelates in plant growth and productivity. *Plant Growth Regulation* **12**, 207-218.

Kinraide, T.B. (1993) Aluminium enhancement of plant growth in acid rooting media. A case of reciprocal alleviation of toxicity by two toxic cation. *Physiol. Plant.* **88**, 619-625.

Kochian, L.V. (1991) Mechanism of micronutrient uptake and translocation in plants. In: Mortvedt, J.J., Fox, F.R., Shuman, L.M., Welch, R.M. (eds) *Micronutrients in Agriculture,* 2nd edn. SSSA, Madison, WIS. pp. 229-296.

Kochian, L.V. (1995) Cellular mechanisms of aluminum toxicity and resistance in plants. *Ann. Rev. Plant Mol. Biol.* **46**, 237-260

Kochian, L.V. and Jones, D.L. (1997) Aluminum toxicity and resistance in plants. In: *Research issues in aluminum toxicity.* R. Yohel, and M.S. Golub (eds.), Taylor and Francis Publishers, Washington DC, pp. 69-89.

Köhl, K.L. (1997) Do *Armeria maritima* (Mill.) Willd. Ecotypes from metalliferous soils and non-metalliferous soils differ in growth response under Zn stress? A comparison by a new artificial soil method. *J. Exp. Bot.* 48, 1959-1967.

Koyama, H., Okawara, R., Ojima, K. and Yamaya, T. (1988) Re-evaluation of characteristics of a carrot cell line previously selected as aluminium-tolerant cells. *Physiol. Plant.* **74**, 683-687.

Krämer, U., Grime, G.W., Smith, J.A.C., Hawes, C.R., and Baker, A.J.M. (1997) Micro-PIXE as a technique for studying nickel localization in leaves of the hyper accumulator plant *Alyssum lesbiacum. Nucl Instrum Methods Phys Res B* **130**, 346-350.

Kumar, V. and Verma, K.S. (1997) Relationship between nutrient element content of the index leaf and cane yield and juice quality of sugarcane genotypes. *Comm. soil sci. plant anal.* **28**, 1021-1032.

Landberg, T. and Greger, M. (1996). Differences in uptake and tolerance to heavy metals in *Salix* from unpolluted and polluted areas. *Appl. Geochem.* **11**, 175-180.

Laperche, V, Logan, T.J., Gaddam, P., and Traina, S.J. (1997). Effect on apatite amendments on plant uptake of lead from contaminated soil. *Environ. Sci. Technol.* **31**, 2745-2753.

Lee, J., Reeves, R.D., Brooks, R.R. and Jaffre, T. (1978) The relation between nickel and citric acid in some nickel-accumulating plants (from New Caledonia). *Phytochemistry* **17**, 1033-1035.

Lichtenberger, O. and Neumann, D. (1997) Analytical electron microscopy as a powerful tool in plant cell biology: examples using electron energy loss spectroscopy and X-ray microanalysis. *Eur J Cell Biol* **73**, 378-386.

Lindberg, S. (1986) Aluminium inhibition of a (Na^+ +K^++Mg^{2+}) ATPase from sugar beet roots. *Biol. Chem.* **367**, 241.

Lindberg, S. (1990) Aluminium interactions with K^+ ($^{86}Rb^+$) and $^{45}Ca^{2+}$ fluxes in three cultivars of sugar beet *(Beta vulgaris). Physiol. Plant.* **79**, 275-282.

Lindberg, S. (2000) Aluminium effects on membrane potentials and cytosolic ion changes in plant cells. Proceeding. International symposium on impact of potential tolerance of plants on the increased productivity under aluminium stress, Kurashiki, Japan, Sept 2000

Lindberg, S. and Strid, H. (1997) Aluminium induces rapid changes in cytosolic pH and free calcium and potassium concentrations in root protoplasts of wheat *(Triticum aestivum). Physiol. Plant.* **99**, 405-414.Lindberg, S., Szynkier, K. and Greger, M. (1991) Aluminium effects on transmembrane potential in cells of fibrous roots of sugar beet. *Physiol. Plant* **79**, 275-282.

Lindberg, S. and Wingstrand, G. (1985) Mechanism for Cd^{2+} inhibition of (K^++ Mg^{2+})ATPase activity and K^+($^{86}Rb^+$) uptake in roots of sugar beet *(Beta vulgaris). Physiol. Plant.* **63**, 181-185.

Lindberg, S., and Griffiths, G. (1993) Aluminium effects on ATPase activity and lipid composition of plasma membranes in sugar beet roots. *J. Exp. Bot.* **44**, 1543-1550.

Lindberg, S., Szynkier, K. and Greger, M. (1998) Aluminium effects on transmembrane potential in root cells of spruce in relation to pH and growth temperature. *J. Mineral Nutr.* **21**, 975-985.

Lindsay, W.L. (1979) Chemical equilibria in soils. John Wiley and Sons, New York, NY.

Liu, D., Jiang, W., Guo, L., Lu, C. and Zhao, F. (1994) Effects of nickel sulfate on root growth and nucleoli in root tip cells of *Allium cepa. Israel. J. Plant Sci* **42**, 143-148.

Liu, J., Reid, R.J. and Smith, F.A. (2000) The mechanism of cobalt toxicity in mung beans. *Physiol. Plant.* **110**, 104-110

Lloyd-Thomas, D.H. (1995). Heavy metal hyperaccumulatoin by *Thlaspi caerulescens J. & C. Presl.* PhD Thesis, University of Sheffield, U.K.

Longnecker, N. and Welch, R.M. (1990) Accumulation of apoplastic iron in plant roots: a factor in the resistance of soybeans to iron deficiency induced chlorosis? *Plant Physiol.* **92**, 17-22.

Lu, Z., Grewal, H.S. and Graham, R.D. (1998) Dry matter production and uptake of zinc and phosphorus in two oilseed rape genotypes under differential rates of zinc and phosphorus supply. *J. of Plant Nutr.* **21**, 25-38.

Marschner, H. and Römheld, V. (1994) Strategies of plants for acquisition of iron. *Plant and soil* **165**, 261-274

Marschner, H (1986) *Mineral nutrition of higher plants.* Academic Press, London

Marschner, H (1995) *Mineral nutrition of higher plants,* 2nd edition, Academic Press, London

Matera, U. and Kneusel, R.E. (1988) Phenolic compounds in plant disease resistance. *Phytoparasit.* **16**, 153-170

Matsumoto, H. and Yamaya, T (1986) Inhibition of potassium uptake and regulation of membrane-associated Mg^{2+}-ATPase activity of pea roots by aluminium *Soil Sci. Plant Nutr.* **32**, 179-188.

McDonald-Stephens, J.L. AND Taylor, G.J. (1995) Kinetics of aluminum uptake by cell suspensions of *Phaseolus vulgaris* L. *J. of Plant Physiol.* **145**, 327-334.

McGrath, S.P. (1998). Phytoextraction for soil remediation. In. (R.R. Brooks, ed.), *Plants that hyperaccumulate heavy metals.* CAB International, Wallingford, pp. 261-288

McGrath, S.P., Baker, A.J.M., Morgan, A.N., Salmon, W.J., Williams, M. (1980) The effects of interactions between cadmium and aluminium on the growth of two metal-tolerant races of *Holcus lantanus* L. *Environ. Pollution,* Series A, **23**, 267-277.

Mengel, K. and Kikby, E.A. (1987) Principles of plant nutrition. International Potash Institute, Bern.

Meredith, C.P. (1978) Selection and characterization of aluminum-resistant variants from tomato cell cultivars. *Plant Sci. Lett.* **12**, 25-34.

Miyasaka, S.C., Kochian, L. V., Shaff, J.E. and Foy, C.D. (1989) Mechanism of aluminum tolerance in wheat. An investigation of genotypic differences in rhizospere pH, K^+, AND H^+ transport, and root-cell membrane potentials. *Plant Physiol.* **91**, 1188-1197.

Mori, S. (1994) Mechanisms of iron acquisition by gramineous (Strategy II) plants. In: J.A. Manthley, D.E. Crowley, and D.G. Luster (eds) Biochemistry of metal micronutrients in the rhizosphere, Lewis Publishers, Boca Raton, F.L. PP. 225-249.

Nagy, L. and Proctor, J. (1997) Plant growth and reproduction on a toxic alpine ultramafic soil: adaptation to nutrient limitation. New Phytol. **137**, 267-274.

Neumann, D., zur Nieden, U., Lichtenberger, O. and Leopold, I. (1995) How does *Armeria maritima* tolerate high heavy metal concentrations? *J Plant Physiol* **146**, 704-717.

Olivetti, G.P., Cumming, J.R. and Etherton, B. (1995) Membrane potential depolarization of root cap cells precedes aluminum tolerance in snapbean. *Plant Physiol.* **109**, 123-129.

Österås, A.H., Ekvall, L. and Greger, M. (2000) Sensitivity to, and accumulation of, cadmium in *Betula pendula, Picea abies,* and *Pinus sylvestris* seedlings from different regions in Sweden. *Can. J. Bot.* **78**, 1440-1449.

Ouzounidou, G., Moustakas, M., Eleftheriou, E.P. (1997) Physiological and ultrastructural effects of cadmium on wheat *(Triticum aestivum* L.) leaves. *Arch Environ Contam. Toxicol,* **32**, 154-160.

Pandolfini, T., Gabrielli, R. and Ciscato, M. (1996) Nickel toxicity to two durum wheat cultivars differing in drought sensitivity. *J Plant Nutr.* **19**, 1611-1627.

Parker, D.R., Kinraide, T.B., Zelazny, l.W. (1989) On the phytotoxicity of polynuclear hydroxy-aluminium complexes. *Am. J. of Soil Sci.* **53**, 789-796.

Pearson, J.N. and Rengel, Z (1997) *In* Basra, A.S. and Basra, R.K. (1997) *Mechanisms of environmental stress resistance in plants,* Harwood academic publishers, the Netherlands, ISBN 905702036X, pp. 213-240.

Pettersson, S. and Strid, H. (1989) Effects of aluminium on growth and kinetics of $K^+(^{86}Rb^+)$ uptake in two cultivars of wheat (*Triticum aestivum*) with different sensitivity to aluminium. *Physiol. Plant.* **76**, 255-261.

Polar, E. (1975) Zinc in pollen and its incorporation into seeds. *Planta* **123**, 97-103.

Poskuta, J.W. and Waclawczyk-Lach, E.(1995) *In vivo* responses of primary photochemistry of photosystem II and CO_2-exchange in light and in darkness of tall fescue genotypes to lead toxicity. *Physiol. Plant.* **17**, 233-240.

Prasad, M.N.V. (1995) Cadmium toxicity and tolerance in vascular plants. *Environ Exp. Bot.* **35**, 525-545.
Prasad, M.N.V. (1999) Metallothioneins and metal binding complexes in plants In: Prasad, M.N.V. and Hagemeyer, J. (eds) *Heavy metal stress in plants. From molecules to ecosystems.* ISBN 3-540-65469-0, Springer-Verlag, Berlin, Heidelberg, New York. pp. 51-72.
Prasad, M.N.V. and Hagemeyer, J. (1999) *Heavy metal stress in plants. From molecules to ecosystems.* ISBN 3-540-65469-0, Springer-Verlag, Berlin, Heidelberg, New York. pp. 236-239.
Quariti, O., Boussama, N., Zarrouk, M., Cherif, A. and Ghorbal, M.H. (1997) Cadmium and copper-induced changes in tomato membrane lipids. *Phytochemistry* **47**, 1343-1350.
Rahimi, A. and Bussler, W. (1979) Die Entwicklung und der Zn- Fe- und P-Gehalt höherer Pflanzen in Abhängigkeit vom Zinkangebot. *Z. Pflanzenernaehr. Bodenkd.* 142, **15**-27.
Rascio, N. (1977). Metal accumulation by some plants growing on zinc mine deposites. *Oikos* **29**, 250-253.
Reddy, G.N. and Prasad, M.N.V. (1990) Heavy metal binding proteins, peptides, occurrence, structure, synthesis and functions review. *Environ. Exp. Bot.* **30**, 252-264.
Reeves, R.D. and Baker, A.J.M. (2000). Metal-accumulating plants. In: (I. Raskin, B.D. Ensley, eds) *Phytoremediation of toxic metals. Using plants to clean up the environment.* John Wiley & Sons, Inc., US, pp. 193-229.
Reeves, R.D., Baker, A.J.M., Borhidi, A. And Berazain, R. (1996). Nickel-accumulating plants from the ancient serpentine soils of Cuba. *New Phytol.* **133**, 217-224.
Rengel, Z. (1992a) Disturbance of cell Ca^{2+} homeostasis as a primary trigger of Al toxicity syndrome. *Cell Plant environ.* **15**, 931-938.
Rengel, Z. (1992b) The role of calcium in aluminium toxicity. *New Phytol.* **121**, 499-513.
Rengel, Z. (1997) Mechanisms of plant resistance to toxicity of aluminium and heavy metals. In: Basra, A.S. and Basra, R.K. (eds.) *Mechanisms of environmental stress resistance in plants,* Harwood Academic Publishers, Amsterdam, pp. 241-276.
Rengel, Z. (1999a) Physiological mechanisms underlying differential nutrient efficiency of crop genotypes. In: Rengel Z. (ed*)* *Mineral nutrition of crops: Fundamental mechanisms and implications*, Haworth Press, New York
Rengel, Z. (1999b) Heavy metals as essential nutrients. In: Prasad, MNV and Hagemeyer J (eds) *Heavy metal stress in plants,* ISBN 3-540-65469-0 Springer-Verlag, Berlin, Heidelberg, New York, pp. 231-251.
Rengel, Z. and Graham, R.D. (1995) Wheat genotypes differ in Zn efficiency when grown in chelate-buffered nutrient solution. II. Nutrient uptake. *Plant and Soil* **173**, 259-266.
Rengel, Z. and Graham, R.D. (1996) Uptake of zinc from chelate-buffered nutrient solutions by wheat genotypes differing in Zn efficiency. *J. Exp. Bot.* **47**, 217-226.
Rengel, Z. Ross, G. and Hirsch, P. (1998) Plant genotype and micronutrient status influence colonization of wheat roots by soil bacteria. *J. of Plant Nutr.* **21**, 99-113.
Reuter, D.J., Robson, A.D., Lonergan, J.F. and Tranthim-Fryer, D.J. (1981). Copper nutrition of subterranean clover (*Trifolium subterranium* L. cv. Seaton Park). *Aust. J. Agric. Res.* **32**, 267-282.
Roberts, B.A. and Proctor, J. (eds) (1992) *The ecology of areas with serpentinized rocks. A world view.*Kluwer, Dordrecht.
Römheld, V. and Marschner, H. (1990) Genotypic differences among gramineaceous species in release of phytosiderophores and uptake of iron phytosiderophores. *Plant and Soil* **123**, 147-153.
Ryan, P.R., Delhaize, E. and Randall, P.J. (1995) Characterization of Al-stimulated efflux of malate from apices of Al-tolerant wheat roots. *Planta* **196**, 103-110.
Ryan, P.R., Shaff, J.E. and Kochian, L.V. (1992) Aluminum toxicity in roots. Correlation among ionic currents, ion fluxes, and root elongation in aluminum-sensitive and aluminum-tolerant wheat cultivars. *Plant Physiol.* **99**, 1193-1200.
Samantaray, S., Rout, G.R. and Das, P. (1998) Differential nickel tolerance of mung bean (*Vigna radiata* L.) genotypes in nutrient culture. *Agronomie* (Paris) **18**, 537-544.
Sattelmacher, B., Horst, W.J., and Becker, H.C. (1994) Factors that contribute to genetic variation for nutrient efficiency of crop plants. *Z. Planzenernähr. Bodenk* **157**, 215-224.
Schwartz, C., Morel, J.L., Saumier, S., Whiting, S.N., and Baker, A.J.M. (1999). Root development of the Zinc-hyperaccumulator plant *Thlaspi caerulescens* as affected by metal origin, content and localization in soil. *Plant Soil* **208**, 103-115.
Sela, M., Tel-Or, E., Fritz, E. and Hüttermann, A. (1988) Localization of toxic effects of cadmium, copper and uranium in *Azolla. Plant Physiol.* **88**, 30-36.

Sivaguru, M., Yamamoto, Y. and Matsumoto, H. (1999) Differential impacts of aluminium on microtubule organization depends on growth phase in suspension-cultured tobacco cells. *Physiol. Plant.* **107**, 110-119.

Soon, Y.K., Clayton, G.W., Clarke, P.J (1997) Content and uptake of phosphorus and copper by spring wheat: effect of environment, genotype, and management. *J. Plant Nutr.* **20**, 925-937.

Steffens, J.C. (1990) The heavy metal-binding peptides of plants. *Annu. Rev. Plant Physiol. Plant Mol.. Biol.* **41**, 553-575.

Stevens, W.B., Jolley, V.D. and Hansen, N.C. (1994). Diurnal rhythmicity of root iron reduction in soybean as affected by various light regimes. *J. of Plant Nutr.* **17**, 2193-2202.

Stoltz, E. and Greger, M. (2001). Wetland plant reduce metal content in dranage water from submersed tailings. *Proceeding. Int. Conf. On mining and the environment. Skellefteå. 2001.*

Strid, H. (1996) Aluminium toxicity effects on growh and on uptake and distribution of some mineral nutrients in two cultivars of spring wheat. Doctoral thesis, Department of Plant Physiology, SLU, Box 7047, ISBN 91-576-5128-0.

Tagliavini, M., Rombola, A.D. and Marangoni, B. (1995) Response to iron-deficiency stress of pear and quince genotypes. *J. of Plant Nutr.* **18**, 2465-2482.

Taiz, L and Zeiger, E. (1998) *Plant Physiology,* Sinauer Associates, Inc., Publishers, ISBN: 0-87893-831-1, pp. 103-124

Tan, K. and Keltjens, W.G. (1995) Analysis of acid-soil stress in sorghum genotypes with emphasis on aluminium and magnesium interactions. *Plant Soil* **17**, 147-150.

Taylor, G.J. (1991) Current views of the aluminium stress response; the physiological basis of tolerance. *Curr. Top. Plant Biochem. Physiol.* **10**, 57-93.

Taylor, G.J. and Foy, C.D. (1985) Mechanisms of aluminum tolerance in *Triticum aestivum* (wheat). I. Differential pH induced by winter cultivars in nutrient solutions. *Am. J of Bot.* **72**, 695-701.

Thornton, F.C., Schaedle, M., Raynal, D.J. and Zipperer, C. (1986) Effects of aluminum on honey locust (*Gleditsia triacanthos* L.) seedlings in solution culture. J. Exp. Bot. 37, 775-785.

Tiller, K.G. (1989) Heavy metals in soils and their environmental significance. *Advances in Soil Sci.* **9**, 113-142.

Tilstone, G.H. and Macnair, M.R. (1997) The nickel tolerance and copper-nickel co-tolerance in *Mimulus guttatus* from copper mine and serpentine habitats. *Plant and Soil* **191**, 173-180.

Tolra, R.P., Poschenrieder, C. and Barcelo, J. (1996) Zinc Hyperaccumulation *in Thlaspi caerulescens.* I. Influence on growth and mineral nutrition. *J. Plant Nutr.* **19**, 1531-1540.

Tong-YiPing, Rengel, Z. and Graham, R.D. (1997) Interactions between nitrogen and manganese nutrition of barley genotypes differing in manganese efficiency (1997) *Annals of Botany* **79**, 53-58.

Turkendorf, A., Lyszcz, S. and Baszynski, B. (1984) Copper binding proteins in spinach tolerant to excess copper. *J. Plant Physiol.* **115**, 351-360.

Van Steveninck, R.F.M., Van Steveninck, M.E., Fernando, D.R., Horst, W.J. and Marschner, H. (1987) Deposition of zinc phytate in globular bodies in roots of *Deschampsia caespitosa* ecotypes: a detoxification mechanism? *J. of Plant Physiol.* **131**, 247-257.

Wagatsuma, T. (1983) Characterization of absorption sites for aluminium in root. *Soil Sci. and Plant Nutr.* **29**, 499-515.

Wagatsuma, T. and Ezoe, Y. (1985) Effect of pH on ionic species of aluminium in medium and on aluminium toxicity under solution culture. *Soil Sci. Plant Nutr.* **31**, 547-561

Wallnöfer, P.R., Engelhardt, G. (1984) Pflanzentoxikologie. In: B. Hock and E.F. Eistner (eds), BI Wissenschaftsverlag, Mannheim, pp. 95-117.

Wei, L.C., Loeppert, R.H. and Ocumpaugh, W.R. (1997) Fe-deficiency stress response in Fe-deficiency resistant and susceptible subterranean clover: Importance of induced H+ release. *J. Exp. Bot.* **48**, 239-246.

Welch, R.M. (1995) Micronutrient nutrition of plants. *CRC Crit. Rev. in Plant Sci.* **14**, 49-82.

Yamamoto, Y., Chang, Y.C., Ono, K. and Matsumoto, H. (1994) Effects of aluminium on the toxicity of iron (II), copper and cadmium in suspension-cultured tobacco cells. *Bulletin of the Research Institute for Bioresources, Okayama University,* **2**, 181-190.

Yang, X., Ye, Z.Q., Shi, C.H., Zhu, M.L. and Graham, R.D. (1998) Genotypic differences in concentrations of iron, manganese, copper and zinc in polished rice grains. *J. of Plant Nutr.* **21**, 1453-1462.

Yu, Q., Osborne, L.D. and Rengel, Z. (1998) Micronutrient deficiency changes activities of Superoxide dismutase and ascorbate peroxidase in tobacco seedlings. *J. Plant Nutr.* **21**, 1427-1437

Zhang, F.S. (1993) Mobilization of iron and manganese by plant-borne and synthetic metal chelators. In: N.J. Barrow (ed) Plant nutrition – from genetic engineering to field practice, Kluwer, Dordrecht, pp. 115-118.

Zhang, G. and Taylor, G.J. (1989) Kinetics of aluminum uptake by excised roots of aluminum-tolerant and aluminum-sensitive cultivars of *Triticum aestivum* L. *Plant Physiol.* **91**, 1094-1099.

Zhang, G. and Taylor, G.J. (1990) Kinetics of aluminum uptake in *Triticum aestivum* L. Identity of the linear phase of aluminum uptake by excised roots of aluminum-tolerant and aluminum-sensitive cultivars. *Plant Physiol.* **94**, 577-584.

Zheng, S.J., Ma, J.F. and Matsumoto, H. (1998) High resistance in buckwheat: I. Al-induced specific secretion of oxalic acid from root tips. *Plant. Physiol.* **117**, 745-751.

CHAPTER 15

GENOTOXICITY AND MUTAGENICITY OF METALS IN PLANTS

BRAHMA B. PANDA AND KAMAL K. PANDA
*Genecology and Tissue Culture Laboratory, Department of Botany,
Berhampur University, Berhampur 760 007, Orissa, India*

1. INTRODUCTION

Metals are ubiquitous in nature. Metals are emitted from a wide spectrum of natural and man-made sources such as volcanic eruption, mining, fossil burning, industrial emissions and automobile exhausts and sewage disposals. Distribution of metals in the environment, however, is uneven (Moore and Ramamoorthy 1984, Nriagu and Pacyna 1988, Nriagu 1990). Upon entering into the environment in a variety of organic and inorganic forms and being neither degradable nor recoverable, metals get incorporated into biogeochemical cycles where they can exert long-term effect (Nriagu 1990). The problem of metal pollution is a global phenomenon. This, however, is accentuated in third world countries that have been offering the dumping grounds for toxic wastes and have become the hot spots of metal pollution because of population explosion coupled with poor economic conditions, use of outdated technologies in industries and lack of stringent anti-pollution laws (Anonymous 1991).

Industrial emissions of metals such as aluminum (Al III), antimony (Sb III), arsenic (As III), cadmium (Cd II), cobalt (Co II), copper (Cu II), chromium (Cr III, VI), iron (Fe II), lead (Pb II), mercury (Hg II), nickel (Ni II), selenium (Se II) vanadium (V II, V) and zinc (Zn II) are primarily responsible for the overall metal pollution (Nriagu and Pacyna 1988, Nriagu 1990). Worldwide input of trace metals from different sources into soil is listed (Table 1). Of the various environmental trace metals, a few such as Co, Mn, Fe, Cu Se, Mo, Ni are essential elements, whereas Al, Cd, Cr, Hg, Pb are nonessential elements.

toxicity from Al, B, Cu, Ni, and Zn while animals are sensitive to As, Ba, Cd, Cr, Cu, Hg, Mo, Ni, Pb, Se etc. Of these, Cd, Hg and Pb have the greatest potential to affect human health (Logan and Traina 1993, Mehera and Farago 1994). A buildup of toxic metals in the food chain results from massive quantities of metals being discharged into the environmental media, in which plants constitute an important link (Wiersma et al 1986, Dudka and Miller 1999). Plants, obviously, can play a crucial role in monitoring and assessment of environmental metal pollution. Phytotoxicity of metals is well documented (Foy et al. 1978, Woolhouse 1983, Prasad 1997). Plants respond to metal toxicity by eliciting a number of bio-markers/endpoints that can be detected and analyzed at various levels of organisation ranging from gross morphology to cellular, biochemical or molecular levels, and thus have been very useful to monitor as well as assess environmental metal pollution (Burton 1986, Briat and Lebrun 1999). Of the various biological effects that environmental pollution may cause, genetic toxicology is one aspect that concerns with the damage to the genome or DNA. The major types of genotoxic effects are 1. mutagenesis refers to gene or point mutation, a change in DNA sequence within a gene; 2. clastogenesis refers to change in chromosome structure, usually resulting in a gain or loss or rearrangement of chromosome pieces within the genome; 3. aneugenesis refers to gain or loss of one or more chromosomes (aneuploidy) or haploid set of chromosomes (euploidy); and 4. recombinogenesis refers to homologous or non-homologous exchange of segments between chromatids or chromosomes. Genetic toxicology of metals has been implicated in human cancer, birth defects and mutations (Flessel 1977, Rossman et al. 1987, Beyersmann 1994, Hartwig 1995). More than 200 short-term assays utilizing microorganisms, insects, plants have been developed over the last 20-25 years to aid identification of agents that pose genetic hazards to human (Waters et al. 1988). Plants provide ideal genotoxicty assays for screening as well as monitoring of environmental mutagens or genotoxins (Grant 1994, Ma 1999). Apart from possible human extrapolation, information on genetic toxicology of environmental agents in non-human or sub-mammalian organisms including plants is very vital from the standpoints of safeguarding the bio-diversity and ecosystem health (Wurgler and Kramer 1992, de Raat et al. 1990, Anderson et al 1994, Gopalan 1999).

2. PLANT GENOTOXICITY ASSAYS

Genetic toxicity bioassays have proved to be very useful in environmental monitoring and assessment of industrial pollution (Houk 1992). The advantages of using plant assays for genetic toxicological testing are that a number of plants having long and low number of chromosomes offer excellent cytogenetic systems with a wide range of genetic endpoints, from gene mutation to mitotic and meiotic chromosome aberrations and DNA-damage (Grant 1999). Genotoxicity test results based on plant assays for a number of standard chemical mutagens and carcinogens have shown positive correlation with that of

Table 1: Worldwide inputs of trace metals into soil, in thousand tons/year (Nriagu 1990)

Source	As	Cd	Cr	Cu	Hg	Pb	Zn
Agriculture and animal waste	5.8	2.2	82	67	0.8	26	31
Wood wastes	1.7	1.1	1.1	10	28	28	7.4
Urban refuge	0.4	4.2	20	26	0.13	40	60
Sewage sludge	0.25	0.18	6.5	13	0.44	7.1	39
Solid waste from metal fabrication	0.11	0.04	1.5	4.3	0.04	7.6	11
Coal ash	22	7.2	298	214	2.6	144	298
Discarded products	38	1.2	458	592	0.68	292	465
Fertilizer and peat	0.28	0.20	0.32	1.4	0.01	2.9	2.5
Atmospheric fallout	13	5.3	22	25	2.5	232	92
Total input	82	22	898	971	8.3	759	1322

mammalian and non-mammalian assays (Grant 1994). It has also been shown that plant genotoxicity assays are highly sensitive, result in few false negatives in predicting carcinogenicity of test agents (Ennever et al. 1986). Plant assays are relatively easy and inexpensive to work with. This aspect is of particular relevance to developing countries where screening or testing of environmental agents for genetic toxicity has to be carried out often with limited resources and supplies in terms of facility, equipment, chemicals, reagents etc. (Plewa 1985). Plant assays are unique in the sense that they can be employed to evaluate genotoxicity of agents under wide range of environmental conditions that include *in situ* monitoring (Sandhu and Lower 1989) as well as testing genotoxicity of complex mixtures (Sandhu et al 1987). Genetic toxicology in plants therefore have profound implications in genetic-ecotoxicology or genecotoxicoly (Wurgler and Kramers 1992, de Raat et al. 1990, Anderson et al. 1994, Gopalan 1999).

The use of higher plants in the study of environmental mutagenesis paralleled the historical developments in the field of chemical mutagenesis. There are about 233 plants that have been used in various aspects of mutagenesis research (Shelby 1976). Of these a few plants namely *Allium cepa* (2n = 16), *Arabidopsis thaliana* (2n = 10), *Crepis capillaris* (2n = 6), *Glycine max* (2n = 40), *Hordeum vulgare* (2n = 14) *Tradescantia* clones (2n = 12), *Vicia faba* (2n = 12) and *Zea mays* (2n = 20) are some of the well worked out assay systems that provide well defined genotoxicity endpoints. These include assays for gene mutation, mitotic and meiotic chromosome as well as cell division aberrations, micronucleus (MNC), sister chromatid exchange (SCE) and the comet assay that evaluates DNA damage (Table 2).

3. GENOTOXICITY OF METALS IN PLANT ASSAYS

Employing some of the aforesaid plant assays it has been possible to screen as well as monitor select metals that show genotoxic effects that include chromosome break or structural aberrations (clastogenesis), spindle malfunction affecting chromosome number (aneugenesis), MNC, SCE, and DNA strand break as evaluated by the Comet assay. Metal ions that have been shown to be clearly positive at least in one of the above assays are listed bellow (Table3).

3.1. Clastogenesis

Induction of chromosome break, clastogenesis, is one of the basic tests for genotoxicity evaluation of any test agent. Depending on the nature of chromosome break whether of chromosome or chromatid type and the time of their appearance during different recovery hours following a brief, 1-2h, treatment of root meristems of *A. cepa* (Kihlman 1971), *V. faba* (Kihlman 1975), *H. vulgare* (Constantin and Nilan 1982a) or *C. Capillaris* (Grant and Owens 1998) the cell-cycle phase (G1, S, and G2) specific action of the test agent may be predicted. It has been shown that agents that act directly on DNA can produce chromosome/chromatid breaks or exchanges seen in metaphase or dicentric chromosomes, acentric chromosomes (Kihlman 1971, 1975). Terminal deletion or the loss of telomere results in the formation of bridges, and sticky chromosomes in ana-telopase Nicoloff and Gecheff, 1976, Rank and Nielson 1993). Induction of chromosomal bridges in ana-telophase is a firm evidence of clastogenicity. Further resolution of such bridges would produce broken ends in the chromosomes that perpetuates bridge formation in ana-telophases of subsequent cell cycles through breakage-fusion-bridge cycle, indicating genomic stress (McClintock 1965, Borboa and De la Torre 1996). Genome may respond to such stress by activating silent genes mediated by the movement of transposable elements.

Chromosomal stickiness is assumed to be via damage to chromosomal peripheral proteins such as DNA topoisomerase II, which may also lead to chromosome breakage aberrations (Gaulden 1987). Of the results available on clastogenicity of metals tested in plant assays employing *A. cepa* (Borboa and De la Torre 1996, Fiskesjo 1979, Fiskesjo 1988, Liu et al. 1992, Liu et al. 1995, Rank and Nielson 1994, Lerda 1992), *C. capillaris* (Ruposhev 1976) or *H. vulgare* (Zhang and Yang 1994); Al, Cd, Cr, Hg, Pb, Se, Zn, have been found to be positive (Table 3). Whereas Cu has been shown to be negative. With *A.. cepa* employing ana-telophase assay, it has been possible to analyse the genotoxicity of waste water sludge that has been partly correlated to the load of heavy metals: Pb, Ni, Cr, Zn, Cu and Cd in the sludge (Rank and Nielsen 1998).

3.2. Aneugenesis

Aneuploidy is an important aspect of environmental mutagenesis and genetic toxiciology (Zimmermann et al. 1979). Agents that interact with spindle function during mitosis or meiosis cause chromosome segregational errors (nondisjunction and noncongression) leading to ploidy. Aneuploidy and polyploidy play a role in plant evolution (Khush 1973,

Table 2: Plants assays with specific genetic toxicological endpoints.

Assay system	Genotoxic endpoint	References
Allium cepa	Mitotic cell division and chromosome aberration	Grant 1982, Fiskesjo 1997, Rank and Nielson 1993
	MNC	Reddy et al 1995, Ma et al. 1995
	SCE	Schvartzman and Cortes 1977, Panda et al 1998
	Comet (DNA strand break)	Navarrete et al 1997
Arabidopsis thaliana	Embryonic and chlorophyll mutations in siliques	Redei 1982, Gichner et al. 1994
Crepis capillaries	Mitotic chromosome aberration	Grant and Owens 1998
Glycine max	Chlorophyll yellow/green twin spots or single spots indicating somatic crossing over or forward /reverse mutations	Vig 1982
Hordeum vulgare	Mitotic and meiotic chromosome aberration	Constantin and Nilan 1982a, Panda et al 1992a)
	Chlorophyll mutations	Constantin and Nilan 1982b
Tradescantia	Pollen tetrad MNC	Ma 1982a, Ma et al 1994a
	Stamen hair specific locus mutation	Van't Hof and Schairer 1982, Ma et al. 1994b
Vicia faba	Mitotic cell division and chromosome aberration	Ma 1982b, Kanaya et al 1994
	MNC	Ma et al. 1995, Degrassi and Rizzoni 1982
	SCE	Kihlman and Kornberg 1975, Templaar et al 1982
	Comet (DNA strand break)	Koppen and Verschaeve 1996
Zea mays	Specific locus mutations	Plewa 1982
	MNC	Wagner and Plewa 1985

Table 3. Genotoxicity of metals in plant assays.

Metal	Clasto-genicity	Aneu-genicity	MNC	SCE	Comet	Gene mutation	References
Al (III)	+ (1,2)	- (1,2)					1. Fiskesjo 1988
As (III)		+ (3)	+ (4)				2. Liu et al 1995
							3. Fiskesjo 1997
							4. Steinkellner et al 1998
Cd (II)	+ (1,2,5-7)	+ (1,5,7)	+ (4)	+ (8)	+ (9)]	+ (10)	5. Borboa and De la Torre 1996
Cu (II)	- (1,2)	- (1,2)	- (4)			+ (10)	6. Lerda 1992
							7. Rupshev 1976
Cr (III	+ (2,11)	+ (2,11)	- (12)		+ (9)		8. Panda et al 1996
							9. Koppen and Verschaeve 1996
Cr (VI)	+ (2,11,13)	+ (2,11)	+ (12)		+ (9,13)		10. Reddy and Vaidyanth 1978
Hg (II)	+ (1,13)	+ (14,15)	+ (16-19)	+ (8)		+ (10)	11. Liu et al 1992
							12. Knasmuller et al 1998
Ni (II)	+ (1)	+ (1,2)	+ (12)	+ (8)			13. Rank and Neilson 1994
Pb (II)	+ (6,13)	+ (6,20)	+ (4)]			± (10)	14. Fiskesjo 1969
Sb (III)		- (12)					15. Ramel 1969
Se (II)	+ (21)	± (21,22)					16. Dash et al 1998
							17. Panda et al 1989
Zn (II)	+ (5)	- (5)	+ (4)	+ (8)			18. Pnada et al 1990
							19. Panda et al 1992b
							20. Ahlberg et al 1972
							21. Fiskesjo 1979
							22. Mukherjee and Sharma 1986

Tested positive (+), weakly positive (±) and negative (-)

Bretaguolle and Thompson 1995). Aneupoidy due to meiotic non-disjunction is the cause of several diseases and constitute a significant factor for fetal loss and early infant mortality in humans (Newcombe 1979). A growing body of evidence from human and animal cancer-cytogenetics indicates that aneuploidy is also associated with carcinogenesis (Oshimura and Barrett 1986). Plant assays have been in use for detection of agents capable of inducing aneuploidy and such agents are termed as aneugens (Sandhu et al. 1986). A good number of metals, those having high affinity for SH groups, preferentially associated with tubulin protein, impair spindle function causing chromosome condensation and aneuploidy (Anderson 1986). Hg is the foremost metals known to induce c-mitotsis (colchicine-mitosis) as well as chromosome condensation in root meristem cells of *A. cepa* (Fiskesjo 1969, Ramel 1969). Among

GENOTOXICITY AND MUTAGENICITY OF METALS IN PLANTS 401

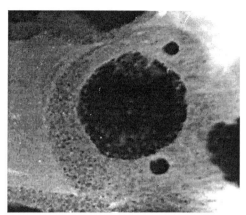 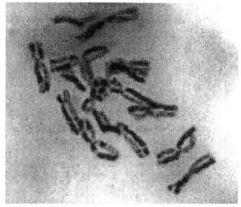

Figure 1. Root meristem cells of A. cepa showing: (A) an interphase cell with two micronuclei and (B) metaphase chromosomes with sister chromatid exchanges.

others As, Be, Cd, Cr, Ni, Pb, Co and Se (Table 3) are aneugenic in plant cells (Ahlberg et al. 1972, Fiskesjo 1979, 1988, 1997, Mukherjee and Sharma 1986, Lerda 1992, Liu et al. 1992, Liu et al. 1995, Borboa and De la Torre 1996).

3.3. Induction of MNC

Plant MNC assay employing root meristems of plants such as *A. cepa* (Reddy 1995, Ma et al. 1995) *Z. mays* (Wagner 1985) and pollen tetrads of *Tradescantia* (Ma 1982a, Ma et al. 1994a) are in vogue. The MNC found in the cytoplasm of mitotic interphase or tetrad cells may either originate from an acentric chromosomal fragment as a result of clastogenesis, or from a lag chromosome due to aneugenesis. The MNC assay therefore has the ability to detect both clastogens and aneugens. On the basis of the size as well as time of appearance of MNC following a brief treatment it may further be possible to resolve whether the test agent is a clastogen or an aneugen (Reddy et al 1995). Plant MNC assay employing the root meristem cells of *A. cepa* (Dash et al. 1988, Panda et al. 1989, Panda et al. 1990, 1992, Steinkellner et al. 1998), *V. faba* (Steinkellner et al. 1998, Minissi and Lombi 1997, Wang 1999, Knasmuller et al 1998) or pollen tetrad cells in *Trdescantia* clone 4430 (Steinkellner 1998, Knasmuller et al 1998) have been used for testing genotoxicity of metals As, Cu, Cd, Cr, Hg, Ni, Pb, Sb, V, Zn or soil and water samples contaminated with some of these metals (Table 3). Whereas *Allium* MNC assay (Figure 1 A) has been demonstrated highly useful for monitoring environmental Hg (Panda et al. 1989, 1990, 1992), *Tradescantia* pollen tetrad MNC assay in comparison to root MNC assays of either *V. fava* or *A. cepa* has been shown to be relatively more sensitive in detection of genotoxicity of heavy metal ions (Steinkellner

1998). As (III), Hg (II), Cd (II), Cr (VI), Pb (II) and Zn (II) were positive in plant MNC assays, whereas no clear positive results were obtained for Cu (II), Cr (III) or Sb (III) with *Tradescantia* pollen tetrad MNC assay (Steinkellner 1998, Knasmuller et al. 1998). The *Tradescantia* pollen MNC test was more sensitive to detect the genotoxicity of Cr (VI), Cd (II), Ni (II) and Zn (II) than *V. faba* root MNC assay, which has been attributed to the higher sensitivity of meiotic chromosomes than mitotic chromosomes to DNA-damage (Knasmuller et al. 1998).

3.4. Induction of SCE

SCE assay has been proved to be one of the most sensitive short-term genotoxicity assays owing to its ability to detect genotoxins at very low concentrations (Tucker et al. 1993). Although the exact mechanism of SCE induction is still a matter of discussion (Cortes et al. 1994), SCEs are widely believed to represent the interchange of DNA replication products at apparently homologous loci, and involve DNA breakage and reunion (Latt et al. 1981). SCE assay has been worked out for a number of plants that include *A. cepa* (Figure 1B), *V. faba, H. vulgare, Picea abies, Nicotiana plumbaginifolia* etc. (Panda et al. 1998, Schvartzman 1987, Schubert 1994). Among a few metals tested for induction of SCE, Cd (II), Hg (II), Ni (II) have shown to be clearly positive and Zn (II) to be weakly positive in *A. cepa* SCE test (Panda et al. 1996).

3.5. Comet Assay

The comet or single cell gel electrophoresis (SCGE) assay as of recent is gaining wide acceptability as a short-term genotoxicity test (Singh et al. 1988, Fairbain 1995, Frenzelli et al. 2000). The Comet assay exclusively using plant cells (Figure 2) has been

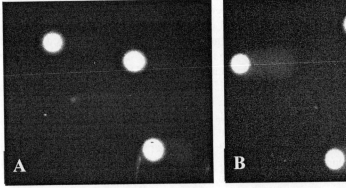

Figure 2: Alkaline Comet Assay in 3 day old germinating Vicia faba roots; comet nuclei in root treated with (A) distilled water (B) 10^{-3} M $CdCl_2$ for 2 h (Courtesy: Dr. G. Koppen).

developed that uses the root or leaf cells of *V. faba* (Koppen and Verschaeve 1996), *A. cepa* (Navarrete et al 1997) and *Nicotiana tabacum* (Gichner and Plewa 1998). The alkaline version (pH > 13) of the comet assay allows quantitative analysis of DNA damage in individual nuclei, including single-strand breaks, alkali-labile sites, incomplete excision repair sites, DNA-DNA or DNA-protein cross-linking (Fairbain et al. 1995, Fortini et al. 1996, Horvathova. et al. 1998). The relative advantages with use of the comet assay over the conventional chromosome aberration, MNC or SCE tests, are that the comet assay is more sensitive for detection of low levels of DNA damage, the requirement for small number of cells or nuclei per sample, its flexibility, low cost, easy application, and short time needed to complete a study (Tice et al. 2000). A number of metals viz. As (III), Cd (II), Cr (III, IV) Hg (II), Ni (II), Pb (II), V (V) have been shown to be positive in mammalian comet tests using human or rat cells *in vitro* (Betti et al. 1993, Pool-Zobel et al. 1994, Hartmann and Speit 1994, Rojas et al. 1996, Anderson et al. 1997, Blasiak and Kowalik 2000). A few metals, Cd (II), Cr (III) and Cr (VI) tested using plant comet assay has been found to be positive (Koppen and Verschaeve 1996, Poli et al. 1999). The comet assay can easily be performed in almost all eukaryotic cells covering a wide range of plants and animals, and therefore possesses wide applicability for assessment of genecotoxicolgy of environmental pollution (Poli et al. 1999, Cotelle and Ferard 1999).

4. METAL-MUTAGEN AND METAL-METAL SYNERGISM AND ANTAGONISM

Organism including plants are exposed to metals in the environment not in isolation but in association with other agents, which include mutagens, metals and a variety of pollutants. Interaction between metals and other environmental agents including mutagens and different metals have been of interest in the overall assessment genotoxicity of xenobiotics in the environment. Of a few metals that were tested for gene-mutation in the chlorophyll deficient system of *Oryza sativa* (Reddy and Vaidyanath 1978); Ba, Cd, Cu, Hg, found to be strongly positive; Sr, Fe, Pb weakly positive; whereas Mn and Ca found to be negative (Table 3). Interestingly, however, when the above metals were administered as post treatment following treatment of gamma ray, Cd, Cu, Hg, Pb and Sr interacted with radiation synergistically yielding high frequency of chlorophyll mutation in M_2 seedlings. Even though Mn, which by it self was not mutagenic, enhanced mutagenicity of radiation (Reddy and Vaidyanath 1978). Likewise combined treatments of seeds with Pb and radiation induced mutation in *Arabidopsis thaliana* synergistically (Dineva et al 1993). In root meristem cells of *A. cepa*, Zn and Cd increased six and eleven times the direct clastogenic effect of β-irradiation produced by the [^{3}H] thymidine-incorporation into DNA, and also increased the cell cycle duration by 1.5 and 2 fold, respectively (Borboa and De la Torre 1996). As, Hg, and Ni likewise prolonged cell cycle in plant (Reddy et al. 1995, Papper et al. 1998) and human cells (Vogel et al. 1986, Jha et al. 1992, Sahu et al. 1995). Mammalian genotoxicity studies established that metals: Cd, Ni, Co, Pb and As when administered in combination enhance genotoxicity of a variety of mutagenic agents such as X-ray, UV, N- methyl- N- nitro- N- nitrosoguanidine (MNNG), mitomycin C etc., possibly

Table 4: Induction of adaptive response by metal salts and oxidative agents to MMCl, MH, EMS or TEM in plant assays (Patra et al. 2000)

Conditioning agent	Challenging agents				Plant assays
	MMCl	MH	EMS	TEM	
Metal salts					
$AlCl_3$	+	-	+	NT	*A. cepa*
	+	-	+	NT	*H. vulgare*
$Cd(NO_3)_2$	NT	+	NT	+	*V. faba*
$CdSO_4$	+	+	NT	NT	*A. cepa*
	+	+	+	NT	*H. vulgare*
$CuSO_4$	NT	+	NT	+	*V. faba*
	+	+	+	NT	*H. vulgare*
$HgCl_2$	+	+	+	NT	*H. vulgare*
CH_3HgCl	+	+	NT	NT	*A. cepa*
	+	+	NT	NT	*H. vulgare*
Residual Hg	+	+	+	NT	*H. vulgare*
$NiCl_2$	NT	+	NT	+	*V. faba*
$NiSO_4$	+	+	+	NT	*H. vulgare*
$Pb(NO_3)_2$	NT	+	NT	+	*V. faba*
	+	+	+	NT	*H. vulgare*
$ZnSO_4$	NT	+	NT	+	*V. faba*
	+	+	NT	NT	*H. vulgare*
Oxidative agents					
H_2O_2	NT	+	NT	-	*V. faba*
	+	+	NT	NT	*A. cepa*
	+	+	NT	NT	*H. vulgare*
PQ	+	-	+	NT	*A. cepa*
	+	-	+	NT	*H. vulgare*

Adaptive response induced (+), not induced (-), not tested (NT).

through interference in DNA repair synthesis, directly or indirectly (Hartwig 1995, Sahu et al. 1989, Roy and Rossman 1992, Hu et al. 1998).

Antagonism of metals against genotoxicity of various mutagens and toxic metals is well documented. Progeny studies employing embryonic shoot ana-telophase or MNC assay with *H. vulgare* (Subhadra et al. 1993) and root MNC assay with *V. faba* (Duan et al. 2000), it has been demonstrated that plants already contaminated with metals like Hg or Cd develop some degree of adaptation/resistance and thereby become less susceptible when challenged by mutagenic or genotoxic treatments subsequently. The ability of plant cells to better resist or adapt to the genotoxic insult caused by mutagenic or genotoxic agents when first pretreated (conditioned) with a metal in non- or sub-toxic doses has been termed as metallo-adaptive response (Subhadra and Panda 1994, Panda et al 1997). In animal cells metals such as Cd induces metallothioneins, a class of specialized proteins that counter metal toxicity (Hamer 1986). Cd-induced metallothioneins protect animal cells from radiation- as well as Cd-genotoxicty (Thornalley and Vasak 1985, Coogan et al. 1994). Like wise a number of metals: Al, Cd, Cu, Hg, Ni, Pb, and Zn are reported to induce adaptive response to genotoxins/chemical mutagens, which include methyl mercuric chloride (MMCl), maleic hydrazide (MH), ethyl methane sulfonate (EMS) or trimethylene melamine (TEM) and the adaptive response was comparable to that induced by oxidative agents like hydrogen peroxide (H_2O_2) or paraquat (PQ) (Table 4, Patra et al. 2000). Possible role of adaptive / detoxifying mechanisms involving metal chelating agents such as glutathione, phytochelatins, metallothioneins, antioxidant responses and/or DNA repair processes underlying the metallo-adaptive response and cross-adaptation to various genotoxins including metals and chemical mutagens in plants has been suggested (Zenk, 1996, Sanità di Toppi and Gabbrieli 1999, Patra et al. 2000, Panda et al. 2000, Cobbett 2000). Many of the metals such as Al, Cd, Cu, Hg, Ni, Pb and Zn induce adapative response against different class of mutagens and share some of the features of oxidative-adaptive response induced by H_2O_2 or PQ (Patra et al 1997, 2000). Metabolic inhibitors namely buthionine sulfoximine an inhibitor of glutathione/phyotchelatin synthesis, and cycloheximide an inhibitor of de novo protein synthesis, inhibited Cd-adaptive response to MH in plant cells (Panda et al 1997, 2000). These studies have pointed to the involvement of glutathione/phytochelatin metabolism in the metallo-adaptive response in plant cells. The foregone metallo-adptive response and cross-adaptation between metals, alkylating, and oxidative mutagens might also involve signal transduction implicating the signaling function of reactive oxygen species (ROS), calcium, abscisic acid and/or jasmonic acid (Xiang and Oliver 1998, Albinsky et al 1999, Bowler and Fuhlr 2000). Activation of some of the signalling pathways may as well delay cell cycle and induce of DNA repair (Albinsky et al 1999).

5. MECHANISMS UNDERLYING METAL INDUCED GENOTOXICITY

Many of the metals have a strong affinity to bind thiol groups of several proteins such as some of the DNA ligases, DNA polymerases, topoisomerase II, glutathione reductase, tubulin or low molecular weight polypeptides e.g. glutathione, and thereby modifying their activity (Gaulden 1987, Albinsky et al. 1999, Moore and Bender 1993). Metal binding to tubulin impairs spindle function resulting in aneugenesis. Binding with chromosomal proteins such as topoisomerase II may result in clastogenesis. Proteins rather than DNA could thus be the target of metal toxicity bringing about genotoxic effects. Besides detoxifying metals directly through direct conjugation, glutathione being an important cellular antioxidant quenches ROS (Alscher 1989, Xiang and Oliver 1998). The ascorbate-glutathione cycle reduces H_2O_2 to water (Dixon et al 1998). Plants are also protected from some of the metals by phytochelatins or class III metalothioneins, which are polypeptides having the general structure (γ-glutmyl-cysteinyl)$_{2-11}$–glycine (Steffens, 1990). Metallothionein of type 1 has been reported in some plants, e. g. *V. faba* (Foley et al 1997). It has been shown that metals such as Cd trigger γ-glutamyl cysteine synthatase (Zhu et al 1999) or phytochelatin synthase (Clemens et al 1999, Vatmaniuk et al 1999) that increase biosynthesis of glutathione and phytochelatin, respectively. The glutathione pool, which is the substrate for phytochelatin synthesis is depleted with the synthesis of the latter causing oxidative stress (De Vos 1992, Dietz et al, 1999). This facilitates lipid peroxidation leading to the formation of free radicals that may ultimately cause DNA strand breaks as well as clstogenesis (Ochi et al, 1983, Ochi and Oshawa 1985, Stohs and Bagcchi, 1995). The fact that certain antioxidants confer protection against Cd, Hg and Cr induced DNA strand break as well as clastogenicity in plant or animal cells proves this point (Snyder 1988, Panda et al 1995, Grillo et al 1999). Furthermore, plants show up antioxidant responses to different metals namely Al, Cd, Cu, Hg, and Zn that points to the metal-induced oxidative stress (Shaw 1995, Chaoui et al 1997, Mazhoudi et al 1997, Patra and Panda 1998, Prasad et al 1999, Ezaki et al 2000, Cho and Park, 2000, Cuypers et al 2001). Metal like Cu, Cr, Fe, and V undergo redox cycling to generate ROS (Stohs and Bagcchi 1995). The other mode of action by which metals such as As, Cd, Co, Ni, and Pb might generate ROS is through participation in Fenton reaction (Roy and Rossman 1992, Luo et al 1996, Lloyd and Phillip 1999). It is important to note that a metal depending on the timing of its treatment, whether pre-, post or in combination with mutagenic/genotoxic agents may either modulate the genotoxicity or act synergistically enhancing the genotoxic effects (Figure 3). Modulation of genotoxicity by metal is possible through metallo-adaptive response involving glutathione, phytochelatins and/or signal transduction pathways (Xiang and Oliver 1998, Albinisky et al 1999, Panda et al 2000). Synergistic or co-mutagenic effect of metals: As, Cd, Ni and Pb has generally been attributed to the inhibition of DNA repair It is important to note that a metal depending on the timing of its treatment, whether pre-, post or in combination with mutagenic/genotoxic agents may either modulate the genotoxicity or act synergistically enhancing the genotoxic effects (Figure 3). Modulation of genotoxicity by metal is possible through metallo-adaptive response involving glutathione, phytochelatins and/or signal transduction pathways (Xiang and Oliver 1998, Albinisky et al 1999, Panda et al 2000). Synergistic or co-mutagenic effect of metals: As, Cd, Ni and Pb has generally been attributed to the inhibition of DNA repair pathways by such metals through competition or interference with the biochemical function of some of the key essential metal ions such as Zn, Ca, Mg etc. (Hartwig 1995).

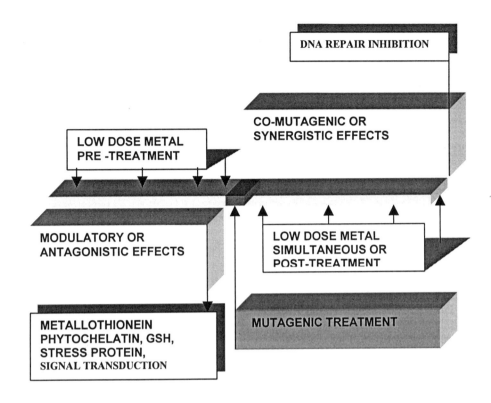

Figure 3: Modulatory and co-mutagenic/synergistic effects of metals

6. CONCLUSION

The foregone review on genetic toxicology of metals in plant that include clastogenicity, aneugenicity, recombinogenicity, gene mutation, DNA damage and repair, and effect on cell cycle did very well agree with results available from mammalian or human systems. The promise that the plant genotoxicity assays lies is its possible application as sentinel bioassay (Lower and Kendeall 1990) in studies pertaining to genetic-ecotoxicology (genecotoxicology) of environmental pollution. Of late efforts are being made to biomonitor environmental genotoxicity employing transgenic plants (Kovalchuk et al 2001) Plants since have a tendency to develop adoptive response to metal or pollution stress, care must be taken that the sentinel species to be employed have already not been exposed to any pollution (Subhadra et al 1993, Hu et al 1998). The role of metallo-adaptive response in evolution of metal tolerance in plants and its implication particularly in genecotoxicology of metals needs better understanding warranting further research.

ACKNOWLEDGEMENTS

The authors are thankful to the authorities of Berhampur University for providing facilities and to Dr. Jita Patra for reading the manuscript. Financial assistance in form of a Major Research Project (No. 3-172/2000) awarded to BBP from the University Grants Commission, New Delhi is gratefully acknowledged.

REFERENCES

Ahlberg, J., Ramel, C., and Wachmeister, C. A. (1972) Organolead compounds shown to be genetically active, *Ambio* 1, 29.

Albinsky, D., Masson, J. E., Bagucki, A., Afsar, A., Vasa, I., Nagy, F. and Paszowski, J (1999) Plant responses to genotoxic stress are linked to an ABA/salinity signaling path way, *The Plant Journal* 17, 73-82.

Alscher, R. G. (1989) Biosynthesis and antioxidant function of glutathione in plants, *Physiologia Plantarum* 77, 457-464

Anderson, O. (1986) Evaluation of spindle inhibiting effects of metals on chromosome length measurement, *Toxicolocgical and Environmental Chemistry* 12, 195-213.

Anderson, S. L., Shugart, L. R. and Suk, W.H. (1994) Napa Conference on Genetic and Molecular Ecotoxicology, *Environmental Health Perspectives* 102 (Supplement 12), 1-108.

Anderson, D., Dobrzynska, M. M. and Basaran, N. (1997) Effect of various genotoxins in human lymphocytes and sperm in the comet assay, *Teratogenesis, Carcinogenesis and Mutagenesis* 17, 29-43.

Anonymous (1991) Toxic terror: dumping of hazardous wastes in the third world, *Third Wold Net Work*, Penang.

Betti, C., Barale, R. and Pool-Zobel, B. B. (1993) Comparative studies an cytotoxic and genotoxic effects of two organic mercury compounds in lymphocytes and gastric mucosa cells of Sprague-Dawley rats, *Environmental and Molecular Mutagenesis* 22: 172-180

Beyersmann, D. (1994) Interaction in metal carcinogenicity, *Toxicology Letters* 72, 333-338Briat, J-F. and Lebrun, M (1999) Plant responses to metal toxicity, *C. R. Academie des Sciences/ Life Sciences*, 322, 43-54.

Blasiak, J. and Kowalik, J. (2000) A comparison of the in vitro genotoxicity of tri- and hexavalent chromoium, *Mutation Research* 469, 135-145

Borboa, L. and De la Torre, C. (1996) The geonotoxicity of Zn (II) and Cd(II) in *Allium cepa* root meristematic cells, *New Phytologist* 134, 481-486.

Bowler, C. and Fuhlr, R. (2000) The role of calcium and activated oxygens as signals for controlling cross-tolerance, *Trends in Plant Science* 5, 241-246.

Bretaguolle, F. and Thompson, J. D. (1995) Tansley Review No. 78. Gamets with the somatic chromosome number: mechanism of their formation and role in evolution of autopolyploid plants, *New Phytologist* 129, 1-22

Briat, J-F. and Lebrun, M. (1999) Plant responses to metal toxicity, *Comptes Rendus de l'Academie des Sciences/ Life Sciences*, 322, 43-54.

Burton, M. A. S. (1986) *Biological monitoring of environmental contaminants, a technical report*, M. A. R. C., London.

Chaoui, A., Mazhoudi, M. H., Ghorbal, S. and Ferjani, E. E. (1997) Cadmium and zinc induction of lipid peroxidation and effect on antioxidant enzyme activities in bean (Paseolus vulgaris L.), Plant Science 127, 139-147

Cho, U.-H and Park, J. O. (2000) Mercury induced oxidative stress in tomato seedlings, Plant Science 156, 1-9

Clemens, S., Kim, E. J., Neumann, D. and Schroeder, J. I. (1999) Tolerance to toxic metals by a gene family of phytochelatin synthases from plants and yeast, *EMBO Journal*. 18, 3325-3333

Cobbett, C. S. (2000) Phytochelatin biosynthesis and function in heavy-metal detoxification, *Current Opinion in Plant Biology* 3, 211-216

Constantin, M. J. and Nilan, R. A. (1982a) Chromosome aberration assays in barley (*Hodreum vulgare*), A report of the US Environmental Protection Agency Gene-Tox Program, *Mutation Research* 99, 13-36.
Constantin, M. J. and R. A. Nilan (1982b) The chlorophyll-deficient mutant assay in barley (*Hordeum vulgare*), A report of the US Environmental Protection Agency Gene-Tox Program, *Mutation Research* 99, 37-40
Coogan, T. P., Bare, R., M. Bjornson, E. J., Walkes, M. P. (1994), Enhanced metallothionein gene expression is associated with protection for cadmium-induced genotoxicity in cultured rat liver cells, *Journal Toxicology and Environmental Health* 41, 233-245.
Cortes, F., Daza, P., Pinero, J., Escalza, P., (1994) Evidence that SCEs induced by mutagens do not occur at the same locus in successive cell cycle: lack of cancellation in three way stained CHO chromosomes, *Environmental Molecular Mutagenesis* 24, 203-207.
Cotelle, S. and Ferard, J. F. (1999) The comet assay in genetic toxicology, *Environmental and Molecular Mutagenesis* 34, 246-255.
Cuypers, A., Vangronsveld, J. and Clijsters, H (2001) The radox status of plant cells (AsA and GSH) is sensitive to zinc imposed oxidative stress in roots and primary leaves of *Phaseolus vulgaris*, *Plant Physiology and Biochemistry* 39, 657-664.
Dash, S., Panda, K. K. and Panda, B. B. (1988) Biomonitoring of low levels of mercurial derivatives in water and soil by Allium micronucleus assay, *Mutation Research* 203, 11-21.
Degrassi, F. and Rizzoni, M. (1982) Micronucleus test in *Vicia faba* root tips to detect mutagen damage in fresh water pollution, *Mutation Research* 97, 19-33.
de Raat, W. K., Vink, G.J. and Hnstveit, A. O. (1990) The significanc of mutagenicity as a criterion in ecotoxicological evaluations, in M. D. Waters (ed.) *Genetic Toxicology of Complex Mixtures*, Plenum Press, New York, pp. 249-267.
De Vos, C.H.R., Vonk, M.J., Vooijs, R. and Schat, H. (1992) Glutathione depletion due to Copper-induced phytochelatin synthesis cause oxidative stress in *Silene cucbalus, Plant physiology* 98, 853-858.
Dietz, K.-J, Bair, M. Kramer, U. (1999) Free radicals and reactive oxygen species as mediators of heavy metal toxicity in plants, in M.N.V. Prasad and J. Hagemeyer (eds.) Heavy Metal Stress in Plants - From molecules to Ecosystems, Springer-Verlag, New York, pp. 73-97.
Dineva, S. B. Abramov, V. I. And Shevchenko, V. A. (1993) Genetic effects of lead nitrate on seeds of chronically irradiated populations of *Arabidopsis thaliana*, *Genetika* 29, 1914-1920
Dixon, D. P., Cummins, I., Cole, D. J. and Edwards, R. (1998) Glutathione-mediated detoxification systems in plants, *Current Opinion in Plants Biology* 1, 258-266.
Duan, C., Hu, B., Guo, T., Luo, M., Xu, X., Chang, X., Wen, C., Meng, L., Yang, L. and Wang, H. (2000) Changes of reliability and efficiency of micronucleus bioassay in *Vicia faba* after exposure to metal contamination for several generations, *Environmental and Experimental Botany* 44, 83-92.
Dudka, S. and Miller, W. P. (1999) Accumulation of potential toxic elements in plants and their transfer to human food chain, *Journal of Environmental Science and Health* B34, 681-708.
Ennever, F. K., Andreano, G. And Rosenkranz, H. S. (1986) The ability of plant genotoxicity assays to predict carcinogenicity, *Mutation Research* 205, 99-105Grant, W. F. (1994) The present status of higher plant bioassays for the detection of environmental mutagens, *Mutation Research* 310, 175-183
Ezaki, B., Gardner, R. C., Ezaki, Y. and Matsumoto, H. (2000) Expression of aluminium-induced genes in transgenic *Arabidopsis* plants can ameliorate aluminum stress and/or oxidative stress, Plant Physiology 122, 657-665.
Fairbain, D. W., Olive, P. L. and O'Neill, K. L. (1995) The comet assay: a comprehensive review, *Mutation Research* 339, 37-59.
Fiskesjo, G., (1969) Some results from Allium tests with organic mercury halogenides, *Hereditas* 62, 314-322
Fiskesjo, G., (1979) Mercury and selenium in a modified Allium test, *Hereditas* 91, 169-178.
Fiskesjo, G. (1988) The *Allium* test - an alternative in environmental studies: the relative toxicity of metal ions, *Mutation Research* 197, 243-260.
Fiskesjo, G. (1997) *Allium* test for screening chemicals; evaluation of cytological parameters, in W. Wang, J. W. Gorsush and J. S. Hughes (eds.) *Plants for Environmental Studies*, Lewis CRC Publishers, Boca Raton, FL, pp.309-333.
Foley, R. C., Liang, Z. M. and Singh, K. B. (1997) Analysis of type 1 metallothionein cDNAs in *Vicia faba*, *Plant Molecular Biology* 33, 583-91.
Flessel, C. P. (1977) Metal as mutagens, *Advances in Experimental Medicine and Biology* 91, 117-128
Fortini, P., Raspaglio, G., Falchi, M. and Dogliotti, E. (1996) Analysis of DNA alkylation damage and repair in mammalian cells by the comet assay, *Mutagenesis* 11, 169-175

Foy, C. D., Chaney, R. L. And White, (1978) The physiology of metal toxicity in plants, *Annual Review of Plant Physiology* **29**, 511-566

Frenzilli, G., Bosco, E and Berale, R. (2000) Validation of single cell gel assay in human leukocytes with 18 reference compounds, *Mutation Research* **468**, 93-108.

Gaulden, M. E. (1987) Hypothesis: some mutagens directly alter specific chromosomal proteins (DNA topoisomerase II and peripherial proteins) to produce chromosomal aberrations, *Mutagenesis* **2**, 357-365.

Gichner, T. Badyev, S. I., Demchenko, J. Relichova, J., Sandhu, S. S., Usmanov, P. D., Usmanova, O. and Veleminsky, J. (1994) Arabidopsis assay for mutagesis, *Mutation Research* **310**, 249-256.

Gichner, T. and Plewa, M. J. (1998) Induction of somatic DNA damage as measured by single cell gel electrophoresis and point mutation in leaves of tobacco plants, *Mutation Research* **401**, 143-152.

Gopalan, H. N. B. (1999) Ecosystem health and human well being: the mission of the international programme of plant bioassays, *Mutation Research* **426**, 99-102.

Grant, W. F. (1982) Chromosome aberration assays in *Allium*, A report of the US Environmental Protection Agency Gene-Tox Program, *Mutation Research* **99**, 273-291.

Grant, W. F. (1994) The present status of higher plant bioassays for the detection of environmental mutagens, *Mutation Research* **310**,175-183.

Grant, W. F. (1999) Higher plant assays for the detection of chromosomal aberrations and gene mutations - a brief historical background on their use for screening and monitoring environmental chemicals, *Mutation Research* **426**, 107-112.

Grant, W. F. and Owens, E. T. (1998) Chromosome aberration assays in *Crepis* for study of environmental mutagens, *Mutation Research* **410**, 291-307

Grillo, C. A. Seoane, A. I and Dulout, F. N. (1999) Protective effect of butylated hydroxytoluene (BHT) agaist the clasogenic activity of cadmium chloride and potassium dichromate in Chinese hamster ovary cells, *Genetics and Molecular Biology* **22**, 59-64.

Hamer, D. H. (1986) Metallothioneins, *Annual Review of Biochemistry* **55**, 913-951

Hartmann, A. and Speit, G. (1994) Comparative investigation of the genotoxic effect of metals in single cell (SSG) assay and the sister chromatid exchange (SCE) test, *Environmental and Molecular Mutagenesis* **23**, 299-305.

Hartmann, A. and Speit, G. (1996) Effect of arsenic and cadmium on the persistence of mutagen-induced DNA lesions in human cells, *Environmental and Molecular Mutagenesis* **27**, 98-104.

Hartwig, A (1995) Current aspects in metal genotoxicity, *Biometal* **8**, 3-11

Horvathova, E., Slameova, D., Hlinikova, L., Mandal, T. K., Gabelova, A. and Collins, A. R. (1998) The nature and origin of DNA single-stranded breaks determined with comet assay, *Mutation Research* **409**, 163-171.

Houk, V. S. (1992) The genotoxicity of industrial wastes and effluents, *Mutation Research* **277**, 91-138.

Hu, Y., Su, L. and Snow, E. T. (1998) Arsenic toxicity is enzyme specific and its effects are not caused by the direct inhibition of DNA repair enzymes, *Mutation Research* **408**, 203-218

Jha, A. N., Noditi, M., Nilsson, R. and Natarajan, A. T. (1992) Genotoxic effects of sodium arsenite on human cells, *Mutation Research* **284**, 215-221.

Kanaya, N. Gill, B. S., Grover, I S., Murin, A. Osiecka, R., Sandhu, S. S. and Andersson, H. C. (1994) *Vicia faba* chromosomal aberration assay, *Mutation Research* **310**, 231-248.

Khus, G. S. (1973) *Cytogenetics of Aneuploids*, Academic Press, New York.

Kihlman, B. A. (1971) Root tips for studying the effects of chemicals on chromosomes, in A. Hollaender (ed.) Chemical Mutagnes Vol. 2, Plenum Press, New York, pp. 489-514.

Kihlman, B. A. (1975) Root tips of *Vicia fava* for the study of the induction of chromosomal aberrations, *Mutation Research* **31**, 401-412.

Kihlman, B. A. and Kornberg, D. (1975) Sister chromatid exchanges in *Vicia faba*, demonstrated by a modified fluorescent plus giemsa technique (FPG) technique, *Chromosoma* **51**, 1-10.

Knasmuller, S., Gottamann, E., Steinkellner, H., Fomin, A., Pickel, C. Paschke, A. God, R. and Kundi, M. (1998) Detection of genotoxicity effects of heavy metal contaminated soils with plant bioassays, *Mutation Research* **420**, 37-48.

Koppen, G. and Verschaeve, L. (1996) The alkaline comet test on plant cells: A new genotoxicity test for DNA strand breaks in *Vicia faba* root cells, *Mutation Research* **360**, 193-200.

Kovalchuk, I., Kovalchuk, O. and Hohn, B. (2001) Biomonitoring the genotoxicity of environmental factors with transgenic plants, *Trends in Plant Science* **6**, 7306-7310.

Latt, S. A., Allen, J., Bloom, S. E., Carrano, A., Falke, E., Kram, D., Schneider, E., Schreck, R., Tice, R., Whitefield, B. And Wolff, S. (1981) Sister chromatid exchange: a report of the Gene-Tox Program, *Mutation Research* **87**, 17-62.

Lerda, D. (1992) The effect of lead on *Allium cepa* L., *Mutation Research* **281**, 89-92.

Liu, D., Jiang, W. and Li, M. (1992) Effect of trivalent and hexavalent chromium on root growth and cell division of *Allium cepa*, *Hereditas* **117**, 23-29.

Liu, D. Jiang, W., Wang, W. and Zhai, L (1995) Evaluation of metal ion on root tip cells by the *Allium* Test, *Israel Journal of Plant Sciences* **43**, 125-133..

Lloyd, D. R. and Phillips, D. H. Oxidative DNA damage mediated by copper (II), iron (II) and nickel (II) Fenton reaction: evidence for site-specific mechanisms in the formation of double-strand breaks, 8-hydroxydeoxyguanosine and putative intrastrand-links, Mutation Research, **424**, 23-36.

Logan, T. J. And Traina, S. J. (1993) Trace metals in agricultural soils, in H. E. Allen, E. M. Perdue and D. S. Brown (eds.), *Metals in Groundwater*, Lewis Publishers, Boca Raton, pp. 309-347.

Lower, W. R. and Kendeall, R. J. (1990) Sentinel species and sentinel bioassay, in J. F. McCarthy and L. R. Shugart (eds.) Biomarkers of Environmental Contamination, Lewis CRC Publishers, Boca Raton, FL, pp. 309-331.

Luo, H., Lu, Y., Shi, X., Mao, Y. and Dalal, N.S. (1996) Chromium (IV)-mediated fenton-like reaction causes DNA damage: implication to genotoxicity of chromate, *Annals of Clinical Laboratory Science* **26**, 185-191.

Ma, T.-H. (1982a) Vicia cytogeneitc tests for environmental mutagens, A report of the US Environmental Protection Agency Gene-Tox Program, *Mutation Research* **99**, 257-271.

Ma, T-H., (1982b) Tradescantia micronuclei (Trad-MCN) test for environmental clastogens, in A. R. Kolber, T. K. Wong, L. D. Grant, R. S. DeWoskin and T. J. Hughes (eds.) *In Vitro Toxicity Testing of Environmental Agents, Part A*, Plenum Press, New York, pp. 191-214.

Ma, T-H. (1999) The international program on plant bioassays and the report of the follow-up study after the hands-on workshop in China, *Mutation Research* **426**,103-106.

Ma, T-H., Cabera, G. L., Chen, R., Gill, B. S. Sandhu, S. S., Vanderberg, A. L. and Salmone, M. F. (1994a) Tradescantia micronucleus bioassay, *Mutation Research* **310**, 221-230.

Ma, T-H., Cabera, G. L. Cebulska-Wasilewaska, A., Chen, R. Loarca, F. Vanderberg, A. L. and Salmone, M. F. (1994b) *Tradescantia* stamen hair mutation bioassay, *Mutation Research* **310**, 211-220

Ma, T-H., Zhidong, X., Xu, C., McConnell, H., Rabago, E. V. Arreola, G. A. and Zhang, H. (1995) The improved *Allium/Vicia* root tip micronucleus assay for clastogenicity of environmental pollutants, *Mutation Research* **334**, 185-195

Mazhoudi, S. and Chaoui, A, Ghorbal, M. H. and Ferjani, E. E. (1997) Response of antioxidant enzymes toexcess copper in tomato (*Lycopersicum esculantum*, Mill), *Plant Science* **127**, 129-137.

McClintock, B. (1965) The control of gene action in maiz, *Brookhaven Symposium in Biology* **26**, 305-309.

Mehera, A. and Farago, M.E. (1994) Metal ions and plant nutrition, in M. E. Farago (ed.) *Plants and the Chemical Elements: Biochemistry, Uptake Tolerance and Toxicity*, VCH, Weinheim, pp. 32-66.

Minissi, S. and Lombi, E. (1997) Heavy metal content and mutagenic activity, evaluated by *Vicia faba* micronucleus test, of Tiber river sediments, *Mutation Research* **393**, 17-21.

Moore, R. C. and Bender, M. A. (1993) Time sequence of events leading to chromosomal aberration formation, *Environmental and Molecular Mutagenesis* **22**, 208-213

Moore, J. W. and Rammoorthy, S. (1984) *Heavy Metals in Natural Waters,* Springer-Verlag, New York.

Mukherjee, A. and Sharma, A. (1986) Effects of cadmium chloride and sodium selenite on plant chromosomes, *Perspectives in Cytology and Genetics* **5**, 325-328

Navarrete, M. H. Carrera, P, de Miguel, M. and de la Torre, C. (1997) A fast comet assay for solid tissue cells. The assessment of DNA damage in higher plants. *Mutation Research* **389**, 271-277.

Newcombe, H. B. (1979) Measuring the public health impact of the aneuploids, *Environmental Health Perspectives* **31**, 3-8.

Nicoloff, H. and Gecheff, K (1976) Methods of scoring induced chromosome structural changes in barley, *Mutation Research* **34**, 233-244.

Nriagu, J. O. (1990) Global metal pollution; poisoning the biosphere, *Environment* 32, 7-32

Nriagu, J. O. and Pacyna, J. M. (1988) Quantitative assessment of worldwide contamination of air, water and soils with trace metals, *Nature* **333**, 134-139.

Ochi, T., Ishigura, T. and Osawa, M. (1983) Participation of active oxygen species in the induction of DNA single strand scissions by cadmium chloride in cultured Chinses hamster cells, *Mutation Research* **122**, 169-175.

Ochi, T. and Oshawa, M. (1985) Participation of active oxygen species in the induction of chromosomal aberrations by cadmium chloride in cultured Chinese hamster cells, *Mutation Research* **143**, 137-142.
Onfelt, A. (1983) Spindle disturbances in mammalian cells. I. Changes in the quantity of free sulfhydril groups in relation to survival and c-mitosis in V79 Chinese hamster cells after treatment with colcemid, diamide, carbaryl and methyl mercury, *Chemico-Biological Interactions* **46**, 201-217.
Oshimura, M. and Barrett, J. C. (1986) Chemically induced aneuploidy in mammalian cells: mechanisms and biological significance in cancer, *Environmental Mutagensis* **8**, 129-159.
Panda, B. B., A. V. Subhadra, and Panda, K. K. (1995) Prophylaxis of antioxidants against the genotoxicity of methyl mercuric chloride and maleic hydrazide in Allium micronucleus assay, *Mutation Research* **343**, 75-84.
Panda, B. B., Patra, J. and Panda, K. K. (2000) Cadmium-induced adaptive response in plant cells *in vivo*: A possible model based on genotoxicity studies, in: M. Yunus, N. Singh and L. J. Dekok Eds., *Pollution Stress: Indication, Mitigation and Eco-conservation*, Kluwer Academic Publication, Amsterdam, pp. 173-184.
Panda, K. K., Lenka, M. and Panda, B. B. (1989) Allium micronucleus (MNC) assay to assess bioavailability, bioconcentration and genotoxicity of mercury from solid waste deposits of a chloralkali plant, and antagonism of L-cysteine, The *Science of the Total Environment* **79**, 25-36.
Panda, K. K., Lenka, M. and Panda, B. B. (1990) Monitoring and assessment of mercury pollution in the vicinity of a chloralkali plant I. Distribution, availability and genotoxicity of sediment mercury in the Rushikulya estuary, India, The *Science of the Total Environment* **96**, 281-296.
Panda, K. K., Lenka, M. and Panda, B. B. (1992a) Monitoring and assessment of mercury pollution in the vicinity of a choralkali plant. II Plant-availability, tissue-concentration and genotoxicity of mercury from agricultural soil contaminated with solid waste assessed in barley (*Hordeum vulgare* L.), *Environmental Pollution* **76**, 33-42.
Panda, K. K., Lenka, M. and Panda, B. B. (1992b) Monitoring and assessment of of mercury pollution in the vicinity of a chloralkali plant III. Concentration and genotoxicity of mercury in the industrial effluent and contaminated water of Rushikulya estuary, India, *Mutation Research* **280**, 149-160.
Panda, K. K., Patra, J. and Panda, B. B. (1996) Induction of sister chromatid exchanges by heavy metal salts in root meristem cells of *Allium cepa* L., *Biologia Plantarum* **38**, 555-561.
Panda, K. K., Patra, J. and Panda, B. B. (1997) Persistence of cadmium-induced adaptive response against genotoxicity of maleic hydrazide and methyl mercuric chloride in root meristem cells of *Allium cepa* L. Differential inhibition by cycloheximide and buthionine sulfoximine, *Mutation Research* **389**,129-139.
Panda, K. K., Sahu, U. K. and Panda, B. B. (1998) A new haematoxylin procedure for differential staining of sister chromatid exchanges in root meristem cells of *Allium cepa*, *Cytobios* **93**, 113-121.
Papper, I. Galaniti, N., Sans, J. and Lopez-Saez, J. F. (1998) Reversible inhibition of root growth and cell proliferation by pentavalent arsenic in *Allium cepa* L., *Environmental and Experimental Botany* **28**, 9-18.
Patra, J. and Panda, B. B. (1998) A comparision of biochemical responses to oxidative and metal stress in seedlings of barley, *Hordeum vulgare* L. *Environmental Pollution* **101**, 99-105.
Patra, J., Panda, K. K. and Panda, B. B. (1997) Differential induction of adaptive response by paraquat and hydrogen peroxide against genotoxicity of methyl mercuric chloride, maleic hydrazide and ethyl methane sulfonate in plant cells *in vivo*, *Mutation Research* **393**, 215-222.
Patra, J., Baisakhi, B., Mohapatro, M. K. and Panda, B. B. (2000) Aluminium triggers genotoxic adaptation to methyl mercuric chloride and ethyl methane sulfonate, but not to maleic hydrazide in plant cells *in vivo*, *Mutation Research* **465**,1-9.
Plewa, M. J. (1982) Specific-locus mutation assays in *Zea mays*, A report of the US Environmental Protection Agency Gene-Tox Program, *Mutation Research* **99**, 317-337..
Plewa, M. J. (1985) Plant genetic assays and their use in studies on environmental mutagenesis in developing countries, in A. Muhammed and R. C. von Borstel (eds.) *Basic and Applied mutagenesis. With Special Reference to Agricultural Chemicals in Developing Countries*, Plenum press, New York, pp. 249-268.
Poli, P Buschini, A, Restivo, M. F., Ficarelli, A, Cassino, F., Ferrero, I. And Rossi, C. (1999) Comet assay application in environmental monitoring: DNA damage in human leukocytes, and plant cells in comparison with bacterial and yeast tests, *Mutagenesis* **14**, 547-555.
Pool-Zobel, B. L., Lotzmann, N., Knoll, M., Kuchenmeister, F., Lambertz, R. Leucht, U. And Schroder, H.-G. And Schmezer, P. (1994) Detection of genotoxic effects of human gastric and nasal mucosa cells isolated from biopsy samples, *Environmental and Molecular Mutagenesis* **24**, 23-45.
Prasad, K. V. S. K., Saradhi, P. P. and Sharmila, P. (1999) Concerted action of antioxidant enzymes and curtailed growth under zinc toxicity in *Brassica juncia*, *Environmental Experimental Botany* **42**, 1-10.

Prasad, M. N. V. (1997) Trace metals, in M. N. V. Prasad (ed.) *Plant Ecophysiology*, John-Wiley, New York, pp 263-273.
Ramel, C. (1969) Genetic effects of organomercury compounds, 1. Cytological investigation on *Allium* root, *Hereditas* 61, 208-230.
Rank, J. and Nielson, M. H. (1993) A modified Allium test as a tool in screening of the genotoxicity of complex mixtures, *Hereditas* 118, 49-53
Rank, J. and Nielson, M. H. (1994) Evaluation of Allium anaphase-telophase test in relation to genotoxicity screening of industrial waste water, *Mutation Research* 312, 17-24.
Rank, J. and Nielsen, M. H. (1998) Genotoxicity of waste water sludge using the *Allium cepa* chromosome anaphase-telophase aberration assay, *Mutation Research* 418, 113-119
Reddy, N. M., Panda, K. K. Subhadra, A. V. and Panda, B. B. (1995) The Allium micronucleus (MNC) assay may used to distinguish clastogens from aneugens, *Biologishes Zentralblatt* 144, 358-368
Reddy, T. P. and Vaidyanath, K. (1978) Synergistic interaction of gamma rays and some metallic salts in the induction of chlorophyll mutations in rice, *Mutation Research* 52, 361-365.
Redei, G. P. (1982) Mutagen assay with *Arabidopsis*, A report of the US Environmental Protecion Agency Gene-Tox Program, Mutation Research 99, 243-255.
Rojas, E., Valverde, M., Herrera, L. A., Altrarmirano-Lozano, M. and Ostrosky-Wegman, P (1996) Genotoxicity of vanadium pentoxide evaluated by the single cell gel electrophoresis assay in human lymphocytes, *Mutation Research* 360, 193-200.
Rossman, T. G., Zelikof, J. T. Agarwal, S. and Kneip, T. J. (1987) Genetic toxilogy of metal compounds: an examination of appropriate cellular models, *Toxioclogy and Environmental Chemistry* 14, 251-262.
Roy, N. K. and Rossman, T. G. (1992) Mutgenicity and comutagenicity by lead compounds, Mutation Research 298, 97-103.
Ruposhev, A. R. (1976) The cytogenetic effect of heavy metal ions on seed of *Crepis capillaris* L., *Genetika* 12, 37-43
Sahu, R. K. Katsifis, S. P., Kinney, P. L. and Christie, N. T. (1989) Effect of nickel sulphate, lead sulphate and sodium arsenite alone and with UV light on sister chromatid exchanges in cultured human lymphocytes, *Journal of Molecular Toxicology* 2, 129-136.
Sahu, R. K., Katsifis, S. P., Kinney, P. L. and Christie, N. T. (1995) Ni (II) induced changes in cell cycle duration and sister-chromatid exchanges in cultured human lymphocytes, *Mutation Research* 327, 217-225.
Sandhu, S. S. and Lower, W. R. (1989) In situ assessment of genotoxic hazard of environmental pollution, *Toxicology and Industrial Health* 5, 61-69.
Sandhu, S. S., Vig, B. K. and Constantin, M. J. (1986) Detection of chemically induced aneuploidy induction with plant test systems, *Mutation Research* 167, 61-69.
Sandhu, S. S., De Marini, D., Mass, M. J., Moore, M. M. and Mumford, J. L. (1987) *Short-Term Assays in the analysis of Complex Environmental Mixtures*, Plenum Press, New York.
Sanità di Toppi, L. and Gabbrieli, R. (1999) Responses to cadmium in higher plants, *Environmental and Experimental Botany* 41, 105-130.
Schubert, I (1994) Sister chromatid exchanges in *Nicotiana plumbaginifilia*, *Biologisches Zentralblatt* 113, 487-491.
Schvartzman, J. B. (1987) Sister chromatid exchanges in higher plant cells: past and perspectives, *Mutation Research* 181, 127-145.
Schvartzman, J. B. and Cortes, F. (1977) Sister chromatid exchanges in *Allium cepa* L. *Chromosoma* 62, 119-131
Shaw, B. P. (1995) Effect of mercury and cadmium on the activities of antioxidant enzymes in the seedling of *Phaseolus aureus*, *Biologia Plantarum* 37, 587-596.
Shelby, M, D. (1976) *Chemical mutagenesis in plants and mutagenicity of plant related compounds*, Environmental Mutagen Information Center, Oak Ridge.
Singh, N.P., McCoy M.T., Tice, R.R. and Schneider, E.L. (1988) A simple technique for quantitation of low levels of DNA damage in individual cells, *Experimental Cell Research* 175, 184-191.
Snyder, R. D. (1988) The role of active oxygen species in metal induced DNA strand breakage in human deploid fibroblasts, *Mutation Research* 193, 237-246.
Steffens, J. C. (1990) The heavy metal binding peptides of plants, *Annual Review of Plant Physiology and Molecular Biology* 41, 553-575.

Steinkellner, H., Mun-Sik, K, Helma, C, Ecker, S., Ma, T.-H., Horak, O., Kundi, M. and Knasmuller, S. Genotoxic effects of heavy metals: Comparative investigation with plant bioassays, *Environmental and Molecular Mutagenesis* 31, 183-191.

Stohs, S. J. and Bagcchi, D. (1995) Oxidative mechanisms in the toxicity of metal ions, *Free Radical Biology and Medicine* 18, 321-326.

Subhadra, A. V. and Panda, B. B. (1994) Metal-induced genotoxic adaptation in barley (*Hordeum vulgare* L.) to maleic hydrazide and methyl mercuric chloride, *Mutation Research* 321, 93-102.

Subhadra, A.V., Panda, K.K. and Panda, B.B. (1993) Residual mercury in seeds of barley (*Hordeum vulgare* L.) confres genotoxic adaptation to ethyl methane sulfonate, Maleic hydrazide, methyl mercuric chloride and mercury contaminated soil, *Mutation Research* 300, 141-149.

Templaar, M. J. De Both, M. T. J. and Verteegh, J. E. G. 1982 Measurement of SCE frequencies in plants: a simple Feulgen-staining procedure of *Vicia faba, Mutation Research* 103, 322-326.

Thornalley , P. J. and Vasak, M., (1985) Possible role of metallothionein in protection against radiation-induced oxidative stress. Kinetics, and mechanism of its reaction with superoxide and hydroxyl radicals, *Biochimica Biophysica Acta* 827, 36-44.

Tice, R. R., Agurell, E., Anderson, D., Burlinson, B., Hartman, A., Kobayashi, H. Miyamae, Y., Rojas, E., Ryu, J.-C. and Sasaki, Y. F. (2000) Single cell gel/comet assay: guidelines for in vitro and in vivo genetic toxicligical testing, *Environmental and Molecular Mutagenesis* 35, 206-221.

Tucker, J. D., Auletta, A., Cimido, M. C., Dearfield, K. L., Jacobson-Kram, D. Tice, R. R. and Carrano, A. V. (1993) Sister chromatid exchange: second report of the Gene-Tox Program, *Mutation Research* 297, 101-180.

Van't Hof, J. and Schairer, L. A. (1982) Tradescantia assay system for gaseous mutagens, A report of the US Environmental Protection Agency Gene-Tox Program, *Mutation Research* 99, 303-315.

Vatamaniuk, O. K., Mari, S., Lu, Y. P. and Rea, P. A. (1999) AtPCS1, a phytochelatin synthase from *Arabidopsis*: isolation and *in vitro* reconstitution, *Proceedings of National Academy of Sciences*, U. S. A. 96, 7110-7115

Vig, B. K. (1982) Soybean (*Glycine max* [L.] Merill as a short-term assay for study of environmental mutagens, A report of the US Environmental Protection Agency Gene-Tox Program, *Mutation Research* 99, 339-347

Vogel, D. G., Rovinvitich, P. S. and Mottet, N. K. (1986) Methyl mercury effects on cell cycle kinetics, *Cell Tissue Kinetics* 19, 95-118.

Wagner, E. D. and Plewa, M. J. (1985) Induction of micronuclei in maiz root tip cells and a correlation with forward mutation at the yg2 locus, *Environmental Mutagenesis* 7, 821-832.

Wang, H. (1999) Clasogenicity of chromium contaminated soil samples evaluated by *Vicia faba* micronucleus assay *Mutation Research* 426, 147-149.

Waters, M. D., Stack, H. F., Brady, A. L., Lohman, P. H. M., Haroun, L. and Vainio, H. (1988) Use of computerized data listing and activity profiles of genetic and related effects in the review of 195 compounds, *Mutation Research* 205, 295-312.

Wiersma, D., van Goor, B. J. and van der Veen, N. G. (1986) Cadmium, lead, mercury, and arsenic concentrations in crops and corresponding soils in The Netherlands, *Journal of Agriculture and Food Chemistry* 34, 1067-1074.

Woolhouse, H. W. (1983) Toxicity and tolerance in the response of plants to metals, in O. Lang, P. S. Nobel, C. B. Osmond and h. Zeigler (eds.) *Encylopedia of Plant Physiology, Vol. 12 Physiological Plant Ecology III*, Springer-Verlag, Berlin, pp. 245-300

Wurgler, F. E. and Kramers, P. G. N. (1992) Environmental effects of genotoxins (eco-genotoxicology), *Mutgenesis* 7, 321-327.

Xiang, C. and Oliver, D. J. (1998) Glutathione metabolic genes coordinately respond to heavy metals and jasmonic acid in Arabidopsis, *The Plant Cell* 10, 1539-1550.

Zenk, M. H. (1996) Heavy metal detoxification in higher plants, *Gene* 176, 21-30.

Zhang, Y. And Yang, X. (1994) The toxic effects of cadmium on cell division and chromosomal morphology of *Hordeum vulgare, Mutation Research* 312, 121-126.

Zhu, Y. L., Pilon-Smits, E. A., Tarun, A. S., Weber, S. U., Jouanin, L. and Terry, N. (1999) Cadmium tolerance and accumulation in Indian mustard is enhanced by overexpressing gamma-glutamylcysteine synthetase, *Plant Physiology* 121, 1169-1178.

Zimmermann, F. K., de Serres, F. J., Shelby, M. D. and Wassom, J. S. (1979) Proceedings of the workshop on systems to detect induction of aneuploidy by environmental mutagens, *Environmental Health Perspectives* 31, 1-167.

CHAPTER 16

METAL DETOXIFICATION PROPERTIES OF PHYTOMASS: PHYSIOLOGICAL AND BIOCHEMICAL ASPECTS

MIGUEL JORDAN
Departamento de Ecología, Facultad deCiencias Biológicas
Pontificia Universidad Católica de Chile, Casilla 114-D, Santiago, Chile

1. INTRODUCTION

The paper pulp industry is a source of various kinds of solid, liquid and gaseous wastes. During the recovery of cellulose, several solid residues of inorganic nature (*ashes, fly-ashes, dregs* and *grits*) or organic wastes derived by sedimentation of sludge mainly composed by fiber debris (*primary sludge* and *brown stock greening rejects*) are generated. The discharges have been significantly reduced during recent years due to new techniques that have been applied both in developed and developing countries. However, the production of residues cannot be completely avoided, and proper techniques for the handling of waste products are required. Although gaseous and liquid residues have greatly decreased, the solid components are still considerable: taking a broad range of mills, an average mill produces about 70.000 hum/tons/year. To date, there is a scarce number of rigorous studies addressing the impact of these wastes on the environment. However, it is likely that some of these residues may be harmful, due to their heavy metals and organochlorides content. At present, waste products are often dried, burnt and disposed off at sites near the mills. This is an expensive strategy excluding consideration of any other possible use. Now, several alternatives to obtain added value from solid residues are being investigated worldwide. They include, among others, anaerobic fermentation, composting, and landspreading. The first two are conducted under controlled conditions - which implies relatively high costs - and are sensitive to the presence of toxic compounds. Therefore, landspreading for the purpose of fertilization and amendment of soils diminishes the environmental impact, and can be used as an adequate alternative.

2. COMPOSITION, CHARACTERISTICS AND TYPES OF WASTE RESIDUES

Solid waste residues are deposited in dumps. A sample of deposits derived from *Pinus radiata* processing shows a mixture of all six residues in proportions that are disposed off by the mill. That is, approx. ashes (46%), primary sludges (23%), fly-ashes (11%), dregs (15%), grits (3.5%), and brown stock greening rejects (2.1%), but proportions vary according to processing and the plant species used as donor trees (Osses, pers. comm.). The ashes are the inorganic residues from the power boiler while dregs and grits correspond to inorganic residues from the caustic treatment area. Primary sludge and brown stock rejects are organic wastes composed primarily of cellulose fibers. The latter organic compounds can reach 97% of dry weight. According to kind, residues contain various levels of essential elements as well as heavy metals and organic matter, in varying amounts, making them valuable as fertilizers (Table 1). (Jordan et al. 2000). Several essential elements are present in residues and their levels are normally much higher than the minerals found in average forest soils. Amounts depend on the plant species used as sources and on the chemical processing at the mill. For example, ashes show high levels of SiO_2 and various elements in the form oxides (Ca, K and Mg), while grits and dregs contain mainly carbonates, Ca and Na. In fresh residues, it is also common to find natural organochlorides (absorbable organic halides: AOX), at present less than 15 mg/Kg, that are further biodegraded by microbial remediation in the soil or dumps. Most residues are alkaline, making them suitable to correct acidity of some soils, i.e. the addition of organic wastes increased soil pH slightly after 16 weeks while fly-ash showed no effect (Simpson et al. 1983). The gradual changes in pH do not seem to affect the natural microflora (Gonzalez, pers. comm.; Jordan et al. submitted) but changes in pH can account for changes in availability of several elements in the soils (Marschner, 1995). For example, the application of sludges and fly-ashes affect the extractable levels of phosphorus and calcium without altering heavy metals uptake by crops (Simpson et al. 1983). Also, depending on waste type, water retention can be noticeable, a positive condition to maintain humidity of the soil and at the root zone. According to Bellote (pers. comm.), at field capacity, ashes applied in the order of 50 ton/Ha can increase water retention about 12-14% besides improving physical, chemical, and biological conditions of soils. Sludge and brown rejects derived from *P. radiata* processing retain approx. 50 and 76% water respectively, ashes and fly-ashes 20-25%, dregs 35%, while grits retained only about 5% water, but values can vary widely.

Estimation of elements in wastes can vary according to mill processes and source trees. Soluble P can be found in residues in ranges of 0.005-0.07%, soluble K within 0.003-0.88 (the last value corresponding to ashes). Ca can range between 0.1-29% (the highest values represented by grits and dregs as a result of processing calcareous sources such as seashells (Waldemar and Herrera, 1986). Also, Mg can reach 10% in dregs. All minor essential elements are found and their contents are higher than levels present in forest soils, especially in the case of Fe (as Fe_2O_3) in fly-ashes showing the highest content (4.0%). Some non-essential elements/heavy metals such as Cd, Cr, and Pb besides Ba, Bi and Sb, were always found in a level close to 0.01% (Jordan et al. not published). Minerals consisting principally of carbonates can increase cathodic exchange and reduce the speed of water percolation in highly drainable soils. However, the amount of hazardous contaminants, which can be leached from the material depends on type of residue used. For example, the total concentrations of Cd, Cr, Cu, Ni, Pb and Zn contained in sludge are less than those contained in a morine - similar to natural geological material (Fällman, 1986) - although available leached concentrations from grits and green liquor dregs are comparatively higher (Gustavsson et al. 1999). This can account for the limited growth responses to both residues when used with seedlings (Jordan et al. 2000)

Table 1. Composition of various solid residues, of two forest soils and of leaves of 10 year-old Pinus radiata trees growing close to a dump.

1Compound	Ashes	Fly-Ash	Dregs	Grits	Prim. Sludge	Brown Stock Rejects	MSS[1]	MTC[1]	Adult Trees
PH	11.2	10.4	10.4	12.0	9.1	10.8	5.67	5.13	----
Humidity	21.5	26.2	35.4	4.5	47.9	75.4	----	----	----
AOX [3]	0.28	2.60	0.17	0.33	9.51	0.0	0.0	0.0	n.e.
Org.matt.[4]	7.0	2.7	0.0	n.e.	76.3	94.6	9.0	7.6	97.8
Phosphorus	0.01	0.005	0.06	0.008	0.012	0.07	<0.001	<0.001	0.03
Potassium	0.88	0.10	0.09	0.19	0.017	0.003	0.007	0.014	0.22
Calcium	9.79	5.43	29.8	29.0	17.49	0.1	0.095	0.045	0.32
Magnesium	1.09	1.06	10.8	0.75	2.02	0.03	0.014	0.011	0.07
Sodium	3.15	1.49	5.7	2.24	1.26	1.24	0.002	0.001	0.24
Manganese	0.23	13.5	1.43	0.01	0.10	<0.01	0.01	0.007	0.02
Iron	2.75	4.01	0.81	0.56	0.87	0.02	0.001	<0.001	0.06
Aluminium	5.91	4.93	0.65	0.23	0.38	0.03	n.e.	n.e.	0.76
Cadmium	0.01	<0.01	0.005	0.005	<0.01	<0.01	n.e.	n.e.	<0.01
Chromium	0.01	0.01	0.03	0.01	0.01	<0.01	n.e.	n.e.	<0.03
Lead	0.01	<0.01	0.001	0.001	0.001	<0.01	n.e.	n.e.	<0.01

[1] MSS, metamorphic sandy soil; MTC, marine terrace clay.
[2] Element content in adult pine trees growing close to the dump.
[3] Absorbable organic halogen (AOX) values are expressed in mg/Kg.
[4] Organic matter, P, K, Mg and Mn available in residues, all expressed in % (d.w.); Ca, Mg and Na extractable in soils as well as heavy metals, also in %; n.e.=not examined (Data from Jordan et al. 2000, modified).

3. BIOREMEDIATION

Metals as well metalloids can be removed from polluted sites through bioadsorption to living and non-living biomass by many microbes since these are needed for structural and or catalytic functions; even dead cells can bind metal ions (Ehrlich, 1997). Although some elements are found in the environment due to industrial processing, as it happens with cellulose mill dumps, many metals including heavy metals of wastes, are normally subjected to bioremediation and mineralization by indigenous microbial flora, affecting selective and non-selective leaching of metal constituents (Ehrlich, 1996; Ehrlich, 1997). For example, after landspread application of sludges it was found that the organic chlorine from NMW-chlorolignins (formed during the bleaching of wood pulp with chlorine and chlorine derivatives), is slowly released to the soil in the form of inorganic chloride (Sherman, 1995). Although relatively slow, the mineralization rate of organic chlorine to chloride seems to occur faster than carbon mineralization (Sherman, 1995). Simultaneously, chlorolignin is rapidly adsorbed or bound to soil and humic matter (Brezny et al. 1993). Both effects, biodegradation and biologically mediated chemical binding into the soil, reduce toxicity during the stabilization of dumps. However, the fate and turnover of heavy metals present in the different wastes of the pulp industry, disposed off in dumps or after landspreading mixed with forest soil, still require more detailed studies.

Regarding AOX degradation (Jordan et al. submitted), the initial contents of fresh residues were relatively low, ranging from 0.28 to 9.51 mg/Kg, (highest levels were found in sludges and none in brown rejects); these levels were further reduced by bioremediation of indigenous microbes. The microbial decomposing activity of microorganisms affecting Kraft bleaching effluent chloroaromatics has been reported in soils (Brezny et al. 1992, González et al. 1995). In solid residues, biodegradation depends on the type of waste, soil characteristics and participation of both phases. The turnover of various added chloroaromatics (2,4-dichlorophenol; 2,4,6-trichlorophenol; 4,5-dichloroguaiacol; and trichlorosyringol; all 1:1 vol/vol) was ca. 80-95 % in presence of sludges after 30 days; intermediate in brown rejects and grits and nil in the mixtures (Jordan et al. submitted). The presence of ash residues in a mixture of wastes that roughly corresponds to the production of each kind of residue in the mill (normal content is approx. 46%) inhibited degradation to a high extent. Bark chips have been used as inoculum with nutrients and reported as useful in enhancing chlorophenol degradation by indigenous microbes (Laine and Jorgensen, 1998).

Growth stimulation of indigenous microorganisms and/or exogenous microbes which can act as degraders, or show bioadsorptive capacity for various heavy metals (Valentine et al. 1996, Lovley and Coates 1997) can be a useful strategy to immobilize toxic compounds in soil, reducing probability of ground water leaching (Joyce et al. 1992), and their incorporation to plants. Comparatively, heterotrophic bacterial or fungal cultivable counts in soil/solid residues mixtures were not affected over 60 days incubation; the counts were typically 10^8 and 10^4 colony-forming units/ml (Gonzalez, not published). Recent studies concerning DNA characterization of microorganisms affecting degradation have been developed (Gonzalez, pers. comm.). Preliminary results, using amplification of the ribosomal sub-unity 16S rDNA region of natural bacterial communities, showed little changes regarding natural microflora of various selected forest soils after addition of cellulose mill residues (Céspedes, pers. comm). The search for adequate microbes is of primary importance; for example, by means of incubation with the fungus *Sporotrichum pulvurulentum* a 40% mineralization of the chlorolignin was observed after 6 weeks (Ericksson and Kolar, 1985).

4. GROWTH EFFECTS OF RESIDUES ON PLANTLETS OF *P. RADIATA*, *E. GLOBULUS* AND OTHER SPECIES IN GREENHOUSE

The amendment, with different levels of each residue mixed with a forest soil, produced different effects on germination and further growth of *P. radiata* and *E. globulus* plantlets in greenhouse conditions. While both organic residues, grits and the mixture of residues, provoked an inhibition on radiata pine seedlings, the use of ashes, fly-ashes and dregs in low proportions (10-20% vol/vol soil, or higher) promoted growth, following development for 1 year (Figure 1; Table 2). *P. radiata* plantlets always showed better growth responses and greater tolerance than *E. globulus* (Jordan et al. 2000)

Table 2. *Various combinations of solid residues and soil substrates that promoted or inhibited growth of P. radiata plants after 210 days*

Proportion Solid Residues / Soil	Germination after 31 days[1]	Plant height (cm) [2]	Humidity %	pH of substrate
100%MSS (Control)	100	17.6	32.0	6.5
10% D+ 90%	46.9 [3]	19.8	37.8	6.7
20% D+ 80%	34.4 [3]	21.3	38.2	7.3
30% D+ 70%	59.4 [3]	21.8	36.9	7.7
40% D+ 60%	59.4 [3]	18.4	36.2	8.1
10%M6+90%	68.8 [3]	16.2	40.0	7.0
20%M6+80%	62.5 [3]	15.8	45.1	7.2
10% G+ 90%	0.0	0.0	30.5	10.0
10% PS+90%	90.6	16.6	41.8	6.7
10% BS+90%	100	15.7	41.6	6.7
20% BS+80%	90.6	13.5	62.0	7.0
100%MTC (Control)	84.3	17.1	28.5	6.8
10% A + 90%	78.1 [3]	24.9	29.2	7.4
20% A + 80%	75.0 [3]	25.9	29.5	7.7
30% A + 70%	75.0 [3]	25.8	35.4	8.0
10% FA+90%	62.5 [3]	27.0	29.0	6.0
20% FA+80%	71.8 [3]	24.8	29.0	6.3
30% FA+70%	62.5 [3]	34.9	29.0	6.6

MSS, metamorphic sandy soil; MTC, marine terrace clay. A, ashes; BS, brown stock screening rejects; D, dregs; FA,: fly-ashes; G, grits; M6, mixed residues in equal parts without grits; PS, primary sludge. [1,2]Germination (%) and growth results determined from 32 seeds or plantlets/treatment. [3]Germination increased in most treatments after 60 days. Plantlets grown in plastic containers vol. 260 ml. (Jordan et al. 2000, mod).

*Figure 1. Growth responses of radiata pine plantlets developing on a soil substrate amended with various levels of solid wastes, in greenhouse conditions. **A,** Left, 100% fly-ashes; center, fly-ash/soil mixtures; right, 100% soil. **B,** Left, 100% soil; center, ash/soil mixtures; right, 100% ashes.*

Plantlet growth was strongly inhibited when using higher levels of residues. In the case of ashes, stems became shorter and thicker in levels over 50% with respect to soil. A common generalized effect was reduction of root branching, compression and thickening of the shoot apex and a reduced amount of axillary buds formed. Cells of the elongation zone were wider, showing less mitosis. However, in a substrate composed of a mixture of 40% vol. of dregs mixed with the two selected forest soils, no negative effects were observed after 300 days. With an increase of wastes, roots and especially root-tips, became darker and thicker due to enlargement of the cortex cells. The meristematic region was stunted and disposed close to the initiation of the central cylinder. Mitosis was strongly reduced. Besides, the root cap was normally damaged and did not recover in the more adverse conditions.

Although in the presence of several residues, growth inhibition, observed both in the roots and shoots was the most common effect; in the presence of brown stock rejects (30% in proportion with soil) shoots were strongly inhibited while root growth was apparently not affected (Jordan, not published). Despite these limitations, germination still reached 37% using a substrate composed of 90% and 100% ashes (not in Table). Inhibition was also strongly exhibited when using grits (10%) or in presence of a mixture of residues (in vol. of 50% with soils). In respect to grits, it is assumed that their toxicity in the substrate is attributed to high pH (ca. pH 10 at 10% vol. in respect to soil) and presence of high levels of Na ions found in this residue (579.3 meq/L). Another limiting factor of grits even when used in the lowest levels tested, as indicated above, is a sealing effect on the surface of the pots, due to the formation of a hard crust that affected gas exchange, damaging plantlets. Despite these conditions, analysis of elements in plantlets growing in presence of various residues did not show major changes in contents or accumulation of heavy metals, compared to controls (Jordan, not published). Substrates composed of (a) ashes/soil, (b) fly-ashes/soil; (c) mixture/sandy soil and (d) dregs/sandy soil (all 1:1 vol/vol) only showed minor variation in metal contents. A higher level of Ca was found in a and d, of Mg in d, and of K (total and soluble) and Na in c. In some of the plants growing on high levels of ashes, TLC revealed the presence of proline in shoot-tips of these plants. (Jordan, not published).

Several experiments carried out in the greenhouse related to germination and plantlet development were performed with other species, i.e rice, peas, oats and *Raphanus* spp. (Jordan, 2000). Preliminary results showed that rice develops especially well in the presence of sludge when added in the substrate, completing its development, while the other species showed to be more sensitive. In the future, various species may become potentially tolerant to harmful conditions of substrates; i.e. by using mutation techniques a single-gene pea mutant, controlled by the genotype of the root, has been found to account for a 10-100 fold higher Fe accumulation compared to its parent plant (Welch and LaRue, 1990).

It is assumed that the detrimental effects observed in some treatments in experiments conducted in greenhouse conditions will not occur in the field. In the former, the direct contact of wastes with seedlings should affect growth of the very sensitive new emerging roots as well as the whole germination process. Under these conditions plantlet development is simultaneously affected by various kinds of stresses, including unbalanced levels and/or toxicity caused by several disolved metals, high pH, less O_2 supply, higher water content, and eventually, competitive microbiota. On the contrary, in the field, older roots are externally protected by the rhizodermis, vascular tissues are organized and filtration mechanisms operate. The residues are applied superficially, diffusing in the soil profile while diluting. This involves a longer time in reaching the root absorption zone. Finally, the root biomass relation vs. residues is much higher compared to seedlings.

5. GROWTH EFFECTS OF RESIDUES ON ADULT PLANTS UNDER FIELD CONDITIONS

Experiments conducted under field conditions with adult trees, testing ashes and residues on *E. grandis* in levels of 10-50 ton/Ha of each with additions of some fertilizers showed an increase of biomass leading to taller trees. The height of the trees was enhanced by 25% (16 to 20 m) and an increase of wood over 80% after 3 years (Bellote et al. 1995). Similar to the results found in *P. radiata* and *E. globulus* trees (Jordan et al. 2001) growing close to dumps (see below, 6.) in E. *grandis* there are no changes, besides, K and Ca were found in the leaves. Therefore, it is assumed that, when used in proper concentrations, in surface applications in the field, residues should not provoke harmful effects in poor soils used for forestry plantation. Ashes definitely help neutralizing soils (Lerner and Utzinger, 1986), providing several nutrients to soils as P and K and water, but care must be taken when applying it alone in sandy soils, since significant leaching has been detected after 1 month (Guerrini and Villa Boas, 1996; Bellote et al. 1995). However, the feasibility of landspreading for the purposes of fertilization and amendment of forest soils, diminishing the environmental impact, has to be assessed in each case. Other residue sources, furnace fly-ashes derived from metallurgic production also showed an increase in dry matter in *E. urophylla* (Novais et al. 1995) and in *E. camaldulensis* with addition of phosphate, especially in sandy soils (Oliveira et al. 1994). Dregs and grits when applied separately in levels of 1 and 4 ton/Ha enhanced growth of *E. grandis* (Valle et al. 1995) and in combination with sludge the former increased plant height in *E. grandis*, *E. saligna* and *Acacia mearnssi* (Bergamin et al. 1994). Fabres et al. (1994) using sludges found growth increase in *E. grandis* determining an optimum level of application in the field equivalent to 38 m3/Ha in combination with KCl (1 kg/m^3 sludges). Sludges are poor in K (Fabres et al. 1994, Jordan et al. 2000). In Finland good results in fertilizers have been reported using pellets composed of 94% ashes of wood bark with 4% sludges applying 8000 ton/Ha/year. Their use helps by raising the pH of highly acidic peatlands and such conditions had a good effect on birch seedlings while improving the growth of forests (Hytonen, 1998). On the other hand, fly-ashes, due to their capacity in adsorption and precipitation of heavy metals, have been considered useful to remove these from sludges favoring land application (Moo-Young and Ochola, 1999). For *P. radiata* 10 year-old plants, levels of applied residues used up to 40 ton/Ha, did not show evidence of an increase in height after 2 years. Besides, no toxicity was observed and only minor changes in the content of some elements, mainly nitrogen, were found in the needles (Jordan et al., not published).

6. NUTRIENT CONTENTS IN LEAVES OF *PINUS* AND *EUCALYPTUS* GROWING CLOSE TO DUMPS

P.radiata and *E. globulus* trees were found growing close to dumps for 10 years. A comparison of the nutrients of leaves of both species with trees developed on soils free of residues at 1 km distance, was performed. The content of essential elements (K, Mg, Ca, Mn, Fe, Cu Zn, Na and Ni) and heavy metals of leaves (Pb, Cr, Cd, Bi, Sn and Al) within species was practically the same. Only Ca and Na were higher in the trees of both species growing close to the residue dump (Jordan et al. 2001). According to analysis, heavy metals are not absorbed by the root system and therefore not incorporated into the wood. On the other hand the trees growing next to the waste disposal showed signs neither of damage or toxicity, nor cumulated heavy metals, in comparison to trees of the same age growing 1 km away. The results are important due to the avoidance of uptake limits, a "feedback effect" in using trees fertilized with the wastes as sources for the mill.

7. TOLERANT SPECIES OCCURRING NATURALLY AND COLONIZING STABILIZED DUMPS

About 30 plant species, found in one site, seem to adapt and are tolerant to the "substrate" conditions of dumps, composed of mixtures of the various residues produced by the paper mills (Jordan and Peña, 2000). In the mixture of residues that constitute the "substrate", a high pH level (>10) high salinity, organic matter and essential elements are found. Although recent deposits remain as nude soils for a couple of years, in the older, more stabilized sites a significant cover of vegetation exist. Stabilized dumps will show a lower pH and decreased leachability partially due to the fact that *sludges* and organic matter may enhance metal adsorption in a solid phase (Xiao et al. 1999). Within the colonizing species in the stabilized sites, European weeds have been found to dominate in older zones, including the shrub *Teline monspessulana* (Papilionaceae) and the thistles *Silybum marianum* and *Cirsium vulgare* (Asteraceae) besides other species from both families and also Poaceae (Jordan et al. 2001). Close to them, associations of *Nolana*, with graminae and compositae with *Avena fatua* are found. In the more humid sites, associations of *Rumex* spp., *Polygonum* spp. and *Azolla filiculoides* Lam. are frequent. *Azolla pinnata* has been described to be a hyperaccumulator for aquatic ecosystems (Duschenkov et al. 1995*);* in the case *of Azolla filiculoides,* the level of heavy essential elements and heavy metals were not higher than those of other species growing on the dump (Jordan et al. 2001). A higher cover and diversity of plants is found in the older more stable areas of the dumps that have had more biotic and abiotic changes making the site better for colonization compared to fresh deposits. Within the many species, which do not appear to behave as "accumulators" the highest levels of some ions detected in leaves (mg/Kg/dw.) were however: *Conyza bonariensis*, (Papilionaceae) [Ca, 2.32; K, 2.49]; *Trifolium sp* (Papilionaceae) [Ca, 4.41; Fe, 0.27]; *Oenothera stricta* (Oenotheraceae) [Ca, 2,06; K 1.83]; *Juncus bufonius* (Juncaceae) [Zn, 0.02]; all others showed a lower level of nutrients. As indicated, *P. radiata* and *E. globulus,* both growing close to the older deposits, tolerate well the levels found in these mixtures. Stabilization as a result of biochemical fixation, precipitation, and physical fixation, besides other processes (Cunningham, et al. 1995) occurring in these complex substrates, can account for a better tolerance of these species developing on the dump and nearby sites. On the other hand, it has been shown that the soluble material in leaches of sludge mixtures, i.e. Na need 1-2 months to move into the first few feet of soil (Smith et al. 1991, cited by Sherman, 1995).

8. SUMMARY AND CONCLUSIONS

A more extensive, possible use of solid residues deriving from a "Kraft" pulp mill as fertilizers in forest soils is explored. Several studies indicate that most waste residues can be re-utilized. Residues such as ashes, fly-ashes, dregs, grits, primary sludge and brown stock rejects used alone or mixed, act as fertilizers in low concentrations depending on plant species. Although various levels of residues can be set in direct contact with germinating seedlings, harm can occur, while superficial application in the field, with older plants, acts rather beneficially. Various reports show an increase in height and biomass of trees, especially in various species of *Eucalyptus*. Responses strongly differ according to species and age of plant material, soil substrate employed, level of residues, and the specific physical and chemical characteristics of each residue. The residues are normally attacked by the natural microbiota reducing toxicity in time. In tropical climates, at higher and constant temperatures around the year, biodegradation (aerobic/anaerobic digestion) and phytoremediation, makes possible the re-conversion of wastes to humus, domestic fertilizers and other sub-products (i.e. ethanol). In more temperate climates, with a lower activity of the microbiota on waste deposits accumulated in dumps, colonization by several species may still occur in a couple of years; although this response is restricted to older, more stable and leached areas of the dumps. The species found growing upon dumps are different from those found in distant sites of the area and adaptation apparently does not involve incorporation of mineral elements into the plant tissues. *P. radiata* and *E. globulus* trees planted close to the dumps and in contact with their leachings grow well in these conditions, not showing any difference with trees of both species located in forests, the dumps not being hazardous for them. Preliminary work has also shown a potential use of residues in some crops (Jordan, et al. 2000a) but specific studies are required to evaluate this effect and to determine mineral content in tissues. It is expected that responses obtained under field conditions can make possible a wider use of these wastes in the long-term. Regarding environmental problems, more research is also needed. Some studies related to ash fertilization of forests in catchment areas of humic lakes in Finland (applying 6400Kg/Ha) indicated however, that there are no significant effects on water quality or plankton communities by leaching of nutrients, concentration of heavy metals and pH changes (Tulonen et al. 2000). According to Sherman (1995), properly applied sludges produce an increase in plant growth of several important crops (i. e. grass, hay, corn) and trees. On the other hand, data obtained from extensive studies (Sherman 1995, and cited therein) do not indicate negative effects on wildlife, health of individuals or on reproductive parameters.

ACKNOWLEDGMENTS

The work from our laboratory was supported by a grant from Conicyt (Fondef D97I1012), Chile. Thanks are given to Mme. Andrée Goreux for her help in revising this manuscript.

REFERENCES

Bellote AFJ, Ferreira CA, Da Silva HD & Andrade GC (1995) Efecto de la aplicación de ceniza de caldera y residuo de celulosa en el suelo y en el crecimiento de *Eucalyptus grandis. Bosque* **16**, 95-100.
Bergamin FN, Gonzaga JV & Bortolas E (1994) Residuo de fábrica de celulose e papel: lixo ou produto? In: Guerrini IA, Bellote AFJ & Büll LT (eds.) Seminario sobre uso de resíduos industriais e urbanos em florestas (pp. 97-120), Botucatu, Brasil.
Brezny R, Joyce TW & González B (1992) Biotransformation in soil of chloroaromatic compounds related to bleach plant effluents. *Wat. Sci. Technol.* **26**, 397-406.
Brezny R, Joyce TW, González B & Slimak, M (1993) Biotransformations and toxicity changes of chlorolignins in soil. *Environ. Sci. Technol.* **27**:1880-1884.
Cunningham SD, Berti WR & Huang JW (1995) Remediation of contaminated soils and sludges by green plants. In: (Hinchee RE, Means JL & Burris DR (eds) Bioremediation of inorganics, pp. 33-54. Battelle Press, Ohio, USA.
Duschenkov V, Kumar B, Motto H & Raskin I (1995) Rhizofiltration: the use of plants to remove heavy metals from aqueous streams. *Environ. Sci. Technol.* **29**, 1239-1245.
Ehrlich (1996) Geomicrobiology. Dekker, New York.
Ehrlich (1997) Microbes and metals. *Appl. Microbiol. Biotechnol.* **48**, 687-692.
Ericksson LE & Kolar MC (1985) Microbial degradation of chlorolignins. *Environ. Sci. Technol.* **19**:1086-1089.
Fabres, AS, Couto C & Conceicao DA (1994) Uso de residuo industrial de celulosa em florestas. In: Guerrini IA; Bellote AFJ & Büll LT (eds). Seminario sobre uso de resíduos industriais e urbanos en florestas (pp 121-140) Botucatu, Brasil.
Fällman AM (1986) Characterization of residues release of contaminants from sludges and ashes. Annual AFR-Report 86, Swedish Environmental Protection Agency, Stockholm, Sweden.
González B, Brezny R, Herrera, M & Joyce T (1995) Degradation of 4,5-dichloroguaiacol in soils. *World J. Biotechnol. Microbiol.* **11**:536-540.
Guerrini IA & Villas Boas RL (1996) Uso de residuos industriais em florestas. In: Aloisi RR & Torrado PV (eds). Congresso Latino Americano de Ciencia do Solo. 13, Agua de Lindoia, 1996. Campinas, Soc. Brasileira de Ciencia do Solo e Sociedade Latino Americana de Ciencia do Solo.
Gustavsson M, Wiberg K & Öberg-Högsta AL (1999) Characterization of pulp and paper waste materials and their field application. Proceedings 3th International Workshop on the Use of Paper Industry Sludges in Environmental Geotechnology and Construction (pp 116-124), June, 1-4, Helsinki, Finland.
Hytonen J (1998) Pellets made of wood ash and other wastes as nutrient sources for silver birch seedlings. *Suo* **49**, 49-63.
Jordan M (2000) Use of solid residues derived from the cellulose paper industry. Proceedings "2nd International Symposium on Biotechnology for Conservation of the Environment. Munster, Germany, 9-12 July.
Jordan M & Peña R (2000) Especies vegetales que prosperan sobre residuos sólidos derivados del proceso productivo de plantas de celulosa, de tipo "kraft". Proceedings 43th Reunión Anual Sociedad de Biología de Chile, Pucón, Chile, 14 -18 Nov. Biological Research 33:79.
Jordan M, Vicuña R, Gonzalez B, Bronfman M, Osses M, Toro J, Balocchi C & Rodriguez E (2000) Growth of *Pinus radiata* in soil containing solid waste from the kraft pulp industry. Proceedings of 4th International Conference on Environmental Impacts of the Pulp and Paper Industry (pp 354-358), 12-15 June 2000, Helsinki, Finland.
Jordan M, Osses M, & Roveraro C (2000a) Efecto de la aplicación de residuos sólidos generados en el proceso de producción de celulosa sobre el desarrollo de *Pinus radiata* en condiciones de invernadero. Proceedings 43th Reunión Anual Sociedad de Biología de Chile, Pucón, Chile, 14-18 Nov. Biological Research 33:63.

Jordan M, Peña RC & Osses, M (2001) Plant biodiversity in a cellulose pulp mill in continuous disposal of solid residues (ISEB 2001 Meeting), Phytoremediation, 15-17 May, Leipzig.

Jordan M, Sanchez MA, Padilla L, Céspedes R, Osses M & González B (2001a) Effect of kraft mill residues-soil mixtures on *Pinus radiata* germination and growth, and soil microbial activity (submitted).

Joyce TW, Brezny R, Gonzalez, B & Palo, N (1992) Degradation in soil of chlorolignin substances from chlorine bleaching of pulp. *Papír a Celulóza* 47, 125-146.

Laine MM & Jorgensen KS (1998) Bioremediation of chlorophenol-contaminated soil by composting in full scale. In: Wickramanayake GB & Hinchee RE (eds) Bioremediation and Phyotoremediation (pp 45-50) Batelle Press, Ohio, USA.

Lerner BR & Utzinger JD (1986) Wood ash as soil limiting material. *HortScience* 21, 76-78.

Lovley DR & Coates JD (1997) Bioremediation of metal contamination. *Current Opinion in Biotechnology* 8:285-289.

Marschner H (1995) Mineral nutrition of higher plants. Academic Press, Cambridge. UK.

Moo-Young HK & Ochola CE (1999) The future of paper industry waste management. Proceedings of the 3rd International Workshop on the Use of Paper Industry Sludges in Environmental Geotechnology and Construction (pp 62-84), 1-4 June, Helsinki, Finland.

Novais RF, Barros NF, Firme DJ, Leite FP, Villani EMA, Texeira JL & Leal PGL (1995) Eficiencia de escória de siderurgia. In: 25th Congreso Brasileiro de Ciencia do Solo, Resumos Expandidos (pp 2282-2284), Campinas, Brasil.

Oliveira AC, Hahne H, Barros NF & Morais EJ (1994) Uso de escoria de alto forno como fonte de nutrientes na adubacao florestal. In: Guerrini IA, Bellote AFJ & Büll LT (eds) Seminario sobre Uso de Resíduos Industriais e Urbanos em Florestas .(pp 77-96), Botucatu, Brasil.

Simpson GG, King LD, Carlile BL & Blickensderfer PS (1983) Paper mill sludges, coal fly ash, and surplus lime mud as soil amendments in crop production. *Tappi Journal* 66, 71-74.

Sherman WR (1995) A review of the Maine "Appendix A" sludge research program. *Tappi Journal* 78, 135-150.

Tulonen T, Arvola L, Ollila S; Mäkinen A, Pihlström M, Rask, M & Karppinen C (2000) Effects of forest fertilization by wood ash on lake ecosystems. Proceedings of the 4th International Conference on Environmental Impacts of the Pulp and Paper Industry, (pp 359-363), 12-15 June, Helsinki, Finland.

Valentine NB, Bolton H Jr, Kingsley MT, Drake, GR, Balkwill DL & Plymate AE (1996) Biosorption of cadmium, cobalt, nickel, and strontium by a *Bacillus simplex* strain isolated from the vadose zone. *J. Indust. Microbiol.* 16:189-196.

Valle CF, Corradini L & Alvarenga SF (1995) Uso de resíduos industriais e urbanos en florestas de eucalipto. Relatorio Técnico. Votorantin de Celullose e Papel Ltda. Luiz Antonio, Brasil.

Waldemar CC & Herrera J (1986) Avaliacao do potencial de utilizacao do "Dregs" e do "Grits" como correctivo de acidez e fertilizante na agricultura. In: 19th Congresso Anual da ABCP; Anais (pp 447-453), Sao Paulo, Brasil.

Welch RM & LaRue TA (1990) Physiological characteristics of Fe accumulation in the 'Bronze' mutant of *Pisum sativum* L., cv 'Sparkle'E107 (brz brz). *Plant Physiology* 93, 723-729.

Xiao CLQ & Sarigumba T (1999) Effects of soil on trace metal leachability from paper mill ashes and sludges. *J. Environ. Engineering* 122: 758-760.

SUBJECT INDEX

α - aminolevulinic acid (ALA), 202
ABC transporters (ATP binding cassette). 9
Accumulation in chloroplasts, 231
Adenosine 5' Phospho Sulphate (APS) reductase
Adsorption of heavy metals, 5
α - aminolevulinic acid (ALA)
- dehydratase (ALAD), 202
- dehydratase activity ,215
- synthesis, 214
Aluminum , 95
- chelating substances, 97
- mineral nutrition, 374
- plasma membrane, 375
- root growth ,374
- signal, 102
- tolerance, 97
- tolerance mechanisms, 381
- tolerance mechanisms in rice, 382
- tolerance mechanisms in wheat, 382
- toxicity, 96
Ammonium uptake, 329
Aneugenesis, 398
Anion channel, 100
Antioxidant, 53
Apophytochelatin, 43
APS reductase, 314
APS-sulphotransferase, 53

Bioavailability, 115
Bioremediation, 418
Biosynthesis of
- chlorophyll, 202
- phytochelatins, 45
- photosynthetic pigments, 201

Buthionine sulfoximine (BSO), 133
Ca-ATPases, 8
Ca-channels, 8

Cadmium, 25
Cadmium tolerance mechanisms, 379
Cation exchange capacity,2
Cell wall proteins, 4
Characteristics of the excretion of organic acids, 98
Chlorophyll a, 220
Chlorophyll accumulation, 206
Chlorophyll b, 220
Chlorophyll
- biosynthesis, 213
- fluorescence, 236
Chlorophyll/Carotenoid ratio, 220
Chlorophyllide, 218
Chlorosis, 209
Circulation of ions in plants, 17
Clastogenesis, 398
Copper and nickel tolerance mechanisms, 380
Copper deficiency, 368
Coproporphyrinogen, 216

Damage of
- active center, 309
- proteins, 309
Dissociation of active complex, 311

Efflux of malic, citric and oxalic acids, 20
Electron transport, 237
Energy yielding metabolism, 124
Enzymatic steps, 213

Enzyme
- inactivation,122
- longevity, 307
Enzymes,303, 333

Enzymes
- activities, 101
- oxidative stress, 315
Essential metal ions, 114
Etioplast inner membranes (EPIM), 203

Excretion of
- malate ,105
- organic acids, 97

Ferritins, 83
Free radical formation, 126
Functions of enzymes in heavy metals treated plants, 313

Gene(s), 105
Genetic background of uptake mechanisms, 10
Genotoxicity, 395
Genotoxicity assays, 396
Genotypic
- differences, 357, 368, 370
- variations, 383
Genotypic differences in toxicity
- cadmium, 370
- copper, 370
- manganese, 371
- zinc, 372
Glutamate dehydrogenase (GDH) activity, 340
Glutamate
- synthase, 338
- synthetase, 338
Glutathione, 37
Glutathione biosynthesis, 37
Glycoproteins, 4
Growth effects of residues, 419, 422
Growth inhibition, 164

Heat shock proteins, 160, 315
Heavy metal stress, 150
Heavy metals affected cells, 303
Heavy metals-dependent down regulation of enzymes, 304

High affinity transport system (HATS), 329
High and low affinity transport systems, 10
Homophytochelatins, 41
Hydroxymethylphytochelatins, 41

Induction of sister chromatid exchange (SCE), 402
Inorganic nitrogen, 336
Intracellular mechanism of Al resistance, 105

Ion uptake, 168
Iron uptake, 15
Isogenic wheat lines, 105
Isophytochelatins, 41
Lead, 23
Lead toxicity, 162
Light phase of photosynthesis, 229
Long distance nitrate transport, 329
Low affinity transport (LAT), 329
Magnesium protoporphyrin, 217
Malate efflux, 100
Manganese as micronutrient, 359
Manganese deficiency ,367
Mechanisms of
- metal toxicity, 121
- excretion of organic acids, 100
- inhibition, 309
Membrane damage, 121
Metabolism of organic acids, 95
Metal
- binding polypeptides, 133
- binding properties of ferritin, 84
- chelating peptides, 59
- deficient conditions, 357
- detoxification, 75, 415
- efflux, 129
- homeostasis, 75
- induced genotoxicity, 405
- mutagen, 403
- permeability, 1

- pollution, 112
- tolerance, 136
- toxicity, 117
- uptake, 231
Metal-metal
- antagonism, 403
- synergism, 403
Metallothioneins, 40, 81, 133, 317
Metals and nucleic acids, 126
Micronucleus (MNC), 397
Micronutrient, 359
Mineral nitrogen forms, 331,332
Molecular chaperones, 315
Molybdenum as micronutrient, 360
Mugineic acid, 17

Nickel
- deficiency, 368
- micronutrient, 361
Nicotianamine, 16, 85
- synthase, 16
- transferase, 16
Nitrate
- reductase (NR) 325, 333
- reductase activity, 334
- uptake, 326
Nitrite
- reductase, 333
- reductase activity, 338
Nitrogen
- fixation 51, 344
- metabolism, 325
Non-vascular plants, 111
Normalizing factor, 16
Nutrient mimicry, 128

Oxygen evolution activity, 238

P-type heavy metal pumps, 9
Patterns of excretion of organic acids, 98
Photoinhibition, 240
Photosynthesis, 165
Photosynthesis and respiration, 372
Photosynthetic electron transport, 232

Photosystem I, 233
Photosystem II, 235
Phycobilin biosynthesis, 222
Phytochelatin
- biosynthesis, 63
- synthase, 70
- synthase activation, 49
- synthase gene, 47
Phytochelatins, 40, 42, 60, 133, 317
Phytoextraction, 385
Phytostabilization, 386
Plasmalemma, 7, 375
Polypeptide composition, 238
Porphobilinogenase (PBGase), 216
Prolamellar bodies (PLB), 203
Proline synthesis, 342
Protein,
- degradation, 306
- folding 306,
Protection to stress, 311
Protochlorophyllide, 218
Protoporphyrin, 202
Redox
- potential 116
- proteins 11
Repair 312

Sister chromatid exchange (SCE), 397
Sites and mode of action, 153
Stimulation of activity, 312
Strategies I and II, 363
Stress proteins, 85
Structural carbohydrates, 1
Structure of cell walls, 1
Symptoms of toxicity, 152

Thiol
- metabolism, 37, 51
- peptides, HMW, 43
- peptides, LMW 43
- peptides, MMW, 43
Tolerance
- index, 129
- mechanisms, 157

- metal deficiency, 366
Toxicity
- aluminum, 373
- heavy metals, 369
Transformed poplars, 46
Transport ATPases, 7
Transport
- interference, 127
- nitrate, 327
Triticale, 359

Ubiquitin, 314
Uptake and accumulation, 162
Uptake of copper, manganese and zinc, 14
Uroporphyrinogen, 216

Vacuolar compartmentalization, 72
Vascular plants, 149

Water relations, 167

Zinc
- deficiency, 368
- micronutrient, 361
- tolerance mechanisms, 380

BIODIVERSITY INDEX

Achlya racemosa, 119
Agaricus bisporus, 133
A. silvicolla, 114
Alatospora acuminata, 119
Allium cepa, 80, 152, 153, 159
Amaranthus hybris, 135
Amphidinium hoefleri, 115
Anabaena, 122, 135
Anabaena cylindrica, 132
A. doliolum, 117, 123, 124, 130, 133, 135
A. flos-aquae, 121, 132
A. inequalis, 117
Anacystis nidulans, 132
Arabidopsis, 8, 14, 21, 24, 26, 67, 69, 77, 78
Arabidopsis thaliana, 12, 15, 46, 47, 49, 76, 82, 100, 159
Armeria maritima, 44
Armoracia rusticana, 40, 41
Aspergillus ocharaceus, 127
Astragalus racemosus, 132

Bacopa monniera, 167
Banana doliolum,127
Brachythecium rivulare, 114
Brassica juncea, 40, 43, 46, 47, 53, 54, 77, 79, 81, 151, 152, 164
*Brassica oleracea,*168
Byoria fuscescens, 116

Caenorhabditis elegans (nematode), 47, 74
Calothrix parietina, 131
Canavalia ensiformis, 84
Candida glabrata, 40, 41, 42, 45, 80
Chironomonas oppositus, 49
Chlorella, 122, 132, 133

Chlorella vulgaris, 123, 124, 127, 128, 137
Chroococcus, 133
Cladonia rangiferina, 121, 124
Cucumis sativus, 163
Cyanidium caldarium, 131

Datura innoxia, 62, 79
Daucus carota, 159
Dicranella varica, 131
Dictylum brightwelli, 135
Dictyostelium discoideum, 49
Diploschistes muscorum, 131
Dunaliella, 136
Dunaliella acidophila, 137

Enteromorpha, 114
Enteromorpha prolifera, 135
Escherichia coli, 16, 46
Euglena, 133
Euglena gracilis, 123, 130

Fagopyrum esculentum, 21, 22
Festuca arundinacea, 153, 166
Fragaria anasassa, 164
Funaria, 115, 120
Fucus vesiculosus, 114, 117

Ganoderma lucidum, 119
Gloeotheca, 131
Glycine max, 39, 51, 68, 84
Gonyaulax polyedra, 135
Grimmia diniana, 131

Hebeloma mesophacus, 119
Hematococcus, 119, 135
Holcus lanatus, 161
Hordeum vulgare 80
*Hydnum repandum,*114

Hydrangea, 20, 105
Hylocomium splendens 131

Laccaria bicolor, 119
Lamanea fluviatilis, 114
Laminaria saccharina, 123
Larrea tridentate, 81
Lecidea, 120
Lemna minor, 153, 165
Lens esculenta, 84
Lotus purshianus, 80
Lycopersicon esculentum, 68, 164

Medicago trancatula, 39
M. truncatula, 68
Minuartia verna, 44

Neurospora crassa, 133
Nostoc, 123
N. calcicola, 130
N. linckia, 122, 124, 127, 128
N. muscorum, 115, 124, 127, 131

Oocystis submarina, 117
Oryza sativa, 41, 68
O. sativa sub sp. Japonica, 68

Paxillus involutus, 119, 120
Peltigera canina, 123
Penicillium digitatum, 127
Phaffia, 136
Phaseolus aureus, 152
P. vulgaris, 39, 51, 54, 99, 101, 152
Picea abies, 119, 168
Pisum sativum, 39, 46, 51, 84, 166, 168
Plectonema boryanum, 121, 127, 132
Pseudomonas aeruginosa, 101

Rauvolfia serpentina, 41, 42, 43, 45, 46, 75
Rhytidiadelphus squarrous, 131
Rubia tinctorum, 75

Russula delica, 114

Saccharomyces cerevisiae, 44, 69, 116, 117, 127, 132, 133
S. exigus, 80
Salvinia natans, 120
Scapania undulata, 115
Scenedesmus armatus, 117
S. bijugatus, 54
S. obliquus, 125
S. quadricauda, 127
Schizosaccharomyces, 41, 69
Schizosaccharomyces pombe, 9, 43, 45, 47, 63, 66, 67, 133
Sesamum indicum, 165
Silene cucubalus, 43-45, 49
S. maritima, 1
S. vulgaris, 42-44, 63, 65, 79, 81, 159
Suillus luteus, 120
Synechococcus, 132-134

Tetraselmis gracilis, 135
Thalassiosira wiessflogii, 127
T. pseudonana, 133
Thlaspi arvense, 79
T. caerulescens, 26, 79, 80, 132, 150
Tricholoma terreum 114
Triticum aestivum, 47, 96, 102
Tritisecale wittmark, 100
Typha latifolia, 68

Ulva lactuca, 114
Urtica dioica, 163
Ulva lactuca, 114

Vigna angularis, 76
V. mungo, 84
V. radiata, 39

Xanthoria parietina, 131

Zea mays 41, 101